Mixed-Mode Chromatography

Qian-Hong Wan

Mixed-Mode Chromatography

Principles, Methods, and Applications

Springer

Qian-Hong Wan
School of Pharmaceutical Science and Technology
Tianjin University
Tianjin, China

ISBN 978-981-16-5484-8 ISBN 978-981-16-5485-5 (eBook)
https://doi.org/10.1007/978-981-16-5485-5

This Springer imprint is published by the registered company Springer Nature Singapore Pte Ltd.
The registered company address is: 152 Beach Road, #21-01/04 Gateway East, Singapore 189721, Singapore

For Hong, Rose, Andrew, and Ada

Preface

Since the FDA approved the first monoclonal antibody drug in 1986, the therapeutic antibody drug market has experienced explosive growth. New antibody drugs have been approved to treat a variety of human diseases, including many cancers, autoimmune, metabolic, and infectious diseases. The manufacture of therapeutic antibody drugs usually involves a lengthy and expensive purification process based on protein A affinity chromatography. However, protein A chromatography has two main problems: ligands are easily cleaved by proteases present in cell lysates, and purified antibodies tend to aggregate under the low pH conditions required for elution. Therefore, considerable efforts have been made to mitigate these adverse effects.

Mixed-mode chromatography has emerged to simulate protein A chromatography for antibody purification. Synthetic ligands with multiple functional groups are used to replace natural biomolecules and bind to target molecules through a combination of hydrophobic, electrostatic, and hydrogen bonding interactions. Binding affinity and selectivity can be optimized by following the principles of molecular recognition, such as multivalency, complementarity, and geometric fitting. This book tends to introduce the principles, methods, and applications of mixed-mode chromatography. It starts from a historical perspective covering the development of chromatography in general and mixed-mode chromatography in particular. Then, theoretical considerations are provided, focusing on the mass transfer and band broadening in the packed bed, and the different separation modes and packing materials are described in detail. The following chapters deal with ligand design, retention mechanisms, stationary phases, and mobile phases in mixed-mode chromatography. A large number of examples are used to explain and illustrate the method development and application in biopharmaceutical manufacturing.

I would like to thank Tianjin University for its continuous strong support and the Springer Nature book project team, especially Lewis Liu and Kavitha Jayakumar, for their suggestions during the writing of this book.

Tianjin, China
July 08, 2021

Qian-Hong Wan

Contents

Chapter 1
Historical Perspectives

Abstract Chromatography is a technique that uses the difference in the partition coefficient of a compound between the stationary phase and the mobile phase to separate the mixture. The stationary phase is a solid-phase carrier that is immobilized during the separation process and its ligands loaded on the surface of the carrier. The mobile phase is a fluid that moves relative to the stationary phase during the separation process and is mainly used to carry solute molecules through the stationary phase. In the course of more than 100 years of development, chromatographic technology has formed two major branches of liquid chromatography and gas chromatography. The mobile phase of the former is liquid and the latter is gas. Liquid chromatography can be divided into normal phase chromatography, reverse-phase chromatography, ion exchange chromatography, and affinity chromatography according to the separation mechanism. In recent years, the rapid development of mixed-mode chromatography has generated considerable interest because it has a different mechanism of action and application characteristics than other chromatographic modes. This chapter will review the birth and development of chromatography focusing on mixed-mode chromatography.

1.1 Introduction

Chromatography is a technique that uses the difference in the partition coefficient of a compound between the stationary phase and the mobile phase to separate the mixture. The so-called stationary phase refers to the solid-phase carrier that is immobile during the separation process and its ligands loaded on the surface of the carrier. The mobile phase refers to the fluid that moves relative to the stationary phase during the separation process, which can be liquid or gas. According to the state of the fluid used, chromatography can be divided into three important branches: liquid chromatography, gas chromatography, and supercritical fluid chromatography. According to the type of interaction involved in the separation process, chromatography can be divided into normal phase chromatography, reverse-phase chromatography, ion exchange chromatography, affinity chromatography, and so on. According to the way the separation process is realized, chromatography can be

Q.-H. Wan, *Mixed-Mode Chromatography*,
https://doi.org/10.1007/978-981-16-5485-5_1

divided into packed-column chromatography, open-tubular column chromatography, electrochromatography, thin-layer chromatography, continuous chromatography, multidimensional chromatography, and so on. According to the scale of application, chromatography can be divided into two major branches: analytical chromatography and preparative chromatography. According to the application field, chromatography can be divided into ion chromatography, chiral chromatography, protein chromatography, nucleic acid chromatography, and so on. Chromatography can also be used in conjunction with other analytical techniques such as mass spectrometry to achieve component separation and structural characterization of analytical samples in one run. In short, the chromatographic technology has a wide variety, convenience, flexibility, and adaptability, making it unique among many separation technologies.

Compared with older separation techniques such as crystallization, solvent extraction, and distillation, chromatography has many advantages. First of all, chromatography is a powerful separation technique. It can separate and purify one or more components of interest from complex multicomponent mixtures without knowing the type, quantity, or relative quantity of the substances present. Secondly, chromatography has a wide range of applications. It can not only deal with molecules of different sizes, from the smallest molecules, which contain only two atoms of hydrogen in the molecule, to viruses composed of millions of atoms, but also deal with molecules of different properties, from simple gaseous molecules to complex and chargeable biological macromolecules such as proteins, antibodies, and nucleic acids; in addition, it can handle materials from trace amounts to bulk quantities. Some forms of analytical chromatography can detect substances present at the level of attogram (10^{-18} g), which makes this method an excellent trace analysis technique widely used to detect residual pesticides in biological materials and the environment, forensic science and detection of therapeutic drugs and drugs of abuse. Preparative chromatography is the main method of downstream processing in the biopharmaceutical process. It is used to separate and purify the recombinant protein produced by genetic engineering to make its purity meet the strict requirements of the pharmaceutical regulatory authority. In the foreseeable future, chromatography will continue to dominate separation science and technology due to its irreplaceable separation capabilities and will play an important role in many industries including pharmaceutical, food, biotechnology, and chemical industry.

In this chapter, we will briefly review the development of chromatography, especially mixed-mode chromatography, and learn from pioneers in the field how to overcome obstacles and create new things.

1.2 Separation Technology Before Chromatography

The separation process is a very old technology, which has already begun almost since human beings entered the civilized age. We know that most elements or compounds in nature are in an impure state, mostly in the form of mixtures, and cannot be used directly without separation. The alchemy or metallurgy created by the ancients was to obtain pure precious metals such as gold from ore raw materials through the calcination process. In the traditional wine making process, the fermentation products are distilled to increase the alcohol content. In the water taken from the pond or river, alum is added and stirred to precipitate impurities and obtain clear and clean drinking water. These examples show that since ancient times, humans have learned to use separation technology to separate raw materials before they can be put into production or used to meet daily needs. This makes separation technology important not only for industrial production but also for daily life.

Industrial separation is usually used for production purposes and is carried out on a large scale by using techniques such as precipitation, crystallization, and distillation to extract effective ingredients from the original mixture. Separation can also be used for analysis and determination, using a small amount of samples to concentrate or to remove impurities in order to determine some of the components. In some cases, separation requires complete purification, such as electrolytic refining of bauxite in aluminum metal ore, crystallization and purification of antibiotic products in fermentation broth, and separation and purification of recombinant protein secreted in cell supernatant. But in other cases, only partial separation is required. A good example of this is the separation of plasma proteins. The Cohn process developed by Edwin J. Cohn is a series of purification steps to extract albumin from plasma, an excellent substitute for human plasma. Using the characteristics of different solubility of plasma protein at different pH, ethanol concentration, temperature, ionic strength, and protein concentration, the plasma protein is divided into several fractions by adjusting the experimental parameters, and each fraction contains several proteins. The cryogenic ethanol precipitation method separates human plasma and produces human serum albumin, serum gamma globulin, fibrinogen, thrombin, and blood group globulin (Surgenor 2002).

Another example of incomplete separation is the refining of crude oil (Gary and Handwerk 1984). In nature, crude oil exists as a mixture of various hydrocarbons and impurities. The refining process breaks down this mixture into other more valuable mixtures, such as natural gas, gasoline, and chemical raw materials. None of them are pure substances, but each must be separated from crude oil. In both cases, a series of separations are required to obtain the desired final product. In the case of oil refining, crude oil is subjected to a series of separate distillation steps, each of which produces different products or intermediates.

Precipitation is the formation of solid precipitates in a liquid solution through reaction. These precipitates aggregate and settle under the action of gravity and separate from the solution (Zumdahl 2005). The chemical substance that causes the solid to form is called a "precipitating agent." There is not enough gravity (sedimentation) to gather the solid particles together, and the sediment stays in suspension. In this situation, centrifugal force can be used to compress it into a compact mass to accelerate the settlement. An important stage of the precipitation process is the beginning of nucleation. Some energy based on the relative surface energy of the solid and the solution is required for the formation of an interface.

Without this energy, and without a suitable nucleation surface, oversaturation will occur.

Distillation is the process of separating components or substances from a liquid mixture by using selective boiling and condensation. Distillation can result in a substantially complete separation (almost pure components), or possibly a partial separation, thereby increasing the concentration of selected components in the mixture (Seader and Henley 1998). In either case, the process takes advantage of the difference in the relative volatility of the components of the mixture. In the chemical industry, distillation is actually a unit operation of universal importance, but it is a physical separation process, not a chemical reaction. Distillation is only suitable for the purification of volatile components. In real life, we have to deal with many substances, which are nonvolatile and cannot be separated and purified by distillation technology, such as many polar molecules such as salts, sugars, and amino acids. In this case, we usually use another purification technique that is crystallization to separate and purify them.

Crystallization is the process of forming solids or crystals (natural or artificial), in which atoms or molecules are highly organized into structures called crystals. Some ways of crystal formation are precipitation from solution, freezing, or rarely directly from gas (McCabe and Smith 2000). The properties of the resulting crystals are largely dependent on factors such as temperature and air pressure, as well as the time for the fluid to evaporate.

Crystallization is a solid–liquid chemical separation technique in which solutes are transferred from a liquid solution to a pure solid crystalline phase. In chemical engineering, crystallization occurs in the crystallizer. Therefore, crystallization is related to precipitation, although the result is not amorphous or disordered, but crystalline.

Recrystallization is a technique used to purify chemical substances. By dissolving both impurities and compounds in an appropriate solvent, the desired compound or impurity can be removed from the solution while the other is retained. It gets its name from the crystals that often form when the compound precipitates. In addition, recrystallization can refer to the natural growth process of large ice crystals, and smaller ice crystals form large ice crystals through the process of dissolution and recrystallization.

Filtration is a physical, biological, or chemical operation that can separate solids and fluids from the mixture through a filter medium with a complex structure (Sparks and Chase 2015). The fluid that can pass through the filter medium is called filtrate. Oversized particles that cannot pass through the filter media may form a filter cake on the top of the filter and be recovered. Solid–liquid separation is a convenient and fast unit operation, the main purpose is to remove solid particles from the solution. The disadvantage of this method is that the recovered solids may be contaminated by some fluids and the filtrate will contain fine particles (depending on the pore size, filter thickness, and biological activity). Filtration occurs both in the natural environment and in engineering systems. There are biological, geological, and industrial forms.

Distillation and crystallization are the most commonly used separation techniques by organic chemists, and many applications have been found in both daily life and industrial processes. However, they also have some obvious shortcomings, for example, time-consuming operation, long separation cycle, and limited applicability. Obviously, with the rapid development of phytochemistry and biochemistry at the end of the nineteenth century, traditional separation and purification techniques such as distillation and crystallization were incapable of meeting the increasing demands of scientific research, prompting scientists to find more efficient separation and purification methods.

1.3 Birth of Chromatography

At the end of the nineteenth century and the beginning of the twentieth century, natural sciences developed rapidly, especially phytochemistry, food chemistry, and biochemistry were in a stage of rapid advancement. Many biologically active compounds with complex structures, strong hydrophilicity, thermal instability, and large molecular weight were discovered. There was an urgent need for better techniques than traditional distillation and crystallization for the complete characterization of these substances. Looking back on history, it is not difficult to find that it is the difficult problems of separating pigments in plants and amino acids from protein degradation products in wool that lead to the birth of adsorption chromatography and partition chromatography, respectively (Wixom 2001).

1.3.1 Adsorption Chromatography

Long before the discovery of chromatography, dye chemists used capillary phenomena to separate dyes. They tested their dye mixtures by dipping string or cloth strips or filter paper into dye buckets. On the inserted material, the dye solution undergoes differential migration through capillary action, dividing the dye components into different colored strips. In the nineteenth century, several German chemists deliberately designed experiments to explore this phenomenon. They observed that by dropping a solution of an inorganic compound into the center of a piece of filter paper, concentric colored circles were formed. This method is named "capillary analysis" (Wixom 2001).

However, the discovery of chromatography is usually attributed to Russian phytochemist Mikhail S. Tswett. Because he first realized the physical and chemical basis of chromatographic separation in 1901 and applied this principle to the separation of complex plant pigments, especially the separation of carotenoids and chlorophyll, which was a major challenge faced by chemists at that time. Tswett's research topic is about the physical and chemical structure of chlorophyll. He needed to separate chlorophyll in a form that is as close to its natural state as possible. In this work, he discovered a strange phenomenon. The chlorophyll in plant leaves can be extracted by polar solvents (such as ethanol), but it cannot be extracted by nonpolar solvents (such as petroleum ether). The chlorophyll isolated from the plant is readily soluble in nonpolar solvents. After thinking about this confusing problem, Tswett came to the conclusion that this behavior of chlorophyll is not due to a simple solubility problem, nor is it caused by the chemical change of chlorophyll in ethanol into a "soluble" form, but it is because of the molecular adsorption force on the plant tissue that the chlorophyll is adsorbed and fixed on the surface of the plant tissue. In order to extract chlorophyll from plant tissues, one need not only to solve the problem of chlorophyll dissolution but also to overcome the adsorption of plant tissues to chlorophyll. The latter may be the main reason for the different extraction abilities of the solvents.

In subsequent work, Tswett studied in detail the interactions of plant pigments with more than 100 organic and inorganic powdery particles that may be used as adsorbents and their relative solvent strength, in order to determine the general adsorption and elution behavior of these substances. Ultimately, these studies led to a new method of separating chlorophyll and carotenoids through stepwise selective adsorption and elution. He named the newly developed separation method chromatography. The packed-column chromatography technique described by Tswett in 1906 is almost identical to the column chromatography technique still widely used today in organic chemistry or phytochemistry laboratories. He packed a vertical glass column with absorbent solid materials (such as calcium carbonate, alumina, silica gel, or powdered sugar), added a plant pigment solution to the top of the column, and then used an organic solvent as an eluent to pass through the column. The pigment is divided into a series of discrete color bands on the chromatographic column, separated by areas that are completely free of pigment. Since Tswett processed colored substances and formed ribbons on the columns, he called it chromatography (derived from Greek, meaning color writing). However, Tswett also specifically pointed out that chromatography is also suitable for the separation of compounds that do not contain chromophores.

Regrettably, the chromatographic method invented by Tswett did not attract enough attention from the scientific community. It may be because his work was mainly published in German and Russian botany journals, which was not easily understood by English-speaking scientists. Another important reason is that Tswett's research work was interrupted by the outbreak of the First World War, and Tswett was transferred to several different universities. In this disturbing period, his health was deteriorating. He died in Voronezh on June 26, 1919, at the age of 47 (Fig. 1.1).

Soon after Tswett invented adsorption chromatography and reported the successful separation of chlorophyll, carotenoids, and lutein from plant leaves, American scientist L. S. Palmer entered the field of food chemistry. The question he faced was why the color of butter made in summer and winter is different. Presumably it is caused by the change of plant pigments in dairy cow feed. The German scientists R. Willstätter and W. Mieg just recently determined the basic composition of "carotene" and "lutein" through a multistep crystallization purification method, but the chemical structure of these compounds was not figured out until the late 1920s.

Fig. 1.1 Mikhail S. Tswett (1872–1919). (Reproduced from Wikipedia at https://en.wikipedia.org/wiki/Mikhail_Tsvet)

Therefore, Palmer had to make a choice in 1910, whether to use the traditional multistep crystallization process or the newly created adsorption chromatography to purify plant pigments. Relying on Tswett's adsorption chromatography, Palmer found carotenoids in his butter extract, which explained the color change of butter caused by cow's diet from summer to winter. His research may represent the first practical application of chromatography in the American continent in addition to Tswett's own work.

Then, it was the excellent work of Richard Kuhn and his colleagues that really made Tswett's adsorption chromatography out of the laboratory and into the public eye. Richard Kuhn is an Austrian–German biochemist. In 1929, he became the head of the Institute of Chemistry of the newly established Kaiser Wilhelm Institute of Medicine (the Max Planck Institute of Medicine in Heidelberg since 1950). Kuhn's research fields include theoretical issues of organic chemistry (stereochemistry of aliphatic and aromatic compounds, synthesis of polyenes and cumene, composition and color, acidity of hydrocarbons), and a wide range of biochemistry (carotenoid, flavins, research on vitamins and enzymes). Specifically, he carried out important work in vitamin B2 and anti-dermatitis vitamin B6. Kuhn, Winterstein, and Lederer published an important paper on the purification of lutein on a $CaCO_3$ adsorption chromatography column in 1931. R. Kuhn was awarded the 1938 Nobel Prize in Chemistry for his "research on carotenoids and vitamins."

In the early and mid-1930s, Paul Karrer (1889–1971) was actively engaged in the research of natural products at the University of Zurich in Switzerland. After learning the results of Kuhn–Lederer, he used adsorption chromatography extensively in his research and introduced the application of chromatography in natural product research in many research papers published in the 1930s. He was awarded the 1937 Nobel Prize in Chemistry for his research on carotenoids, flavins, and vitamins A and B2. Although Tswett himself did not receive the Nobel Prize for his invention of chromatography, there is no doubt that his chromatography played an irreplaceable role in the pioneering work of many Nobel Prize winners.

Kuhn's assistant Edgar Lederer (1908–1988) moved to France in 1933 and began an active and independent research career. Edgar Lederer's research involves the separation and identification of a variety of biologically active substances, including carotenoids (aminopterin in red-boiled lobster shells, shrimp xanthine in goldfish skin); vitamin A2, perfume, and pheromone from queen bee; Lisopin in crown gallbladder; aryl alcohol in parasitic nematodes; and microbial lipids, glycolipids, and peptide lipids and murine amido peptides, and these studies mostly rely on chromatography to extract biologically active compounds from complex biological samples.

1.3.2 Partition Chromatography

The adsorption chromatography invented by Tswett provides a powerful tool for the separation and purification of pigments in plants, which greatly promotes the development of phytochemistry and food chemistry. However, since the biologically active molecules of strongly polar compounds or amphiphilic compounds cannot be dissolved in typical nonpolar organic solvents, the application of adsorption chromatography in the separation and purification of biologically active molecules is limited. However, biochemistry developed rapidly in the 1940s, and there was an urgent need for new and applicable separation and purification methods. Partition chromatography was established to solve the separation of natural amino acids (Ettre 2001). In the late 1930s, in the laboratory of the Wool Research Association in England, A. J. P. Martin and R. L. M. Synges began to study the amino acid composition of wool. Initially, they used a liquid–liquid countercurrent extraction process to extract acetyl amino acids from wool. The acetylated amino acid is partitioned between the chloroform and water phases. Liquid–liquid countercurrent extraction was the most advanced separation method at that time. They designed, constructed and operated a countercurrent liquid–liquid extraction tower composed of 40 units to continuously separate acetyl amino acids. However, the separation device was primitive and required adjustment of the device continuously, which was laborious; the working environment was harsh, and the chloroform vapor escaped in large quantities and polluted the entire space; a separation took a week to complete (Martin and Synge 1941a). It was urgent to find alternative technologies. About 40 years later, Martin recalled this experience (Martin 1979):

> In 1940 it occurred to me that the crux of the problem was that we were trying to move two liquids in opposite directions simultaneously. Equilibrium had to be established rapidly or the experiment took far too long, but this meant converting the liquids to very fine droplets and if the droplets were too small they would not settle out or move into the required

> direction within any reasonable period of time. This meant that the machine was bound to be a compromise unless I could either introduce centrifugal force to speed up the movement of droplets or think of a completely different system. Then I suddenly realized that it was not necessary to move both liquids; if I just moved one of them the required conditions were fulfilled. I was able to device a suitable apparatus the very next day, and a modification of this eventually became the partition chromatograph with which we are now familiar.

In their partition chromatography research, they used a device similar to Tswett adsorption chromatography, but used a solid-phase carrier gel containing water as the column packing. The principle of this method is to firmly bond a liquid such as water to the fine particulate solid contained in a glass tube and then infiltrate a second liquid immiscible with the first liquid such as chloroform as a mobile phase. Silica gel is a granular solid, which was chosen by Martin and Synge as a stationary phase composed of water tightly bound to silica particles. They used this technique to successfully isolate acetyl amino acids. Although the principle of their invention is similar to Tswett's method, Martin and Synge's innovation involves the concept of liquid–liquid separation in a continuous flow mode with one liquid (water) fixed on an inert solid (silica gel) and another immiscible solvent (chloroform) flowing through so that the solute molecules continuously distribute between the two liquids on the way moving through the column. This technique is called partition chromatography. In a paper published in 1941 (Martin and Synge 1941b), Martin and Synge detailed the experimental results and theoretical considerations of partition chromatography. They not only demonstrated the feasibility and potential of this new technology but they also developed the theory of chromatographic separation process and introduced the concept of theoretical plates in distillation into chromatography, laying a foundation for the study of chromatographic process kinetics. In addition, a famous prediction was made, that is, "The mobile phase does not have to be a liquid, but can be vapor... Therefore, in a chromatographic column that allows permanent gas to flow through a gel impregnated with a nonvolatile solvent, it should be possible to separate volatile substances very finely." Perhaps because of World War II, this innovative idea was not adopted for nearly a decade until 1952 when Martin and James published their work on gas–liquid partition chromatography (James et al. 1952; James and Martin 1952). In 1952, Martin and Synge won the Nobel Prize in Chemistry for their work on partition chromatography (Fig. 1.2). The partition chromatography established by them and the theoretical plate model proposed by them have a meaning far beyond the separation of amino acids and peptides and have had a lasting impact on scientific research.

1.4 Development of Chromatography

1.4.1 Thin-Layer Chromatography

The initial partition chromatography system had experienced some difficulties because of the lack of reproducibility of silica gel performance, the uneven packing

Fig. 1.2 Archer John Porter Martin and Richard Laurence Millington Synge. (Reproduced from Wikipedia at https://en.wikipedia.org/wiki/Archer_Martin and https://en.wikipedia.org/wiki/Richard_Laurence_Millington_Synge, respectively)

of the chromatographic column, and the large amount of sample required. Partly for this reason, Martin and his colleagues proposed a new partition chromatography method based on cellulose filter paper, in which the fixed medium is filter paper, the stationary phase is water combined with cellulose on the filter paper, and the mobile phase is organic solvents immiscible with water (Consden et al. 1944). This technique provides the required reproducibility, and since the 1940s, paper chromatography has been widely used to analyze biologically important compounds such as amino acids, steroids, carbohydrates, and bile pigments. At the same time, they also demonstrated the great potential of two-dimensional chromatography in separating complex mixtures. The filter paper developed in one organic solvent was rotated through 90° and placed in another organic solvent for elution to produce the two-dimensional chromatogram generated by the compound moving in two orthogonal directions. Two-dimensional chromatography overcomes the peak capacity limitation of one-dimensional chromatography and further improves the separation ability of chromatography.

Possibly for the same reason, in order to overcome the abovementioned shortcomings of column chromatography, two Soviet pharmacists Nikolay A. Izmaylov and Maria S. Shrayber coated the carrier material in the form of a thin film on the glass plate. The plate and carrier material are then manipulated in a manner similar to paper chromatography. They reported the results of their research in the Soviet Union in 1938, but it was not until 1956 that the German chemist Egon Stahl began to conduct in-depth research on its application, that the potential of the method was not widely understood. This system is now called thin-layer chromatography (TLC) (Fried and Sherma 1999).

1.4.2 *Gas Chromatography*

As Martin and Synger predicted in 1941, chromatography can also use gas as the mobile phase for separation, which is gas chromatography. As early as 1944, the German chemist Erika Cremer made a preliminary attempt using a chromatographic method for the separation of gases. Martin and his colleagues first reported the elution of organic acids and amines by gas chromatography (James et al. 1952; James and Martin 1952) in 1952. After the small particle carrier material is coated with a nonvolatile liquid as a stationary phase, it is packed into a glass tube to make a packed chromatographic column. The column is heated to the set temperature. The sample mixture is injected into the tube inlet and driven by compressed gas through the chromatographic column. The components in the mixture appear in well-separated zones according to their boiling points. The potential of this technology was immediately recognized by petrochemists as it provides a simple and fast method for the analysis of complex hydrocarbon compounds in petroleum products.

In 1957, when conducting theoretical research on gas chromatography columns, Marcel Golay (1979), a consultant of PerkinElmer Corporation, concluded that the inner wall of a column with a length of 90–180 m and an inner diameter of 0.25 mm coated with a liquid film on it will produce an excellent separation effect. Fortunately, around the same time, a flame ionization detector with extremely low detection limits was introduced, which can detect the small sample volumes required by these new columns. These capillary columns (or Golay columns), now called open-tubular chromatographic columns, characterized by an open tube design and an inner diameter of less than 1 mm, have caused an explosive development of high-resolution chromatographic methods. It is now possible to separate hundreds of mixture components in one single chromatographic run.

1.4.3 *Size Exclusion Chromatography*

Molecular sieves are porous materials that capture mobile-phase gases. When a chromatographic column is packed with molecular sieve materials for the separation of compounds of different molecular sizes, the macromolecules cannot enter the micropores and largely bypass the granular materials and flow through the pores in the interstitial spaces unhindered. However, when small molecules are introduced into the column, they meander into and out of the micropores due to diffusion and their migration rate is delayed. Medium-sized molecules show different migration rates according to their size. In 1959, Sweden's Per Flodin and Jerker Porath developed a cross-linked dextran gel material, which acts as a molecular sieve for dispersing substances of different molecular sizes dissolved in a liquid (Porath and Lindner 1961). This extends the molecular weight range of chromatography to peptides, proteins, and high-molecular-weight polymers. When this separation method is applied to biological macromolecules, it is also called gel filtration chromatography; when it is applied to synthetic polymers, it is called gel permeation chromatography; now it is collectively called size exclusion chromatography.

1.4.4 Ion Exchange Chromatography

Ion exchangers are natural substances (such as certain clays) or synthetic resins containing fixed negative ions (cation exchangers) or fixed positive ions (anion exchangers), which can adsorb oppositely charged solute ions from the solution. The adsorbed solute ions can be desorbed by competing with the ions of the same charge in the solution. This method is called ion chromatography or ion exchange chromatography (Walton 1976). T.I. Taylor and Harold C. Urey first reported this chromatographic separation method using zeolite in 1938. In 1942, during the Manhattan Project, this method received widespread attention as a method for separating rare earth and transuranium elements, fission products of uranium, and other elements produced by thermonuclear explosions. After the end of World War II, the focus of scientists shifted to the biochemical applications of ion exchange chromatography. Stanford Moore and William Stein made the next major advance in ion exchange chromatography (Moore and Stein 1951). In the late 1950s, quantitative amino acid analysis was carried out by ion exchange chromatography, and an automatic amino acid analyzer with key functions of high sensitivity, precision, automation, and quantification was developed. They first improved the chromatographic method using starch derivative columns to separate amino acids and then invented the photoelectric droplet counting fraction collector, which is a vital equipment that is still widely used in laboratories around the world. Later, they used a sulfonated polystyrene-divinylbenzene cross-linked polymer chromatography matrix and together with Daryl Spackman (Daryl Spackman) invented an automatic amino acid analyzer. In the early 1960s they determined the amino acid composition and amino acid sequence of ribonuclease. This groundbreaking research led them to win the 1972 Nobel Prize in Chemistry (Fig. 1.3). The work of Sober and Peterson (Sober and Peterson 1954; Peterson and Sober 1956) on the

Fig. 1.3 Stanford Moore and William H. Stein. (Reproduced from Wikipedia at https://en.wikipedia.org/wiki/Stanford_Moore and https://en.wikipedia.org/wiki/William_Howard_Stein, respectively)

use of cellulose ion exchange chromatography for protein separation opened up the application of polysaccharide media in the purification of biological macromolecules.

1.4.5 Affinity Chromatography

Affinity chromatography is a method for separating compounds based on the phenomenon of biomolecular recognition (Wilchek and Chaiken 2000). Molecular recognition is the specific binding reaction between specific molecular pairs observed in nature, such as the binding between enzyme and substrate, antibody and antigen, biotin and avidin, and lectin and polysaccharide. Through a chemical reaction, a partner, such as a substrate, of the biological recognition reaction pair is immobilized on a solid carrier, usually agarose or dextran gel. During the separation process, another partner biomolecule, such as an enzyme, binds to the substrate on the solid phase, while other components are eluted in flowthrough. The adsorbed molecules can then be eluted by changing the mobile-phase composition. Since Cuatrecasas, Wilcheck, and Anfinsen (1968) first used the term affinity chromatography in 1965, affinity chromatography methods based on biomolecular recognition have quickly gained popularity in protein separation and purification.

Affinity chromatography is a variant of liquid chromatography. This technology takes advantage of the unique biological specificity of protein–ligand interactions. This concept is realized by combining the ligand with an insoluble support and using the prepared substance as the column packing. Ligands can have low or high-molecular weights and consist of nucleic acids, enzymes, or many other compounds, while the support can be cellulose derivatives, cross-linked dextran, polyacrylamide, or agarose. Using the prepared material as the column packing will retain certain proteins that have significant affinity for specific ligands. These compounds can then be eluted by changing the composition or pH of the mobile phase to facilitate dissociation and weaken ligand binding.

1.4.6 Chiral Chromatography

Many attempts have been made to use natural chiral biopolymers as a chiral chromatographic stationary phase to separate enantiomers (Allenmark 1991). In 1938, Henderson and Reid first reported the separation of the racemate *p*-phenylene-bis-iminocamphor (+)-lactose on a chromatographic column. Dalgliesh proposed the famous “three-point action” model of chiral separation based on the paper chromatography experiment of phenylalanine derivatives in 1952. A few years later, Krebs separated the chiral isomers of amino acids and hydroxy acid derivatives on the starch stationary phase, confirming the feasibility of the “three-point interaction” concept. Hesse and Hagel made an interesting discovery in 1973, that is, when microcrystalline cellulose is acylated under heterogeneous conditions, it retains its chiral crystal structure and proved that multiple enantiomers can be separated.

By reacting cross-linked polyacrylic acid chloride with the secondary hydroxyl group of quinine, Grubhofer and Schleith prepared a chiral ion exchange resin in 1953, which can partially resolve racemic mandelic acid. Subsequently, a large number of chiral polymers were synthesized and used as stationary phases for chiral separation, such as condensation reaction of L-tyrosine and formaldehyde, acrylic and methacrylic derivatives of optically active amines and amino acids, and amino acid bonding to chloromethylated polystyrene.

A major breakthrough in chiral chromatography occurred in the 1970s. Davankov introduced chiral ligand exchange chromatography in 1970, Cram reported the high enantioselectivity of bonded chiral crown ethers in 1974, and Gil-Av and colleagues demonstrated the enantiomeric separation ability of silica bonded charge transfer selectors in 1976. At present, almost all types of organic compounds can be resolved into enantiomers on a preparative chromatographic scale. The separation of several racemic compound mixtures shows the great potential of chiral chromatography in the analysis of enantiomers.

1.5 High-Performance Liquid Chromatography

As mentioned above, after entering the 1960s, with the development of phytochemistry, biochemistry, and clinical chemistry, the scope of application of liquid chromatography has continued to expand, from routine laboratory preparation and separation into the field of microanalysis and testing. As an analytical technology, gas chromatography, especially capillary gas chromatography, has extremely high column efficiency. It can separate hundreds of compounds in one chromatographic experiment, and it plays an irreplaceable role in process detection and product analysis in the petrochemical industry. As gas chromatography requires the separated substances to be thermally stable and volatile, this greatly limits the application of this technology in the separation and analysis of biologically active substances. Liquid chromatography can overcome the limitations of gas chromatography, but the separation efficiency of liquid chromatography at that time was far from that of gas chromatography. In 1964, Calvin Giddings and John H. Knox (Fig. 1.4) carried out groundbreaking research in the theory of liquid chromatography, and clearly pointed out the way to improve the separation efficiency of liquid chromatography, to be comparable with that of gas chromatography. Since the diffusion coefficient of the solute in the liquid is about 10^5 smaller than the diffusion coefficient in the gas phase, the particle sizes of the stationary phase used in the liquid chromatography should be reduced accordingly to achieve the equivalent column efficiency of the gas chromatography. Therefore, using very small particles such as 5–10 μm in liquid chromatography and coating or bonding a very thin liquid film on solid particles are the key to obtaining high column efficiency.

Fig. 1.4 J. Calvin Giddings (September 26, 1930–October 24, 1996) (Reproduced from Giddings 1997). John H. Knox (October 21, 1927–October 15, 2018)

The technology now called high-performance liquid chromatography (HPLC) is the result of the integration of multiple innovative technologies. Its successful development depends on (1) the development of a pump that can deliver a stable high-pressure liquid stream to the chromatographic column to force the liquid to pass through the narrow gap channel formed in the chromatographic column of the packing and perform sample detection at a reasonable rate; (2) the development of high-performance chromatographic packing materials, wherein high-efficiency packing uses a rigid porous material with a small particle size as a carrier to bond the organic stationary phase to the carrier through a chemical reaction to avoid loss during run; and (3) a high-sensitivity detector which can detect the required small sample volume. Among them, the development of high-performance chromatographic stationary phases with high-pressure resistance and small particle size is the most critical. Initially, only adsorptive solids were used as the stationary phase, because the liquid coating was easily dissolved by the mobile phase and lost. Drawing lessons from the chemical bonding method used in gas chromatography to reduce adsorption activity, István Halász first used these silanization reactions to bond molecular layers on the surface of silica gel. These molecules act as a stationary phase in the liquid system and interact with solutes. The chemically bonded phase has good stability, low mass transfer resistance, and versatility. It has become the most important method for the preparation of stationary phases in high-performance liquid chromatography.

1.6 Mixed-Mode Chromatography

The adsorption or retention of solutes on chromatographic media is achieved through non-covalent molecular interactions between solutes and stationary-phase ligands. The mobile phase affects the interaction between the solute and the ligand through solvation effect or ionization effect, thereby regulating solute retention. Intermolecular interactions can be roughly divided into van der Waals forces, electrostatic forces, hydrogen bonds, dipole–dipole forces, metal coordination, and hydrophobic interactions. In liquid chromatography with organic solvent as the mobile phase, polar forces, including electrostatic interaction, hydrogen bonding, metal coordination, and dipole interaction, are the main driving force for solute binding; while in the aqueous mobile phase, hydrophobic interaction, electrostatic interaction, hydrogen bond, and metal coordination play a major role. The contribution of van der Waals forces is small and can usually be ignored.

According to the main types of intermolecular interactions involved in the chromatographic retention process, we can divide chromatography into normal phase chromatography, reversed-phase chromatography, ion exchange chromatography, hydrophobic interaction chromatography, hydrophilic interaction chromatography, and metal coordination chromatography. In these chromatographic modes, there is usually only one type of intermolecular interactions that dominate the entire retention process. For example, in normal phase chromatography, solute retention is determined by polar interactions such as hydrogen bonds and dipole–dipoles, which are only related to the polar part of the solute. In reversed-phase chromatography, the situation is just the opposite. Solute retention is dominated by hydrophobic interaction. Solute retention in ion exchange chromatography is determined by the Coulomb force between the solute and the ionic groups in the stationary phase and has nothing to do with the hydrophobic part of the solute. Because the retention of solutes in these chromatograms is based on a single retention mechanism, these chromatographic formats are also called single-mode chromatography. In these single-mode chromatography, the retention mechanism is relatively well understood, and the factors that control retention are well defined. Therefore, it brings great convenience to method development and optimization and is the most important mode of use in chromatography.

Mixed-mode chromatography has received increasing attention in the past two decades. Mixed-mode chromatography refers to the fact that the retention of solute on stationary phase in the process of chromatographic separation involves two or more different types of interactions. This means that the solute must be amphiphilic, that is, it contains a hydrophobic and a hydrophilic/ionic parts; and the chromatographic ligand must be multifunctional, containing a hydrophobic group and a hydrophilic/ionic group complementary to the solute. In strict sense, the intensity of the different types of intermolecular interactions that cause solute retention should be comparable in mixed-mode chromatography. For example, in reversed-phase ion exchange mixed-mode chromatography, the retention of ionic solutes is controlled jointly by reversed-phase retention mechanism and ion exchange retention mechanism. Their contributions to the retention are comparable in quantitative terms, and the absence of any kind will result in a substantial decrease in the retention of solutes or even not being retained. In fact, this mode of action can also be understood as the simultaneous action of multiple sites between the solute and the stationary-phase ligand, that is, multivalent action. In this case, the bonding strength is not only related to the complementarity of the two group properties but also related to the matching of their geometric configurations.

Multimodal chromatography usually means that the stationary phase containing multifunctional ligands can function in different separation modes under different mobile-phase conditions. The multimodal chromatographic stationary phase can be a mixed-mode chromatographic stationary phase. In this case, the multimodal chromatogram and the mixed-mode chromatogram are equivalent and can be used interchangeably. However, it should be noted that the multimodal chromatographic stationary phase can also be a mixture of different single-mode stationary phases, such as reversed-phase and ion exchange stationary phases. Under the conditions of reversed-phase chromatography, the solute is retained by the hydrophobic mechanism, while under the conditions of ion exchange chromatography, the solute is retained by the Coulomb electrostatic mechanism. Since the distance between the two stationary-phase ligands is much greater than the size of the solute, the solute cannot act simultaneously with the two. Therefore, it can only be bound either in hydrophobic mode or in ionic mode. In this case, multimodal chromatography cannot be interchanged with mixed-mode chromatography, but should be treated separately.

In bioaffinity chromatography, the binding of solute molecules on affinity ligands is based on specific interactions between biomolecule pairs, such as antibodies and antigens, enzymes and substrates, lectins and glycoproteins, avidin and biotin, etc. The molecular recognition between these biomolecules involves the interaction of multiple amino acid residues between different proteins at the molecular binding interface. The contribution of each amino acid residue may not be large, but a combination of many small effects can produce a very large affinity. This multivalent effect is fully utilized in nature and can be found from protein folding, nucleic acid replication to virus invasion, cell signal transduction, and other processes. The molecular recognition involved in bioaffinity chromatography is generally very complicated, and the genetic engineering of ligands to improve performance requires expertise in molecular biology and biotechnology. Mixed-mode chromatography can be regarded as a simplified version of bioaffinity chromatography, which mimics the behavior of biologically active ligands by tailored synthesis of ligands, thereby providing chromatographic packings of high selectivity, high stability, and low cost for the separation and purification of complex biomacromolecules.

Mixed-mode chromatography is also different from multidimensional chromatography. Multidimensional chromatography refers to the separation of solute mixtures using chromatographic modes that are orthogonal to each other in terms of retention mechanisms. For example, Martin uses paper chromatography for two-dimensional chromatographic separation of amino acid mixtures. In recent years, multidimensional chromatography has been widely used in proteomics research to separate a large number of proteomes in complex biological samples. The most typical two-dimensional chromatography uses ion exchange chromatography for the first-dimensional separation and then uses reversed-phase chromatography for the second-dimensional separation of the fractions pooled from ion exchange chromatography, which greatly increases the peak capacity of column chromatography. In the separation and purification of biological macromolecules, three-step chromatographic purification protocols of capture, intermediate purification, and polishing are often used. The chromatographic methods used in these three steps are often based on different retention mechanisms. Therefore, this cascaded purification chain can also be regarded as a multidimensional chromatogram operated offline. Mixed-mode chromatography can be used in any step of the three-step purification scheme, making it a powerful tool for biological separation and purification. As can be seen from the following description, mixed-mode chromatography was actually developed mainly to meet the challenges of separation and purification of biological macromolecules.

1.6.1 Mixed-Mode Chromatography of Nucleic Acids

Nucleic acids can be separated by anion exchange chromatography. The mechanism of nucleic acid retention is based on the electrostatic interaction of phosphodiester with cationic sites on the chromatographic matrix. Nucleic acids bind to these matrices through electrostatic interactions and negatively charged ions in the mobile phase compete with nucleic acids for binding sites, so the adsorbed nucleic acids can be eluted by changing the ionic strength of the mobile phase. (Diethylamino)-ethyl (DEAE)-modified cellulose or agarose has been successfully used to separate and purify various nucleic acids. Although it is generally believed that on such chromatographic materials, anion exchange is the main driving force leading to retention, but the functional groups on the surface of the material can also bind to nucleic acids through hydrophobic and hydrogen bond interactions. Therefore, the separation of nucleic acids on DEAE-modified materials may be facilitated by a mixed-mode retention mechanism. Supporting this idea is experimental observation. If the nucleic acid retention is dominated by pure ion exchange theory, then the nucleic acid retention is only determined by the number of phosphates in its molecule and has nothing to do with the base sequence. However, experimentally it was observed that nucleic acid fragments can be separated not only by size (using electrostatic interaction) but also by nucleobase sequence (using hydrophobic and hydrogen bond interactions). The mixed functionality present on the stationary phase makes separation possible (McLaughlin 1989).

Inspired by these experimental observations, researchers tried to introduce hydrophobic groups into ion exchange materials to prepare amphiphilic chromatographic materials containing ionic and hydrophobic interaction sites for the separation and purification of nucleic acids. The indirect method is to adsorb cationic surfactant on the surface of the reversed-phase chromatography packing so that the packing contains both hydrophobic and ion exchange groups. The direct method is to simultaneously bond hydrophobic groups and ionic groups on the surface of the carrier or introduce ion exchange groups by further modifying the bonded chemically active groups. In short, through chemical bonding, the surface of the inert carrier is functionalized, so that it has the necessary multifunctional properties.

The direct method can be exemplified by work that began in the 1960s. Mandell and Hershey (1960) prepared a nucleic acid separation material by converting the carboxylic acid residue of albumin into the corresponding methyl ester and adsorbing the resulting substance on silica gel. The chromatographic column packed with methylated albumin can effectively separate DNA, soluble RNA, and ribosomal RNA according to molecular size, hydrogen bond, and base content. There are side chains of lysine and arginine on the surface of methylated albumin, which act as an anion exchange matrix, and the methyl group added by esterification reaction acts as a hydrophobic interaction matrix, so the chromatographic packing has features of dual hydrophobic/ionic action.

Gillam et al. (1967) introduced benzoyl or naphthoyl into DEAE cellulose by esterification to enhance the hydrophobic property of the anion exchange matrix. They observed the separation of transfer RNA (tRNA) on benzoylated DEAE cellulose, partly based on molecular size and partly due to the hydrophobic interaction between tRNA bases and benzoyl groups. Due to the increase in the strength of its hydrophobic effect, some ethanol needs to be added to the mobile phase to elute some tRNAs from this modified cellulose, especially tRNAs specific to the amino acids phenylalanine and tyrosine. This is because both tRNAs contain very hydrophobic super-modified bases in the anticodon loop.

The indirect method can be illustrated by using the RPC-5 column as an example (Pearson et al. 1971, 1974). The RPC column is prepared with Plascon 2300 (a nonporous spherical polymer, poly(chlorotrifluoroethylene)) as a hydrophobic medium and coated with trimethyloctylammonium chloride. Because the materials have strong hydrophobic properties, they are one-sidedly described as reversed-phase chromatographic materials. In fact, the quaternary ammonium salt adsorbed on the surface will produce a material with both hydrophobic and ionic interaction sites, which is a typical amphiphilic chromatographic material. The RPC-5 column has been used in the separation and purification of oligodeoxynucleotides and tRNA and DNA fragments. McLaughlin and coworkers (Bischoff et al. 1983) used octadecyl bonded silica gel as a substrate and adsorbed methyl trioctyl ammonium chloride on the surface, which greatly improved the chromatographic separation efficiency and can be applied to the purity analysis of tRNAs crude extracts.

Crowther et al. (1982) reported an improved method for determining the chain length, leading and/or terminating nucleotides and base composition of deoxyribose and ribose oligonucleotides. The method is based on phosphodiesterase-catalyzed oligonucleotide hydrolysis, ion pair high-performance liquid chromatography with tetrabutylammonium phosphate as the counter ion to separate oligonucleotide hydrolysates. On this basis, they developed a method for preparing silica-based mixed-mode stationary phase by chemical bonding, which simulates ion pair separation in terms of retention characteristics (Crowther et al. 1982). These stationary phases consist of a mixture of two ligands with different chemical properties that are chemically bonded to a silica support. The stationary phase thus obtained is reversed phase to a large extent, but also contains significant ion exchange properties. By bonding ionic and hydrophobic groups in an appropriate ratio, a mixed retention mechanism can be formed, resulting in unique selectivity. At the same time, the excellent stability and efficiency of the bonded phase are maintained. The ratio of hydrophobic properties to ionic properties can be controlled during the synthesis process, and it can be used as a tool to change the stationary phase instead of only changing the mobile phase to achieve the desired separation. The synthesis and behavior of anions/reversed-phase and cations/reversed-phase materials were described and applied to the simultaneous separation of nucleosides and nucleotides and the separation of catecholamines. They used a mixture of chloropropyl- and octyl-dimethylchlorosilane to modify the silica support. After this initial step, the chloropropyl group is then converted to an amino functional group by reaction with benzyldimethylamine. For reversed-phase/cation exchangers, the alkyl group and the alkylphenyl group are bonded, and then the aromatic ring is sulfonated.

Crowther et al. (1983) bonded various proportions of ionic and hydrophobic ligands to the surface of silica support to prepare a new type of stationary phase. The ratio of hydrophobicity to ionicity can be reproducibly controlled by changing the volume ratio of monochlorosilane reagent in the bonding mixture. In this study, octyl- and 3-chloropropyldimethylmono-chlorosilane ligands were bonded, and then the chloro group was converted into a quaternary amine by nucleophilic substitution of benzyldimethylamine. The obtained hydrophobic ion exchanger was found to be very suitable for separating oligonucleotides with a chain length of about 20, and the resolution improved compared with the existing ion exchange or reversed-phase separation methods. The selectivity for polynucleotides of these mixed-mode packings was found to be comparable to that observed for RPC-5 and similar resins.

Aromatic cation exchange packings are synthesized by sulfonating chlorodimethyl(3-phenylpropyl)silane bonded to 10 μm irregular silica. The sulfonation can be performed before or after attaching the silane to the support, each producing phases with different exchange capacities and selectivities. The use of elemental analysis, titration data, and capillary gas chromatography analysis of the acid-hydrolyzed phase is to monitor the progress of the reaction (Crowther et al. 1984a, b).

Hydrophobic anion exchangers are formed by co-bonding ionic and hydrophobic ligands to silica gel. These phases are used to separate single-stranded oligonucleotides and double-stranded DNA restriction fragments. By changing the ratio of *n*-octyldimethylsilane to 3-chloropropyldimethylsilane or 4-chlorobutyldimethylsilane added in the silylation process, a series of mixed ligand stationary phases can be obtained. A new gas chromatography fluorination method was used to determine the concentration and ratio of bonded ligands. For *n*-octyldimethylsilane, the total ligand coverage was found to be close to 2.1 ligands/nm^2. The bonding reproducibility of the mixed-mode phase is very good. By using the ligand density as a variable that can be easily modified, nucleic acid separation is achieved under mild mobile-phase conditions (Floyd et al. 1986).

McLaughlin et al. (Bischoff et al. 1983) reported the separation of tRNA in baker's yeast by high-performance liquid chromatography with a reversed-phase support (ODS-Hypersil) treated with trioctylmethylammonium chloride. Although it is likely that both anion exchange and reversed-phase mechanisms play a significant role in causing retention and separation on this column, the hydrophobic interaction is largely responsible for the separation of one tRNA from another. Chromatographic separation of large amounts of tRNA produced many fractions with different amino acid receptor activities and almost no cross-contamination. In some cases, the column resolved several single nucleotide modifications of $tRNA^{Phe}$. Using a 250 × 6.2 mm chromatographic column, at least 100 A_{260} units of tRNA can be chromatographically separated without serious loss of resolution. As the last step of the purification step, the tRNA separated from the column has high amino acid receptor activity.

The commercially available aminopropylsilyl bonded phase was modified by amidation with three different organic acids. Each anion exchanger contains an unreacted amine group and a hydrophobic amide group (Bischoff and McLaughlin 1983). Two oligonucleotides $(Ap)_3A$ and $(Up)_3U$ with the same chain length but different hydrophobic properties were used as probes to examine the chromatographic properties of these carriers. The separation of these two solutes on a modified carrier with a pH range of 4.5–6.5 and an acetonitrile concentration of 0–48% was achieved. The most effective carrier contains the phenylalanine moiety (APS-PHE) bound to a commercial carrier via an amide bond. This support has been used to separate oligonucleotides of the same size (6–12 residues) but with different sequences. The difficulties inherent in the optimization of mixed-mode chromatographic parameters were discussed.

Then, they reported on the preparation methods of two mixed-mode (ionic-hydrophobic) carriers for high-performance liquid chromatography (Bischoff and McLaughlin 1984a). First, the hydrophobic group was covalently bound to the anion exchange carrier (ASP-Hypersil). In the second method, the ionic amine was combined with the reverse-phase carrier (ODS-Hypersil) through hydrophobic interaction. In the former case, the hydrophobicity of the carrier is controlled by adding alkyl groups of various chain lengths, and the ionic interaction is controlled by the pH of the mobile phase. The analysis of two simple solutes $(Ap)_3A$ and $(Up)_3U$ shows that there are ionic and hydrophobic interactions. In the latter case, it has been shown that the packing can be used to isolate large oligonucleotides, such as $(Up)_{90}$ $(Up)_{90}$. Hydrophobic interactions are important in the observed separation. The combination of nonpolar amino acids with specific tRNAs will cause a large change in the retention volume, which is due to the stronger adsorption on the carrier.

About 65% of the amine groups of the aminopropylsilyl bonded phase silica anion exchanger (APS-Hypersil) are modified by organic acids containing *n*-alkyl moieties of different chain lengths, resulting in mixed-mode chromatography media with different hydrophobicity (Bischoff and McLaughlin 1984b). These stationary phases can achieve high-resolution transfer ribonucleic acid (tRNA) separation when used with a decreasing ammonium sulfate gradient. The observed resolution seems to be mainly a function of interfacial precipitation, which is affected by the high initial salt concentration of the mobile phase and the hydrophobic nature of the stationary phase, and then selectively redissolves during the decreasing salt gradient. Compared with the pure reverse-phase carrier, the residual amine group reduces the hydrophobicity of the stationary phase. Mixed-mode media can be used to separate aminoacylated tRNAs from non-aminoacylated tRNAs.

C18 reversed-phase high-performance liquid chromatography carrier (octadecylsilyl-Hypersil) coated with tetraalkylammonium salt (methyl trioctyl ammonium chloride) produces a chromatographic matrix with ionic and hydrophobic properties (Bischoff and McLaughlin 1985, 1990; McLaughlin 1986; McLaughlin and Bischoff 1987). Using this material, oligonucleotides and tRNAs can be separated with high resolution. The observed resolution is partly due to the apparently no diffusion process taking place during chromatographic separation with this matrix. After a single chromatographic step, some tRNA can be obtained in high purity from a large tRNA mixture. Generally, it is more effective to use the matrix as the last step of a specific tRNA purification procedure. A two-step procedure is described, which in some cases allows the separation of small amounts of specific tRNA isoforms.

1.6.2 Mixed-Mode Chromatography of Proteins

The first use of mixed-mode chromatography in protein purification is usually attributed to A. Tiselius. As early as 1956, he reported the preparation method for calcium phosphate (hydroxyapatite) columns suitable for protein chromatography and specific elution conditions for proteins on this column (Tiselius et al. 1956). The protein adsorbed on the chromatographic column can be eluted by gradually or continuously increasing the concentration of the eluent (phosphate buffer with constant pH) or by displacement. Chromatography separates many proteins with a molecular weight of about 10,000 to millions. In several cases, multiple components from a single peak were resolved in gel electrophoresis, indicating that the purified protein was not homogeneous. Some features of this method were discussed, especially the narrow elution range and mutual displacement between proteins. And it was found that low-molecular-weight substances hardly adsorbed on these chromatographic columns. It is now generally accepted that hydroxyapatite (HA) with the chemical structural formula $Ca_{10}(PO_4)_6(OH)_2$ is a crystalline material, with positively charged calcium-based sites and negatively charged phosphate-based sites on the surface. When they are placed in contact with proteins, metal coordination and electrostatic interaction occur, leading to protein adsorption by a dual metal affinity/cation exchange mechanism.

Kremzner and Wilson (1963) first reported the use of hydrophobic/ion exchange mixed-mode chromatography for protein purification. They synthesized benzyldiethylaminoethyl cellulose as a specific adsorbent to capture acetylcholinesterase from the extract. The enzyme extract was loaded on a benzyldiethylaminoethyl cellulose column that had been equilibrated with 0.01 M sodium phosphate buffer (pH 7.0) and rinsed with 0.01 M sodium phosphate buffer (equilibration buffer) at pH 7.0. When the protein concentration in the effluent dropped sharply, a gradual gradient with increasing salt concentration was used for elution. Acetylcholinesterase was eluted at 0.50 M NaCl. This procedure can purify the enzyme by more than tenfolds, the specific activity is 20–40, and the recovery is more than 70%.

R. J. Yon (1972), in the process of studying the adsorption behavior of protein affinity chromatography, found that the protein was irreversibly adsorbed due to the hydrophobic interaction with the alkyl spacer. In order to overcome the strong adsorption of alkyl groups, a hybrid adsorbent containing hydrophobic and ionic groups was proposed. By selecting an appropriate ionic group and pH, the charge repulsion between the adsorbent and the protein is introduced to weaken the strong hydrophobic interaction between the adsorbent and the protein. This adsorbent may help to distinguish between lipophilic proteins and weaken the ability of hydrophobic stationary phases to inactivate proteins. Using the cyanogen bromide activation process of agarose, the aminodecyl group and *N*-(3-carboxypropionyl)aminodecyl group were respectively bonded to the agarose gel microspheres to prepare cation and anion exchanges containing hydrophobic groups. The agent, through the C10 alkyl chain and the ionic group of the terminal group, has hydrophobic and electrostatic interaction with the protein. Using bovine serum albumin with an isoelectric point of 4.7–4.9 as a model protein, the adsorption behavior under different pH conditions was studied. At pH 5 and 6.5, the protein is almost completely absorbed by hydrophobic interaction. However, when the pH is increased to 8.0 and 9.5, the charge repulsion effect is enhanced, the non-adsorbed protein component increases, and the hydrophobically bound protein component decreases. The behavior of bovine serum albumin is completely consistent with the mechanism of charge repulsion and can be used for the selective elution of hydrophobic proteins in complex protein mixtures. This technique was used for the purification of aspartate transaminase serum carbamate from wheat germ (Yon and Simmonds 1975) and the purification of human erythrocyte glycoprotein (Simmonds and Yon 1976).

Shaltiel et al. used the cyanogen bromide activation method to bond a series of homologous α-alkylamine compounds to agarose gel microspheres and investigated the use of these stationary phases in protein purification (Er-el et al. 1972). The retention behavior of glycogen phosphorylase changes significantly as the length of the alkyl side chain increases: from no retention in the methyl group to a slight retention in the propyl group and to a strong binding in the hexyl group. On the preparation scale, one-step purification of phosphorylase can be accomplished by passing the crude muscle extract through a butyl sepharose column. This work was extended to include the homologous series of ω-diaminoalkyl sepharose. On ω-aminobutyl agarose, a 350-fold of the synthase purification was achieved. In the α-alkyl agarose series, the butyl side chain is sufficient to retain glycogen phosphorylase, while in the ω-aminoalkyl agarose series, the hexyl side chain is required to retain the enzyme, indicating that the protein binding is subject to the dual effect of hydrophobicity and electrostatic interactions.

Shaltiel et al. continued to develop this technology and purify histone binding proteins and glutamine synthetase in rapid succession (Shaltiel and Er-el 1973). Each purification method uses the coupling of aminoalkyl or diaminoalkyl with cyanogen bromide-activated agarose. We now know that the agarose coupled in this way can form positively charged isourea groups, thereby generating amphiphilic adsorbent with cations. The influence of the ionic portion of the adsorbent can be seen from the fact that the enzyme can be eluted by using a gradient of increasing salt concentration, and the hydrophobic interaction can be enhanced by increasing the ionic strength. Since the matrix with a large amount of methylene can bind many enzymes with strong hydrophobicity, the hydrophobic effect is also obvious. Obviously, this is a mixed-mode adsorbent with both hydrophobic and ionic groups.

Almost at the same time, Hofstee also noticed the synergistic effect of hydrophobic and electrostatic forces on protein purification (Hofstee 1973a). He used a hydrophobic alkyl and aryl amine linked to agarose activated by cyanogen bromide to adsorb the enzyme and then eluted the enzyme through a salt gradient with the addition of ethylene glycol. Serum albumin, β-lactoglobulin, and ovalbumin are strongly bound to phenylbutylamine (PBA)-agarose at pH 8 and ionic strength of about 0.05 M. However, when 1 M NaCl solution is used as an eluent, the degree of protein elution varies. In each case, adding a polarity modifier such as ethylene glycol to the eluent can enhance the elution capacity. These findings indicate that the binding of these proteins to PBA-Sepharose is caused by the combined effect of hydrophobic and electrostatic forces (which may have mutually reinforcing effects).

Hofstee studied the binding of a series of proteins to agarose (Sepharose 4B) substituted with *n*-alkylamines of different C chain lengths (C1, C4, or C8) (Hofstee 1973b). At pH 8 and 0.05 M ionic strength, positively charged proteins (chymotrypsinogen A, α-chymotrypsinogen, lysozyme) are weakly retained on these phases. In contrast, negatively charged proteins (chymotrypsinogen X, serum albumin, ovalbumin, and β-lactoglobulin) show a strong affinity for longer C-chain (C4, C8) adsorbents. The binding seems to depend on the synergy between hydrophobic and electrostatic forces, the latter involving positive charges on the adsorbent, which are introduced by the coupling process.

Hofstee uses *n*-octylamine agarose to immobilize xanthine oxidase, lactate dehydrogenase, DNAse I, and alkaline phosphatase, all of which are in active form (Hofstee 1974). The adsorption of the enzyme on octylamine-agarose is irreversible, but if *n*-butylamine is used instead of *n*-octylamine, the enzyme can be easily removed by using high ionic strength (such as 1 M NaCl). Similar immobilization by hydrophobic adsorption has been used for many other enzymes. Through the above work, Hofstee came to a conclusion. The chromatographic separation with an amphiphilic adsorbent with both a charge and a hydrophobic group may be more advantageous than an ion exchange method based solely on the charge effect. The separation of proteins may be carried out according to the two dimensions of their molecular size and electric charge.

Wilchek et al. systematically studied the effect of cyanogen bromide activation method on the chromatographic behavior of organic amine coupled agarose (Jost et al. 1974; Wilchek and Miron 1976). The coupling of alkyl or aryl amines with agarose activated by cyanogen bromide results in the formation of isourea with an apparent pK_a of about 10, which is a type of strong ion exchanger. In order to reduce the alkalinity of the isourea, they carried out an acylation reaction on the isourea to obtain almost neutral hydrazide derivatives. At physiological pH, the coupling of similar acylates produces substantially uncharged agarose derivatives. They found that leucine aminopeptidase has no affinity for alkylamine agarose, but it binds firmly to tyramine agarose, partially inactivating the enzyme. Leucine aminopeptidase has no adsorption on the hydrazide analog of tyramine-agarose, that is, 4-hydroxyphenyl acetylhydrazine derivative. Assuming that the alkyl or aryl amine coupled with activated agarose has a cationic detergent-like mode of action, the binding of the protein to the adsorbent can occur in two steps. The first step involves the binding of protein anions to cationic sites on the column. The second step involves the binding of the hydrophobic amino acid residues of the protein to the hydrophobic part of the chromatographic column. Therefore, the binding mode of "detergent-like" amphiphilic chromatography is significantly different from that of "pure" hydrophobic interaction chromatography.

Sasaki et al. (1979) reported the application of hydrophobic ion chromatography in the purification of porcine pancreatin. They used Amberlite CG-50, a weakly acidic cation exchanger based on polystyrene-divinyl cross-linked resin, as the adsorbent. Because the pK_a of the carboxylic acid immobilized on the surface is about 6, under the condition of pH 4, these ion exchange groups are basically not dissociated. All tested porcine pancreatic enzymes, regardless of their pI, were adsorbed on Amberlite CG-50 at pH 4. Adsorption is hardly affected by ionic strength. By increasing the pH from 4 to 6, the adsorbed enzyme can be effectively eluted. It is speculated that the adsorption–elution mechanism of the protein on Amberlite CG-50 undergoes a process of hydrophobic and electrostatic interaction. At pH $\leq$ 4.5, the protein is adsorbed on the surface of the resin by hydrophobic force, and each site contains at least one carboxyl group. As the pH value increases, the carboxyl groups dissociate, resulting in an increase in the negative charge on the resin surface and a decrease in hydrophobicity. At this time, the negative charge of the protein also increases with the increase in pH. When the increased electrostatic repulsion between the stationary phase and the protein is greater than the remaining hydrophobic force, the protein is eluted. Therefore, they call this technique "hydrophobic ionic chromatography."

Hydrophobic ion chromatography is also used to purify microbial glucose oxidase, hyaluronidase, cholesterol oxidase, and cholesterol esterase (Sasaki et al. 1982). When the pH is $\leq$4.5, the enzyme is adsorbed and eluted by increasing the pH, where the ion exchange group dissociates. Through single-step chromatography, glucose oxidase, hyaluronidase, cholesterol oxidase, and cholesterol esterase were purified 30-, 12-, 45-, and 20-folds, with the yield of 82%, 83%, 80%, and 90%, respectively. This shows that hydrophobic ionic chromatography on Amberlite CG-50 column is effective for purifying various enzymes as long as they are stable at acidic pH.

The general applicability of hydrophobic ionic chromatography has been confirmed by the work of Kasche et al. (1990). They compared the adsorption behavior of penicillin amidase on Eupergit C, agarose, and phenyl agarose bonded with *p*-phenylbutylamine (PBA). Under certain pH conditions, phenylbutylamine does not dissociate, so proteins can be adsorbed to the phenylbutylamine adsorbent through hydrophobic interactions. By lowering the pH value of the eluent, the secondary amine on the adsorbent is dissociated to generate a positive charge, and the positive charge on the protein is electrostatically repelled, which can selectively desorb the hydrophobically adsorbed protein. On the PBA-Eupergit, 1500 U penicillin amidase is purified from *E. coli* homogenate per unit of wet weight gel, and a purification factor of 5.3 and a yield of over 90% can be obtained.

This technique is also called affinity exclusion chromatography by Teichberg (1990), which focuses on electrostatic repulsion to promote desorption for protein separation. In this process, the protein is loaded on the affinity matrix in the form of a high-salt solution, and the protein binds to the matrix through hydrophobic interaction. It is eluted with a low-ionic-strength solution or deionized water, and the hydrophobic interaction is partially offset by the electrostatic repulsion, which reduces the possibility of protein inactivation due to strong adsorption.

Harding et al. reported the use of various chemical methods to prepare mixed-mode adsorbents based on cellulose, agarose, and methacrylate matrix and applied them to the one-step purification of chymosin (Burton and Harding 1997; Burton et al. 1997). These matrices contain hydrophobic (aliphatic and/or aromatic) and ionic (carboxylate or alkylamine) groups. Hydrophobic amine ligands are attached to agarose gel activated by epichlorohydrin (mixed-mode amine ligands). Other mixed-mode matrices are produced by combining the same amine ligand with the carboxylate matrix in different proportions (0–80%). The best results can be obtained with unmodified or partially modified aminocaproic acid agarose and cellulose. The use of high ligand density results in high capacity. In addition, chymosin is adsorbed by mixed-mode media at both high and low-ionic strength, which reduces the requirements for sample preparation. The purified chymosin can be eluted and recovered by a small pH change. The described method is simple, effective, and inexpensive and provides very good resolution of chymosin from crude recombinant sources. In contrast, the carboxylate matrix has the best combination of capacity and regeneration performance but requires lower ionic strength for adsorption.

Burton and Harding (1998) introduced a new technique called hydrophobic charge induction chromatography in 1998. The mixed-mode matrix prepared by coupling a weak acid or weak base ligand to a suitable carrier has no charge at the adsorption pH. At low ligand densities, lyotropic salts are usually added to the mobile phase to promote protein adsorption. At higher ligand densities, the adsorption of chymotrypsin, chymotrypsinogen, and lysozyme is independent of ionic strength. Changes in pH will cause ionization of ligands adsorbed on the matrix, release the electrostatic potential of the matrix, and weaken hydrophobic interactions, leading to elution. According to the requirements of purified protein, the appropriate ligand and density should be selected and the hydrophobicity and dissociation range of the matrix should be adjusted, so that both adsorption and elution can be performed in the range of pH 5–9. Compared with ion exchange chromatography, the advantage of hydrophobic charge induction chromatography is that protein adsorption is hardly affected by the ionic strength of the solution, eliminating the need for dialysis and desalination of the crude protein solution.

Johansson et al. (2003a) reported the preparation and characterization of multimodal anion exchangers designed to capture negatively charged biomolecules under high-salt conditions. These multimodal anion exchangers are selected by screening a diverse library of multimodal ligands, from which the test protein is eluted at high ionic strength. They found that nonaromatic multimodal anion exchange ligands based on primary or secondary amines (or both) are most suitable for capturing proteins under high-salt conditions. In addition, these new multimodal anion exchange ligands not only use hydrophobic and electrostatic interactions but also use hydrogen bond interactions to selectively adsorb proteins. By modifying the ligand, a hydroxyl group is introduced near the ionic group. Compared with strong anion exchangers (such as Q Sepharose Fast Flow), the new multimodal weak anion exchanger has 20–30 times higher BSA breakthrough capacity when the ionic strength of the mobile phase is 28 mS/cm and pH 6.0. The new multimodal anion exchanger can also be used under normal anion exchange conditions, and the elution of the protein can be completed with a salt step or by adjusting the pH to acidic. Many multimodal anion exchange ligands based on aromatic amine and showing high breakthrough capacity of BSA have been discovered. Due to the strong non-electrostatic interaction, the recovery of proteins is usually very low. However, for anion exchange media coupled with phenol derivatives, the recovery of active protein can be improved by using charge repulsion to facilitate desorption at high pH.

Johansson and coworkers continued to expand the scope of research and prepared several aromatic (Ar) and nonaromatic (NoAr) ligand-based cation exchangers on Sepharose 6 Fast Flow, which are used to capture proteins from the mobile phase with high conductivity (about 30 mS/cm) (Johansson et al. 2003b). These multimodal cation exchangers were discovered by screening a diverse library of multimodal ligands and selecting cation exchangers that elute the test protein at high ionic strength. The recovery of the protein was also tested by applying a salt step or a pH step. They found that carboxylic acid-based aromatic multimodal cation exchange ligands seem to be most suitable for capturing proteins under high-salt conditions. The amide group on the α-carbon is essential for capturing proteins under high-salt conditions. Compared with strong cation exchangers (such as SP Sepharose Fast Flow), the breakthrough capacity of BSA, human IgG, and lysozyme on the new multimodal weak cation exchanger is 10–30 times higher under high-salt conditions. The new multimodal cation exchanger can also be used under normal cation exchange conditions, and the protein can be eluted through a salting step or a pH step (to reach the negatively charged pH of the protein). For BSA, cation exchangers based on high-capacity nonaromatic carboxylic acid ligands under high-salt conditions can also be designed. The common feature of these ligands is that they contain a hydrogen bond acceptor group close to the carboxyl group. In addition, if the phenyl group is connected to the beads of a strong cation exchanger (SP Sepharose Fast Flow), a mixed ligand ion exchange medium can be prepared to increase the breakthrough capacity of lysozyme and BSA. Changing the ligand ratio (SP/phenyl) can be used to optimize the function of the mixed ligand ion exchange medium.

1.6.3 Mixed-Mode Chromatography of Small Molecules

The development of mixed-mode chromatography for small molecules can be traced back to the pioneering work of Moore and Stein in 1951 on the use of sulfonated polystyrene resins for the separation of amino acids (Moore and Stein 1951). They described in detail the chromatographic separation of amino acid mixtures on a sulfonated polystyrene resin Dowex-50 column. On a 0.9 × 100 cm, a resin chromatography column operating in the form of sodium ion binding, the amino acids are eluted by using a buffer with a gradually increasing pH value (from 3.4 to 11). In experiments using synthetic amino acid mixtures to simulate protein hydrolysates, a series of buffers were developed to generate a single chromatogram. In the elution curve of amino acids, each component appears as a discrete peak (Fig. 1.5).

When discussing the chromatographic behavior of amino acids, they pointed out:

> It is clear that the separation obtained in Fig. 1.5 cannot be ascribed solely to differences in the ionic natures of the amino acids. The rate of travel on a column of a sulfonated polystyrene resin is a function both of the charge possessed by an amino acid and the nature of its side chain; it is a result of the affinity of the resin for both the ionic and the non-ionic portions of the molecule. Accordingly, columns of Dowex-50 exhibit a resolving power for mixtures of amino acids which compares favorably with that obtained on starch.

Moore and Stein have actually revealed that the separation of amino acids on synthetic resin ion exchangers is due to the mixed-mode chromatographic retention mechanism, that is, the ionic and hydrophobic parts of amino acids and the corresponding parts of the medium undergo electrostatic and hydrophobic interactions at the same time. The elution order of solutes is determined by their electrostatic and hydrophobic properties. As far as we know, this is the first unambiguous statement about mixed-mode chromatography in the literature.

In 1981, Hancock and Sparrow (1981) first reported the mixed-mode retention behavior of peptides and proteins in reversed-phase high-performance liquid chromatography and introduced the term "mixed-mode high-performance liquid chromatography" for the first time. They mentioned that the recently introduced packing used in radial compression columns has low octadecyl (C18) coverage (5% w/w) and has not been treated with silanol end capping. Such a carrier contains a large amount of free silanol together with alkyl groups and can be used for mixed-mode separations through electrostatic and reversed-phase retention mechanisms. In any given separation, the main mechanism depends on the nature of the sample and the mobile phase. In order to effectively separate peptides and proteins, most silanol-based interactions must be inhibited by using mobile-phase additives. For example, the mobile phase contains a high concentration of phosphate amines, such as 0.17 M triethylammonium phosphate, pH 3.2. In addition, it is necessary to further deactivate the silanol groups by an initial column wash of at least 20 column volumes of

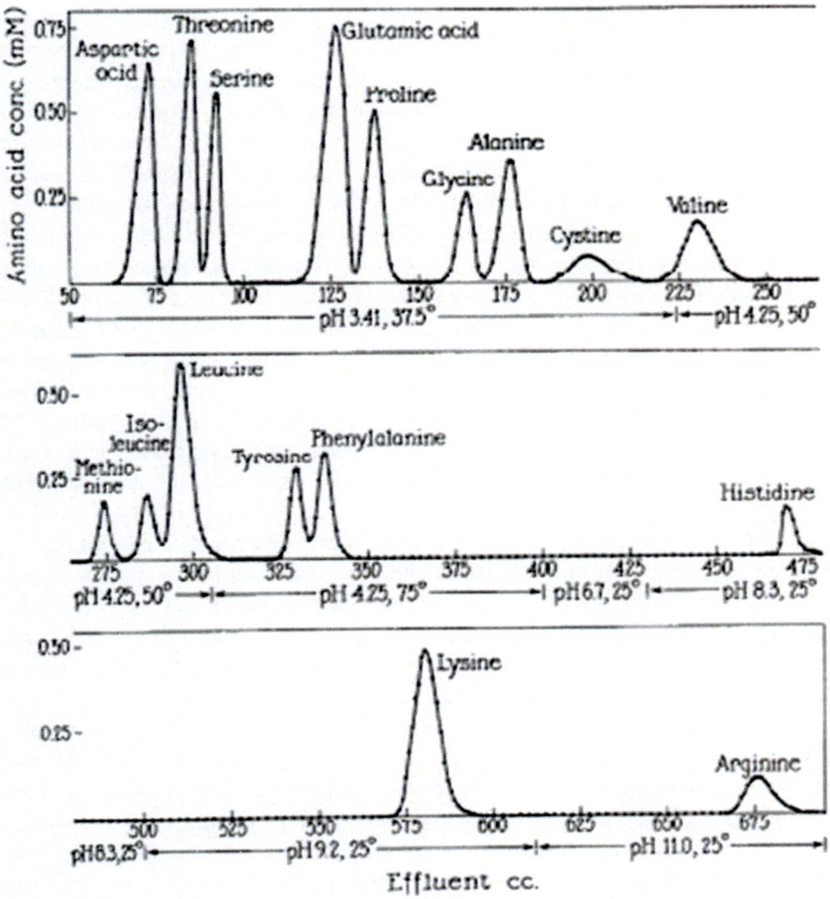

Fig. 1.5 Separation of amino acids from a synthetic mixture simulating the composition of a protein hydrolysate. The column of Dowex-50, 0.9 × 100 cm, was operated in the sodium form, with buffers of the pH and temperature indicated as eluents. A sample of about 6 mg of amino acids was chromatographed. The position of an amino acid peak is reproducible, on the average, to better than 5%. (Reproduced from Moore and Stein 1951)

methanol. Samples containing strong basic groups (such as the guanidine group of arginine) still show poor separation efficiency on this carrier. However, by using isopropanol as an organic modifier, these problems can be overcome to a large extent. If these precautions are followed, the packing will have excellent selectivity and increased sample capacity when compared with other packing materials with similar properties. Obviously, the mixed-mode chromatographic separation mechanism caused by the electrostatic interaction between the sample and the residual silanol groups has been identified as an unfavorable factor affecting the separation efficiency of reversed-phase chromatography, and measures need to be taken during the separation process to eliminate these unfavorable effects as much as possible.

In high-performance liquid chromatography, the initial efforts to improve the separation of small-molecule compounds using mixed-mode chromatography may be attributed to Hartwick and colleagues. Perhaps inspired by the separation of amphiphilic molecules by ion pair chromatography, they reported in 1982 the preparation of a chemically bonded silica-based mixed-mode stationary phase, which mimics the retention characteristics of ion pair reversed-phase chromatography (Crowther and Hartwick 1982). These stationary phases consist of a mixture of ligands with different chemical properties loaded on a silica gel support through chemical bonding. For example, the ligand mixture can contain alkyl groups and ionic groups, so that the prepared mixed-mode stationary phase can contain both reversed-phase retention characteristics and significant ion exchange characteristics. By adjusting the ratio of bonded ionic groups and hydrophobic groups, a dual retention mechanism of ionic interaction and reversed-phase partition can be formed, resulting in unique selectivity while maintaining excellent stability and column efficiency of the bonded phase. The synthesis and behavior of anionic/reversed-phase and cationic/reversed-phase materials were described and applied to the simultaneous separation of nucleosides and nucleotides and the separation of catecholamines. They modified the silica gel carrier with a mixture of chloropropyl- and octyl-dimethylchlorosilane. After this initial step, the chloropropyl group is then converted to an amino functional group by reaction with benzyldimethylamine (Crowther et al. 1983; Floyd et al. 1986).

The potential of mixed-mode chromatography in the separation of small-molecule compounds is best demonstrated in the separation of TNT and its metabolites. Caton and Griest (1996) conducted the separation of 15 kinds of 2,4,6-trinitrotoluene (TNT) explosives, by-products, and metabolites on a commercial C18/anion exchange mixed-mode column (Alltech) using the following ternary gradient: (1) 90:10 (v/v) 0.015 M potassium phosphate, pH 5.1: methanol; (2) methanol; and (3) acetonitrile. TNT, hexahydro-1,3,5-trinitro-1,3,5-triazine (RDX), octahydro-1,3,5,7-tetranitro-1,3,5,7-tetrazoxazine (HMX), six reduced metabolites, and three oxidized metabolites of TNT have been separated and quantified in one single run. Due to the anion exchange characteristics of the stationary phase, the anionic TNT metabolite 2,4,6-trinitrobenzoic acid can also be separated and quantified using this method. The systematic validation process has established a limit of quantification that is usually less than 1 ppm. The versatility of the mixed-mode chromatography combined with diode array ultraviolet absorption detection has been demonstrated by analyzing samples from biological (tissue and physiological fluids) and environmental (soil, compost, and leachate) sources.

1.7 Summary

At the end of the nineteenth century, the development of science and industrial technology promoted the rapid development of separation techniques. Extraction, distillation, adsorption, crystallization, and other separation and purification technologies had been widely used. However, for phytochemists, these techniques were too primitive to meet the needs of separating and purifying plant pigments with complex and diverse structures from complex matrices, so as to identify their structures. At the beginning of the last century, Tswett was the first to realize that by combining solid-phase adsorption and liquid flow, a new separation technique called chromatography could be formed to separate pigments such as chlorophyll and carotene from plant extracts. In the following 30–40 years, chromatography had become a powerful tool in the fields of phytochemistry and food science and has played an irreplaceable role in revealing the structure and function of a large number of complex natural products. In the early 1940s, Martin and Synge combined solid-phase adsorption, liquid flow, and liquid–liquid extraction to create a new separation technology called partition chromatography for hydrophilic compounds such as amino acids. Partition chromatography enables the application of chromatography to the separation of biologically interested compounds, thus contributing significantly to the development of life sciences. Since then, chromatographic techniques including thin-layer chromatography, gas chromatography, and high-performance liquid chromatography have emerged, making chromatography unprecedented progress in expanding the scope of application, improving separation efficiency and operating automation.

Conventional liquid chromatography mostly operates on a single binding mode, such as reversed-phase chromatography based on the hydrophobic interaction between the solute and the ligand, and ion exchange chromatography based on the electrostatic interaction between the solute and the ligand. These separations are characterized by interacting with a single type of functional group on the surface of the solute. Mixed-mode chromatography is a technique that binds and separates solutes through the coupling of two or more different interactions. Compared with single-mode chromatography, mixed-mode chromatography can act simultaneously with different functional groups of the solute, such as hydrophobic groups and ionic groups. Therefore, it can identify the surface characteristics of the solute and provide higher selectivity. Obviously, if the solute contains only a single type of functional group, mixed-mode chromatography will retain the solute as a single-mode chromatography and will not show any advantages. However, if the solute is a compound with multiple functional groups, such as amino acids, nucleic acids, peptides, proteins, etc. that are frequently observed in biological samples, mixed-mode chromatography will show completely different behavior from single-mode chromatography. In fact, the mechanism of mixed-mode chromatography was first discovered by Moor and Stein in the process of separating amino acids by ion exchange chromatography in the early 1950s. In the 1980s, mixed-mode chromatography was widely used in nucleic acid separation. After entering the twenty-first century, the application of mixed-mode chromatography in the separation and purification of recombinant proteins and monoclonal antibodies has been rapidly developed. With the deepening of the understanding of its mechanism of action and the continuous improvement of the method development process, the application of mixed-mode chromatography will become more and more extensive, and it will become a powerful tool in separation and purification of biological compounds.

References

Allenmark S (1991) Chromatographic enantioseparation: methods and applications, 2nd edn. Ellis Horwood, Ltd, England

Bischoff R, McLaughlin LW (1983) Chemically synthesized hydrophobic anion-exchange high-performance liquid chromatography supports used for oligonucleotide resolution by mixed mode chromatography. J Chromatogr 270:117–126

Bischoff R, McLaughlin LW (1984a) Nucleic acid resolution by mixed-mode chromatography. J Chromatogr 296:329–337

Bischoff R, McLaughlin LW (1984b) Mixed-mode chromatographic matrices for the resolution of transfer ribonucleic acids. J Chromatogr 317:251–261

Bischoff R, McLaughlin LW (1985) Isolation of specific tRNAs using an ionic-hydrophobic mixed-mode chromatographic matrix. Anal Biochem 151:526–533

Bischoff R, McLaughlin LW (1990) HPLC of transfer RNAS using ionic-hydrophobic mixed-mode chromatography and hydrophobic-interaction chromatography. J Chromatogr Library 45:A73–A102

Bischoff R, Graeser E, McLaughlin LW (1983) tRNA separation by high-performance liquid chromatography using an aggregate of ODS-Hypersil and trioctylmethylammonium chloride. J Chromatogr 257:305–315

Burton SC, Harding DRK (1997) High-density ligand attachment to brominated allyl matrices and application to mixed mode chromatography of chymosin. J Chromatogr A 775:39–50

Burton SC, Harding DRK (1998) Hydrophobic charge induction chromatography: salt independent protein adsorption and facile elution with aqueous buffers. J Chromatogr A 814:71–81

Burton SC, Haggarty NW, Harding DRK (1997) One step purification of chymosin by mixed mode chromatography. Biotechnol Bioeng 56:45–55

Caton JE, Griest WH (1996) Determination of explosives and some metabolites of TNT in biological and environmental samples by liquid chromatography on a mixed-mode C18-anion column. J Liquid Chromatogr Related Technol 19:661–677

Consden R, Gordon AH, Martin AJP (1944) Qualitative analysis of proteins: a partition chromatographic method using paper. Biochem J 38:224–232

Crowther JB, Hartwick RA (1982) Chemically bonded multifunctional stationary phases for high-performance liquid chromatography. Chromatographia 16:349–353

Crowther JB, Caronia JP, Hartwick RA (1982) An improved method for the determination of oligonucleotide chain length using phosphodiesterase hydrolysis and ion-pair high-performance liquid chromatography. Anal Biochem 124:65–73

Crowther JB, Fazio SD, Hartwick RA (1983) High-performance liquid chromatographic separation of oligonucleotides and other nucleic acid constituents on multifunctional stationary phases. J Chromatogr 282:619–628

Crowther JB, Griffiths P, Fazio SD et al (1984a) Synthesis of aromatic silica-based cation-exchange packings for HPLC. J Chromatogr Sci 22:221–224

Crowther JB, Fazio SD, Schlksnis R et al (1984b) Analysis of the acid hydrolysis products of monofunctional chemically bonded stationary phases for high-performance liquid chromatography using capillary gas chromatography. J Chromatogr 289:367–375

Cuatrecasas P, Wilchek M, Anfinsen CB (1968) Selective enzyme purification by affinity chromatography. Proc Natl Acad Sci U S A 61:636–643

Er-el Z, Zaidenzaig Y, Shaltiel S (1972) Hydrocarbon-coated Sepharoses. Use in the purification of glycogen phosphorylase. Biochem Biophys Res Commun 49:383–390

Ettre LS (2001) The birth of partition chromatography. LC-GC 19:506–512

Floyd TR, Cicero SE, Fazio SD et al (1986) Mixed-mode hydrophobic ion exchange for the separation of oligonucleotides and DNA fragments using HPLC. Anal Biochem 154:570–577

Fried B, Sherma J (1999) Thin-layer chromatography, 4th edn. Marcel Dekker, New York

Gary JH, Handwerk GE (1984) Petroleum refining technology and economics, 2nd edn. Marcel Dekker, Inc., New York

Giddings MC (1997) Tributes to J. Calvin Giddings. Special Issue: In honor of J. Calvin Giddings. J Microcolumn Sep 9:123–124

Gillam I, Millward S, Blew D et al (1967) The separation of soluble ribonucleic acids on benzoylated diethylaminoethylcellulse. Biochemistry 6:3043–3056

Golay MJE (1979) In: Ettre LS, Zlatkis A (eds) 75 years of chromatography—a historical dialogue. Elsevier, Amsterdam, pp 104–114

Hancock WS, Sparrow JT (1981) Use of mixed-mode, high-performance liquid chromatography for the separation of peptide and protein mixtures. J Chromatogr A 206:71–82

Hofstee BHJ (1973a) Hydrophobic affinity chromatography of proteins. Anal Biochem 52:430–448

Hofstee BHJ (1973b) Protein binding by agarose carrying hydrophobic groups in conjunction with charges. Biochem Biophys Res Commun 50:751–757

Hofstee BHJ (1974) Non-specific binding of proteins by substituted agaroses. In: Dunlap RB (ed) Immobilised biochemicals and affinity chromatography. Plenum Press, New York

James AT, Martin AJP (1952) Gas-liquid partition chromatography: the separation and micro-estimation of volatile fatty acids from formic acid to dodecanoic acid. Biochem J 50:679–690

James AT, Martin AJP, Smith GH (1952) Gas-liquid partition chromatography: the separation and micro-estimation of ammonia and the methylamines. Biochem J 52:238–242

Johansson BL, Belew M, Eriksson S et al (2003a) Preparation and characterization of prototypes for multi-modal separation media aimed for capture of negatively charged biomolecules at high salt conditions. J Chromatogr A 1016:21–33

Johansson BL, Belew M, Eriksson S et al (2003b) Preparation and characterization of prototypes for multi-modal separation aimed for capture of positively charged biomolecules at high-salt conditions. J Chromatogr A 1016:35–49

Jost R, Miron T, Wilchek M (1974) The mode of adsorption of proteins to aliphatic and aromatic amines coupled to cyanogen bromide-activated agarose. Biochim Biophys Acta 362:75–82

Kasche V, Löffler F, Scholzen T et al (1990) Rapid protein purification using phenylbutylamine-Eupergit: a novel method for large-scale procedures. J Chromatogr A 510:149–154

Kremzner LT, Wilson IR (1963) A chromatographic procedure for the purification of acetylcholinesterase. J Biol Chem 238:1714–1717

Mandell JD, Hershey AD (1960) A fractionating column for analysis of nucleic acids. Anal Biochem 1:66–77

Martin AJR (1979) In: Ettre LS, Zlatkis A (eds) 75 Years of chromatography—a historical dialogue. Elsevier, Amsterdam, pp 285–296

Martin AJR, Synge RLM (1941a) Separation of the higher monoamino-acids by counter-current liquid-liquid extraction: the amino acid composition of wool. Biochem J 35:91–121

Martin AJP, Synge RLM (1941b) A new form of chromatogram employing two liquid phases. A theory of chromatography. 2. Application to the micro-determination of the higher monoamino-acids in proteins. Biochem J 35:1358–1368

McCabe W, Smith J (2000) Unit operations of chemical engineering. McGraw-Hill, New York

McLaughlin LW (1986) Resolution of nucleic acids by high-performance liquid chromatography. Trends Anal Chem 5:215–219
McLaughlin LW (1989) Mixed-mode chromatography of nucleic acids. Chem Rev 89:309–319
McLaughlin LW, Bischoff R (1987) Resolution of RNA using high-performance liquid chromatography. J Chromatogr B Biomed Sci Appl 418:51–72
Moore S, Stein WH (1951) Chromatography of amino acids on sulfonated polystyrene resins. J Biol Chem 192:663–681
Pearson RL, Weiss JF, Kelmers AD (1971) Improved separation of transfer RNA's on polychlorotrifuoroethylene-supported reversed-phase chromatography columns. Biochim Biophys Acta 228:770–774
Pearson RL, Weiss JF, Holladay DW et al (1974) Purification of calf liver tRNA's. Methods Enzymol 29:514–522
Peterson EA, Sober HA (1956) Chromatography of proteins. I. Cellulose ion-exchangers. J Am Chem Soc 78:751–755
Porath J, Lindner EB (1961) Separation methods based on sieving and ion exclusion. Nature 191: 69–70
Sasaki I, Gotoh H, Yamamoto R et al (1979) Hydrophobic-ionic chromatography. Its application to purification of porcine pancreas enzymes. J Biochem 86:1537–1548
Sasaki I, Gotoh H, Yamamoto R et al (1982) Hydrophobic-ionic chromatography: its application to microbial glucose oxidase, hyaluronidase, cholesterol oxidase, and cholesterol esterase. J Biochem 91:1555–1561
Seader JD, Henley EJ (1998) Separation process principles. Wiley, New York
Shaltiel S, Er-El Z (1973) Hydrophobic chromatography: use for purification of glycogen synthetase. Proc Natl Acad Sci U S A 70:778–781
Simmonds RJ, Yon RJ (1976) Protein chromatography on adsorbents with hydrophobic and ionic groups. Chromatography of dodecyl sulphate-solubilized proteins of the human erythrocyte membrane on N-(3-carboxypropionyl)aminodecyl-sepharose. Biochem J 157:153–159
Sober HA, Peterson EA (1954) Chromatography of proteins on cellulose ion-exchangers. J Am Chem Soc 76:1711–1712
Sparks T, Chase G (2015) Filters and filtration handbook, 6th edn. Butterworth-Heinemann, Oxford
Surgenor DM (2002) Edwin Cohn and the development of protein chemistry. Harvard University Press, Cambridge, MA
Teichberg VI (1990) Affinity-repulsion chromatography. J Chromatogr A 510:49–57
Tiselius A, Hjertén S, Levin Ö (1956) Protein chromatography on calcium phosphate columns. Arch Biochem Biophys 65:132–155
Walton HE (ed) (1976) Ion-exchange chromatography, benchmark papers in analytical chemistry (early papers reprinted). Dowden, Hutchinson Ross, Stroudsburg, PA, p 440
Wilchek M, Chaiken I (2000) An overview of affinity chromatography. Methods Mol Biol 147:1–6
Wilchek M, Miron T (1976) On the mode of adsorption of proteins to "hydrophobic columns". Biochem Biophys Res Commun 72:108–113
Wixom RL (2001) The beginnings of chromatography the pioneers (1900–1960). J Chromatogr Library 64:1–38
Yon RJ (1972) Chromatography of lipophilic proteins on adsorbents containing mixed hydrophobic and ionic groups. Biochem J 126:765–767
Yon RJ, Simmonds RJ (1975) Protein chromatography on adsorbents with hydrophobic and ionic groups. Some properties of N-(3-carboxypropionyl)aminodecyl-sepharose and its interaction with wheat-germ aspartate transcarbamoylase. Biochem J 151:281–290
Zumdahl SS (2005) Chemical principles, 5th edn. Houghton Mifflin, New York

Chapter 2
Theoretical Background

Abstract Chromatography theory can be divided into three branches: dynamics, thermodynamics, and separation optimization. Chromatographic dynamics is mainly concerned with the peak shape. In the chromatographic separation process, solute molecules move through the chromatographic column along with the mobile phase. Due to molecular diffusion and fluid flow, as well as the dynamics of interaction with the stationary phase, the solute zone gradually widens. In this chapter, we will provide an introduction to chromatographic dynamics. Starting from the chromatographic process, we will define basic concepts, such as retention time and retention factor. Then, we will introduce the theory of zone migration based on the principle of mass conservation, flow and diffusion in porous media, and finally, the theoretical plate height equation in chromatography.

2.1 Introduction

Chromatography theory studies the law of molecular motion in the process of chromatography and explores the internal relationship between microscopic molecular motion and chromatographic separation. It mainly involves three basic aspects:

1. Chromatographic process dynamics: In the chromatographic separation process, solute molecules move through the chromatographic column along with the mobile phase. Due to molecular diffusion and fluid flow, as well as the dynamics of interaction with the stationary phase, the solute zone gradually widens. Studying the mechanism and influencing factors of chromatographic zone broadening is the main content of chromatographic kinetics.
2. Thermodynamics of the chromatographic process: As the solute moves forward in the chromatographic column, its migration rate changes due to the interaction with the stationary phase, and the migration rate of the solute with strong interaction becomes slower, thereby separating it from the faster solute. The study of solute retention mechanism, chromatographic selectivity, and its influencing factors is the main content of chromatographic thermodynamics.

Q.-H. Wan, *Mixed-Mode Chromatography*,
https://doi.org/10.1007/978-981-16-5485-5_2

3. Chromatographic separation theory: The actual chromatographic separation requires that the chromatographic peaks of each component are narrow enough, and the retention time/volume of each component must be sufficiently different. Changing the chromatographic conditions may lead to changes in both, that is, the resolution varies with chromatographic conditions. When solving the problem of separation of multicomponent mixtures, it is required not only to achieve a certain degree of separation between the components but also to achieve fast separation of components as many as possible. This is a comprehensive problem related to chromatographic kinetics and thermodynamics. The screening of operating conditions and method development are mainly based on this.

Chromatographic separation is a process of fluid molecular movement. Various column chromatography methods, although operated on different modes, all have basically the same kinetic laws. These studies will help to predict the shape of the elution curve, elucidate the mechanism of chromatographic peak broadening, and define the factors that affect peak broadening. The solute retention mechanism is related to the specific separation mode and is determined by the intermolecular interaction or molecular recognition of the specific system. In gas chromatography, the mobile phase only serves to carry solute molecules through the chromatographic column and does not participate in the solute retention process. However, in liquid chromatography, the mobile phase is a key factor in determining solute retention, and the composition and properties of the mobile phase play an important role in solute retention. In this chapter, we will mainly discuss chromatographic kinetics. Starting from the chromatographic process, we will define basic concepts such as retention time and retention factor. Then, we will introduce the theory of zone migration based on the principle of mass conservation, flow and diffusion in porous media, and finally, the theoretical plate height equation in chromatography. Chromatographic thermodynamics and separation theory will be discussed separately in the chapters of retention mechanism, stationary phase, mobile phase, and method development.

2.2 Chromatographic Process

Except for capillary gas chromatography, most chromatographic separations are carried out in packed chromatographic columns. The chromatographic column is usually a cylindrical tube made of glass, plastic, or stainless steel, with porous screen plates fixed at both ends, and the tube is packed with a stationary phase based on porous particles. Many tortuous flow channels are formed between the gaps of the particles, allowing the mobile phase to pass through. When the solute molecules are introduced into the chromatographic column, a solute zone is formed, and the solute zone is driven by the mobile phase toward the outlet end of the chromatographic column. Solute molecules diffuse into the pores of the packing, interact with the stationary-phase ligands, and are adsorbed on the stationary phase, and the migration

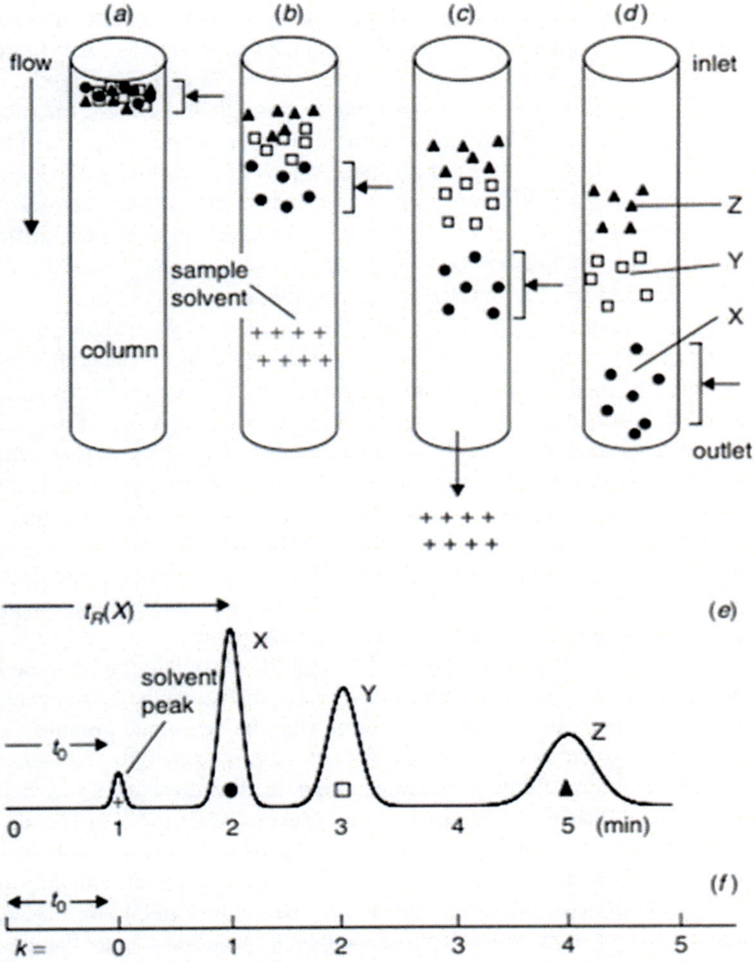

Fig. 2.1 Schematic diagram of HPLC separation process. (**a–d**) The sequential separation in the column (i.e., as a function of time); (**e**) the final chromatogram; (**f**) estimate the value of k from the chromatogram (**e**). Solute molecules X, Y, and Z are represented by black dot, open square, and filled triangle, respectively; sample solvent molecules are represented by plus symbol. (Reproduced from Snyder et al. 2010)

speed of the solute zone will decrease. The stronger the interaction with the stationary phase, the more the migration rate of the zone decreases. At the same time, because the flow path in the packed-column bed is tortuous and constrictive, the zone will continue to expand.

Figure 2.1 schematically illustrates the process of chromatographic separation of mixtures (Snyder et al. 2010). The mixture contains three sample compounds (or solutes), in which a single sample molecule is represented by black dot solute X, open square solute Y, and filled triangle solute Z. For clarity, mobile-phase molecules are not shown, and solvent molecules dissolved in the sample are represented by plus symbol. The sample is applied to the chromatographic column (Fig. 2.1a) and is carried by the mobile phase through the column in successive stages (Fig. 2.1b–d), and finally, the sample leaves the column (Fig. 2.1e) to reach the detector. A graph of the detector response versus time is obtained, which is also known as chromatogram. As the separation proceeds in Fig. 2.1a–d, the molecules of sample components X, Y, and Z exhibit two characteristic behaviors: differential migration and molecular diffusion. According to Fig. 2.1d, the solutes X, Y, and Z have been separated from each other in the chromatographic column.

Differential migration (X, Y, and Z solute molecules move or migrate through the column at different average speeds) constitutes the basis of chromatographic separation. If there is no difference in the migration rate of the two compounds, they co-elute in the chromatogram, and separation between them will not occur. In this example, the X (black dot) molecule moves the fastest, the Y (open square) molecule is in the middle, and the Z (filled triangle) molecule moves the slowest; the three solute molecules are baseline separated. The solvent molecules in the sample or the solvent molecules in the mobile phase are usually not retained on the stationary phase but pass through the column at the fastest flow rate and then leave the column before the separated solute molecules. In Fig. 2.1b–e, plus symbol represents a part of the solvent molecules that make up the injected sample.

When a given solute molecule passes through the chromatographic column, the solute molecules in the high-concentration area will diffuse to the low-concentration area, thereby occupying a larger volume in the chromatographic column. The volume of a given solute molecule in the column defines the so-called band. The width of the solute volume is measured along the flow direction and is defined as the bandwidth, as shown by the arrows and brackets along the X molecule (black dot) movement direction in Fig. 2.1a–d. When a band leaves the column and is recorded in the chromatogram (Fig. 2.1e), it is called a peak. The identity of each peak can be determined from the time it leaves the column (retention time t_R), and the concentration of each solute in the sample is proportional to the peak size (measured by area or height). For samples that are small enough (low ng to μg injection volume usually used in HPLC analysis procedures), the peak retention time does not change with the sample concentration (and the size of the peak produced).

2.3 Solute Retention

The retention time t_R of each solute is the time from injection to the appearance of the peak in the chromatogram. In Fig. 2.1e, the retention times of solutes X, Y, and Z are 2 min, 3 min, and 5 min, respectively. The retention time of the solvent peak at 1 min is called the column dead time t_0 (sometimes t_m is used instead of t_0 to indicate the column dead time). The migration rate or velocity u_x of the solute X through the chromatographic column is determined by the fraction R of solute molecules present in the mobile phase at any time. On average, u_x is equal to R multiplied by the migration rate or velocity u of solvent molecules:

$$u_x = Ru \tag{2.1}$$

For example, if half of the X molecules are in the mobile phase ($R = 0.5$) and the other half are in the stationary phase, only half of the molecules are in motion at a given time, so the average mobility of X will be half of the solvent migration rate.

The fraction R of the solute molecule X in the mobile phase is determined by the solute distribution equilibrium process between the stationary phase and the mobile phase. The equilibrium constant K can be expressed as

$$K = C_s/C_m \tag{2.2}$$

where C is the molar concentration of solute molecules and the subscripts s and m represent stationary phase and mobile phase, respectively. This equilibrium and the migration rate of a given solute are affected by the molecular structure of the solute, the chemical composition of the mobile phase and stationary phase (solvent and column), and temperature. The influence of the average pressure in the column on the retention of the sample is usually negligible.

For a given solute, the retention factor k (sometimes still called the capacity factor k) is defined as the amount of solute in the stationary phase (s) divided by the amount in the mobile phase (m). The amount of solute in each phase is equal to its concentration (C_s or C_m, respectively) multiplied by the phase volume (V_s or V_m, respectively), and then

$$\begin{aligned} k &= C_sV_s/(C_mV_m) = C_s/C_m/(V_s/V_m) \\ &= K\phi \end{aligned} \tag{2.3}$$

where K is the distribution equilibrium constant of the solute between the two phases and $\phi = (V_s/V_m)$ is the phase ratio, that is, the ratio of the volume of the stationary phase to the mobile phase in the chromatographic column. Solute molecules must exist in the mobile phase or stationary phase. Therefore, if the ratio of molecules in the mobile phase is R, the ratio in the stationary phase must be $(1 - R)$; therefore, according to Eq. (2.3), we get

$$k = (1 - R)/R \tag{2.4}$$

or

$$R = 1/(1 + k) \tag{2.5}$$

The retention time (t_R) of the solute molecule X can be defined as the distance divided by the band speed, where the distance is the column length L and the band speed is u_x:

$$t_R = L/u_x \tag{2.6}$$

Similarly, the retention time of the solvent peak is

$$t_0 = L/u \tag{2.7}$$

where u is the average linear velocity of the mobile phase. Equation (2.6) is divided by Eq. (2.7), and L is eliminated to give

$$t_R = t_0 u/u_x \tag{2.8}$$

In the case of $R = u_x/u_0$ (Eq. (2.1)) and Eq. (2.5), we obtain

$$t_R = t_0(1 + k) \tag{2.9}$$

Equation (2.9) can also be expressed by retention volume $V_R = t_R F$, where F is the volume flow rate of the mobile phase (mL/min):

$$V_R = V_m(1 + k) \tag{2.10}$$

Here, V_m is the column dead volume, which is equal to $t_0 F$ (see further discussion of V_m and Eq. (2.10) below).

Equation (2.9) can be rearranged as

$$k = (t_R - t_0)/t_0 \tag{2.11}$$

In this way, the k value of each peak in the chromatogram can be calculated. As shown in Fig. 2.1f (corresponding to the chromatogram in Fig. 2.1e), the distance t_0 can be used to mark the approximate value of k from time t_0; therefore, for compounds X, Y, and Z, k is equal to 1, 2, and 4, respectively. The column dead time can be estimated from the solvent peak, or it can be accurately measured using unretained compounds as markers. Alternatively, the t_0 value can be estimated from the column size and flow rate (for a column packed with fully porous particles):

$$t_0 \approx 5 \times 10^{-4} L d_c^2/F \tag{2.12}$$

L is the length of the column in millimeters, d_c is the inner diameter of the column in millimeters, F is the volumetric flow rate in mL/min, and t_0 is in minutes. For hundreds of different RPC columns, it was found that Eq. (2.12) was consistent with the experimental value of t_0 and the average error was only ±10% (1 SD), which was also accurate enough for practical purposes. The column dead volume V_m is related to t_0 because

$$V_m \approx 5 \times 10^{-4} L d_c^{\,2} \tag{2.13}$$

The unit of L and d_c is mm. The dead volume V_m represents the total volume of the mobile phase inside and outside the column. For example, if $V_m = 2$ mL and $F = 0.5$ mL/min, then $t_0 = V_m/F = 2/0.5 = 4$ min; otherwise, t_0 can be considered as the time required to empty the mobile phase originally present in the column. For the common case of column inner diameter ≈4.6 mm, we can easily estimate the value of V_m (by combining Eqs. (2.12) and (2.13)):

$$V_m(\text{mL}) \approx 0.01L \quad (\text{For columns with an inner diameter of } 4-5 \text{ mm}, L \text{ is in mm}) \tag{2.14}$$

Then, the value of t_0 can be obtained from Eq. (2.14), where $t_0 = V_m/F$.

2.4 Solute Zone Migration

The above discussion on the migration and retention of solute bands pointed out the simple relationship between solute retention and solute distribution equilibrium between the two phases. But it does not involve the type of solute distribution equilibrium. Wilson (1940) proposed a theoretical model of band migration in the chromatographic process. The migration rate of the solute is related to the band shape and the adsorption type of the solute. The transport process of solute in a chromatographic column is a typical transport problem in porous media. Strict mathematical treatment should be based on the solute migration process in the column and various influencing factors, the corresponding partial differential equations should be listed, and the equations describing the chromatographic band migration should be solved. However, it is quite difficult to solve this type of equations, and certain assumptions are often used in actual dynamics research to simplify mathematical processing. Wilson's assumptions include the following: (1) the distribution equilibrium of the solute between the mobile phase and the stationary phase can be achieved instantaneously throughout the chromatographic process; (2) the influence of molecular diffusion and flow resistance on the equilibrium is negligible; and (3) during the migration process in the chromatographic column, within a certain period of time, the change of the solute mass of each segment of the chromatographic column conforms to the principle of mass conservation, that is, the difference between the solute mass entering the small segment of the column and the solute mass exiting the small segment of the column is equal to the total changes of the solute mass in the mobile and stationary phase of the small

Mobile phase: $C_m \varepsilon audt$	$\Delta = \varepsilon adx(\frac{\partial C_m}{\partial t})_x dt$	$(C_m - dC_m)\varepsilon audt$
Stationary phase:	$\Delta = (1-\varepsilon)adx(\frac{\partial C_s}{\partial t})_x dt$	

Fig. 2.2 Schematic diagram of the mass balance between the mobile phase and the stationary phase in the chromatographic column

segment within the corresponding time. Based on this, the equations describing the migration rate of the chromatographic band can be derived.

Figure 2.2 illustrates the change of the solute mass in a short period of time *dx*. The following assumptions are made:

- The loading area of the chromatographic column is *a*, and then, the volume of this small segment is equal to *adx*.
- The fraction of column loading area occupied by fluid is ε, and the fraction of loading area occupied by the stationary phase is $(1 - \varepsilon)$.
- The linear velocity of the mobile phase is *u*.
- When the mobile phase flows into *dx*, the solute concentration is C_m, and the outflow concentration is $(C_m - dC_m)$, then the solute mass of inflow into *dx* within a short period of time is $C_m\varepsilon audt$, and the solute mass of outflow from *dx* is $(C_m - dC_m)\varepsilon audt$.

In the *dt* time, the change rate of the solute concentration in the mobile phase in *dx* is $\left(\frac{\partial C_m}{\partial t}\right)_x$, the change rate of the solute concentration in the stationary phase is $\left(\frac{\partial C_s}{\partial t}\right)_x$, and then the change in solute concentration in the mobile phase and stationary phase are $\varepsilon adx\left(\frac{\partial C_m}{\partial t}\right)_x dt$ and $(1-\varepsilon)adx\left(\frac{\partial C_s}{\partial t}\right)_x dt$, respectively.

The mass balance (the concentration in the incoming mobile phase minus the outgoing concentration is equal to the sum of the concentrations in the two phases in *dx*) can be expressed by

$$C_m\varepsilon audt - (C_m - dC_m)\varepsilon audt = \varepsilon adx\left(\frac{\partial C_m}{\partial t}\right)_x dt + (1-\varepsilon)adx\left(\frac{\partial C_s}{\partial t}\right)_x dt \quad (2.15)$$

Eliminate the common factor *adt* from the above equation and divide by *dx* to get

$$\varepsilon u\left(\frac{\partial C_m}{\partial x}\right)_t = \varepsilon\left(\frac{\partial C_m}{\partial t}\right)_x + (1-\varepsilon)\left(\frac{\partial C_s}{\partial t}\right)_x \quad (2.16)$$

Solute adsorption equilibrium constant *K* between two phases:

$$K = \left(\frac{\partial C_s}{\partial C_m}\right) \tag{2.17}$$

The change rate of the solute concentration in the stationary phase can be transformed as follows (from Eqs. (2.16) and (2.17)):

$$\left(\frac{\partial C_s}{\partial t}\right)_x = \left(\frac{\partial C_s}{\partial C_m}\right)\left(\frac{\partial C_m}{\partial t}\right)_x \tag{2.18}$$

Substituting Eq. (2.18) into Eq. (2.16),

$$\begin{aligned} \varepsilon u\left(\frac{\partial C_m}{\partial x}\right)_t &= \varepsilon\left(\frac{\partial C_m}{\partial t}\right)_x + (1-\varepsilon)\left(\frac{\partial C_s}{\partial C_m}\right)\left(\frac{\partial C_m}{\partial t}\right)_x \\ &= \left(\frac{\partial C_m}{\partial t}\right)_x\left[\varepsilon(1-\varepsilon)\left(\frac{\partial C_s}{\partial C_m}\right)\right] \end{aligned} \tag{2.19}$$

The change rate of solute concentration in the mobile phase can be transformed as follows:

$$\left(\frac{\partial C_m}{\partial t}\right) = \left(\frac{\partial C_m}{\partial x}\right)\left(\frac{\partial x}{\partial t}\right) \tag{2.20}$$

Substitute Eq. (2.20) into Eq. (2.19),

$$\varepsilon u\left(\frac{\partial C_m}{\partial x}\right)_t = \left(\frac{\partial C_m}{\partial x}\right)\left(\frac{\partial x}{\partial t}\right)\left[\varepsilon + (1+\varepsilon)\left(\frac{\partial C_s}{\partial C_m}\right)\right] \tag{2.21}$$

When the linear velocity of the mobile phase is a certain value, the time t and the length x coordinate are equivalent, so

$$\frac{\partial x}{\partial t} = \frac{\varepsilon u}{\varepsilon + (1-\varepsilon)\left(\frac{\partial C_s}{\partial C_m}\right)} \tag{2.22}$$

or

$$\frac{\partial x}{\partial t} = \frac{u}{1 + \left(\frac{1-\varepsilon}{\varepsilon}\right)K} \tag{2.23}$$

Equations (2.22) and (2.23) are the band migration rate equations derived from equilibrium chromatographic theory. The band migration rate $(\partial x/\partial t)$ is proportional to the linear velocity of the mobile phase u and the column porosity $\varepsilon/(1-\varepsilon)$.

When $((1 - \varepsilon)/\varepsilon)K \gg 1$, then Eq. (2.23) can be simplified to

$$\frac{\partial x}{\partial t} = \frac{u\varepsilon}{(1-\varepsilon)\,K} \tag{2.24}$$

Because the distribution coefficient K of each component is different, the migration rate of the components is different, so as to realize the separation.

Component retention time, t_R:

$$t_R = \frac{L}{\left(\frac{\partial x}{\partial t}\right)} = \frac{L}{u}\left[1 + \left(\frac{1-\varepsilon}{\varepsilon}\right)K\right] \tag{2.25}$$

where L is the length of the chromatographic column.

Component retention volume, V_R:

$$V_R = t_R \varepsilon a u = L\varepsilon a\left[1 + \left(\frac{1-\varepsilon}{\varepsilon}\right)K\right] \tag{2.26}$$

If K can be expressed by the Langmuir adsorption equation and the adsorption isotherm is a straight line, then the equilibrium constant K is a constant, the band migration rate remains unchanged, and the resulting band is a symmetrical shape curve; if the adsorption isotherm is a convex curve, K decreases as the concentration of the solute in the mobile phase increases, the band migration rate $(\partial x/\partial t)$ increases, so the K value of the solute in the low-concentration area is large, and the solute migration rate is low, while the solute migration rate is high in the high-concentration area, and an asymmetric trailing peak is formed. The retention time decreases as C_m increases (K decreases); conversely, if the adsorption isotherm is a concave curve, the resulting band is a forward-tilted peak.

The equation of band migration derived from the equilibrium chromatography theory and the corresponding equations for retention time and retention volume reveal the relationship between the adsorption equilibrium of the solute in the stationary phase and the mobile phase and its migration rate in the chromatographic column. The asymmetric chromatographic peaks, especially the tailing peaks, observed experimentally can be explained by nonlinear adsorption isotherms. However, due to the assumption that the effect of band broadening caused by diffusion and flow is negligible, the model fails to provide insights into the factors that affect the chromatographic peak spreading during the chromatographic process.

2.5 Flow Dynamics

Up to date, most liquid chromatographic separations have been carried out in chromatographic columns packed with porous particles. The solute molecules are driven by the mobile phase through the chromatographic column. The solid particles used as the chromatographic medium restrict, deflect and disperse the flow and diffusion of the solute in the mobile phase, making the solute zone continue to expand. Therefore, it is useful to have a detailed knowledge about the packing structure and its relation to flow patterns in the column.

The main function of fluid flow in chromatography is mass transfer. It provides a transport mechanism for the movement of the chromatographic zone along the column bed toward the column outlet. However, the flow process is not limited to the simple and rather passive role of the transfer zone. It is an integral part of chromatographic dynamics. In order to improve the performance of chromatographic separation, it is often necessary to carefully design an experimental system to precisely control the flow effect and minimize the peak broadening caused by the flow.

A careful examination of the liquid flow that occurs in the column helps us understand its important role in dynamics. All known flows in confined spaces are different and distinguishable in nature (Geankoplis 2003). Different points in the liquid stream have different speeds. The distinguishable nature of flow is that all fluids have different degrees of viscosity, and the viscous force between fluids can either stop the flow of fluids or continue the flow. Let us consider the passage of a liquid through a pipe under pressure. At the wall of the pipe, the fluid is in contact with the fixed surface, and the movement of the fluid layer at the interface is stopped due to viscous friction. However, the outer fluid above the stationary layer can continue to flow. This is because under the action of an external force, shear is generated between the different flow layers of the fluid, and the viscous force of the fluid overcomes the shear force and causes the liquid to move forward as a whole. At the same time, uneven flow patterns are produced. The flow velocity of each point on the cross section of the pipeline is parabolic, the flow velocity of the fluid at the pipe wall is zero, and the fluid has the maximum flow velocity at the center of the pipe.

The flow pattern in the packed column is more complicated. It is not difficult to imagine that the seepage of liquid in porous media occurs in the flow space formed by the interstices of particles. The flow space is fixed by the packed structure, which is determined by the orientation, arrangement, and relative position of the packing particles. Therefore, the precise form of the flow pattern in the packed structure is determined by the geometry of the flow space between the particles. There are two extreme difficulties in developing a quantitative understanding of percolation in a chromatographic column. First of all, the packed structure is very complicated and cannot be accurately described by mathematical methods. Second, even if the geometry of the structure is easy to handle, the flow pattern can only be approximated by numerical analysis. Therefore, our actual understanding of flow patterns must be limited to inaccurate but intuitive concepts supported by experience. Simple models and empirical parameters are used to obtain mathematical tools necessary for theoretical development in order to better understand the relationship between flow phenomena and structural factors in chromatographic separation.

2.5.1 Packing Structure

The majority of chromatographic separations is carried out on packed bed columns which consist of two parts: (1) the solid part provided by the particle skeleton and (2) the free space between the particles and in the pores. Porosity ε_T is defined as the fraction of free (nonsolid) space within a certain volume element of a porous material. It is a measure of the space available for the mobile phase. If these particles are porous, some spaces may appear within the particles themselves (e.g., perfusion chromatography packing). However, fluid flow mainly occurs in the free space composed of gaps and vacancies between particles. Therefore, for some purposes, we may only consider the interparticle porosity, which gives only a portion of the free space between the particles. This basic parameter, the interparticle porosity, will be identified by ε. This interparticle porosity is less than or at most equal to the total porosity ε_T. So, we can write

$$\Phi = \varepsilon/\varepsilon_T \tag{2.27}$$

where Φ is the fraction of the free space existing between particles in the total free space.

The interparticle porosity ε (measured for bulk materials) of the hardest particles is in the range of 0.35–0.50. A well-packed column usually has $\varepsilon = 0.4 \pm 0.03$. Of course, if the particles are porous, the real porosity ε_T changes more and higher value is expected. For example, the porous carrier used in liquid chromatography may confine half of its total pore space within the particle, giving $\Phi \sim 0.5$, so $\varepsilon_T = \varepsilon/\Phi = 0.4/0.5 = \sim 0.8$.

The porosity of granular materials is affected by many factors, such as external force, particle shape, size distribution, and wall effect. Agitation and compaction will reduce the porosity of the hard particulate material to a moderate level (about 20%). On the other hand, soft or fragile materials may suffer substantial compaction under moderate stress. Under the same degree of compaction, the angular particles will maintain a higher porosity than the round particles. This is due to particle bridging. The extreme case of bridging occurs in fibrous materials, where the porosity may be equal to or greater than 0.9. The mixture of large particles and small particles is packed more densely than any single kind of particles. This is due to the fact that fine particles can be embedded in the voids of the coarse material. A related phenomenon is an increase in the porosity of the column bed in a capillary column, whose diameter is reduced to only a few particle diameters, which is the well-known tube wall effect. This is due to the obstructive effect of the rigid tube wall on the particles, which limits the configuration of the particle arrangement, so that the particles cannot be assembled at the maximum density.

The packing structure of the granular material is the result of a compromise between two relative forces. The gravitational (or pressure) effect that causes the particle packing in the first place will cause the state of maximum density, which leads to the greatest order. On the other hand, the process of mixing particles together is highly random, resulting in a low-density, disordered structure. The final structure will consist of dense and loose regions. The particles in the dense region have reached almost the maximum order and density, while the particles in the loose region are imperfect in the packing structure and cannot form a dense packing. The loose region can be visually imagined as a bridging effect between the particles, the particles cannot be close to each other, thus forming a high porosity region. Therefore, the local porosity of the particle-packed structure is different, the dense region has a lower porosity, and the loose region has a higher porosity. The distribution of the two depends entirely on the shape of the particles, the physical properties such as rigidity, the packing method, and the ratio of the diameter of the chromatographic column to the particle diameter of the chromatographic medium.

By considering the packing of equal-diameter spherical particles, the concept and magnitude of bridging can be better understood (Scheidegger 1957). The regular cubic packing of spheres forms an array with six coordination numbers (the number of particles in contact with other particles), and the interparticle porosity ε is 0.48. The coordination number of orthorhombic packing is 8, and the interparticle porosity is 0.40. The rhombohedrons are formed by an array with 12 coordination numbers with the maximum possible density. The interparticle porosity ε is only 0.26, while the natural stacking of spheres will have $\varepsilon \sim 0.40$. The porosity of the tightly packed regions in the natural aggregates of spheres may be close to the minimum value of 0.26. However, the overall porosity increases due to the bridging effect, and the average (or overall) porosity is close to a larger value of 0.40. The difference between the average porosity and the densest porosity indicates that the bridging region occupies a considerable volume fraction of the aggregate and therefore may have a greater impact on the flow in the column bed.

2.5.2 Permeability of the Packing Structure

In the study of the flow phenomenon of porous materials, Darcy's law is of great value (Whitaker 1986). Darcy's law is an equation that describes the flow of fluid through porous media. It was originally proposed by Henry Darcy based on the experimental results of water flowing through the sand bed. In essence, it is similar to Fourier's law in heat conduction, Ohm's law in electric current, and Fick's law in diffusion theory. It points out that the fluid flux per unit cross section (unit: $m^3/s/m^2$) is proportional to the pressure drop per unit length (unit: Pa/m):

$$q = -K\frac{dp}{dz} \tag{2.28}$$

The proportional constant K is an empirical parameter (the minus sign is used to indicate that flow occurs in the direction of decreasing pressure). Morris Muskat first improved the Darcy equation for single-phase flow by including viscosity in Darcy's single-phase (fluid) equation (Muskat 1937). It can be understood that fluids with high viscosity are more difficult to penetrate through porous media than fluids with low viscosity. In this case, the flow resistance is caused by the viscous shear between the different fluid layers. The fluid flux permeating through the porous medium is proportional to the pressure gradient and inversely proportional to the fluid viscosity η:

$$q = -\frac{K_0}{\eta}\frac{dp}{dz} \tag{2.29}$$

where K_0 is the specific permeability and has a unit of length squared (m^2). (K_0 is usually given in Darcy units, where 1 Darcy is 9.87×10^{-9} cm^2.)

In chromatography, we are usually more interested in the volumetric flow rate F of the fluid rather than the fluid flux q per unit area. In porous media, fluid flow only occurs in the pores. Therefore, through the porosity, the flow velocity can be simply linked to the fluid flux, $F = q/\varepsilon$. We get

$$F = -\frac{K_0}{\eta\varepsilon}\frac{dp}{dz} \tag{2.30}$$

It correlates flow rate with pressure gradient, viscosity, porosity, and specific permeability.

2.5.3 *Capillary Model of Seepage in Porous Media*

The analogy between a flow channel in a porous medium and a bundle of capillaries is based on the following assumption: the flow through the pores of a packed bed of particles is similar to the flow through a thin capillary tube. Therefore, many capillary models have been designed. In the simplest case, the granular material is depicted as being arranged along the axis of flow, and the pores between them form a bunch of capillaries with uniform diameter and equal length. Other models assume capillaries with unequal lengths (tortuosity), capillaries with uneven diameters (pore throat model), capillaries with interconnected channels, nonparallel capillaries, etc. The detailed discussion of these models is beyond the scope of this book, because our main purpose is only to have a general understanding of the packing structure and the flow pattern in the column.

Naturally, the multi-capillary flow method is built from a single capillary flow based on Poisson's law (Sutera and Skalak 1993). Because the flow in the capillary is relatively simple, the Poiseuille's equation can be used for accurate calculations. This law states that the fluid flux Q (fluid volume per unit time, m^3/s) passing through a single capillary is given by

$$Q = -\frac{\pi d_c^4}{128\eta}\frac{dp}{dz} \tag{2.31}$$

or

$$q = \frac{4Q}{\pi d_c^2} = -\frac{d_c^2}{32\eta}\frac{dp}{dz} \tag{2.32}$$

where d_c is the diameter of the capillary and dp/dz is the pressure gradient along the net flow or z-axis.

The most direct application of the capillary model is to correlate the diameter of the capillary with the size of the particle that constitutes the particle-packed structure to give an estimate of the size of the channel in the porous medium.

2.5.4 *Flow Channels in Porous Media*

Many researchers have obtained expressions for the dependence of permeability on porosity. By far, the most famous expression is the Kozeny–Carman equation (Carman 1956). These approximate equations can be derived from the capillary model using empirical parameters. The specific permeability given by the Kozeny–Carmen equation is

$$K_0 = \frac{d_p^2}{180\eta}\frac{\epsilon^3}{(1-\epsilon)^2} \tag{2.33}$$

Comparing the flow equation of the capillary model with the Kozeny–Carman equation, we can get

$$\frac{d_c^2}{32} = \frac{d_p^2}{180}\frac{\epsilon^3}{(1-\epsilon)^2} \tag{2.34}$$

After rearrangement, the relationship between d_c and particle diameter d_p is obtained as

$$d_c = d_p\sqrt{\frac{8}{45}\frac{\epsilon^3}{(1-\epsilon)^2}} \tag{2.35}$$

It can be seen from the above equation that the size of the flow channel in the particle-packed structure is proportional to the particle size and increases with the increase of porosity. For the natural packing structure, the porosity is 0.4, and the diameter of the "circular" channel in the porous structure is approximately 18% of the particle diameter. For the densest regularly packed area, the porosity is 0.26, and the channel diameter is approximately 7.5% of the particle size. For the loosest regular packing, the porosity is 0.48, and the channel diameter is 27% of the particle size.

2.5.5 Flow Velocity Distribution

As mentioned earlier, the simplest capillary model assumes that the flow in the porous material occurs along a bundle of parallel, uniform diameter capillaries and equal length capillaries. Therefore, each interstitial flow "channel" is equivalent to a capillary in the model in a sense. However, this simple model has many disadvantages. First, each capillary is uneven in diameter along its length. The contact between the particle and the particle forms a narrow part, and the gap between the particles forms an open part. The entire flow channel is composed of the narrow part and the open part arranged alternately. The narrow part is the main obstacle to flow, often forcing liquid into the adjacent capillary. This means that there are a large number of interconnections in the particle packing structure, which cannot be described by a simple model. Second, the lengths of the capillaries are not equal. If the alignment between the particles is poor and there are bridging areas, much greater flow will occur. Finally, the flow paths in the packed column are not parallel but tortuous. Therefore, for the detailed study of the flow pattern in the packing structure, the simplest capillary model seems inadequate and cannot provide more quantitative information. Even so, the proper application of the capillary model to evaluate the various types of flow rate deviations that may occur in the chromatographic column will help us better understand the complex problem of flow rate distribution in the chromatographic column.

First, consider the flow rate distribution of the fluid in the capillary. Because the viscous force is dominant, there is friction along the wall, resulting in a well-defined streamline with a parabolic radial velocity distribution:

$$u = 2u_0\left(1 - \frac{r^2}{R^2}\right) \tag{2.36}$$

where r is the radial coordinate and R is the radius of the capillary. The average linear velocity u_0 is given by the Hagen–Poiseuille equation (Muskat 1937; Carman 1956):

$$u_0 = \frac{R^2}{8\eta}\frac{dp}{dz} \tag{2.37}$$

As mentioned earlier, at the center of the tube, $r = 0$, the flow layer has a maximum velocity, which is twice the average velocity. At the pipe wall, $r = R$, the flow layer is in a static state, and the flow velocity is 0.

For the flow velocity distribution of the fluid in the packing structure, we can use the capillary model to estimate it. In the flow channel composed of each particle gap, the fluid presents a parabolic flow velocity distribution. However, due to the different diameters of different channels, their average flow rates are also different. The different channel sizes produce flow rate deviations. In addition, fluids are exchanged between adjacent flow channels, thereby creating a cross-channel effect.

For porous chromatographic packing materials, the mobile phase usually occupies the internal pore space accessible by solvent molecules. This part of the mobile phase is usually considered stagnant because the internal pore structure is so fine (as mentioned earlier, the average velocity of the fluid increases with the square of the channel diameter). Therefore, there is usually a strong velocity difference between the fluids within and between particles, leading to trans-particle effects. However, when there are larger internal pores, the internal fluid may not stagnate. The normal passage between the particles is so narrow and tortuous that certain fluids can choose any reasonable alternative route, thereby enhancing their movement.

Second, the speed deviation between important short-range channels is also increasing, which can be understood as the difference in flow velocity between the tightly packed area and the loosely packed area. If we assume that the packing of the bridging area is similar to a simple cubic structure ($\varepsilon = 0.48$) and that the tight area is a tightly packed structure ($\varepsilon = 0.26$), then the Kozeny–Carman equation (to some extent for this local flow disturbance is valid) tells us that the fluid velocity in the bridging area is 13 times the fluid velocity in the tight area. Due to the wide interconnection between the flow channels, or possibly the wide interconnection between the bridging regions themselves, it is entirely possible that there are large velocity changes in the cross section of the chromatographic column.

Another source of flow rate variation across the column is related to the nature of the flow channel formed by the packing near the wall surface. Due to the existence of the rigid wall, the arrangement of particles on the surface of the wall is restricted, and it has a higher porosity, which is usually called the wall effect. This wall effect decreases as the distance from the wall increases. When the ratio of pipe diameter to particle size is greater than 50, this wall effect is generally considered to be negligible.

2.6 Molecular Diffusion

In addition to the mass transfer caused by flow, molecular diffusion also plays an important role in mass transfer in the chromatography process. Molecular diffusion is caused by the thermal movement of molecules. The molecular diffusion itself is random and nondirectional. However, since the number of molecules in the high-concentration area is greater than the number of molecules in the low-concentration area, the movement of molecules from the high-concentration area to the low-concentration area is actually observed. As long as there is a difference in concentration, diffusion may occur, and there is molecular diffusion of solutes in both the mobile phase and the stationary phase.

2.6.1 *Diffusion in Solution*

The main rule describing the precise nature of diffusion broadening is Fick's law (Giddings 1965; McGuffin 2004). Fick's first law states that the flux of a given component (such as a solute) on a fixed unit area in the medium J is

$$J = -D\frac{\partial c}{\partial z} \tag{2.38}$$

where D is the diffusion coefficient and $\partial c/\partial z$ is the concentration gradient. The equation shows that the mass flux is proportional to the concentration gradient but in the opposite direction. In other words, solute molecules will migrate from a higher-concentration area to a lower-concentration area. The diffusion coefficient D is a characteristic of the solute and its diffusion medium. The flux of J will be equal to the number of moles per unit area per unit time (perpendicular to the concentration gradient). This equation is generally applicable to true molecular diffusion, but it has been found to be applicable to a large variety of natural migration phenomena (e.g., horizontal flow of water in unsaturated soil, mixing of air masses in the atmosphere, light penetrating into snow or clouds, reactor Neutron flux in neutrons, etc.). These processes can be called effective diffusion processes, and each process has an effective diffusion coefficient, simply because they follow Fick's first law of diffusion.

The mass transfer process that obeys Fick's first law also obeys Fick's second law:

$$\frac{\partial c}{\partial t} = \frac{\partial}{\partial z}D\frac{\partial c}{\partial z} \tag{2.39}$$

Generally, D can be considered as a constant, and the equation can be written in a more common form:

$$\frac{\partial c}{\partial t} = D\frac{\partial^2 c}{\partial z^2} \tag{2.40}$$

where t is the time and all other variables are as previously defined. This is a partial differential equation, whose solution is consistent with appropriate initial conditions and describes how the concentration curve changes over time. For example, if the initial profile is a sharp peak (corresponding to a narrow sample injection area), the solution of this equation is in the Gaussian form:

$$c = \frac{M}{\sqrt{t}} \exp\left(-\frac{z^2}{4Dt}\right) \tag{2.41}$$

where M is a constant proportional to the total mass of the diffused solute. By substituting it into the differential equation, it can be easily verified as the solution of the equation. It can be seen that the maximum concentration change when $z = 0$ is $1/t^{1/2}$. This shows that as diffusion progresses (e.g., during column migration), the flattening of the solute zone will occur. In chromatography, at a given flow rate, the time t is proportional to the migration distance. Therefore, the maximum concentration of the zone decreases with the increase of the column length L according to the relationship of $L^{-1/2}$. If the flow rate changes, this simple rule no longer applies, because various effective diffusion coefficients are sensitive to the flow rate.

The Gaussian curve can usually be written in the form of standard deviation σ:

$$c \propto \exp\left(-\frac{z^2}{2\sigma^2}\right) \tag{2.42}$$

A direct comparison of the last two equations shows

$$\sigma^2 = 2Dt \tag{2.43}$$

This is the famous Einstein relationship. This equation shows the relationship between σ, which is a quarter of the width of the solute zone, and the molecular (or effective) diffusion process that causes the band to expand. And, of course, since $H = \sigma^2/L$, the plate height is related to the effective diffusion coefficient, $H = 2Dt/L = 2D/Ru$, where the zone displacement speed Ru replaces L/t.

This relationship is shown in Fig. 2.3, where the dispersion of the concentration increases with the diffusion coefficient or time.

If the initial solute concentration is a continuous source, like the feed we encounter in front chromatography, the solution to Fick's second law is

$$c = \frac{1}{2}\left(1 - \operatorname{erf}\left(\frac{z}{\sqrt{4Dt}}\right)\right) \tag{2.44}$$

Therefore, diffusion from a continuous source causes the breakthrough curve to have an S-shaped distribution. This relationship is shown in Fig. 2.4, where the concentration dispersion increases as the diffusion coefficient or time increases.

Various molecular diffusion processes occur in liquid chromatography. Liquid molecules are more or less constrained in place by relatively strong intermolecular attractive forces, and a large amount of heat energy must be obtained to break the constraints of these bonds and achieve displacement. Therefore, logically speaking, diffusion and transport is an activation process and has an exponential dependence on temperature:

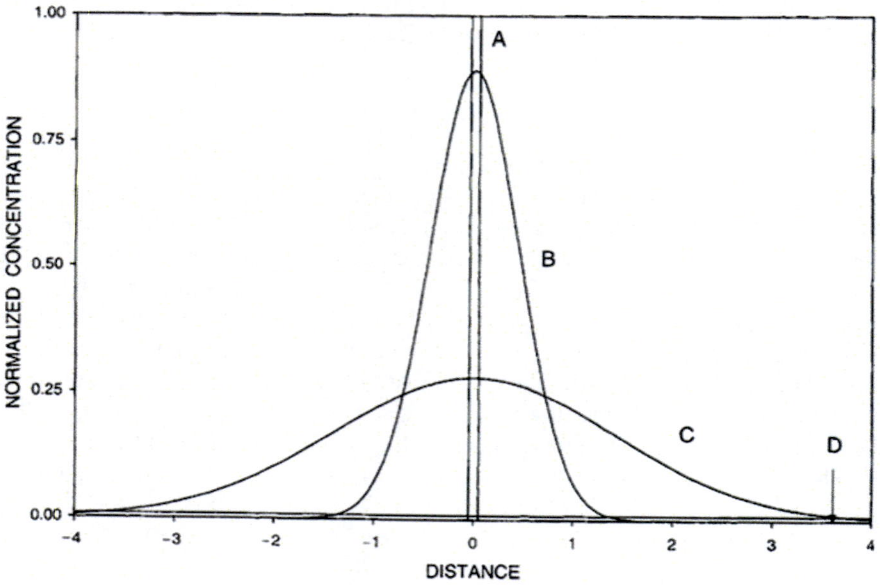

Fig. 2.3 With the diffusion of sharp peaks in the elution chromatogram, the solute concentration is flattened as the value of the product of the diffusion coefficient D and time t increases. (A) $Dt = 0.00$, (B) $Dt = 0.01$, (C) $Dt = 1.00$, (D) $Dt = \infty$. (Reproduced from McGuffin 2004)

$$D_l = D_0 \exp\left(-\frac{W_l}{RT}\right) \tag{2.45}$$

where W_l is the activation energy required for molecular displacement, R is the gas constant, T is the absolute temperature, and D_0 is the diffusion coefficient that hardly changes with temperature. This equation can be used to calculate the temperature dependence of liquid diffusion in a chromatographic system.

It is well-known that the viscous shear of liquid is an activation process, and $1/\eta$ shows an exponential temperature dependence similar to D. Interestingly, the experiment found that the activation energy W_l is almost equal for D_l and l/η. Since viscosity data is more common (easier to obtain) than diffusion data, we can usually obtain W_l in this way. If η is known at a range of temperatures, the graph of log $(1/\eta)$ versus $1/T$ will produce a straight line with a slope of $-W_l/2.3R$.

The diffusion in the binary mixture (solute and solvent) shows the same activation energy as the process in the pure solvent (viscous shear), which indicates that the molecular displacement in the diffusion is controlled to some extent by the slip of solvent molecules. In the hole theory of liquids (Eyring et al. 1964), it is assumed that the rate-controlling process of diffusion is to form a pore space next to the solute, and then, the solute can easily fall into the pore. The energy W_l is approximately equal to the energy required to create the hole.

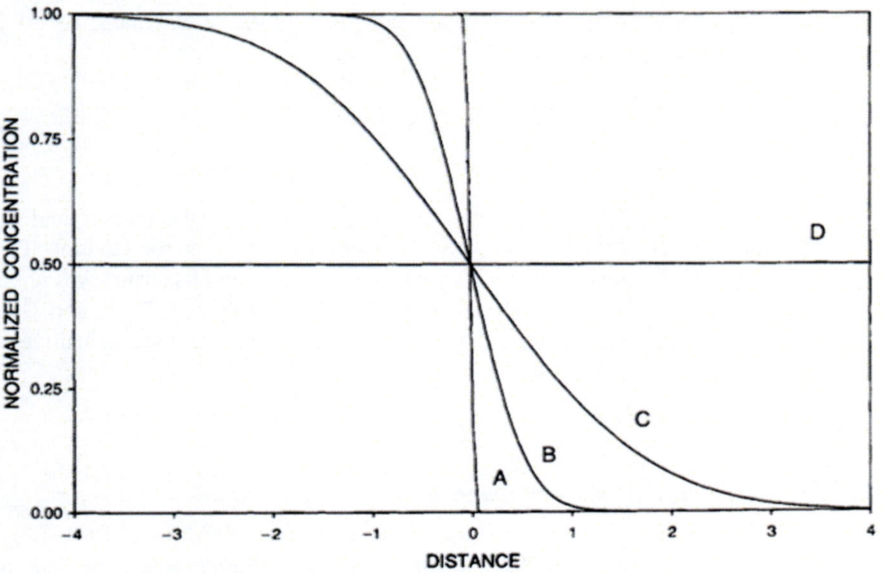

Fig. 2.4 The diffusion of the solute breakthrough curve in the frontier chromatogram varies with the value of the product of the diffusion coefficient D and the time t. (A) $Dt = 0.00$, (B) $Dt = 0.01$, (C) $Dt = 1.00$, (D) $Dt = \infty$. (Reproduced from McGuffin 2004)

If you want to obtain the value of D_l itself, not just the value of W_l, you can call a suitable equation, such as the Stokes–Einstein equation.

$$D_l = \frac{RT}{6\pi r_1 \eta N} \text{ (Stokes-Einstein)} \tag{2.46}$$

where r_1 is the "radius" of the solute molecule and N is the Avogadro's number. (Hereafter, subscript 1 applies to solutes and 2 applies to solvents.) Assuming that the molar volume of spherical molecules is V_1, the equation can be simplified as

$$D_l = \frac{10^{-7} T}{\eta V_1^{0.33}} \tag{2.47}$$

where D_l is inversely proportional to the cube root of the volume of the solute and, therefore, inversely proportional to the molecular weight. In fact, there should be a form factor in this equation. In contrast to the spherical shape, for rod-shaped molecules, D_l will be reduced.

The Stokes–Einstein equation is derived for large Brownian particles (colloids) and does not apply to solute molecules with a molecular weight less than 1000. Diffusion coefficients for small molecules can be estimated through the use of empirical expressions. The Wilke–Chang equation (Wilke and Chang 1955) is perhaps the most commonly used:

$$D_{\mathrm{l}} = 7.4 \times 10^{-8} \frac{(\Psi_2 M_2)^{0.5} T}{\eta_2 V_1^{0.6}} \tag{2.48}$$

where M_2 is the molecular weight of the solvent, ψ_2 is the association factor, and 1 is for nonpolar solvent, 2.6 for water, 1.9 for methanol, and 1.5 for ethanol. This equation is suitable for small- and medium-sized molecules, and its accuracy is about 10%. Please note that D_{l} in the Wilke–Chang equation varies with $V_1^{-0.6}$ and D_{l} in the Stokes–Einstein equation varies with $V_1^{-0.3}$. The former is more sensitive to changes in the molecular size of the solute.

2.6.2 *Diffusion in Porous Media*

When migrating through a granular medium, the chromatographic zone will be partly expanded due to molecular diffusion and partly expanded due to various imbalances and differential flow phenomena (which of course may involve molecular diffusion). There is also a situation where the solute molecules interact with the stationary-phase ligands through the liquid retained in the pores of the particles. Due to the complexity of the geometry of the micropores, the effective diffusion coefficient of the solute will be lower than the diffusion coefficient in the bulk solution. In the following sections, we will discuss this phenomenon in more detail.

2.6.2.1 Diffusion in the Retained Mobile Phase

For most liquid chromatography packings, the pore size of the medium is about 100–300 Å, and the mobile phase is retained after entering these pores. It can be assumed that all flows have stopped. The solute molecules diffuse along the stationary fluid in the pores inside each particle, and the solute diffusion may be reduced by the hindrance of the medium. This situation can be compared with the impeding effect of porous media on the electrophoresis rate of particles. Boyack and Giddings (1963) proposed that the obstacle to electrophoresis may come from two effects: (1) the tortuous path of the particle through the medium and (2) the alternating contraction and widening of the particle motion path, resulting in a potential gradient along the path. The potential gradient is not a fixed value but constantly changes as the particles move. It can be inferred that the diffusion of molecules in porous media will also be affected by the same two obstructive effects. First, the pores in the porous material exist in a tortuous manner, thereby preventing the diffusion of molecules along the shortest direct path. Second, the size of the cross-sectional area of the tunnel varies from wide to narrow alternately. The narrow throat greatly

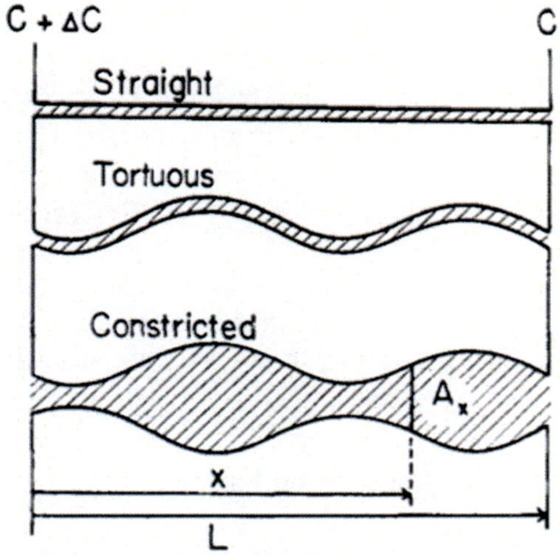

Fig. 2.5 The nature of tortuous and narrow paths in chromatographic packing. (Reprinted with permission from Knox JH, McLaren L (1964) A new gas chromatographic method for measuring gaseous diffusion coefficients and obstructive factors. Anal Chem 36: 1477–1482. Copyright {1964} American Chemical Society)

slows solute diffusion. Therefore, the hindrance factor composed of the tortuosity factor and the constrictive factor reduces the diffusion of solute through the chromatographic material.

The tortuosity and shrinkage of pores in porous media are shown in Fig. 2.5. The uniform and parallel paths of the channels will not cause any hindrance to diffusion. The curved path slows the spread due to the increased path length. The ink bottle-like channel is composed of a wide-open area and a narrow constriction. The constriction forms a "traffic bottleneck" and hinders diffusion. There are two obstacles in the combined path.

2.6.2.2 Tortuosity Factor

According to Giddings (1965), the influence of channel bending properties on diffusion can be determined as follows. If the solute peak is allowed to diffuse outward, the root mean square displacement (or standard deviation) σ is given by the Einstein relationship:

$$\sigma^2 = 2Dt \tag{2.49}$$

where D is the effective diffusion coefficient of the solute in the retained mobile phase. However, because the solute molecules diffuse along a tortuous path, the length of which is l_T, which is greater than the length of the linear path, l. Therefore, the time required for solute diffusion to reach the end point is τ times that of the linear path ($\tau = l_T/l$). From a molecular point of view, the actual displacement of the solute is ($\tau\ \sigma$) instead of the apparent displacement σ. Therefore,

$$(\tau\sigma)^2 = 2D_{\mathrm{m}}t \tag{2.50}$$

where D_{m} is the molecular diffusion coefficient in the mobile phase, not the apparent diffusion coefficient D. When dividing Eq. (2.49) by (2.50), we get

$$D/D_{\mathrm{m}} = \tau^{-2} \tag{2.51}$$

where, τ, called the tortuosity factor, is the average ratio of the actual path of the solute through the porous medium to the shortest path length. Of course, due to the complexity of the gap geometry, the actual path length is very difficult to determine.

2.6.2.3 Constriction Factor

As mentioned earlier, there is a narrow part of the diffusion channel in the micropore, which constitutes the bottleneck of molecular diffusion. According to Boyack and Giddings (1963), the influence of the constricted part of the channel on the diffusion can also be obtained in a relatively simple way. We first consider an idealized hole model, which has the property that the size of the cross-sectional surface of the hole varies sequentially along its length. Without considering the twists and turns, we can write the solute flux J_{c} through the cross section of a single channel according to Fick's first law:

$$J_{\mathrm{c}} = -D_{\mathrm{m}}\frac{\partial c}{\partial z}A \tag{2.52}$$

where A is the cross section of the channel, and its value changes from small to large or from large to small due to constriction. We can rearrange the equation as $\frac{\partial c}{\partial z} = -\frac{J_{\mathrm{c}}}{D_{\mathrm{m}}A}$ and then integrate the channel segment length I:

$$\Delta c = \frac{J_{\mathrm{c}}}{D_{\mathrm{m}}}\int_0^l \frac{dz}{A} \tag{2.53}$$

where Δc is the concentration difference over length l. The integral in this equation is only the product of the length l and the average value of $1/A$, that is, $\overline{(1/A)}l$. Therefore,

$$\Delta c = -\frac{J_{\mathrm{c}}}{D_{\mathrm{m}}}\overline{(1/A)}l \tag{2.54}$$

It is assumed that the solute flux J_{c} is constant throughout the fragment, because in the steady state (almost always valid for short fragments), the flux through each cross section along the fragment must be the same. Now, we can use the effective diffusion coefficient D and the average cross-sectional area A to express the flux through the entire fragment length l as

$$J_c = -D\frac{\Delta c}{l}\overline{A} \tag{2.55}$$

where $\Delta c/l$ is the concentration gradient of the solute along the fragment direction. Solve the equation to get Δc:

$$\Delta c = -\frac{J_c}{D}\overline{\left(\frac{1}{A}\right)}l \tag{2.56}$$

Now, we have an equation for Δc similar to Eq. (2.54). When the two equations are set equal, we get

$$\frac{D}{D_m} = \frac{1}{\overline{A}\overline{\left(\frac{1}{A}\right)}} \tag{2.57}$$

We can denote the factor reflecting the influence of pore shrinkage as constriction factor χ, that is, $\chi = \left[\overline{A}\overline{\left(\frac{1}{A}\right)}\right]^{-1}$, which is the reciprocal of the product of the average area $\overline{A}$ and the average reciprocal area $\overline{\left(\frac{1}{A}\right)}$. For a uniform channel, $1/[\overline{A}]$ and $\overline{\left(\frac{1}{A}\right)}$ are equal, and the value of constriction factor is 1. For highly contracted pores, A becomes almost 0 at certain points, so $1/A$ becomes almost infinite, causing the average value of $1/A$ to become disproportionately larger, which further causes the constriction factor χ to become smaller. The tendency of $1/A$ to become very large in contraction is a mathematical reflection of the fact that this contraction excessively hinders diffusion. The constriction factor that causes the diffusion coefficient to become smaller can be compared with the increase in resistance on a wire with a narrow part. The high-pressure drop caused by the narrow part cannot be compensated by the low-pressure drop of the part with a larger cross section.

2.6.2.4 Obstruction Factor

In summary, the diffusion of solute in porous media is reduced by the influence of two geometric factors. One is related to the tortuosity of the diffusion path, and the other is related to the cross-sectional change of the diffusion path. Therefore, the obstruction factor, γ, is the product of the two independent terms, the tortuosity factor and the constriction factor:

$$\gamma = \chi\tau^{-2} \tag{2.58}$$

The value of each factor in the above equation can be calculated using various cell models. Therefore, it is not difficult to understand that their relative contribution to the total hindrance factor also varies with the type of the model selected (Knox and McLaren 1964; Giddings and Boyack 1964; Boyack and Giddings 1963); in particular, whether there are some straight paths or only staggered paths when passing through the medium will have a greater impact on the calculation results (Giddings 1965). The complexity of the pore structure of porous media makes it impossible to determine their respective contributions through experimental methods. So far, experimental studies have almost focused on the determination of the total hindrance factor γ. Knox and McLaren (1964) used non-retained solute molecules as probes and used the arrested elution method to determine the obstruction factor of a column filled with solid glass beads. They found that the γ value was 0.60 ± 0.02, while the theoretical value was 0.63. For most nonporous, densely packed particulate materials used in chromatography, this value of 0.60 may be a good approximation. However, for other porous materials, the experimentally determined value deviated from the theoretical value considerably.

The theory of tortuosity and contraction shows that the hindering factor γ is essentially a structural constant, which is mainly determined by the geometry of the pore structure in the porous medium, and has nothing to do with the specific mobile phase or solute considered. However, the flow rate of the mobile phase may have a certain impact on the experimentally determined obstruction factor, which makes the obstruction factor exhibit flow rate dependence. When the flow rate is large, the porous medium will be more compacted, which will cause the pore structure to deform, resulting in an increase in the obstruction factor. In addition, the retained mobile-phase layer at the contact between the particles becomes thinner, which changes the diffusion path of the solute. Therefore, both the contraction effect and the tortuous effect may change. Although the results about tortuosity and contraction are approximate, they clearly show the importance of the geometry of the pore structure in solute diffusion. As will be mentioned in the discussion below, optimizing the pore structure of the chromatographic packing and improving the solute diffusion behavior are important ways to improve the efficiency of the chromatographic column.

2.7 Chromatographic Peak Spreading

In the chromatographic process, solute molecules have to undergo multiple adsorption and desorption processes or repeated distribution between the mobile phase and the stationary phase. When they are in the mobile phase, they migrate at the speed of the mobile phase; when they are adsorbed on the stationary phase, they are fixed relative to the mobile phase. Due to the collision between molecules in the process of motion, the path of motion is different, and the path of solute molecules is completely irregular, so random models can be used to describe their behavior. Random process always leads to the Gaussian distribution, so standard deviation σ

or variance σ^2 can be used as a measure of the dispersion of solute molecules in the column. The total dispersion should be the sum of the molecular dispersion per unit column length and is proportional to the column length:

$$\sigma^2 \propto L \tag{2.59}$$

If the scale factor is assumed to be H, the above formula can be written as

$$\sigma^2 = H\,L \tag{2.60}$$

where H is the molecular dispersion of unit column length:

$$H = \sigma^2/L \tag{2.61}$$

Traditionally, we call H the theoretical plate height, which is also known as the height equivalent to a theoretical plate (HETP) and is used to express the variance of the solute concentration distribution curve in chromatography, that is, the degree of molecular dispersion.

There are a variety of irregular molecular motion states in the chromatographic process, which causes the chromatographic zone to expand. According to stochastic theory, the variance of the sum of a finite number of independent random variables is equal to the sum of their variances. Therefore, the total chromatographic zone expansion of the chromatographic process is equal to the sum of the variances of each independent factor, that is, the discrete terms of each independent molecule are additive:

$$\sigma^2 = \sigma_1^2 + \sigma_2^2 + \sigma_3^2 + \cdots\cdots + \sigma_i^2 = \sum_{i=1}^{n} \sigma_i^2 \tag{2.62}$$

or

$$H = H_1 + H_2 + H_3 + \cdots\cdots + H_i = \frac{\sigma_1^2}{L} + \frac{\sigma_2^2}{L} + \frac{\sigma_3^2}{L} + \cdots\cdots + \frac{\sigma_i^2}{L} = \sum_{i=1}^{n} H_i \tag{2.63}$$

The total plate height is equal to the sum of the contributions of various factors and is the statistical concept of molecular dispersion.

2.7.1 Plate Height Equation

In 1956, the Dutch physicist J. J. van Deemter and his colleagues (van Deemter et al. 1956) reported their research on the influencing factors of peak broadening in gas-liquid chromatography. According to the material balance, they established partial differential equations for the chromatographic process involving eddy diffusion, axial molecular diffusion and mass transfer between two phases. By solving these equations, they derived the general rate equation in chromatography, also known as van Deemter equation. For a packed column, the van Deemter equation is expressed as

$$H = 2\lambda d_{\mathrm{p}} + \frac{2\gamma D_{\mathrm{g}}}{u} + \frac{8}{\pi^2} R(1-R) \frac{d_{\mathrm{f}}^2 u}{D_{\mathrm{l}}} \tag{2.64}$$

where d_p is the average particle diameter, u is the average gas flow velocity, D_g and D_l are the diffusion coefficients of solutes in gas and liquid respectively, R is the ratio of the peak velocity of the solute to the gas velocity, and d_f is the effective film thickness on the particle surface. The two terms λ and γ are empirical constants.

The equation includes three main contributions that cause the dispersion of a chromatographic peak: (1) eddy diffusion; (2) longitudinal molecular diffusion; and (3) mass transfer resistance, including mobile-phase mass transfer resistance and stationary-phase mass transfer resistance.

The simplified form of the rate theory equation is as follows:

$$H = A + \frac{B}{u} + Cu \tag{2.65}$$

where A is the coefficient of eddy diffusion term, B is the coefficient of longitudinal molecular diffusion term, and C is the coefficient of mass transfer resistance.

For an open-tubular column, A is 0. B/u is caused by the longitudinal diffusion of the gas phase, while Cu is caused by the failure to reach a complete equilibrium between the gas phase and the liquid phase. In the original van Deemter equation (2.64), only the mass transfer resistance in the liquid phase was considered important for the formation of C. However, Golay showed that the mass transfer resistance in the gas phase should also be considered when dealing with open-tubular columns (Golay 1963). The Golay equation for the plate height in capillary chromatography is as follows:

$$H = \frac{2D_{\mathrm{g}}}{u} + \frac{2}{3} R(1-R) \frac{d_{\mathrm{f}}^2 u}{D_{\mathrm{l}}} + \frac{1}{96}\left(6R^2 - 16R + 11\right) \frac{d_{\mathrm{c}}^2 u}{D_{\mathrm{g}}} = \frac{B}{u} + C_{\mathrm{l}} u + C_{\mathrm{g}} u \tag{2.66}$$

where d_c is the inner diameter of the capillary and d_f is the thickness of the liquid film coated on the inner wall of the capillary. The mass transfer resistance is divided into gas phase ($C_g u$) and liquid phase ($C_l u$).

Golay pointed out that, due to the irregularities of the packing, the resistance of gas-phase mass transfer in a packed column has a greater impact on column efficiency than in an open-tubular column. A large number of experimental results show that the mass transfer resistance C_g in the gas phase is usually comparable to the mass transfer resistance C_l in the liquid phase, and in some cases, it is even higher. For non-retained solutes, the liquid-phase mass transfer resistance C_l should theoretically be 0. However, it is observed in the experiment that there is a clear upward trend in the HETP and u curve, so C_g must have an obvious positive value, which can be completely attributed to the slow mass transfer in the gas phase.

Giddings (1961) put forward a different opinion on the van Deemter equation, arguing that it did not correctly explain the role of structural factors in packed columns. He pointed out that the tortuous nature of the streamlines in the packed column will help mass transfer. Therefore, the molecules in any streamline will randomly walk from the low-speed area to the high-speed area (even without lateral diffusion) and will be very close to the surface of the particle from time to time. Therefore, there are two alternative ways, in which the molecules can move in the gas flow and thus promote the mixing of the gas or in close proximity to the particle surface to promote mass transfer. They spread in the airflow and move along the tortuous streamline. Since these two effects are complementary, they should be compounded like parallel resistors instead of series resistors, and their combined contribution to H should be written as

$$A' = \left(\frac{1}{A} + \frac{1}{C_g u}\right)^{-1} \tag{2.67}$$

It can be seen from the above equation that the terms of mass transfer and eddy diffusion in the mobile phase are not completely independent. Solute molecules can exchange in these two processes, which reduces the molecular dispersion. In other words, eddy diffusion is beneficial to reduce the mass transfer resistance of solute molecules in the mobile phase. Giddings coupling equation is not only applicable to gas chromatography but also to liquid chromatography.

$$H = \frac{B}{u} + C_s u + \frac{1}{\frac{1}{A} + \frac{1}{C_m u}} \tag{2.68}$$

The subscripts m and s in the equation denote mobile phase and stationary phase, respectively.

2.7.1.1 Eddy Diffusion

Eddy diffusion is also known as multipath effect. In a particle-packed column, the fluid flow channel is composed of free space in the interstices of the particles. Due to the irregular geometry of the flow channel, the fluid can pass through the chromatographic column at quite different flow rates. When solute molecules are loaded on the chromatographic column, they move toward the column outlet with the mobile phase. Solute molecules can take many different paths within the chromatographic column. Some paths are long and some are short; some are wide and some are narrow. Solute molecules will take a different time to reach the column outlet and leave the chromatographic column, resulting in a widening of the solute zone. In the earlier literature, this process is often referred to as "eddy diffusion," focusing on turbulence and mixing as the cause of eddy dispersion. Nowadays, eddy diffusion is also referred to as the multipath effect, emphasizing that the unevenness of the packing structure of the chromatographic column is responsible for band broadening. The average linear velocity of the mobile phase determines the average migration velocity of the solute, but the difference in the flow velocity of the different paths of the mobile phase leads to the dispersion of the solute zone, which is expressed by the following equation:

$$H_{\mathrm{A}} = 2\lambda d_{\mathrm{p}} \tag{2.69}$$

where λ is an empirical constant in the range of 0.5–1.0, which reflects the uniformity of the packing structure. Therefore, as the first approximation, the peak broadening caused by the multipath effect is only related to the uniformity and particle size of the packed bed and has nothing to do with the linear velocity.

2.7.1.2 Longitudinal Diffusion

Longitudinal diffusion is also known as molecular diffusion. The random thermal motion of molecules also causes the solute zone to disperse. As mentioned earlier, molecular diffusion causes solute molecules to move from a high-concentration area to a low-concentration area, resulting in a symmetrical solute distribution curve, which can be described by the Gaussian distribution equation. The band dispersion is proportional to the square root of the residence time of the solute in the system, that is, proportional to the square root of the distance, the solute moves in the column bed. The peak broadening caused by the diffusion of molecules in the mobile phase can be expressed as

$$H_{\mathrm{B}} = 2\gamma_{\mathrm{m}} D_{\mathrm{m}}/u \tag{2.70}$$

where D_{m} is the diffusion coefficient of the solute in the mobile phase and γ_{m} is an empirical constant in the range of 0.5–0.7, called the obstruction factor, which reflects the hindrance of the path curvature and width changes in the packing structure to molecular diffusion. For an empty column, the obstruction factor is 1.00; for a column packed with nonporous spherical particles, the obstruction factor is 0.6. It can be seen that the tortuosity factor is inversely proportional to the degree of curvature of the flow path in the packed column. In liquid chromatography, the solute diffusion coefficient is usually 10^{-5} $\mathrm{cm^2/s}$, which is 4 to 5 orders of magnitude smaller than the diffusion coefficient in the gas phase. Therefore, the molecular diffusion term is usually negligible.

2.7.1.3 Mass Transfer Resistance

Mass transfer resistance refers to the tendency of molecules to remain in their current position due to inertial forces. In liquid chromatography, mass transfer resistance is composed of three contributions:

$$H_C = H_m + H_s + H_{sm} \tag{2.71}$$

The subscripts *m*, *s*, and *sm* represent the mass transfer resistance in the mobile phase, the stationary phase, and the mobile phase retained in the stationary phase, respectively.

The mass transfer resistance in the mobile phase is due to the fact that the flow velocity near the surface of the stationary phase is slower than that in the middle of the flow path, so the solute molecules near the surface of the stationary phase diffuse shorter than the molecules in the middle of the mobile phase, causing zone dispersion. As mentioned earlier, the laminar flow in the open-tubular column has a parabolic radial velocity distribution, the velocity of the tube wall is 0, and the velocity of the tube center is twice the average velocity. The mass transfer resistance phase in the mobile phase can be expressed as

$$H_{C.m} = \frac{1 + 6k + 11k^2}{96(1+k)^2}\frac{d_c^2 u}{D_m} \tag{2.72}$$

where k is the solute retention factor and d_c is the inner diameter of the open-tubular column.

The flow pattern of laminar flow in a packed column is relatively complicated, and so far, it has not been possible to derive an expression similar to that in an open-tubular column. The mass transfer resistance in the mobile phase is usually expressed as

$$H_{C.m} = \omega\frac{d_p^2 u}{D_m} \tag{2.73}$$

where ω is a dimensionless empirical constant between 0.01 and 10, which is related to the column inner diameter, shape, and packing properties.

The mass transfer resistance of the stationary phase is caused by the different residence time of the solute molecules in the stationary phase, which can be expressed as

$$H_{C.s} = q\frac{k}{(1+k)^2}\frac{d_p^2 u}{D_s} \tag{2.74}$$

where q is the structure factor, which depends on the shape of the stationary phase and D_s is the diffusion coefficient of the solute in the stationary phase.

The mass transfer resistance in the stagnant mobile phase is due to the mass transfer of solutes between the mobile phase and the mobile phase retained in the particles, which must pass through the static mobile-phase zone in the pores. If the pores of the chromatographic packing are small and deep, the mass transfer rate will decrease. This part of the contribution can be expressed as

$$H_{\mathrm{C.sm}} = \frac{(1-\varepsilon+k)^2}{30(1-\varepsilon)(1+k)} \frac{d_\mathrm{p}^2 u}{2\gamma D_\mathrm{s}} \tag{2.75}$$

where ε is the porosity of the stationary phase and γ is the obstruction factor for solute diffusion in the stationary phase.

Giddings (1965) described in detail the contribution of other mass transfer resistance terms in the stationary phase. These contributions are proportional to the linear velocity of the mobile phase.

Similar to mass transfer process in gas chromatography, the mass transfer in liquid chromatography also involves coupling of eddy diffusion and mass transfer resistance in the mobile phase. Solute molecules can migrate from one flow path to another flow path with completely different flow rates through lateral diffusion or convective diffusion, so the eddy diffusion H_e and the mobile-phase mass transfer resistance H_m are coupled with each other to form a coupled phase H_{em}:

$$H_{\mathrm{em}} = \frac{1}{\frac{1}{H_\mathrm{e}} + \frac{1}{H_\mathrm{m}}} = \frac{1}{\frac{1}{2\lambda d_\mathrm{p}} + \frac{D_\mathrm{m}}{\omega d_\mathrm{p}^2 u}} \tag{2.76}$$

At high flow rates, H_{em} is close to the H_e value, that is, H_{em} is controlled by the eddy diffusion term. At low flow rates, H_{em} is close to H_m, which is controlled by the mass transfer resistance of the mobile phase.

2.7.2 Reduced Plate Height Equation

As proposed by Giddings, the use of the so-called reduced parameters is most convenient when comparing the plate heights under different chromatographic conditions.

The dimensionless reduced flow velocity v is defined as

$$v = \frac{u d_\mathrm{p}}{D_\mathrm{m}} \tag{2.77}$$

The dimensionless reduced plate height h is defined as

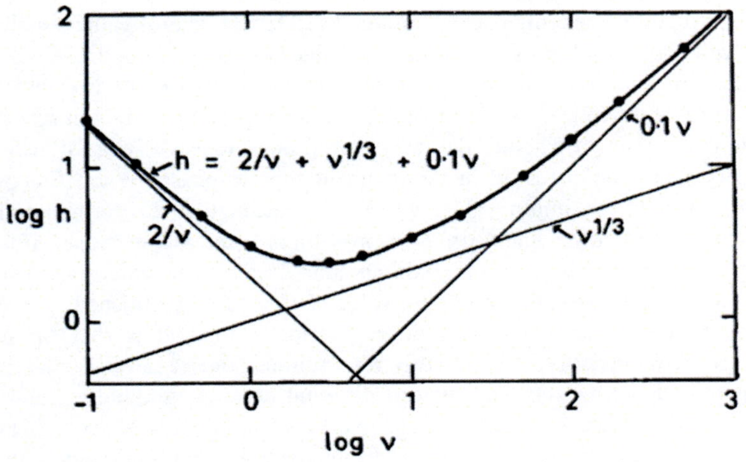

Fig. 2.6 Log h is plotted against log v, where $h = \frac{2}{v} + v^{1/3} + 0.10v$. (Reproduced from Bristow and Knox 1977)

$$h = \frac{H}{d_\mathrm{p}} \tag{2.78}$$

The reduced flow rate is the normalization of the solute diffusion rate within the particle size range of the packing, that is, the ratio of the product of the linear velocity and the diameter of a particle of the packing to the solute diffusion rate. The reduced plate height is measured by the particle diameter of the packing and expressed by the number of packing particles.

Knox and coworkers (Knox and Saleem 1969a, b) derived a semiempirical reduced-parameter plate height equation, describing the relationship between chromatographic zone dispersion and flow velocity based on the reduced parameters, which is also known as the Knox equation:

$$h = \frac{B}{v} + Av^{1/3} + Cv \tag{2.79}$$

The general form of Eq. (2.79) is shown by the log–log plot in Fig. 2.6. In practice, the v value for the best column performance should be slightly higher than the minimum value, but it is better not to be too high.

Like h and v, the coefficients A, B, and C are dimensionless. For a good column, it has approximate values: $A = 1$, $B = 2$, and $C < 0.1$ (Bristow and Knox 1977). B reflects the geometry of the mobile phase in the column and the extent to which solute diffusion is hindered due to the presence of the packing. In an ideal situation, when the hindering effect is small, B should be close to 2. This means that the solute molecules diffuse freely in the packed bed, as they do in the bulk solution. A is a weak function of the retention factor, but it mainly depends on the uniformity of the packing structure: unevenly packed columns will have higher A values, such as 2–5, while well-packed columns should have A values between 0.5 and 1. C reflects the efficiency of mass transfer in the stationary-phase region (Knox and Saleem 1972; Knox and Scott 1983). In theory, from the dependence of C on k, we should be able to distinguish whether the rate-determining process is mass transfer in the stationary phase or diffusion in the stagnant mobile phase. When k approaches 1, C reaches its maximum value, which may mean that the former dominates. Otherwise, C should gradually rise to its maximum value as k increases, if the latter is dominant. Obviously, for non-retaining solute, no matter which rate determines the process is dominant, the minimum value of C should be obtained.

An important feature of the reduced-parameter method is that for different packings, different solvents, and different solutes, the same h–v curve can be obtained. In this way, the efficiency of different chromatographic systems can be compared on the same basis. Knox and coworkers (Bristow and Knox 1977) used h–v logarithmic diagrams to compare gas chromatography and liquid chromatography and proposed methods to determine the best operating conditions for the two. From the log h–log v graph, a lot of meaningful information about the column performance can be obtained. If the curve is straight and the lowest point is too high, such as $h_{\min} = 10$, the A value is too large, which means that the column is not uniformly packed. If the rising part of the right end of the curve is steep, such as v is about 20 and $h = 10$, it means that the C value is too large and the mass transfer resistance is large, which means the quality of the packing is poor.

2.8 Summary

Chromatography is a technique that uses the difference in the partition coefficients of solutes between the mobile phase and the stationary phase to separate them. In this process, if the solute is adsorbed on the stationary phase, its migration rate is 0. If the solute is desorbed from the stationary phase and returns to the mobile phase, it will move forward at the same migration rate as the mobile phase. Therefore, the rate of solute migration is a function of the fraction of solute residence time in the mobile phase. The relationship between the solute zone shape and the adsorption isotherm is given by Wilson's equilibrium theory. When the adsorption isotherm is a straight line, the solute zone is a symmetrical peak; when the adsorption isotherm deviates from the linearity, the solute zone is a tailing peak or a fronted peak. In a chromatographic column formed by a packed structure, the movement of the solute zone is dispersed due to the hindrance of particles. The structure of the column bed can have a great influence on the band broadening, which is mainly manifested in two aspects. One is that the path length of the liquid in the column bed is different, forming a multipath effect; the other is the width of the path in the column bed is different and the paths cross each other, forming an eddy effect. In addition, the diffusion and mass transfer of the solute in the column bed also affect the band broadening of the solute. The diffusion of solute between particles is also called cross-channel diffusion, and it is the main contribution of mass transfer resistance in the mobile phase. The mass transfer resistance of the solute in the particle can be divided into two: the diffusion in the retained mobile phase and the adsorption kinetics on the surface of the stationary phase. The diffusion of solutes in the retained mobile phase is affected by the porous structure of the particles, especially the obstructive effect caused by the bending and narrowing of the pores.

The band broadening of the solute in the chromatographic column bed can be measured by the height equivalent to a theoretical plate (HETP). The theoretical plate height can be expressed by the ratio of the standard variance of the Gaussian distribution curve to the column length. Comprehensively considering the band spreading effect of the solute in the chromatographic column bed, the plate height equation used to characterize the chromatographic efficiency can be derived. These equations include the van Deemter equation for packed column gas chromatography, the Golay equation for open-column capillary chromatography, the Giddings equation containing the coupled term of eddy diffusion and mobile-phase mass transfer, and the Knox equation containing reduced parameters. The Knox equation is not limited by chromatographic operating conditions and can be used to compare different chromatographic systems composed of different chromatographic packings and different mobile phases, such as gas chromatography and liquid chromatography. Therefore, the Knox equation has universal applicability.

References

Boyack JR, Giddings JC (1963) Theory of electrophoretic mobility in stabilized media. Arch Biochem Biophys 100:16–25

Bristow PA, Knox JH (1977) Standardization of test conditions for high performance liquid chromatography columns. Chromatographia 10:279–289

Carman PC (1956) Flow of gases through porous media. Butterworths, London

Eyring H, Henderson D, Stover BJ et al (1964) Statistical mechanics and dynamics. Wiley, New York

Geankoplis CJ (2003) Transport processes and separation process principles, 4th edn. Pearson, London

Giddings JC (1961) Lateral diffusion and local nonequilibrium in gas chromatography. J Chromatogr 5:61–67

Giddings (1965) Dynamics of chromatography: principles and theory. Marcel Dekker, Inc., New York

Giddings JC, Boyack JR (1964) Mechanism of electrophoretic migration in paper. Anal Chem 36: 1229–1233

Golay M (1963) Height equivalent to a theoretical plate of tubular gas chromatographic columns lined with a porous layer. Nature 199:370–371

Knox JH, McLaren L (1964) A new gas chromatographic method for measuring gaseous diffusion coefficients and obstructive factors. Anal Chem 36:1477–1482

Knox JH, Saleem M (1969a) Kinetic conditions for optimum speed and resolution in column chromatography. J Chromatogr Sci 7:614–622

Knox JH, Saleem M (1969b) A comparison of plate efficiencies in gas and liquid chromatography using various supports. J Chromatogr Sci 7:745–751

Knox JH, Saleem M (1972) Mobile and stationary phase contributions to peak dispersion in gas chromatography. J Chromatogr Sci 10:80–88
Knox JH, Scott HP (1983) B and C terms in the Van Deemter equation for liquid chromatography. J Chromatogr A 282:297–313
McGuffin VL (2004) Theory of chromatography. In: Heflmann E (ed) Chromatogray, 6th edn. Elsevier, Amsterdam
Muskat M (1937) Flow of homogeneous fluids. McGraw-Hill, New York
Scheidegger AE (1957) Physics of flow through porous media. Macmillan, New York
Snyder LR, Kirkland JJ, Dolan JW (2010) Introduction to modern liquid chromatography, 3rd edn. Wiley, Hoboken, NJ
Sutera SP, Skalak R (1993) The history of Poiseuille's law. Annu Rev Fluid Mech 25:1–19
van Deemter JJ, Zuiderweg FJ, Klinkenberg A (1956) Longitudinal diffusion and resistance to mass transfer as causes of nonideality in chromatography. Chem Eng Sci 5:271–289
Whitaker S (1986) Flow in porous media I: a theoretical derivation of Darcy's law. Transp Porous Media 1:3–25
Wilke CR, Chang P (1955) Correlation of diffusion coefficients in dilute solutions. AlChE J 1:264
Wilson JN (1940) A theory of chromatography. J Am Chem Soc 62:1583–1591

Chapter 3
Separation Modes

Abstract Liquid chromatography is a versatile and powerful separation technique that can operate in different modes to separate various solutes, including ionic, ionizable, polar, nonpolar, and polymer compounds. According to the retention mechanism involved, the separation mode of liquid chromatography can be divided into normal phase, reversed-phase, ion exchange, size exclusion, hydrophilic interaction, hydrophobic interaction, affinity, chiral, and mixed-mode chromatography. In this chapter, we will review each of the commonly used chromatographic methods in terms of retention mechanism, stationary phase, mobile phase, and application.

3.1 Introduction

One of the main reasons for the widespread acceptance of liquid chromatography as a separation and purification technology is that it can use a variety of separation mechanisms (or modes), which makes it possible to separate a wide range of solutes, including ionic, ionizable, polar, nonpolar, and polymer compounds. As mentioned in Sect. 1.4, the chromatographic methods can be classified according to the actual mechanism by which the separation occurs. In this chapter, we define the chromatographic separation modes according to the nature of the separation method: normal phase, reversed-phase, ion exchange, size exclusion, hydrophilic interaction, hydrophobic interaction, affinity, chirality, and mixed-mode chromatography (Table 3.1).

3.2 Normal Phase Chromatography

Normal phase chromatography, also known as adsorption chromatography or liquid–solid chromatography, represents the oldest mode of chromatographic separation. It was used to separate plant pigments in the form of packed columns more than one century ago. The stationary phase of normal phase chromatography is a polar solid adsorbent, such as calcium carbonate, alumina, and silica gel, and the mobile phase is a nonpolar organic solvent, such as *n*-hexane and petroleum ether.

Q.-H. Wan, *Mixed-Mode Chromatography*,
https://doi.org/10.1007/978-981-16-5485-5_3

Table 3.1 Liquid chromatographic methods, principles, and applications

Chromatographic method	Retention mechanism	Applicable compounds
Normal phase	The solute molecules are bound to the polar adsorbent through intermolecular forces such as hydrogen bonds and dipoles and are separated due to the difference in their interaction with the adsorbent	Medium, weak to nonpolar compounds, such as fat-soluble cellulose, steroids, fatty compounds, plant pigments, herb medicine ingredients, etc. The sample should be dissolved in organic solvent
Reversed-phase	The solute molecules are bound to the hydrophobic stationary phase through hydrophobic interaction and separated due to the difference in hydrophobic interaction with the stationary phase	Most organic compounds; biological macromolecules, small molecules, such as polypeptides, proteins, nucleic acids, carbohydrate compounds. The sample should generally be dissolved in an organic–water mixed solvent
Ion exchange	The solute molecules are bound to the ion exchanger through Coulomb action and separated due to the difference in the electrostatic interactions with the ion exchanger	For ionic compounds or dissociable compounds, such as basic drugs, amino acids, polypeptides, proteins, nucleic acids, etc., the sample should generally be dissolved in an aqueous solution or an organic–water mixed solvent
Size exclusion	Solute molecules are retained by diffusion into the pores of the packing and separated according to their residence times because of their different sizes	Biological macromolecules or synthetic polymers. Generally, it should be dissolved in organic solvents or aqueous solutions
Hydrophilic interaction	Solute molecules are retained and separated by hydrogen bond or dipolar interaction with the polar adsorbent in a highly polar organic solvent	Highly polar and highly water-soluble compounds, such as sugars, nucleotides, peptides, metabolites in the body, etc. generally, it should be soluble in aqueous solution
Hydrophobic interaction	Solute molecules are retained and separated by hydrophobic interaction with hydrophobic stationary phases in high-concentration salt solutions	Weakly hydrophobic compounds, such as proteins, nucleic acids
Coordination interaction	Solute molecules are bound and separated by coordination with metal ions embedded in or bound on the stationary phase. Metal ions are bound and separated by coordination with ligands covalently bonded on the stationary phase	Metal ions, proteins, nucleic acids, sugars
Affinity	Specific molecular recognition between the solute molecule and the ligand on the stationary phase	Peptides, proteins, nucleic acids, sugars, and other biomolecules and small molecule compounds that can bind with biomolecular ligands
Chiral	Molecular recognition phenomenon between chiral compounds and chiral ligands	Resolution of optical isomers with chiral centers

(continued)

Table 3.1 (continued)

Chromatographic method	Retention mechanism	Applicable compounds
Mixed mode or multimodal	Simultaneous interaction between different components of the solute molecule and multiple functional groups on the stationary phase	Amphiphilic compounds, such as basic drugs, peptides, proteins, nucleic acids, and other biological molecules

Obviously, in normal phase chromatography, the polarity of the stationary phase is greater than that of the mobile phase. The sample components are retained on the stationary phase through the interaction of the polar groups on the solute molecule and the polar binding sites on the stationary phase. In this separation mode, the greater the polarity of the solute, the stronger the retention; increasing the polarity of the mobile phase will result in a decrease in solute retention.

Retention mechanism In normal phase or adsorption chromatography, there are many discrete active sites on the surface of the polar stationary phase, that is, the adsorption center, which adsorbs solutes and solvents through hydrogen bonding, dipole–dipole, and other polar interactions. In the chromatographic process, the polar solvent molecules in the mobile phase are adsorbed on the surface of the adsorbent. After sample injection, the solute molecules and the solvent molecules adsorbed on the active center undergo competitive adsorption to form the adsorption and desorption equilibria of different solute molecules on the surface of the adsorbent:

$$L + S \rightleftharpoons LS \tag{3.1}$$

$$LS + A \rightleftharpoons LA + S \tag{3.2}$$

where L is the adsorption site of the adsorbent, S is the polar solvent molecule, and A is the solute molecule. In this equilibrium, it is assumed that the adsorption site forms a 1:1 complex with solvent and solute molecules, and the adsorption of solvent molecule S can be described by the Langmuir equation:

$$[LS] = \frac{Q_m K_1 [S]}{1 + K_1 [S]} \tag{3.3}$$

where $[S]$ and $[LS]$ are the molar concentrations of free polar solvent molecules and adsorbed polar solvent molecules, respectively. Q_m is the monolayer saturated adsorption capacity of solvent molecules, and K_1 is the equilibrium constant of solvent adsorption. The competitive equilibrium constant K_2 of solute molecules is given by the following equation:

$$K_2 = \frac{[LA][S]}{[LS][A]} \tag{3.4}$$

The partition coefficient D of the solute molecule in the two phases:

$$D = \frac{[LA]}{[A]} = \frac{K_2[LS]}{[S]} = \frac{Q_m K_1 K_2}{1 + K_1[S]} \tag{3.5}$$

According to the definition, the retention factor k of the solute is related to the distribution coefficient D of the solute by the following equation:

$$k = \phi D = \phi \frac{Q_m K_1 K_2}{1 + K_1[S]} \tag{3.6}$$

where ϕ is the ratio of the surface area of the stationary phase to the volume of the mobile phase in the chromatographic column and has the unit of the reciprocal of the length. It can be seen from the above equation that the retention factor of the solute is directly proportional to the concentration of the adsorption site of the adsorbent and inversely proportional to the concentration of the polar solvent in the mobile phase. Solute molecules that interact strongly with the adsorbent have a larger adsorption coefficient, which leads to stronger retention, while the weaker component has a smaller adsorption coefficient and is eluted from the column first. In the region of low concentration of polar solvents, $1 + K_1[S] \approx 1$, $k = \phi Q_m K_1 K_2$; in the region of high concentration of polar solvents, $1 + K_1[S] \approx K_1[S]$, $k = \phi Q_m K_2/[S]$. This equation can be written in logarithmic form as

$$\log k = \log k_0 - \log [S] \tag{3.7}$$

where k_0 is the solute retention factor when the concentration of solvent molecules approaches one. The interaction with the active center of the polar adsorbent will increase with the increase of the polarity of the functional groups on the solute molecule or the increase of the number of functional groups, thereby increasing the retention of the solute on the stationary phase. In addition, the solute retention value is related to steric hindrance. If a large alkyl group is adjacent to the polar functional group, the retention value will be reduced; the *cis* isomer with polar substituents has stronger retention than the *trans* isomer. In addition, solute retention is also related to the surface structure of the adsorbent, that is, the geometric arrangement of adsorption centers. When the functional groups with a certain geometric shape of the solute and the active center on the surface of the adsorbent are arranged in parallel, the two have the largest contact area. In this case, the adsorption effect is the strongest. Therefore, adsorption chromatography has good selectivity for structural isomers and geometric isomers and also shows good separation ability for aromatic isomers and halogenated alkane isomers.

Stationary phase Adsorption chromatography can be performed on a variety of stationary-phase materials, including cellulose, starch, silica gel, Florisil, magnesium oxide, hydroxyapatite, and alumina. However, so far, silica gel is the preferred polar stationary phase. The surface of silica gel is rich in silanol groups, Si–OH. These silanol groups can exist in various forms such as free, hydrogen bonding, and gemini, and the ratio depends on the treatment method of the silica gel before use. Among them, the free silanol group has the highest acidity, while the gemini silanol group has the lowest acidity. Due to their acidic nature, free silanols lead to strong retention of basic solutes and weak retention of gemini silanol groups, causing chromatographic peak tailing and irreversible adsorption of basic compounds. These unfavorable factors of silica gel can be eliminated by surface alkylation. Silane reagents containing polar functional groups can be used to modify the surface of silica gel, such as reacting with aminopropyl silane to generate aminopropyl-bonded silica gel, reducing the adverse effects of silanol groups. In addition, there are glycol-based and cyano-functionalized silica gels to replace bare silica gel. The polarity of these replacement functional groups is significantly lower than that of silanol groups.

Mobile phase Adsorption separation is usually carried out on a silica gel stationary phase. The change of solute retention can be achieved by changing the composition of the mobile phase. The elution strength of the solvent in the normal phase system is directly related to the polarity, and a solvent with a large polarity has a strong elution strength. Two solvents with different elution strengths are usually selected, wherein solvent A is a nonpolar solvent such as *n*-hexane and the other solvent B is a polar solvent such as isopropanol. According to the need, the two solvents are mixed in a certain ratio to achieve the required elution strength. Gradient elution can also be used to continuously change the elution strength from pure A to pure B.

Application Normal phase chromatography has been widely used in organic synthesis laboratories, mainly as thin-layer chromatography for the monitoring of synthesis reactions, and flash chromatography for the purification of synthetic products. Normal phase chromatography is most commonly used to analyze samples that are soluble in nonpolar solvents (such as hexane), while solutes that are more polar, water-soluble, and/or ionic are best separated using reversed-phase chromatography. As mentioned earlier, adsorption chromatography is generally more suitable for isomer separation than reversed-phase chromatography. Adsorption chromatography also shows a unique ability to distinguish between solutes with different numbers of electronegativity atoms (such as oxygen or nitrogen) or solute molecules with different functional groups. Both fat- and water-soluble vitamins have been analyzed by normal phase chromatography. For example, 1% isopropanol in hexane as a mobile phase can be used to analyze vitamin D in milk on a silica gel column. For more than 50 years, this method has successfully analyzed less polar plant pigments, such as carotenoids and porphyrins. Natural oils and spice extracts are also easy to analyze by adsorption chromatography. Normal phase chromatography is still widely used to determine nonpolar additives in various commercial products and pharmaceutical preparations, for example, to separate nonpolar components in the nonionic surfactant Triton X-100. Finally, one of the more important uses of normal phase chromatography is the separation of isomers. This chromatographic mode can separate a variety of isomers, including the six isomers of trinitrotoluene and the separation of the 1, 2, and 1, 6 isomers of 7-hydroxytetrahydrocannabinol.

3.3 Reversed-Phase Chromatography

Reversed-phase chromatography (also known as reversed-phase bonded-phase chromatography) is the most widely used method in liquid chromatography. It is reported that more than 75% of all HPLC separations are performed using this method. This separation mode uses a combination of a polar mobile phase and a nonpolar stationary phase, which is the opposite of normal phase (adsorption) chromatography. The most commonly used stationary phases in reversed-phase chromatography can be divided into silica-based and polymer-based stationary phases. The silica-based stationary phase used for reversed-phase chromatography is obtained by bonding hydrophobic groups on the surface of the silica gel, and the polymer matrix such as polystyrene can be directly used for reversed-phase chromatography due to its strong hydrophobicity. In reversed-phase chromatography, sample components are retained through nonspecific hydrophobic interactions with the stationary phase. In this separation mode, the greater the polarity of the solute, the lower the retention. Increasing the polarity of the mobile phase will increase solute retention.

Retention mechanism Compared with the simple polar interaction responsible for solute retention in normal phase chromatography, the retention mechanism in reversed-phase liquid chromatography (RPLC) is more complicated. In reversed-phase chromatography, the polar mobile phase usually consists of a mixed solution of a polar organic solvent and water. It is well known that water is a solvent with a strong hydrogen bond effect, forming a hydrogen bond network in the bulk solution, showing strong cohesion. In the polar mobile phase, when the solute is in contact with the hydrophobic ligand, a molecular complex is formed through intermolecular forces, and the solvent at the interface between the two contacts is released at the same time, which increases the entropy of the system. In addition, the cohesion of the solvent is equivalent to a repulsive force, which promotes a stronger bond between the solute and the ligand. Therefore, the simple displacement relationship between solvent molecules and solutes observed in normal phase chromatography is no longer applicable. A reasonable reversed-phase chromatography retention model must consider the influence of the cohesion of polar solvents, including water, on solute retention.

Sinanoglu and colleagues developed the solvophobic theory to describe the interaction of nonpolar molecules in polar solvents (Sinanoglu and Abdulnur 1965). When the nonpolar solute molecules are moved into the polar solvent, the solvent will repel the solute molecules, and the nonpolar solute molecules tend to gather together to reduce the contact area with the polar solvent and thus the surface tension and system energy. The solvent is then oriented and rearranged in an orderly manner around the solute, forming a cavity. The nonpolar regions of solute molecules also have a tendency to reduce contact with polar solvents. This effect is called "hydrophobic" or "solvophobic effect." The formation and area size of cavities are related to solvent properties such as surface tension, dielectric constant, etc. and solute properties such as surface area and dipole moment. These properties can be related to the thermodynamic process of the solvophobic effect.

Horvath and coworkers were the first to apply this principle to a reversed-phase chromatography system (Horvath et al. 1976; Horvath and Melander 1977). It is assumed that there is an association equilibrium between solute A and the nonpolar ligand L of the stationary phase of reversed-phase chromatography:

$$L + A \rightleftharpoons LA \tag{3.8}$$

The relationship between the free energy change ΔG of the association process and the corresponding equilibrium constant K is

$$\Delta G = -RT \ln K \tag{3.9}$$

where R is the universal gas constant and T is the absolute temperature.

According to the solvophobic theory, the solvation process of a solute can be seen as a two-step process, as shown in Fig. 3.1. The first step is to form a cavity in the solvent equivalent to the size and shape of the solute. The energy required for this process is recorded as cavity energy, ΔG_{C}. The second step is to move the solute into the cavity, and the interaction energy between it and the surrounding solvent, ΔG_{int}, is accompanied by a change in "free volume"

$$\Delta G = \Delta G_{\mathrm{c}} + \Delta G_{\mathrm{int}} + RT \ln \frac{RT}{P_0 V} \tag{3.10}$$

where P_{o} is the atmospheric pressure and V is the molar volume of the solvent.

When any substance forms a cavity in the solvent, it must overcome the cohesive force of the solvent, increase the surface area, and increase the free energy of the system. Therefore, the energy required for cavity formation can be approximately expressed by the following equation:

$$\Delta G_{\mathrm{c}} = NA_{\mathrm{s}}\gamma(\chi^e - 1) + NA\gamma \tag{3.11}$$

where N is Avogadro's constant, A is the cavity area, γ is the surface tension of the solvent, A_{s} is the surface area of the solvent molecule, and χ^e is a factor that converts the macroscopic surface tension into the molecular size.

The solute–solvent association process is equivalent to the filling of cavities, resulting in a decrease in surface area and free energy. The free energy change of this process, $\Delta G_c^{\mathrm{assoc}}$, is given by

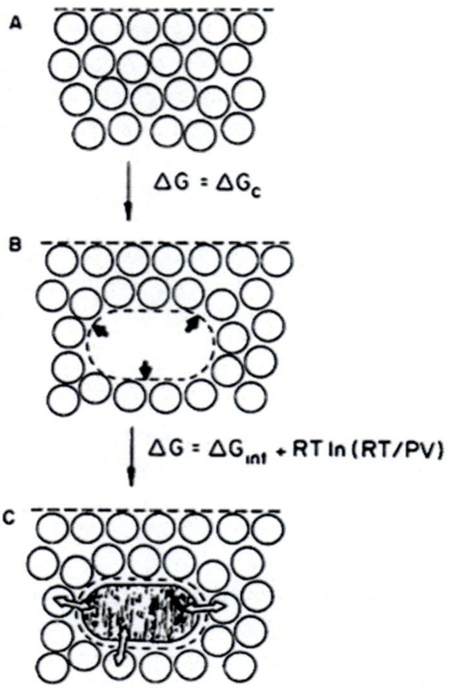

Fig. 3.1 Schematic diagram of putting the solute in the solvent shown in **a**. ΔG_c gives the energy required to form the appropriate cavity shown in **b**. Its size is basically determined by the surface area of the cavity and the appropriate surface tension of the solvent and is greater than zero. As shown in **c**, the solute in the cavity interacts with the solvent, and the energy of the interaction is $\Delta G_{int} + RT \ln (RT/PV)$, which is less than zero. (Reproduced from Horvath and Melander 1977)

$$\Delta G_c^{assoc} = -NA_s\gamma(\chi^e - 1) - N\Delta A\gamma \tag{3.12}$$

where ΔA is the area of the interface between the solute and the ligand, or contact area, given by $\Delta A = A_A + A_L - A_{AL}$, where A_A, A_L, and A_{AL} are the surface areas of the solute, ligand, and complex, respectively. Equation (3.12) shows that during the chromatographic association process, the cavity term increases with the increase of the contact area ΔA between the solute and the hydrocarbon ligand. The nonpolar surface area of the solute molecule increases, and the retention increases; on the other hand, the cavity term increases with the increase of the solvent surface tension. Since water has the highest surface tension among common solvents, the cavity term has the largest value for pure aqueous eluents.

The interactions involved in the solute association process in the solvent cavity are attributed to van der Waals forces and electrostatic interactions. Therefore, the interaction energy ΔG_{int} is the sum of the van der Waals contribution (ΔG_{vdw}) and the electrostatic contribution (ΔG_{es}).

$$\Delta G_{int} = \Delta G_{vdw} + \Delta G_{es} \tag{3.13}$$

Van der Waals contribution ΔG_{vwd} can be approximated by the following equation

$$\Delta G_{\text{vdw}} = W + aA \tag{3.14}$$

where W and a are solvent-dependent parameters and A is the surface area of the solute molecule. Therefore, the contribution of van der Waals energy to the associative energy can be expressed as

$$\Delta G_{\text{vwd}}^{\text{assoc}} = -W - a\Delta A \tag{3.15}$$

where ΔA is the contact area between the solute and the ligand.

The free energy change associated with electrostatic interaction, ΔG_{es}, is simplified as

$$\Delta G_{\text{es}} = \frac{Z}{\varepsilon} \tag{3.16}$$

where ε is the dielectric constant of the medium and Z is a parameter related to the solute charge. The exact expression depends on the charge distribution, molecular size, and other factors. The net electrostatic contribution $\Delta G_{\text{es}}^{\text{assoc}}$ produced by the interaction between the solvent and the substances involved in the association process can be expressed in a simplified form as

$$\Delta G_{\text{es}}^{\text{assoc}} = -\frac{\Delta Z}{\varepsilon} \tag{3.17}$$

where

$$\Delta Z = Z_{\text{AL}} - Z_{\text{A}} - Z_{\text{L}} \tag{3.18}$$

where Z_{A}, Z_{L}, and Z_{AL} are the Z values of solute, ligand, and association, respectively.

Combining Eqs. (3.12), (3.15), and (3.17) to obtain the total free energy change of the association reaction:

$$\begin{aligned}\Delta G^{\text{assoc}} &= \Delta G_{\text{c}}^{\text{assoc}} + \Delta G_{\text{vwd}}^{\text{assoc}} + \Delta G_{\text{es}}^{\text{assoc}} + RT\ln\frac{RT}{P_0V} \\ &= -NA_{\text{s}}\gamma(\chi^e - 1) - \Delta A(N\gamma + a) - W - \frac{\Delta Z}{\varepsilon} + RT\ln\frac{RT}{P_0V}\end{aligned} \tag{3.19}$$

According to the chromatographic principle, the relationship between the retention factor k and the association equilibrium constant K is

$$k = \phi K \tag{3.20}$$

where ϕ is the ratio of the volume of the stationary phase to the mobile phase in the chromatographic column.

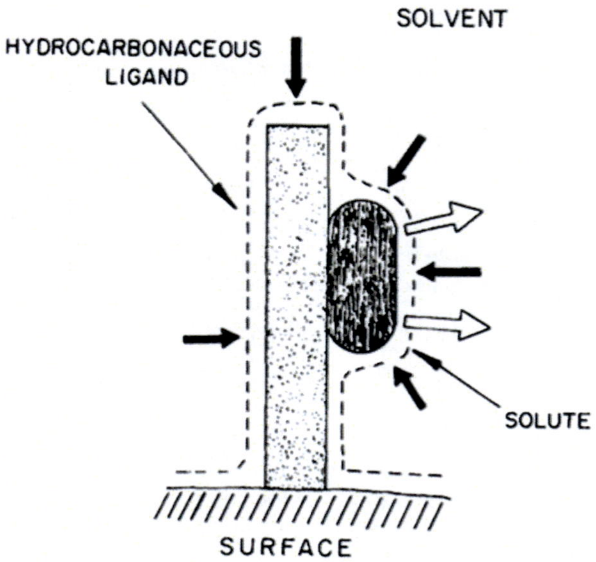

Fig. 3.2 Association between solutes and nonpolar ligands on the surface of the stationary phase in reversed-phase chromatography. The binding is basically due to the solvophobic effect. As the solid arrows suggest, by forming a complex, the reduction of the molecular surface area exposed to the solvent promotes the association of the two substances. Open arrows indicate that the interaction with the solvent has a counteracting effect on the association. The nonpolar interaction between the solute and the ligand increases solute retention. The polar part of the solute enhances the interaction with the polar solvent, thus reducing the binding force. On the other hand, when the contact area in the complex is larger and/or the surface tension of the solvent is higher, the bond is stronger. (Reproduced from Horvath and Melander 1977)

$$
\begin{aligned}
\ln k &= \varphi - \frac{\Delta G^{\text{assoc}}}{RT} \\
&= \varphi + \frac{\Delta A(N\gamma + a) + NA_{\text{s}}\gamma(\chi^e - 1) + W + \frac{\Delta Z}{\varepsilon}}{RT} - \ln \frac{RT}{P_0 V}
\end{aligned}
\tag{3.21}
$$

where $\varphi = \ln \phi$.

In Eq. (3.21), every term except φ is a function of the physical properties of the solute and the eluent and can be measured at least in principle. Therefore, it can be used to explain or predict the relationship between solute retention and mobile-phase properties in a given chromatographic system. Figure 3.2 schematically illustrates the phenomenon quantitatively represented by Eq. (3.21).

According to Eq. (3.21), the mobile-phase parameters that affect solute retention are mainly surface tension and dielectric constant. If the solute is a simple dipole, the effect of the change in the dielectric constant is small and usually negligible. The biggest influence on the retention factor comes from the change of surface tension and its conjugate function $A_{\text{S}}\gamma(\chi^e - 1)$. Consistent with experimental observations,

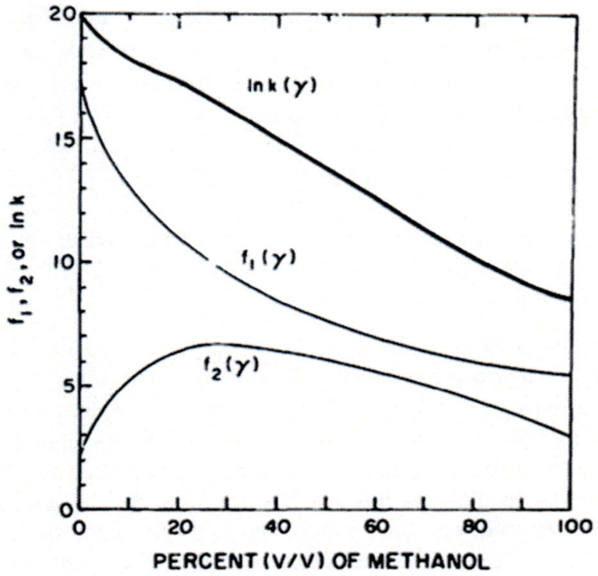

Fig. 3.3 The logarithm of the retention factor and the logarithm of the two surface tension-related terms in Eq. (3.21) change with the composition of the water–methanol mixture. The function $f_1(\gamma)$ is given by $N\Delta A\gamma/RT$, where ΔA is the surface area change. When the change occurs, the solute is bound by the nonpolar ligand of the stationary phase, and γ is the surface tension of the eluent. The value of ΔA is taken as 100 Å^2. The bottom curve $f_2(\gamma)$ represents the energy phase $NA_s(\chi e - 1)/RT$, where A_s is the average surface area of solvent molecules and χ^e is the factor that converts the macroscopic surface tension into molecular size. The top line is $f_1(\gamma) + f_2(\gamma)$, which almost reproduces the solvent dependence of the logarithm of the retention factor obtained by the solute binding experiment. (Reproduced from Horvath and Melander 1977)

Eq. (3.21) predicts that the retention factor will decrease as the surface tension of the eluent decreases. For example, the increase of the organic solvent content in the aqueous organic solution leads to a decrease in surface tension and a decrease in retention factor.

Figure 3.3 shows the two terms of Eq. (3.21) including surface tension displayed as a function of the percentage of organic solvent volume in the aqueous mobile phase. Assuming that the ΔA of the solute is equal to 100 Å^2, the theoretical curve has been calculated for methanol. It can be seen from the change curve of $f_2(\gamma)$, as the concentration of organic solvent increases, the contribution of χ^e first rises and then decreases, while the surface tension itself (see $f_1(\gamma)$) drops sharply. However, the sum of these two contributions is an approximately linear function of the percentage of organic solvent (v/v) (see ln$k(\gamma)$). Similar results were obtained for the water–acetonitrile mixture.

Different research groups have reported experimental observations on the dependence of the logarithm of the retention factor on the solvent composition (Abbott et al. 1973; Horvath et al. 1976; Karger et al. 1976), which greatly simplifies the prediction of the effect of unionized solute retention. In view of the quasi-linear relationship shown in Fig. 3.3, the following relationship can be used to approximate the relationship between the solute retention factor and the volume percentage of methanol or acetonitrile in the aqueous eluent (v/v), X:

$$\ln k = \ln k_{\mathrm{w}} - SX \tag{3.22}$$

where k_{W} is the retention factor of the solute in the pure water eluent and S is the parameter related to the ligand, solute, and solvent. For a given solute, chromatographic column, and solvent, S is a constant.

Stationary phase Reversed-phase stationary phases are mainly divided into two categories, one is hydrocarbyl-bonded silica gel and the other is polymer matrix. Most reversed-phase chromatographic separations are carried out on bonded silica gel stationary phases, while polymer reversed-phase chromatographic columns are mainly used for the separation and purification of peptides and other biomolecules. The silica gel matrix packing has high mechanical strength, and silica gel with a pore diameter of 10 nm can withstand a pressure of 60–70 MPa and is the preferred stationary phase in high-performance liquid chromatography. As the pore size increases, the pressure resistance of silica gel will decrease, and the specific surface area will also decrease. There are three types of ligands for the silica-based reversed-phase chromatography stationary phase, including linear alkanes, aromatic hydrocarbons, and weakly polar groups. Straight-chain alkanes are the main ligands used, from C18, C12, C8, to C4, with decreasing hydrophobicity. C18 is the most widely used ligand for the separation of small organic molecules. The most widely used for protein separation is C8 and C4. Aromatic ligands, such as benzene rings, have intermediate hydrophobicity. Some polar groups, such as cyanopropyl, can be used as a stationary phase for normal phase chromatography or reversed-phase chromatography, depending on the nature of the solvent used. When the solvent polarity is greater than the ligand, it is a reversed-phase chromatography system, and when the solvent polarity is less than the ligand, it is a normal phase chromatography system. There are two main limitations of silica-based packings. One is the limited range of usable pH, which is between pH 2 and 8. Use beyond this range will cause degradation of the silica backbone or loss of chemically bonded ligand. The second is the residual silanol groups on the surface of the solid phase, causing irreversible adsorption of basic compounds.

The most widely used polymer based reversed-phase packing is cross-linked polystyrene microspheres. The surface of this type of resin has the nonpolar characteristics of the alkyl-bonded phase, and it can be directly used as a reversed-phase chromatography packing without chemical modification. In addition, there are some other types of polymer matrix reverse phase packings, such as vinyl acetate copolymers, polyacrylamides with alkyl side chains, alkyl esters of polymethacrylic acid, esters of polyvinyl alcohol, alkyl group-bonded polyvinyl alcohol, alkyl-derived cross-linked polystyrene, and phenyl- or alkyl-derived hydroxylated polyether resin, etc. These packings can be used in a wide range of pH (1–14) and usually do not contain acidic residues, so they are widely used in the separation of complex biological macromolecules. The mass transfer performance and pressure tolerance of polymer matrix packing are not as good as silica gel, which limits their application in high-performance liquid chromatography.

Mobile phase The mobile phase of reversed-phase chromatography is a mixture of polar organic solvents and water. The most commonly used are water and water-miscible organic solvents, such as methanol, acetonitrile, and isopropanol. As mentioned earlier, the surface tension and dielectric constant of the solvent play an important role in controlling solute retention. In addition, solvent viscosity is related to the flow resistance in the chromatographic column and the solute diffusivity in the mobile phase, so it has an important influence on the efficiency of the chromatographic column. Among the commonly used reversed-phase chromatography solvents, water has the highest polarity (dielectric constant) and surface tension. Therefore, water is the weakest eluent in reversed-phase chromatography. Methanol and acetonitrile are the most commonly used organic solvents because they have a relatively low UV cutoff point and viscosity and are easy to obtain in sufficient purity. In most applications, pure methanol or acetonitrile is the strongest eluent used in reversed-phase chromatography. A medium-strength eluent is usually obtained by mixing one of these solvents with water or an aqueous buffer. Both surface tension and dielectric constant decrease monotonously with the increase of organic solvent content. On the other hand, the viscosity first increases with the composition of the organic solvent and then decreases after reaching the maximum value. In the case of a methanol–water mixture, this effect is particularly obvious. The viscosity of an aqueous solution containing 40% (v/v) methanol is approximately twice the viscosity of pure water and three times the viscosity of pure methanol.

In many applications, especially when separating ionic solutes, a buffer must be used to control the pH of the eluent. The choice of an appropriate pH value depends on the dissociation constant of the solute, because the ionization of the solute will greatly reduce its reverse phase retention. Maintaining a high buffer concentration is conducive to quickly establishing a proton equilibrium and avoiding asymmetric peaks or band splits due to the slow secondary equilibria in the chromatographic process. As mentioned earlier, the interaction between basic compounds and residual silanol groups on the surface of the silica gel matrix will cause chromatographic peak tailing and irreversible adsorption. In order to reduce or eliminate the adverse effects of residual silanol groups, appropriate acidic or basic compounds can be added to the mobile phase. For example, trifluoroacetic acid (0.05–0.1%) is usually added to inhibit the dissociation of silanol groups and triethylamine (0.1–1.0%) to compete for silanol groups and reduce peak tailing of amine solutes on such columns.

Application Reversed-phase chromatography (RPC) is the most widely used separation method in all chromatographic separation modes. It is the best choice for separating neutral solutes that are soluble in water or other relatively polar solvents and have a molecular weight of less than 2000–3000 Da. In addition, many ionic solutes can be separated by using pH control or adding ionic interaction reagents to the mobile phase, which greatly expands the scope of application of reversed-phase chromatography. Reversed-phase chromatography is more and more widely used in the field of food science and technology. The separation of fatty acids is one of the first applications of reversed-phase chromatography. These acids are resolved according to their carbon chain length and saturation. Typical applications include the analysis of coconut oil, tallow, and cosmetics. The separation and purification of natural products is another important application of reversed-phase chromatography, such as phenolic compounds and colored substances in plants, digoxigenin, and sapogenin esters. Trace analysis of carcinogens of aflatoxins and other contaminants in food makes reversed-phase chromatography an important tool for food and environmental quality control.

Reversed-phase chromatography is widely used in the pharmaceutical industry. It is not only used for the separation and purification of pharmaceutical intermediates or final products but also widely used in pharmaceutical quality control. It is the main method in pharmaceutical analysis. Pharmaceutically active compounds include parabens, vitamins, beta blockers, alkaloids, steroids, tetracyclines, prostaglandins, etc., which are usually analyzed by reversed-phase chromatography. The technology of reversed-phase chromatography combined with mass spectrometry has been widely used in clinical chemistry, drug metabolism, disease diagnosis, metabolomics, and other fields.

Reversed-phase chromatography also has a wide range of applications in the separation and purification of biomolecules. Typical applications include the separation and purification of amino acids, peptides, proteins, nucleic acids, nucleosides, nucleotides, and oligosaccharides. The molecular weight range of these compounds is relatively wide, and the properties are quite different. Therefore, gradient elution is usually required to shorten the separation time and improve the chromatographic peak shape. The significant advantage of reversed-phase chromatography for protein separation is its strong separation ability. When other chromatographic separation modes such as ion exchange and hydrophobic chromatography fail, reversed-phase chromatography may be tried to provide ideal separation. Secondly, reversed-phase chromatography can be used to separate some poorly water-soluble proteins, such as membrane proteins, which are poorly water-soluble but easily soluble in organic solvents such as formic acid, or add surfactants to increase the solubility of proteins. In recent years, reversed-phase chromatography has played an important role in protein sequence analysis. Under the action of proteolytic enzymes (such as trypsin) or chemical reagents, the protein is partially decomposed. Reversed-phase chromatography can effectively separate the peptide mixture. The obtained reversed-phase chromatogram is called the "peptide profiling" of the protein and used to determine the primary structure of a protein. Peptide mapping analysis has been used for recombinant protein drugs such as human growth hormone and human granulocyte colony-stimulating factor to determine their primary structure, also used for protein modification products such as polyethylene glycol-human growth hormone to determine the modification site. RPC-based peptide mapping has played an indispensable role in the quality control of biopharmaceuticals.

3.4 Size Exclusion Chromatography

Size exclusion chromatography (SEC) is also known as gel chromatography or molecular sieve chromatography. Generally, the use of organic solvents as mobile phase is suitable for separating oil-soluble substances (such as organic polymer polyethylene, polyvinyl chloride, polystyrene, etc.), and in this case the technique is also known as gel permeation chromatography (GPC). Aqueous solution is used for the separation of water-soluble substances (such as proteins, enzymes, polysaccharides, nucleic acids, etc.), and this mode of SEC is also known as gel filtration chromatography (GFC) (Hong et al. 2012). In size exclusion chromatography, solutes of small volumes are retained by diffusion into the pores of the column packing, while solutes of large volumes are excluded from the pores and flow out with the mobile phase. Therefore, the separation of size exclusion chromatography is based on the difference in the size of the solute molecules, and the peaks appear in the order of molecular size from large to small. The chromatographic column is like a screen to sieve the components in the solution by size, so it is also called molecular sieve chromatography.

Retention mechanism The separation principle in size exclusion chromatography is different from other chromatographic modes and does not involve the interaction of solutes with the stationary-phase surface. In fact, in size exclusion chromatography, the adsorption of solutes on the surface of the packing is regarded as an interference phenomenon, which should be avoided by adjusting the composition of the mobile phase. The composition of the mobile phase is not a variable that controls or changes solute retention.

In size exclusion chromatography, the liquid volume (V_t) in the chromatographic column is composed of the pore volume (V_i) in the particles and the pore volume (V_o) between the particles. Solute components can be divided into three categories according to their volume: full exclusion, selective permeation, and full permeation. Some sample components with large molecular volume cannot enter the inner pores of the packing particles at all. They can only flow out from the gaps between the packing particles with the mobile phase, and their retention volume is the same as the interparticle pore volume of the packing. Some molecules are small in size and can enter the entire space of the particle pores, and their retention volume is the same as the total volume. This type of solute is called a fully permeable component. The retention volume of the third type of components is between the full exclusion and full permeability limits. They can enter part of the pore volume of the packing, and their retention volume is between the pore volume of the particles and the total volume of the packing. Solute retention can be regarded as the result of the distribution of the solute between the mobile phase retained in the pores of the packing particles and the mobile phase outside the particle pores. The partition coefficient, K_d, is the retention value of the component in the size exclusion chromatography. K_d can vary from 0 to 1. Obviously, $0 \leq K_d \leq 1$, which represents the fraction of the component that can reach the particle pore volume. When the solute molecules are so small that they can diffuse freely in the entire pore volume of the packing, $K_d = 1$; on the contrary, when the solute molecules are too large to enter any pores, $K_d = 0$. The retention volume of the solute, V_e, can be expressed as

$$V_e = V_0 + K_d V_i \tag{3.23}$$

Obviously, K_d is related to the pore size and pore size distribution of the packing in size exclusion chromatography; it is related to the molecular weight, molecular shape, and molecular volume of the component and has nothing to do with the geometric size of the chromatographic column. Different molecular weight components have different K_d values, so they can be separated.

The measurement of interparticle pore volume usually uses fully exclusion molecules as probes. The most commonly used fully exclusion molecule is blue dextran 2000, which has a molecular weight of about 2 million Daltons. The intraparticle pore volume can be determined by using fully permeable small molecules as probes and subtracting the elution volume of the total exclusion probe from the elution volume of the component.

It is worth noting that when there is adsorption between the separated component and the gel matrix, it will lead to a $K_d > 1$ situation. This interaction includes (1) hydrophobic interaction, where there is a hydrophobic interaction between the hydrophobic group such as the aromatic group in the sample composition and the cross-linking agent in the gel, which makes the elution of the solute lag behind; (2) ionic effect, where the opportunistic charge on the surface of the gel causes electrostatic interaction with the charged groups of the components, resulting in an increase in K_d; and (3) biological specificity, where the biological specific interaction between the sample components and the gel matrix, such as the specific interaction between cellulase and the cellulose matrix, leads to an increase in K_d. According to the cause of adsorption, corresponding countermeasures should be taken to eliminate the interference factors in size exclusion chromatography.

Stationary phase Stationary phases for size exclusion chromatography can be divided into two categories: cross-linked rigid polystyrene divinylbenzene gels traditionally used for polymer characterization and hydrophilic gels used to analyze water-soluble solutes. The column packing used for water-based size exclusion chromatography includes polydextran, polyvinyl alcohol gel, hydroxylated methacrylate gel, and silica gel. Then, they are further classified according to the pore size of the exclusion chromatography column. The pore size of the chromatographic columns matches the expected range of solute sizes, so they show selective permeation on the gel.

The smallest pore chromatographic column (usually 60–100 Å) can be used to separate small molecules (molecular weight <5000 Da). Solutes with a molecular weight range of 10^3–10^5 Da can be separated on a 300 Å column. A series of chromatographic columns with different pore diameters are usually used to analyze solutes with a wide molecular weight distribution (such as polymers). In addition, mixed-bed chromatography columns that mix materials with different pore diameters can also be used. These "linear" columns are very suitable for initially indicating the molecular weight distribution of unknown samples.

Mobile phase As mentioned earlier, the mobile phase in size exclusion chromatography is not normally used to control retention or selectivity. The main consideration in selecting a mobile phase is the solubility of the sample. Ideally, the mobile phase should have a low viscosity at the separation temperature and should be compatible with the selected detection method. The mobile phase must also be selected based on its compatibility with the chromatographic column. Polar solvents, including methanol, ethanol, isopropanol, and dimethyl sulfoxide, should not be used with polystyrene packing because they can cause excessive shrinkage, which can permanently damage the chromatographic column bed. These packings are most commonly used with low-polarity solvents such as toluene, tetrahydrofuran, chloroform, dichloromethane, dimethylformamide, *m*-cresol, and trichlorobenzene, although the last three solvents must be operated at high temperatures to reduce viscosity. Silica columns can be used with a variety of solvents, including water, although their pH range is limited to 2–8.

In addition to considering sample solubility, viscosity, and column compatibility, the mobile phase is also selected to eliminate (or minimize) the interaction between the solute and the stationary phase. Ideally, the solute does not interact with the stationary phase. Interactions such as hydrophobic association, adsorption, ion repulsion, or ion exchange are undesirable because they affect the retention time and calibration curve. Therefore, solvents that reduce hydrophobic interactions (such as THF or toluene) are usually used for polymer size exclusion chromatography, and salt solutions (i.e., 0.1–0.5 M sodium sulfate) are used for aqueous size exclusion chromatography to separate polyelectrolytes. When separating polyethersulfone, other reagents, such as lithium bromide and dimethylformamide (DMF), can also be added to the solvent to reduce unwanted interactions.

When separating polymers, another consideration in GPC is that the injection concentration may be limited by its viscosity. As a rule of thumb, the viscosity of the sample solution injected into the column should be no more than twice the viscosity of the mobile phase. Injecting a sample with a viscosity higher than this value will cause a "viscous fingering" effect. These effects are due to the limited time it takes for the mobile phase to dilute high-viscosity samples, which leads to peak broadening and extended elution time, resulting in incorrect molecular weight distribution. Generally, if the average weight of the polymer exceeds 1×10^6 Da, a 0.01% solution should be used; between 20,000 and 10^6 Da, a 0.1% solution should be used; between 800 and 20,000 Da, a 0.2% solution should be used; and when it is below 800 Da, a solution with a concentration of 0.5% can be used.

Application Traditionally, the main application of SEC has been to characterize polymers using a chromatographic column packed with rigid polystyrene divinylbenzene particles. The molecular weight distribution is a critical parameter for characterization of bulk polymers, including tensile strength, brittleness, and hardness, melt viscosity, etc. The subtle differences between materials and the properties of the final product can be determined by the molecular weight distribution obtained using SEC, which is why this chromatographic mode has been widely used in the polymer industry.

Polymers commonly characterized by SEC include acrylic acid, alkyd resin, carboxymethyl cellulose, ethylene–ethyl acrylate, nitrocellulose, nylon, polybutadiene, polycarbonate, polyethersulfone, polyethylene, and polyisoprene diene, polypropylene, polystyrene, and silicone. Similarly, water-soluble ionic, nonionic amphoteric polymers can be characterized by water-based SEC. These polymers include polysaccharides, starch, dextran, polyacrylamide, hyaluronic acid, carrageenan, sulfonated lignin, collagen, and gelatin.

Although SEC has limited resolving power compared with other separation modes, it is still a useful method for analyzing matrices containing high levels of triglycerides and other hydrophobic substances. Samples such as edible oils, creams, ointments, or lotions usually require extensive cleaning before reversed-phase chromatographic analysis to prevent column fouling. However, after simple dilution in the mobile phase, small molecules in such samples can be extracted by SEC and then analyzed by RPC.

SEC is commonly used in the separation of biomolecules, such as proteins. It provides a fast and simple method for the separation, purification, and characterization of proteins. Because it is separated based on the size of the sample molecule, there is no interaction between the components and the stationary-phase matrix, so the separation process will not cause protein loss or inactivation, and the total protein can be recovered with high yield and its biological activity can be retained. In the purification process of extracting recombinant protein from cell culture fluid, three steps of capture, purification, and polishing are usually adopted, and size exclusion chromatography can be used in any step. The most common application is in the final purification step to remove high molecular weight impurities such as dimers and aggregates in the product.

3.5 Ion Exchange Chromatography

Ion exchange chromatography (IEC) achieves separation through competition between sample ions and mobile-phase ions for charged binding sites on the surface of the stationary phase. It is mainly used to separate ionic compounds. Ion exchange chromatography is a liquid chromatography method developed and applied earlier. In the 1940s, polystyrene ion exchange resins with stable ion exchange properties were introduced into markets. In the early 1950s, polystyrene ion exchange resins were used in ion exchange chromatography for separation of amino acids. In the mid-1950s, carboxymethyl cellulose and diethylaminoethyl cellulose were reported for the separation of protein by ion exchange chromatography, which greatly promoted the application of ion exchange chromatography in the separation of biological macromolecules. At present, ion exchange chromatography has become the most commonly used method in protein separation and purification as it accounts for more than half of all the chromatographic steps surveyed (Karlsson and Hirsh 2011).

Retention mechanism The process of ion exchange chromatography is mainly controlled by the electrostatic interaction between the sample ions and the charged groups on the stationary phase, and the retention order depends on the magnitude of the electrostatic force of the sample components. The electrostatic force between the sample ion and the ligand, F, can be approximated by Coulomb's law:

$$F = \frac{q_1 q_2}{\epsilon r^2} \tag{3.24}$$

where q_1 and q_2 represent the point charges of the interacting ions, ? is the dielectric constant of the medium, and r is the distance between two point-charges. The electrostatic force is proportional to the charge number of the sample ions and inversely proportional to the square of the distance.

The retention mechanism of ion exchange chromatography is simple ion exchange between solute ion, A, and counter ion, E, in the mobile phase with the charged ligand, L, of the stationary phase. For monovalent ions, the exchange equilibria can be expressed by the following equations:

$$\text{Cation exchange}: A^+ + E^+L^- \rightleftharpoons E^+ + A^+L^- \tag{3.25}$$

$$\text{Anion exchange}: A^- + E^-L^+ \rightleftharpoons E^- + A^-L^+ \tag{3.26}$$

Ion exchange equilibrium constant:

$$K_A = \frac{[AL][E]}{[EL][A]} \tag{3.27}$$

where $[A]$ and $[AL]$ are the concentration of solute, A, in the mobile phase and the stationary phase, respectively, and $[E]$ and $[EL]$ are the concentration of the counter ion, E, in the mobile phase and the stationary phase, respectively. For the sake of clarity, the ionic charges in the formula are omitted.

Assuming that the adsorption of the counterion E on the stationary phase can be described by the Langmuir equation, then

$$[EL] = \frac{Q_m K_E [E]}{1 + K_E [E]} \tag{3.28}$$

where Q_m is the monolayer adsorption amount of the counterion, E, on the surface of the stationary phase and K_E is the adsorption constant of the counterion on the stationary phase.

The distribution coefficient, D, of solute ions can be expressed by the following equation:

$$D = \frac{[AL]}{[A]} = \frac{K_A[EL]}{[E]} = \frac{K_A Q_m K_E}{1 + K_E[E]} \tag{3.29}$$

Under the conditions of ion exchange chromatography, it is reasonable to assume $(1 + K_E[E]) \approx K_E[E]$, and Eq. (3.29) can be simplified as

$$D = \frac{K_A Q_m}{[E]} \tag{3.30}$$

Assuming that the mass of the ion exchanger in the chromatographic column is W and the volume of the mobile phase is V, the number of moles of solute on the stationary phase is $W\,[AL]$, and the number of moles of solute in the mobile phase is $V\,[A]$. By convention, W/V is phase ratio and expressed by ϕ, then the retention factor k of the solute is

$$k = \phi D = \frac{\phi K_A Q_m}{[E]} \tag{3.31}$$

It can be seen from Eq. (3.31) that the solute retention increases with the ion exchange equilibrium constant and decreases with the increase of the counterion concentration. If the retention factor k is plotted against the reciprocal of the counter ion concentration, a straight line passing through the origin can be obtained, and the slope of the straight line is $\phi K_A Q_m$. However, experimentally, it has also been observed that the straight line does not pass through the origin, but with a positive intercept on the k-axis. This can usually be attributed to the fact that solute retention is not completely controlled by electrostatic interactions. Hydrophobic and polar interactions may also participate in the solute retention process, and their contribution to solute retention will not be cancelled by the increase in counterion concentration.

Stationary phase According to matrix materials used, ion exchange chromatography packing materials are classified into three types: (1) rigid ion exchangers based on inorganic materials such as silica gel and porous glass, which have high-pressure resistance and are available in small particle size for high resolution and are generally used for analysis; (2) semirigid ion exchangers based on high-molecular polymers, such as polystyrene resins and composite polymers, in which this type of packing have good mechanical strength and are available in small particle size for high resolution and are generally used for analysis and preparation of small organic molecules; and (3) polysaccharides and modified polysaccharide gel-based ion exchangers, which have low pressure tolerance and are supplied in large particle sizes. They are generally used for the preparation of biological macromolecules due to their large adsorption capacity, low nonspecific adsorption, and high activity recovery.

According to the functional groups bonded on the matrix framework, ion exchange chromatography packing can be divided into strong ion exchangers and weak ion exchangers. The former retains the charge on the bonded ions in a wide pH range, while the latter can only be ionized in a narrow pH range. Therefore, ion exchangers can be further divided into four types, namely, (1) strong cation exchangers, where the surface groups are charged anions, such as sulfopropyl groups; (2) strong anion exchangers, where the surface groups are charged cations, such as trimethylaminoethyl groups; (3) weak cation exchange, where the surface groups can be dissociated into anions under certain pH conditions, such as carboxymethyl groups; and (4) weak anion exchanger, where the surface groups can be dissociated into cations under certain pH conditions, such as diethylaminoethyl groups.

Despite the availability of various ion exchange functional groups, most separations using silica and organic polymer ion exchangers are carried out on sulfonic acid-type strong acid cation exchangers and quaternary ammonium-type strong base anion exchangers. These strong cation exchangers and strong anion exchangers are usually labeled SCX and SAX, respectively. Correspondingly, weak cation exchangers and weak anion exchangers are labeled WCX and WAX.

An important characteristic of ion exchangers is their ion exchange capacity, which depends on the number of functional groups per unit weight of the resin, which can be expressed in different units, the most common of which is milliequivalent (charge) per gram of dry resin or milliequivalent per milliliter of wet resin. Ion exchange capacity is usually measured by the following method: immersing a known weight of resin in a solution of specific ions, then washing the resin, and then quantitatively replacing the ions. Then the number of moles of replacement ions can be determined by a titration procedure. It should be noted that the capacity measured by this method is usually higher than the applicable capacity when the resin is used as a stationary phase in the column.

Mobile phase The mobile phase (also known as the eluent) used in ion exchange chromatography is usually composed of an aqueous solution made of a suitable salt or salt mixture. Sometimes a small amount of organic solvent is added. The salt mixture itself can be a buffer, or if needed, a separate buffer can be added to the eluent. The main component of the eluent is a competitive ion, which has the function of eluting the sample component through the chromatographic column in a reasonable time. The three most important characteristics of the eluent that affect the retention of solute ions are (1) the pH of the eluent, (2) the nature of the competing ions, and (3) the concentration of the competing ions. The pH value of the eluent will affect the ionization of functional groups on the ion exchange matrix and the ion exchange equilibrium constant between the competing ions and the specific solute ions, which will determine the extent to which the competing ions can displace the solute from the stationary phase. Since different competing ions have different equilibrium constants, the nature of the competing ions will be the main factor that determines whether the solute ion is easily eluted or not. Finally, it can be seen that the concentration of competing ions plays a major role. By affecting the position of ion exchange equilibrium, as shown in Eq. (3.31). The higher the concentration of competing ions in the eluent, the higher the efficacy of the eluent to displace solute ions from the stationary phase and therefore the faster the solute is eluted from the chromatographic column. In addition to the above three factors, the elution of the solute is also affected by the flow rate and temperature of the eluent. A faster flow rate results in a lower elution volume because solute ions have fewer opportunities to interact with bonded ions. The effect of temperature is small, and it depends to some extent on the type of ion exchange material used. Elevated temperature increases the rate of diffusion within the ion exchange matrix, usually resulting in increased interaction with bonded ions and therefore an increase in elution volume. Chromatographic efficiency usually increases at higher temperatures.

Elution mode In the separation and purification of protein by ion exchange chromatography, it is generally necessary to change the ionic strength or pH value for elution. In order to ensure the adsorption of proteins on the chromatographic column, the salt concentration of the initial equilibrium buffer is relatively low. The adsorbed proteins are then rinsed and eluted by increasing the salt concentration in the eluent. The contaminants are eluted either in flowthrough or retained on the column and stripped off after the target proteins are recovered. Generally, the neutral salt is added to the eluent to increase the ionic strength, the most commonly used is sodium chloride. The greater the salt concentration, the stronger the elution strength. The salt concentration can be increased in steps or continuously. The former is known as step gradient elution and the later as linear gradient elution.

If the elution is performed by changing the pH instead of increasing the ionic strength, neutral salt is not added to the eluate, but the salt concentration is kept low and constant, and the pH value is changed to elute the adsorbed protein.

Application Ion exchange chromatography plays an important role in the separation of various organic and inorganic solutes. A large number of examples of ionic separation can be found in the literature, from metal ions to organic ions, from amino acids to proteins, from metabolites to virus and cells. Today, ion exchange chromatography is indispensable tool for the separation and purification of biomolecules such as amino acids, peptides, proteins, and nucleic acids. It is said that in the field of bioprocessing, more than 70% of the applications are completed on ion exchange chromatography, which is equivalent to the absolute dominance of reversed-phase chromatography in high-performance liquid chromatography applications.

3.6 Hydrophobic Interaction Chromatography

Hydrophobic interaction chromatography (HIC) is a liquid chromatography method that uses hydrophobic interactions to separate biological macromolecular compounds (Eriksson and Belew 2011; Queiroz et al. 2001; Tomaz 2017). With a mobile phase containing a high concentration of salt, biomolecules will be bound on the surface of a moderately hydrophobic stationary phase and then eluted by reducing the salt concentration in the eluent. This mode of separation is often incorporated in a multistep chromatographic purification protocol because it provides good resolution and high recovery of biologically active proteins. In reversed-phase chromatography, the use of organic solvents as polarity modifiers may cause denaturation and inactivation of proteins. In contrast, in hydrophobic interaction chromatography, the use of aqueous buffers avoids any denaturation of biopolymers (proteins). Because hydrophobic interaction occurs only under conditions of high ionic strength, reconditioning of the raw materials is sometimes necessary to facilitate protein adsorption. In addition, hydrophobic interaction chromatography allows large-volume samples to be loaded at high ionic strength, so it is most commonly used in the subsequent steps of salting-out precipitation. For example, after the target protein is precipitated by ammonium sulfate, it can be directly loaded into the hydrophobic interaction chromatography for intermediate or fine purification.

Retention mechanism Hydrophobic interaction chromatography takes advantage of the differential hydrophobic interaction between the sample component and the hydrophobic ligand of the packing in the brine system to achieve the purpose of separation. Horvath and colleagues (Melander and Horvath 1977; Melander et al. 1984) proposed a retention model for hydrophobic chromatography based on Sinanoglu's solvophobic theory (Sinanoglu and Abdulnur 1965). This model is built on promise that a molecular complex is formed between the protein and the ligand through hydrophobic interaction. The driving force for the formation of the complex comes from the tendency of the protein molecule to reduce the nonpolar surface in contact with water. This process causes the free energy of the system to decrease. This change in free energy is not only related to the free energy of cavity formation ΔG_{cav}, the free energy of electrostatic interaction ΔG_{es}, and the free energy of van der Waals interaction ΔG_{vdw} but also related to the free energy of the interaction between the protein and the ligand in the presence of the solvent ΔG_{red} and the free energy of the interaction between protein and ligand in the absence of solvent ΔG_{assoc}. The free energy change of the entire system has the following relationship with the logarithm of the protein retention factor $\ln k$:

$$\ln k = -\frac{\Delta G_{\text{cav}} + \Delta G_{\text{es}} + \Delta G_{\text{vdw}} + \Delta G_{\text{red}} + \Delta G_{\text{assoc}}}{RT} + \ln \frac{RT}{PV} + \ln \phi \qquad (3.32)$$

where R is the gas constant, T is the absolute temperature, ϕ is the phase ratio, P is the pressure, and V is the molar volume of the solvent.

At high salt concentration, each free energy change is proportional to the molar concentration of salt under certain conditions. Under the conditions of constant pressure P and temperature T, the above relationship can be simplified as

$$\ln k = \ln k_w + Sm \tag{3.33}$$

where k_w is the retention factor of the solute when the salt concentration approaches zero, S is a constant, and m is the molar concentration of neutral salt in the mobile phase. When ln k is plotted against salt concentration m, a straight line should be obtained, with the intercept for ln k_w and the slope for S, which is a constant determined by the contact area between the protein and the ligand.

In an aqueous solution of salt, the molar concentration of the salt is directly proportional to the surface tension of the solution. Therefore, after adding salt to the solution, the surface tension of the solution will increase, which means that the cohesion will increase, which is conducive to the contact between proteins and between proteins and ligands. Therefore, the retention value of protein increases as the salt concentration in the solution increases.

This model is only applicable for salts that have a strong salting-out effect with proteins without specific effects. For those salts with specific effects such as magnesium chloride, calcium chloride, and salt-in salts such as potassium bromide, guanidine salt, thiocyanate, etc., the rule is not followed.

In addition to the solvophobic theory, the entropy increase principle is also used to explain the retention mechanism in hydrophobic interaction chromatography. The classic hydrophobic interaction of nonpolar substances in aqueous solutions was proposed by Kauzmann (1959), based on promise that the hydrophobic interaction was driven by the increase in entropy. Jennissen and coworkers used the principle of entropy increase to explain the process of protein in HIC (Jennissen and Heilmeyer 1975; Jennissen 1976, 1986). When the protein and ligand are present separately in the solution, the water molecules are bound to the nonpolar part of the protein and the ligand in an orderly manner and form a highly ordered hydration structure. When the protein comes to contact with the ligand to form a molecular complex, the ordered water in the affected part is released and returns to the more disordered bulk solution, resulting in an increase in the entropy of the system. In this process, the enthalpy change, ΔH, is small (because the energy that destroys the ordered hydration structure of the nonpolar part is very close to the energy that water recombines to the nonpolar part, so ΔH is small), and the entropy change, $T\Delta S$, is much larger and becomes a driving force for the binding interaction between protein and ligand. It appears likely that both the increase in surface tension and the increase in entropy are responsible for the adsorption of proteins on hydrophobic ligands in hydrophobic interaction chromatography.

Stationary phase Conventional reversed-phase chromatographic columns are too hydrophobic, and when used together with pure water mobile phase, protein retention will be too strong to be eluted. Adding organic modifiers to the mobile phase may cause protein denaturation. Therefore, a new material is specially designed for the HIC of biopolymers. The stationary-phase material of HIC has a large pore size (e.g., 1000 Å), which is prepared by low-density bonding of hydrophobic groups such as butyl, phenyl, and octyl on a hydrophilic matrix. The hydrophobicity of the ligand increases in the order of butyl, phenyl, and octyl. The large pore size of the stationary phase allows the protein to diffuse quickly and the controlled functionalization results in low to moderate hydrophobicity, allowing the adsorbed protein to be eluted in an aqueous solution with low salt concentration. HIC has the advantages of high sample capacity and excellent protein activity recovery. It is best to use resin-based HIC chromatographic columns because they can be cleaned in situ with sodium hydroxide to disinfect the packing. The main limitation of HIC is the extra cost due to disposal of a large amount of lyotropic salts.

Mobile phase In hydrophobic interaction chromatography, proteins are adsorbed in the mobile phase with high salt concentration and eluted in the mobile phase with low salt concentration. There are at least two types of mobile phases for hydrophobic interaction chromatography: one is the high salt concentration buffer *A* used before and after loading, also called the equilibration solution; the other is the low salt concentration buffer *B* used for elution, also called eluent. According to the different effects on protein, salt can be divided into three categories: salting-out salt, salt-in salt, and specific action salt. Salting-out salts include sodium sulfate, ammonium sulfate, etc., which can stabilize the conformation of the protein in the solution, promote the hydrophobic interaction between the protein itself and precipitate, or promote the hydrophobic interaction between the protein and the ligand. The increase in molar surface tension of the salting-out salt is also large. Salt-in salts include thiocyanate and guanidine salt, which can improve the water solubility of protein and denature the protein at the same time. Salt-in salt has a small increase in molar surface tension. Specific action salts include magnesium chloride and calcium chloride, which specifically interact with proteins to promote the solubility of proteins in aqueous solutions. However, their molar surface tension increases are also large, which is conducive to the retention of proteins on the chromatographic column. The elution strength of salts and ions and their ability to disrupt the orderly arrangement of water are consistent with the Hofmeister sequence.

In hydrophobic interaction chromatography, the mobile phase should use salting-out salts, which can stabilize the protein and have a high recovery of activity. The most used salt is ammonium sulfate, because ammonium sulfate has a large solubility, a wide range of concentration changes, and a strong salting-out ability, which is beneficial to protein retention and will not cause protein inactivation.

The pH of the mobile phase has an important effect on protein retention. Most proteins are stable in the pH range of 4 ~ 8. Strong acid and alkali media can cause protein denaturation. Therefore, the pH value of the mobile phase should be selected within the range that protein remains active, generally near the isoelectric point of the protein. The farther away from the isoelectric point of the protein, the more the surface charge of the protein, the weaker the hydrophobicity, the lower the adsorption force, and the less likely it is to be retained.

When the retention of hydrophobic proteins on the chromatographic column is too strong, a small amount of organic compounds, such as urea, ethylene glycol, sucrose, and surfactants, can be added to the mobile phase to reduce the polarity of the mobile phase and weaken the retention of proteins. When adding these regulators, care should be taken to control their influence on protein activity.

Elution mode In hydrophobic interaction chromatography, proteins are adsorbed at high salt concentrations and eluted at low salt concentrations. There are basically three ways to reduce the salt concentration: step elution, segment elution, and gradient elution. In step elution, the adsorbed protein is washed with the equilibration solution and then eluted directly with the eluent. Segment elution is to stepwisely reduce the salt concentration in the eluent to selectively elute the adsorbed protein. In gradient elution, the salt concentration in the eluent is gradually reduced to sequentially elute proteins with different hydrophobicities.

Application Hydrophobic interaction chromatography is a widely used chromatographic method in the separation and purification of biological macromolecules. In addition to separation and purification, it can also be used for the concentration of biological macromolecules and the renaturation of proteins. The protein purification process can be roughly divided into capture, intermediate purification, and polishing stages. The hydrophobic interaction chromatography can be used in the appropriate stages according to its characteristics and its connection with other methods in the process.

HIC is often used for the separation and purification of proteins, and its resolution is much higher than that of size exclusion chromatography. And its separation mechanism is based on the hydrophobic interaction between the component molecules and the packing. HIC can provide solutions to the challenging separation problems of other chromatographic methods. With mild chromatographic conditions and high activity recovery, it is a common method for separating biological macromolecules and is usually used in the first steps of the purification process. The carboxypeptidase B (CPB) is a zinc-containing pancreatic exopeptidase that can specifically hydrolyze the basic amino acids at the C-terminus of the peptide chain: arginine, lysine, or ornithine. CPB contains 307 amino acids. The molecular weight is 35,000 Da. The naturally occurring CPB is produced by procarboxypeptidase B, and its N-terminal contains a fragment of 108 amino acids (13 amino acids are the signal sequence and 95 amino acids constitute the active peptide part), which is connected to the C-terminal CPB. After the procarboxypeptidase B is transported to the endoplasmic reticulum, the signal peptide is cleaved off to obtain procarboxypeptidase B (pCPB), which has no enzymatic activity, which is then digested into enzymatically active CPB by trypsin. CPB is one of the essential tool enzymes in the production of recombinant insulin and has important uses in protein sequencing and polypeptide preparation. CPB is mostly extracted from porcine pancreas. The crude porcine pancreas extract is treated by ammonium sulfate salting out and precipitation and then loaded on the octyl-agarose gel, washed with equilibration solution, eluted with low salt buffer, and collected. The collected fractions were pooled and dialyzed with deionized water to remove ammonium sulfate. The product was further purified by DEAE anion exchange chromatography and size exclusion chromatography and electrophoretic pure CPB were obtained (Wang et al. 2007).

HIC has unique advantages as a polishing step in many monoclonal antibody purification processes and is an alternative to high-cost and low-capacity affinity media (Lu et al. 2009). HIC is particularly useful for removing aggregates. It can be used to remove other process-related impurities such as host cell proteins, leached protein A, and endogenous viruses (Shukla and Thömmes 2010). HIC is a reference technology to determine the relative hydrophobicity of mAbs and separate different antibody/drug conjugate (ADC) populations. Conjugation of antibody with drug molecules produces a heterogeneous mixture as the number of drug molecules attached to the antibody varies. This is usually called DAR (drug–antibody ratio) species (Wiggins et al. 2015; Haverick et al. 2014). HIC can separate natural ADCs of different DARs according to their physiological-like conformations, such as DAR0, DAR2, DAR4, DAR6, and DAR8 (Ouyang 2013). It also proved that HIC can be used to purify other biological molecules besides proteins, such as plasmid DNA or DNA vaccines for gene therapy (Bo et al. 2013).

3.7 Hydrophilic Interaction Chromatography

Hydrophilic interaction chromatography (HILIC) achieves the retention and separation of sample components through the polar interaction between the solute and the polar stationary phase (Cavazzini et al. 2017). There are similarities and differences between hydrophilic interaction chromatography and normal phase chromatography. They both use polar stationary phases, and both require the polar interaction between the solute and the stationary phase to promote solute retention. However, hydrophilic interaction chromatography uses polar organic solvents while normal phase chromatography uses nonpolar organic solvents as the main component of the mobile phase. Polar organic solvents are conducive to the dissolution of polar compounds. Therefore, hydrophilic interaction chromatography is a powerful tool for the analysis of strongly polar compounds, which overcomes the deficiencies of normal phase chromatography, reversed-phase chromatography, ion exchange chromatography, and other methods. In recent years, with the rapid development of genomics, proteomics, metabolomics, and drug research and development, more and more polar small molecules have become the object of attention, and the demand for separation and analysis of the biologically active substances (amino acids, peptides, nucleobases, analysis of nucleosides, monosaccharides, polysaccharides, etc.), physiological metabolites (body fluids, biological fluids, drug metabolites, etc.), food components, and environmental pollutants has increased dramatically, posing new challenges to chromatographic technology. Hydrophilic interaction chromatography meets the needs of the abovementioned fields and has become the preferred method for the separation and analysis of strongly polar small molecules.

Retention mechanism In hydrophilic interaction chromatography, polar solute molecules interact with a polar stationary phase in a mobile phase rich in organic polar solvents. Different solute molecules pass through the column at different rates due to different forces. Polar interactions mainly include hydrogen bonds, dipole–dipole interactions, and electrostatic interactions. The hydrogen bond is formed by the interaction between the proton donor/acceptor in the solute molecule and the corresponding proton acceptor/donor on the stationary phase. The strength of the hydrogen bond can be adjusted by the type and content of the polar organic solvent in the mobile phase. Acetonitrile and tetrahydrofuran are typical polar aprotic solvents, while methanol and ethanol are typical polar protic solvents, which are proton donors and proton acceptors. For example, methanol can compete with solute molecules for active sites on the surface of the stationary phase, thereby reducing the retention of polar solutes. Electrostatic interaction is the interaction between the dissociative group in the solute molecule and the charged group on the stationary phase. Charges of the opposite signs attract each other and charges of the same signs repel each other. When the solute molecule is charged, the electrostatic interaction can be superimposed on the hydrogen bond and the dipole–dipole interaction to form a complex with stronger binding force. Taking cytosine as an example, its hydrophilic interaction chromatographic retention on the silica gel stationary phase arises from the hydrogen bond and electrostatic interaction between the amino group on the cytosine and the silanol group, that is, the electrostatically assisted hydrogen bond. The hydrophobic interaction between cytosine and the stationary phase is inhibited by the high concentration of organic solvent in the mobile phase. In recent years, electrostatic repulsion hydrophilic interaction chromatography has been used in proteomics research. When the hydrophilic solute molecules are charged and have the same signs as the charge on the hydrophilic chromatography ligand, electrostatic repulsion will occur between them. This repulsion is superimposed on the hydrogen bond and other polar interactions, making solute retention weakened. By adjusting the ionic strength in the mobile phase, charge repulsion can be suppressed, thereby enhancing solute retention.

Stationary phase The stationary phase used in hydrophilic interaction chromatography is mainly bare silica gel or chemically bonded silica gel with polar functional groups. As mentioned earlier, the surface of silica gel is rich in silanol groups, which can be dissociated under the commonly used conditions of pH 4 ~ 5 and deprotonated to generate negatively charged active sites. Aminopropyl-bonded silica is the most commonly used stationary phase for the separation of monosaccharides in hydrophilic interaction chromatography. Other polar bonded phases such as diol- and cyanopropyl-bonded silica are less used because of their weaker polarity and weak retention in hydrophilic interaction chromatography. Zwitterion-bonded silica gel is a more commonly used stationary phase in hydrophilic interaction chromatography. Zwitterionic functional groups are mainly sulfobetaine, phosphocholine, and 3-P,P-diphenylphosphoranesulfonic acid (3-P,P-diphenyl phosphonium propyl sulfonate). The stationary phase with zwitterions has obvious positive and negative charge centers, but its net charge is zero, so it has a strong ability to adsorb water and does not have a strong electrostatic interaction ability. The zwitterionic stationary phase can be used for the separation of a variety of polar molecules and has versatility. Amide-bonded silica gel is another commonly used stationary phase for hydrophilic interaction chromatography. Typical ligands are carbamoyl and asparagine, which are neutral, nonionic polar functional groups, have a permanent dipole moment and have strong hydrogen bond forming ability.

Mobile phase In hydrophilic interaction chromatography, the mobile phase is composed of a high content (volume ratio of about 60–95%) polar organic solvent mixed with an aqueous solution. Commonly used organic solvents are divided into two categories, one is polar aprotic solvents, such as acetonitrile, tetrahydrofuran, and *p*-dioxane, which are only hydrogen bond acceptors, and the other is polar protic solvents, such as methanol, ethanol, isopropanol, and acetic acid, which are both hydrogen bond donors and hydrogen bond acceptors. Among the most commonly used are acetonitrile, tetrahydrofuran, methanol, and isopropanol. Their elution strength in hydrophilic interaction chromatography decreases in the order of methanol > isopropanol > tetrahydrofuran > acetonitrile. In addition to polar organic solvents, the mobile phase of hydrophilic interaction chromatography also contains a portion of aqueous solution, usually a buffer with a certain pH. The pH of the mobile phase can affect the retention of dissociable solutes. In addition, the salt concentration in the mobile phase is another factor that affects the retention of dissociable solutes. They cause selectivity changes by affecting the electrostatic interaction between the solute and the stationary phase.

Application HILIC is a very powerful tool for the analysis of antibiotics, most of which are natural or derived hydrophilic substances. It is used to detect aminoglycosides, β-lactams, and tetracyclines, to name a few, and can be used in a variety of sample matrices, including plasma, serum, environmental water, food, etc. (Kahsay et al. 2014). HILIC separation has also been applied to drugs of abuse and their metabolites, such as the analysis of cocaine (van Nuijs et al. 2011) and the separation of opium, amphetamine (i.e., MDMA), and other biological samples, such as human plasma, serum, hepatocyte solute, urine, and fingerprint (Kahsay et al. 2014; Li et al. 2016; Ares and Bernal 2012).

The application of HILIC separation in the field of agriculture and food analysis includes the separation and identification of several types of compounds, such as amino acids, peptides and their sugar conjugates, alkaloids, and toxins. In the agricultural sector, HILIC has recently been used to determine nine selected fungicides (Noga et al. 2012). The food matrix analyzed in HILIC comes from different sources, including wheat gluten, Parmesan cheese, potatoes, carbohydrate-rich energy drinks, drinking water, etc. (Zywicki et al. 2005; Schlichtherle-Cerny et al. 2003; Bernal et al. 2011). A recent application is the separation of phenolic compounds which are a class of food ingredients with anticancer, antioxidant, and anti-inflammatory activities (Pyrzynska and Sentkowska 2015).

The separation of carbohydrates by HILIC deserves special attention. The analysis of carbohydrates covers a broad range of applications: the analysis of sugars and nucleosides from the most diverse matrix, the determination of natural glucosinolates and saponins, the analysis of glycans, and the characterization of low-molecular weight and high-molecular-weight polysaccharides and their derivatives in the food and cellulose industries, analysis of synthetic glycopolymers, and applications to characterize complex glycan structures of low-abundance glycoproteins in the field of "glycoproteomics" or "glycomics" (Zhou and Lucy 2008; Treadway et al. 2011; Ruhaa et al. 2008).

3.8 Coordination Chromatography

Coordination chromatography refers to the chromatographic process that leads to the separation of mixtures through the coordination of central ions. It is also known as complex chromatography, ligand exchange chromatography (Davankov et al. 1988), chelating ion chromatography (Nesterenko et al. 2011), and immobilized metal affinity chromatography (Kagedal 2011). The separation generally involves the rapid and reversible formation of a complex between the metal ion and the ligand. If metal ions are immobilized on the stationary phase, this chromatographic mode can be used to separate organic compounds and biological macromolecules, for example, the separation of amino acid optical isomers in ligand exchange chromatography; the separation of histidine-labeled proteins in immobilized metal affinity chromatography; the separation of proteins in hydroxyapatite chromatography; the separation of monosaccharides and disaccharides by the coordination with calcium, zinc, or lead cations immobilized on ion exchangers; and the separation of tetracycline residues in concentrated milk on metal-chelated agarose. If a coordination ligand is immobilized on the stationary phase, coordination chromatography can also be used to separate metal ions by forming complexes with metal ions, such as the separation of rare earth elements in chelated ion chromatography.

Retention mechanism The coordinate complex is formed between a ligand ion or molecule that can give a lone pair of electrons and a central atom or ion that accepts a lone pair of electrons. The stability of the formed complex varies depending upon the composition and spatial configuration. Similar to the hydrogen bond, the coordination bond has covalent properties, and the bond is directional, and the strength of the coordination bond is hardly affected by the ionic strength in the solution. On the hydroxyapatite matrix, the protein can be bound by the formation of coordination bonds between the vicinal carboxyl groups and the calcium ions in the matrix or by the electrostatic interaction between the amino groups and the phosphate groups in the matrix. Under the condition of high ionic strength, the electrostatic effect is shielded, and the coordination bond is not affected. Therefore, the bound acidic protein cannot be eluted with a linear gradient with increasing salt concentration, but through the competition of phosphate or changing the pH of the mobile phase to control the elution behavior of the protein. In immobilized metal affinity chromatography media, nickel ions are usually bound on the matrix through metal coordination, and proteins are adsorbed through the coordination of histidine with the immobilized nickel ions. The adsorption strength of protein is proportional to the number of histidine residues in the protein. Similarly, the adsorbed protein cannot be eluted by increasing the ionic strength of the solution, but by the competitive binding of imidazole to selectively elute the protein. Coordination chromatography is used for the separation of metal ions, also known as chelation ion chromatography, because different metal ions bind to the same coordination ligand with different stability constants, which are similar to the separation mechanism in ion exchange chromatography. Detailed discussion on the binding and elution mechanism of chelation ion chromatography can be found in the monograph by Nesterenko et al. (2011).

Stationary phase Metal coordination chromatography stationary phases can be divided into two categories: natural sources and synthetic sources. The natural source stationary phase is represented by hydroxyapatite, which will be discussed in detail in Chap. 5. The synthetic stationary phase can be directly used for the separation of metal ions by bonding metal ion chelating groups on a conventional chromatographic support medium. If used for the separation of organics or biomolecules, metal ions can be loaded on the chelating group to obtain an immobilized metal ion stationary phase. Many coordination stationary phases have been synthesized using support materials such as silica gel, synthetic polymers, and polysaccharides. Examples of chelating ligands immobilized by chemical bonding include iminodiacetate (IDA), propylenediaminetetraacetate, diketones (such as trifluoroacetoacetate), 8-hydroxyquinoline, hydroxamic acid, maleic acid, dithiocarbamate, and dithizone (Nesterenko et al. 2011). In the separation and application of biological macromolecules, the coordination ligands should be able to form complexes with higher stability with metal ions to reduce the loss of metal ions during the separation process. For the separation of metal ions, the chromatographic stationary phases should show higher formation and dissociation rates to avoid the peak broadening due to slow mass transfer.

Compared with monodentate and multidentate ligands, macrocyclic ligands can provide higher stability and selectivity. They can recognize appropriate guest molecules based on the size of the central ion, molecular configuration, and chirality. As a representative of macrocyclic ligands, crown ethers have been chemically bonded to silica gel or coated on hydrophobic bonded silica to form a highly selective coordination stationary phase. Crown ethers are cyclic compounds with internal cavities, which are usually composed of oxygen atoms connected by ethylene bridges. The chromatographic uses of such stationary phases depend on their ability to complex cations of a specified size. Therefore, changing the size of the cavity in the crown ether will change the selectivity of the stationary phase. The binding of metal ions on a specific crown ether stationary phase depends on the size of the cation, the nature of the relevant anion, and the content of the organic modifier in the mobile phase. In this chromatographic format, water is usually used as the mobile phase, in combination with conductivity detection.

Mobile phase In coordination chromatography, the mobile phase has several different functions. The first is to ensure that the coordination in the stationary phase is the main retention and separation mechanism. Because uncharged coordination ligands are rarely used in coordination chromatography, the undesirable electrostatic interaction between solute molecules and charged ligands always exists. Negatively charged carboxyl, phosphate, and arsenate groups attract positively charged cations without forming any donor–acceptor coordination bonds. Therefore, the presence of coordination groups in the stationary phase does not mean that coordination is the only and main retention mechanism. For the same reason, when the ligand group contains an amino group or other basic group, the protonated amino group can retain the acidic form of the solute molecule in the sample through electrostatic interaction. These electrostatic side effects can usually be eliminated by adjusting the ionic strength of the eluent, for example, adding 0.5–1 M sodium chloride or sodium perchlorate. The presence of this high concentration of ions forms an effective counter-ion shielding layer around the charged coordination groups, thus competitively inhibiting the electrostatic interaction with the solute molecules. At the same time, the increase of electrolyte concentration in the eluent has little effect on the coordination effect. This can be reasonably inferred from the fact that the complex formation constant of several metal ions with iminodiacetic acid (IDA) hardly changes with the change of ionic strength from 0.1 to 2.0 (Martell and Smith 2004).

The effect of mobile phase pH on solute retention in coordination chromatography depends on several factors, including the pH-dependent secondary equilibrium that occurs in the mobile phase, which can affect separation selectivity. These reactions may involve partial hydrolysis, formation of water-complexes, solvation of organic solvents, or most commonly complexation with organic additives in the mobile phase. The decrease in pH induces the protonation of the ionized groups in the coordination ion exchanger, thus reducing the effective concentration of the active form in the immobilized ligand, thereby causing the elution of the adsorbed solute.

Polar organic solvents that are miscible with water such as methanol, ethanol, isopropanol, and acetonitrile are often used as mobile-phase additives for coordination chromatography. The presence of organic solvents can improve the solubility of hydrophobic solutes in the mobile phase and, more importantly, can inhibit the hydrophobic interaction between the solute and the ligand, so that the solute retention is based on a pure coordination binding mechanism rather than the mixed retention mechanism of coordination together with hydrophobic interaction.

Application The main application areas of coordination chromatography include the separation and purification of biomolecules, the separation of metal ions, and the separation of chiral compounds. Coordination chromatography is applied to biomolecules under the name of hydroxyapatite chromatography and immobilized metal affinity chromatography and applied to metal ions under the name of chelation ion chromatography and applied to chiral compounds under the name of crown ether chromatography and ligand exchange chromatography.

Immobilized metal affinity chromatography is an important technique to purify histidine-labeled recombinant proteins from fermented cells. The histidine tag (His-Tag) composed of six histidine peptides has a strong affinity for metal ions (such as Co^{2+}, Ni^{2+}, Cu^{2+}, and Zn^{2+}). One of the earliest packing materials is the use of immobilized iminodiacetic acid, which can bind metal ions and allow the formation of coordination complexes with His-tagged proteins. One of the difficulties of iminodiacetic acid carriers is the possibility of metal ions leaching, leading to protein contamination. Modern metal coordination support media, including nickel trinitrotriacetate (Ni-NTA) and cobalt carboxymethyl aspartate (Co-CMA), show limited leaching and can therefore improve protein purification efficiency. Once the labeled protein is bound by the immobilized chelating agent, it can be eluted by introducing a chelating group competitor (imidazole) or other metal chelating agent (EDTA). Histidine tagging technology makes the process of separating and purifying recombinant proteins from expressing cells easier and simpler and can often achieve one-step purification of tagged molecules.

One advantage of using His-tags for protein purification includes the small size of the affinity ligand. Due to its small size, it has little effect on protein folding. In addition, if the His-tag is placed at the N-terminus of the protein, it can be easily removed with endoprotease. Another advantage of using the His-tag purification method is that the polyhistidine tag can bind to the immobilized metal under natural and denatured conditions. When a protein is found in inclusion bodies and must be denatured to dissolve it, it becomes important to use denaturing conditions. The disadvantages of using His-tag protein purification include the potential degradation of His-tag, the formation of dimers and tetramers, and the co-elution of other histidine-containing proteins. First, when some histidine residues are proteolytically degraded, the affinity of the tagged protein is greatly reduced, resulting in a decrease in protein yield. Secondly, once His-tag is added to the structure of the protein, it may form dimers and tetramers in the presence of metal ions. Although this is usually not a big problem, it can lead to inaccurate estimates of the molecular weight of the tagged protein. The third disadvantage of using His-tag to purify proteins is co-elution with proteins that naturally have two or more adjacent histidine residues.

3.9 Affinity Chromatography

Affinity chromatography is a liquid chromatography method that uses specific interactions between biomolecules for separation, and is mainly used for the separation and purification of proteins (Hage et al. 2017). As is well known, there are many selective interactions in biological systems, and organisms use them to perform various biological functions, such as protein synthesis, nucleic acid replication, cell proliferation, signal transduction, virus invasion, and so on. Examples of these interactions include interactions between enzymes and substrates, hormones and receptors, and antibodies and antigens (Cuatrecasas et al. 1968). In affinity chromatography, one of a pair of biological interaction partners is immobilized on a chromatographic carrier to selectively adsorb the other partner in a complex mixture sample, and the impurity in the sample that does not form an interaction pair with the ligand is not retained and washed away in flowthrough, thereby separating the target biomolecule from the complex mixture.

Different from conventional liquid chromatography, affinity chromatography usually uses a "on and off" switch mode to adsorb and elute target molecules in one step. This is because the affinity of biological interactions is relatively strong, the association equilibrium constant is usually higher than 10^5–10^6 M^{-1} and some even as high as 10^{12} M^{-1}. Under conventional chromatographic conditions, the target protein is retained on the column, and other impurities are removed by washing. Then, more severe conditions are needed to dissociate the interaction pair. The loading buffer usually simulates the pH and ionic strength that are most conducive to the binding of the fully active affinity ligand to the target, so that in the presence of the loading buffer, the target is retained and rinsed. Then the strong mobile phase or elution buffer is applied to the chromatographic column to release the target for collection or analysis.

Affinity chromatography has the characteristics of high binding affinity, strong selectivity, and high enrichment factor. It is mostly used to separate a small amount of specific protein from a large volume of complex solutions. This purification method also has a concentration effect. Sometimes through the one-step separation process of affinity chromatography, fast and effective protein purification can be achieved. The most important application of affinity chromatography in biopharmaceuticals lies in the separation and purification of monoclonal antibody drugs (Batista-Viera et al. 2011). Due to the rapid development of upstream processes, the titer of recombinant proteins in cell culture has been greatly increased, making the bottleneck of biopharmaceuticals manufacturing move to downstream processes, and separation and purification has become a limiting factor for cost control. Affinity chromatography based on protein A has achieved great success in the purification process of monoclonal antibody drugs and is the preferred technique for capturing targets from cell lysates or supernatants. However, affinity chromatography also has some shortcomings, such as the high cost of chromatography packing, mainly because some biological ligands such as protein A itself are expensive, the service life of chromatography packing is short, and the biological ligands can be degraded by proteases in cell lysates. In addition, chromatographic packing has strong specificity and can only be used in specific applications, which limits the versatility of affinity chromatography.

Retention mechanism Affinity chromatography is a separation method based on the specific interaction between immobilized ligands and target biomolecules. The interaction between the ligand and the solute is very complicated. It is generally believed that this is the result of the combined effect of van der Waals, hydrophobic, electrostatic interactions, π–π-stacking and hydrogen bond between the ligand and the solute. This type of interaction is also known as molecular recognition. Ligand and solute must meet the requirements of space and energy matching in order to maximize the interaction energy and achieve high stability, high selectivity, and high specificity binding.

It is generally believed that the three-dimensional structure of a protein contains binding sites that can match those of the ligand to produce high-affinity binding. These binding sites are composed of a certain sequence of amino acid residues and have a specific three-dimensional structure with a certain shape. If the three-dimensional structure and active sites of the ligand are complementary to the binding sites of the protein, a highly stable complex can be formed. This recognition between biomolecules can be understood by the classic lock-and-key model and the induced fit model.

A large amount of structural data indicates that the binding of small molecule compounds such as drug molecules to the target occurs in the binding pockets located in the recessed areas of the protein surface. The ligand molecule is embedded in the binding pocket, and different functional groups in the molecule interact with the amino acid residues in the binding pocket. Obviously, the size of the ligand molecule and the distribution of functional groups must match the size of the binding pocket and the distribution of residues for high-affinity binding to occur. If the ligand molecule is too large or too small, the binding affinity will decrease or even disappear. At the same time, the orientation and distribution of functional groups in the ligand molecule are also important factors affecting the binding affinity. Any mismatch will lead to the dissociation of the ligand–target complex.

The specific interaction between protein and protein plays a key role in almost all biological processes, including signal transduction, cell regulation, immune response and others (Wells and McClendon 2007). The molecular structure of the protein–protein interaction interface has been clarified (Kastritis and Bonvin 2013). Just as there are active sites in enzymes, there are also active (or "warm") sites for interactions on the surface of proteins. The surface area involved in the protein-binding interface is relatively large, each protein is about 700–1500 Å^2, and the interacting surfaces are relatively flat (Janin 1999). Some interface residues contribute more to the binding, while the contribution of other residues is almost negligible. Residues that contribute more are called hot spots and residues that do not contribute are called cold spots. The protein-binding interface is composed of hot and cold spots. The hot spot is surrounded by weakly bound residues, forming an O-ring, excluding the solvent. Different hot spots in the bonding interface can act independently or synergistically. By studying the three structures of insulin dimer, trypsin–trypsin inhibitor complex, and oxyhemoglobin, Chothia and Janin (1975) found that the surface area of 1130–1720 Å^2 on the protein was removed from contact with water, indicating that hydrophobicity is the main factor for the stability of protein–protein interactions. The specificity of protein recognition determines the need for a high degree of shape and electrostatic complementarity and accuracy in the docking of the binding site (Jones and Thornton 1996). The binding constant of protein A-monoclonal antibody is around 10^8 M^{-1} while the binding constant of biotin and streptavidin is around 10^{12} M^{-1}.

Stationary phase Affinity chromatography packing consists of three parts: matrix, spacer, and ligand. Since affinity chromatography is mainly used for the separation and purification of proteins, the materials which are rigid, hydrophilic, and macroporous are selected as matrices. Commonly used affinity chromatography matrices can be divided into two categories. One type is organic and polysaccharide polymers, such as agarose, cellulose, dextran, and so on. The polysaccharide matrix has a large surface hydroxyl group density, good biocompatibility, low nonspecific adsorption to proteins, and stable physical and chemical properties and can withstand strong alkaline disinfection treatment. However, the mechanical strength is not high and cannot withstand high operating pressures. The second type of matrix is porous glass and silica gel. This type of matrix has high mechanical strength and can withstand higher column pressures. However, the nonspecific adsorption caused by the surface silanol groups is difficult to completely overcome; silica gel degrades under alkaline conditions, which limits the pH range of the mobile phase. The spacer arm is used to extend the distance between the ligand and the matrix, improve the accessibility of the ligand to the protein, and overcome the steric hindrance of the ligand. The length and structure of the spacer arm have an influence on the performance of the packing. Polar linear compounds with a length of 4–6 methylene groups can be used as spacer arms.

Ligand is connected to the matrix through the spacer arm and interacts with the protein. It is also the core part of the affinity chromatography packing. Ligands can be divided into two categories: one is a specific ligand, which has a binding ability to only one type of protein, and the other is called a group-specific ligand, which has a binding ability to a family of proteins. Specific ligands are generally complex biomolecules, which only specifically interact with certain biomolecules in the sample. For example, immunoaffinity chromatography with monoclonal antibodies as ligands can adsorb the corresponding antigens with high selectivity. Group-specific ligands can be simple small molecules, such as immobilized metal ions, which have affinity for proteins with histidine tags, or biological macromolecules, such as protein A, which can bind to a large class of monoclonal antibodies, or lectins, which recognize glycoproteins with specific glycan structures. In addition, other ligands widely used in affinity chromatography include antibiotics, hormones, plant hormones, steroids, substrates, coenzymes, amino acids, sugars, starch, tRNA, enzymes, proteins, antibodies, etc.

Proteins A, G, and L are natural or recombinant proteins derived from microorganisms and can specifically bind to immunoglobulins (including immunoglobulin G (IgG)) (Cope and McNeely 2009). IgG accounts for 80% of serum immunoglobulins. Protein A is a 42 kDa surface protein originally found in the cell wall of the bacteria *Staphylococcus aureus*. Protein G is an immunoglobulin-binding protein expressed in group C and G *Streptococcal* bacteria much like Protein A but with differing binding specificities. Protein L was first isolated from the surface of bacterial species *Peptostreptococcus magnus* and was found to bind immunoglobulins through L chain interaction, from which the name was suggested. All three proteins bind extensively to the IgG subclass. Generally, protein A is more suitable for cat, dog, rabbit, and pig IgG, while when purifying mouse or human IgG, protein G is usually preferred. The combination of protein A and G can also be used to purify various mammalian IgG samples. Since protein L binds to the κ light chain of immunoglobulin and these light chains are present in other immunoglobulins (i.e., IgG, IgM, IgA, and IgE), protein L is suitable for purifying different kinds of antibodies.

Lectins are carbohydrate-binding proteins that contain two or more carbohydrate-binding sites and can be divided into five categories based on their specificity for monosaccharides (Sharon and Lis 2004). They show the highest affinity for the following substances: mannose, galactose/*N*-acetylgalactosamine, *N*-acetylglucosamine, fucose, and *N*-acetylneuraminic acid. In this affinity technique, the protein binds to the immobilized lectin through its sugar moiety (N-linked or O-linked). Once the glycosylated protein is bound to the affinity carrier, unbound contaminants are washed away and the purified protein is eluted. Currently, many immobilized lectins are used to separate glycoproteins, glycolipids, polysaccharides, subcellular granules, and cells. Among them, concanavalin A (Con A) agarose and wheat germ agglutinin (WGA) agarose are most commonly used for glycoprotein purification.

Mobile phase In affinity chromatography, the choice of appropriate loading and elution buffers is very important. It is necessary to control the pH and ionic strength of the loading buffer so that the target protein has the maximum adsorption, while the impurities are basically not adsorbed. The elution buffer should ensure the high activity recovery of the target protein. The elution buffer can be applied by using a step change or a linear gradient. The release of the target by changing the pH, ionic strength, or mobile-phase composition is called nonspecific elution. Another elution method is to use a competitor, which displaces the target through mass action (i.e., biospecific elution).

Application Affinity chromatography is one of the most important techniques in biological separation and purification and can be applied to various stages of the downstream process of biological processing, such as capture, intermediate purification, and polishing. At present, as a platform technology for the purification of monoclonal antibodies in the biopharmaceutical industry, protein A affinity chromatography is used to directly capture the target antibody from the supernatant of mammalian cells. After one or two steps of ion exchange chromatography, high-purity antibody drugs that can meet the requirements of drug regulatory authorities can be obtained.

3.10 Chiral Chromatography

Chiral chromatography is a chromatographic method used to separate optical isomers or enantiomers with chiral centers (Wahab et al. 2018). There are many different chromatographic methods that can be used to separate such compounds, for example, by derivatizing the analytes to form diastereomeric products of two chiral compounds and then separating them by conventional chromatography, or use chiral additives in the mobile phase to form instantaneous diastereomeric products with the enantiomers in the sample and then separate them by conventional liquid chromatography or direct separation of enantiomers using chiral bonded stationary phases. At present, a large number of chiral bonded stationary phases are commercially available. They can be used for routine separation and analysis and are suitable for large-scale preparation. They have become the preferred method for chiral chromatography (Wahab et al. 2018).

Chiral compounds are widespread in nature. In fact, living organisms including cells are composed of many chiral biomolecules, such as L-amino acids, D-sugars, proteins, and nucleic acids. Chiral compounds are indistinguishable in most inanimate environments, but they exhibit different characteristics in biological systems. Two molecules that are mirror images of each other are called enantiomeric pairs. Under isotropic conditions, they have exactly the same physical and chemical properties. In anisotropic biological systems, the two enantiomers of chiral drugs may have completely different pharmacological effects. The molecular recognition between drug molecules and targets forms the basis of chiral chromatography: enantiomeric separation always means interaction between purely chiral compounds (selective agents or chiral ligands).

Chiral chromatography packing is usually silica gel with bonded optically active functional groups. Optically active substances used as chiral ligands contain molecules of various structures, from small amino acids, to medium-molecular-weight cyclic peptides, to biological macromolecules such as proteins. For example, (D or L) phenylurea, naphthylurea, phenylglycine, or leucine-bonded silica stationary phase can be used to separate the enantiomers including mandelic acid analogs, alkyl aryl sulfoxides, and propranolol isomer. Cyclodextrin-bonded silica gel can separate hydrophobic optical isomers by forming inclusion compounds in the cyclodextrin cavity. The D and L enantiomers of many compounds form clathrates with different binding strengths in the cavity and are retained to varying degrees to achieve separation. The enantiomers of tryptophan, ibuprofen, atenolol, and norgestrel and other pharmaceutically active compounds can be resolved using a cyclodextrin-functionalized silica stationary phase. Polysaccharides, such as cellulose and starch, are the most abundant optically active substances in nature. Its chiral recognition ability is greatly enhanced by derivatization reactions such as 3,5-dimethylphenyl carbamate. They are coated or bonded on the surface of silica gel to prepare the most widely used chiral chromatographic stationary phases. According to reports in the literature, about 90% of chiral liquid chromatography separations are done on polysaccharide chiral stationary phases.

Retention mechanism The key step of chiral recognition is the formation of diastereomeric complexes between enantiomers and chiral ligands. The binding constants of the two diastereomeric (enantiomer–selector) complexes are different, leading to their different migration speeds in the chromatographic system for separation (Berthod 2006). In order to explain the different pharmacological activities of chiral drugs, Easson and Stedman (1933) first proposed that at least three contact points between an asymmetric drug and its target are required, which is usually called a three-point interaction model. Dalgliesh (1952) applied it to the chiral separation of amino acids observed in paper chromatography. This model explains the differential binding of the chiral stationary phase (cellulose) to the three sites around the chiral centers of the two enantiomers. The three substituents on the chiral atom of one of the optical isomers can interact with the stationary phase at three positions at the same time, while the other isomer cannot complete the three-point interactions due to steric hindrance. Obviously, the direct mutual attraction and steric repulsion between the solute and the ligand are the key to the chiral separation. The key point in the three-point interaction model is that at least three interactions are required at the same time, and they should occur on three different substituents connected to the stereogenic center. Two different interactions with the same substituents only increase the solute--ligand-binding energy but do not increase the efficiency of chiral separation.

Stationary phase With the continuous development of chiral chromatography methods and applications, many different types of chiral stationary phases have emerged. At present, there are more than 200 commercially available stationary phases for high-performance liquid chromatography. According to their separation mechanism, chiral stationary phases can be divided into two categories: independent type and cooperative type. Chiral molecules used as chiral ligands in independent chiral stationary phases have independent chiral recognition capabilities. Brush-type, ligand exchange, cyclodextrins, crown ethers, and macrocyclic antibiotics chiral stationary phases all belong to this type. Synergistic chiral stationary phases are mainly chiral polymers, such as polysaccharides, proteins, and synthetic high-molecular polymers. The chiral recognition mechanism of this kind of stationary phase is more complicated, which is related to its supramolecular structure. Chiral stationary phases can also be classified according to the structure of chiral ligands: polysaccharides, brush type, or Pirkle type, proteins, cyclodextrins, crown ethers, ligand exchanges, macrocyclic antibiotics, and synthetic polymers.

Polysaccharide chiral stationary phase Polysaccharide derivatives are mainly prepared by esterification or etherification of the hydroxyl groups on the sugar unit. Acid chloride and ethylcyanate are the most commonly used derivatization reagents. In addition to the most commonly used cellulose and starch, chitin, chitosan, galactosamine, xylan, dextran, curdlan, and inulin are also used in the preparation of polysaccharide chiral stationary phases. In the past, polysaccharide stationary phases were mostly coating types, which were prepared by physically coating polysaccharide derivatives on a substrate such as silica gel. The biggest disadvantage of this type of stationary phase is poor solvent tolerance. Some commonly used organic solvents, such as chloroform, tetrahydrofuran, acetone, dichloromethane, etc., cannot be used because they can dissolve away the polysaccharide derivatives. In order to overcome this shortcoming, a variety of methods have been developed to chemically bond polysaccharide derivatives to chromatographic substrates, especially silica gel. Bonded polysaccharide stationary phases with good dissolution tolerance are commercially available and are widely used in the analytical and preparative scale separation of chiral compounds.

Cyclodextrin chiral stationary phase Cyclodextrin (CD) is a group of cyclic oligosaccharides that are slightly cone-shaped and are formed by connecting D-

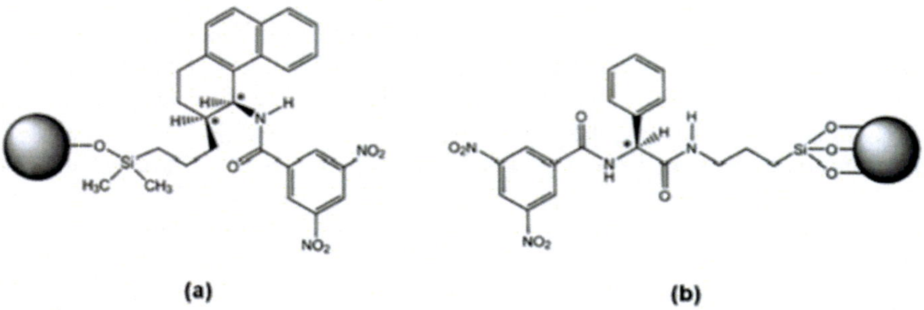

Fig. 3.4 Structural characteristics of the π-complex stationary phase. For example (**a**) *S,S* Whelk-O1, and (**b**) L-phenylglycine. (Reproduced from Fernandes et al. 2013)

glucopyranose units through α-1,4-glycosidic bonds. The most common α-CD, β-CD, and γ-CD contain 6, 7, and 8 glucose units, respectively. The primary hydroxyl group at the 6-position of the glucose unit is located at the end with the smaller opening, while the secondary hydroxyl groups at the 2-position and the 3-position are located at the other end with the larger opening. Since the polar hydroxyl groups are located on the outside of the CD cavity, the CD has a hydrophobic inner cavity and a hydrophilic outer surface. The unique structure of CD enables it to form inclusions with many compounds. At the same time, the CD molecule itself has chirality (there are five chiral centers on each sugar unit of CD). This dual role makes CD have good chiral recognition ability. CD is bonded to the chromatographic substrate such as silica gel through an amine bond, an amide bond, or an ether bond formed by the reaction of the hydroxyl group on the edge of the opening.

Brush-type chiral stationary phase Brush-type chiral stationary phase is a monolayer of chiral small organic molecules bonded to silica gel and other chromatographic substrates through suitable spacer arms. The Pirkle group has conducted a comprehensive study of such stationary phases (Pirkle and Pochapsky 1989). Therefore, they are also called Pirkle-type chiral stationary phases. The structural feature of the brush-shaped chiral stationary phase is that it contains at least one of the following groups near the chiral center: π-acid or π-basic aryl group; polar hydrogen bond donor–acceptor; polar groups that form dipole interactions; and bulky nonpolar group, providing steric hindrance, van der Waals effect, or structure control effect (Fig. 3.4). The brush-type chiral stationary phase has a definite chemical structure, a wide range of applications, and a well-defined chiral recognition mechanism, which can be used to predict the resolution and elution order of enantiomers and determine its configuration. It is suitable for analytical separation and preparation separation.

Protein chiral stationary phase Proteins and glycoproteins are composed of chiral amino acids and sugar units and have chiral recognition capabilities. Similar to the preparation of cyclodextrin stationary phases, protein stationary phases are bonded to chromatographic substrates such as silica gel through the reaction of amino, carboxyl, and sulfhydryl groups of protein surface residues to form a protein chiral

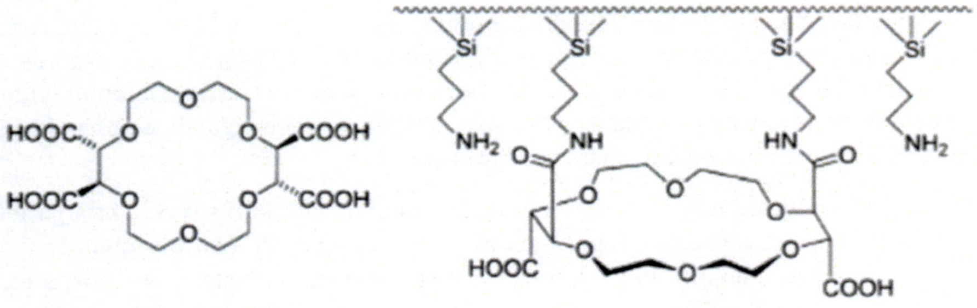

Fig. 3.5 (**a**) Tetracarboxylic crown ether. (**b**) Carboxylic crown ether chiral stationary phase. (Reproduced from Hyun and Cho 2005)

stationary phase. This type of stationary phase has the following characteristics: (1) the use of a buffer containing a low concentration of organic solvent as the mobile phase and (2) has multiple interaction sites and a wide range of chiral recognition capabilities. Its disadvantages include low sample loading capacity, poor durability, and chiral recognition ability easily affected by the mobile-phase composition. Currently, only a few proteins are used as chiral ligands, mainly bovine serum albumin, human serum albumin, α1-acid glycoprotein, and chicken egg mucin.

Crown ether chiral stationary phase Crown ether is a kind of macrocyclic compound with a certain size cavity. The outer side of the ring is a lipophilic methylene group, and the inner side is an electron-rich oxygen atom. It can form inclusion complexes with metal ions, amine cations, etc. Chiral units such as (1,1-′-binaphthol), tartaric acid, etc. are incorporated into the cyclic structure of crown ethers to obtain chiral crown ethers with chiral recognition capabilities. The chiral amine cation forms an inclusion complex with the crown ether. The hydrophobic interaction, electrostatic interaction, steric hindrance, and other non-covalent interactions between the solute and the substituents of the chiral ligand strengthen the stability of the complex and provide the three-point interaction necessary for chiral recognition. Crown ether chiral stationary phases are mainly used for the separation of primary amine chiral compounds, including amino acids and their derivatives, amino alcohols, as well as some secondary amines and non-amine chiral compounds (Fig. 3.5).

Ligand exchange chiral stationary phase Ligand exchange chromatography is a chromatographic method that uses metal ions immobilized on a stationary phase to form coordination complexes with solute molecules for separation. Chiral ligand exchange stationary phases are prepared by coating or bonding a chiral ligand coordinated with a metal ion, such as an amino acid, on a chromatographic matrix. In the process of chromatographic separation, the metal ion and the chiral ligand and the enantiomers to be separated form two ternary complexes that are diastereomers of each other. Due to the steric hindrance effect, their stability constants may differ, leading to their separation. Among the various central metal ions that can form complexes, Cu(II) complexes are usually the most stable, and Cu(II) can form highly stereoselective ternary complexes with a variety of receptor ligands and has been widely used in chiral ligand exchange chromatography.

The immobilized chiral ligands are usually amino acid derivatives. Through its amino group and carboxyl group, the chiral ligand occupies two positions in the copper ion coordination sphere. Small water molecules occupy two positions, leaving two positions for the enantiomeric solute. Enantiomeric solutes must be able to form bidentate chelates, which are α- or β-amino acids, amino alcohols, hydroxy acids, diamines, amino amides, and dicarboxylic acids. The two interactions described are necessary, but not enough. The third interaction required for chiral recognition is provided by the hydrophobic, polar, or spatial interaction with the chiral ligand. The large group and/or rigid group located near the chiral center of the solute will greatly enhance the chiral recognition ability. Ligand exchange chromatography is particularly suitable for the separation of enantiomers such as amino acids and their derivatives, hydroxy acids, and polypeptides.

Ion exchange chiral stationary phase Two cinchona alkaloid derivatives (mainly carbamate compounds of quinine and quinidine) can be used to study specific chiral recognition mechanisms. Quinine is a natural alkaloid extracted from the bark of Cinchona tree in South America and is commonly used as an antimalarial drug. It has 8*S* and 9*R* configurations. Quinidine, the stereoisomer of quinine, also exists in cinchona bark. It has 8*R* and 9*S* configurations. These two alkaloids can be easily derivatized to prepare two chiral stationary phases with opposite configurations.

Both quinine and quinidine carbamate derivatives contain tertiary amine groups, which are positively charged under neutral pH conditions and can interact with the acidic groups of the solute through ion exchange mechanisms. The quinolinyl group in the ligand molecule interacts with the aromatic ring of the solute, and the carbamoyl substituent has a dipole–dipole interaction and hydrogen bond or steric hindrance with the polar group of the solute. *N*-3,5-Diaminotribenzoyl (DNB) derivatized amino acid enantiomers are well separated on the quinine stationary phase. The analysis of the three-point action required for chiral recognition is as follows. The first interaction that leads to molecular docking is the ionic attraction between the negative charge of the carboxylic acid of the DNB amino acid and the positive charge of the quinine ammonium group. The second interaction is the interaction between the DNB π-acid group and the quinoline π-basic group of quinine. The third type of interaction is the repulsive steric hindrance between amino acid substituents (e.g., the phenyl group of phenylalanine, the isobutyl group of leucine, and the methyl group of alanine) and the bulky tert-butyl substituent of carbamate group on the quinine derivative (Fig. 3.6).

The introduction of acidic groups such as carboxylic acid or sulfonic acid into quinine derivatives can prepare amphoteric ion exchange chiral stationary phases. Such amphoteric chiral stationary phases show extensive chiral recognition ability

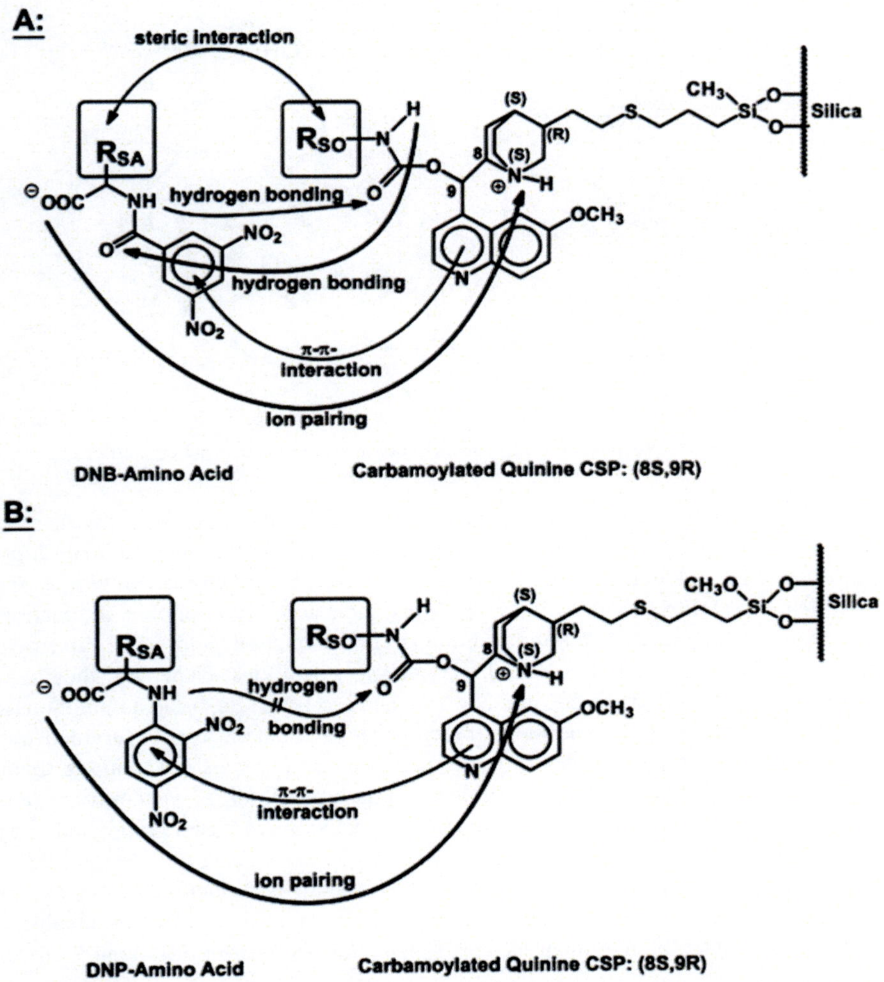

Fig. 3.6 Tentative chiral host–guest complexation model. Intermolecular interactions of (**a**) DNB amino acids and (**b**) DNP amino acids with the carbamoylated quinine selector (RSO: carbamate substituent, e.g., tert.-butyl, RSA: amino acid side chain). (Reproduced from Mandl et al. 1999)

for ionic chiral compounds, including chiral acids, chiral amines, and amino acids and small peptides. Weakly polar organic mobile phases are recommended for use with amphoteric ion exchange chiral stationary phase. Under this condition, three ion exchange modes can be realized: the anion exchange mode is used to separate chiral acids, cation exchange mode is used to separate chiral amines, and amphoteric ion exchange mode is used to separate amphoteric chiral compounds.

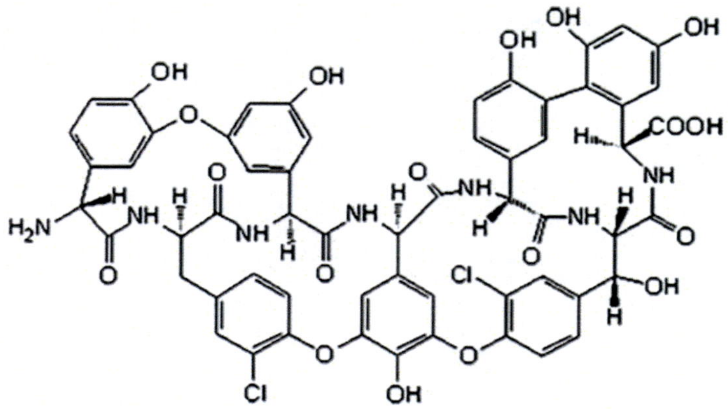

Fig. 3.7 The structure of teicoplanin aglycone. (Reproduced from Vadinská et al. 2008)

Macrocyclic antibiotics chiral stationary phase The macrocyclic antibiotic molecules contain cyclic structures of different numbers and shapes and contain multiple chiral centers and multiple functional groups, which can produce multiple and complex interactions with solutes, such as inclusion, hydrophobic interaction, dipole–dipole interaction, hydrogen bond, ionic interaction, and steric repulsion, thus showing a wide range of chiral recognition capabilities. Generally, they have molecular weights between 600 and 2000 Da and can be acidic, basic, or amphoteric compounds. Macrocyclic antibiotics chiral stationary phases can be divided into three categories: glycopeptides, mycotoxins, and aminoglycosides. Among them, glycopeptide antibiotic chiral stationary phases including vancomycin and teicoplanin (Fig. 3.7) are the most important. Vancomycin-based chiral stationary phases have been used to separate pyridone enantiomers, aryl dihydropyrimidine carboxylates, citalopram and its demethylated metabolite enantiomers in human plasma, α-amino acids and DNS-amino acids, cyclic imide, ergot alkaloids, diclofenac, and tetalol in human plasma. Teicoplanin-based chiral stationary phases have been successfully used in the separation and determination of natural amino acids and dipeptides, unnatural amino acids, and many other enantiomers, amino acid derivatives, secondary aromatic amino acids, 2-amino analogues of indole phytotoxins, salbutamol, and its metabolites in biological samples and other enantiomers.

Mobile phase In chiral chromatography, there are many kinds of stationary phases, and the appropriate chiral stationary phase and corresponding elution mode should be selected according to the nature of the solute. The elution modes can be divided into three categories: normal phase, reverse phase, and polar organic. The normal phase and reverse phase modes are similar in composition to the mobile phase used in normal phase and reversed-phase chromatography, consisting of *n*-hexane-polar modifier and polar organic solvent-aqueous solution, respectively. The polar organic mode is similar to the mobile phase used in hydrophilic interaction chromatography, mainly polar organic solvents such as methanol and acetonitrile, and a small amount of organic acid or base is added to it. For example, the separation on the polysaccharide stationary phase is usually carried out in the normal phase chromatography mode, and the nonpolar solvent *n*-hexane should be selected as the mobile phase; the separation on the cyclodextrin stationary phase should be carried out in the reversed-phase chromatography mode, and the water–polar solvent mixture should be selected as the mobile phase; while the separation on the macrocyclic antibiotic stationary phase is carried out in the hydrophilic interaction chromatography mode, a polar organic solvent should be selected as the mobile phase; the separation on the cinchona alkaloid-based amphoteric stationary phase is carried out in a hydrophilic/ion exchange mixed-mode, and the corresponding polar organic solvent and a buffer containing a certain salt concentration are used as the mobile phase.

Today, there are a large number of chiral stationary phases (CSP) commercially available. Each CSP has unique operating conditions. Information about all mobile-phase modes available for CSP is essential for successful chiral separation screening. The comparison between different CSPs of analytes using the same mobile phase is not the best screening technique. Several different mobile-phase modes are used in chiral liquid-phase separation. The correct choice of mobile-phase mode depends on the physical and chemical properties of the CSP and racemic mixture. These modes include reverse phase, normal phase, polar organic, and polar ion modes.

The reverse phase (RP) mode uses organic modifiers (such as methanol, acetonitrile, or tetrahydrofuran) and aqueous buffers (pH 3.0–7.0). The correct choice of buffers and organic components is essential for the separation of enantiomers. Buffer type and pH value can cause the difference between no separation and separation of enantiomers. Separation differences such as elution strength and peak efficiency can be obtained by using methanol, ethanol, or acetonitrile. The composition of the RP solution can contain high or low organic content. The optimal mobile-phase composition depends on the solubility of the analyte and the chemistry of the column. The main interactions used for chiral recognition in RP are ionic interactions, hydrogen bonds, steric effects, hydrophobic interactions, and in some cases also include pi-pi interactions. RP is used for all types of compounds. CSP categories that use the RP mode elution include macrocyclic glycopeptides and polysaccharide phases, cyclic oligosaccharides, and crown ethers.

Traditionally, the normal phase (NP) eluent consists of nonpolar hydrocarbons (such as hexane, heptane, or methyl tert-butyl ether (MTBE)) and polar components (such as ethanol or isopropanol). Due to safety issues related to hexane toxicity, heptane is becoming the hydrocarbon of choice. The interactions that control chiral recognition are hydrogen bonding, π–π, steric, and dipole–dipole. Neutral compounds are most suitable for NP elution. CSP categories that use the NP mode elution include π-complex phases, crown ethers, derivatized cyclic oligosaccharides, macrocyclic glycopeptides, and derivatized polysaccharide phases.

Polar organic mode (POM) elution uses only organic solvents for the mobile-phase composition. Solvents such as methanol or ethanol are added to acetonitrile in different amounts. For example, the screening can start with a high acetonitrile content and use MeOH to adjust the retention factor. In POM, hydrogen bond and π–π, steric hindrance, and dipole–dipole play a vital role in chiral recognition.

Neutral, acidic, or basic compounds are suitable for this mode (usually two hydrogen bond groups are beneficial). The types of CSP that can be run in POM are π-complex phase, cyclic oligosaccharide, macrocyclic glycopeptide, and polysaccharide phase. Polar ion mode (PIM) eluent consists of pure methanol and 0.01–0.50% acid and base. The most commonly used acid is acetic acid and the base is triethylamine. Acids and bases play a key role in controlling the ionic effects that contribute to the chiral recognition mechanism. By using volatile substances such as ammonium acetate or ammonium formate, both POM and PIM eluents can be transferred to electrospray ionization mass spectrometry (ESI-MS) technology. In PIM, the main interaction between analyte and CSP is ionic effect (with ionizable analyte), but hydrogen bonding, steric hindrance, and π–π interaction can play a role in chiral recognition.

Application Three-quarters of the US Food and Drug Adminstration (FDA)-approved small molecule drugs are single enantiomers, some of which are separated in multiton levels by chiral chromatography during the manufacturing process. The FDA stipulates: "The concentration of each isomer must be determined, and the limits of all isomer components, impurities and contaminants must be determined for the compounds to be used in clinical trials before clinical trials." These requirements make chiral chromatography indispensable. Modern separation sciences in the fields of pharmaceuticals, agricultural chemistry, petrochemicals, nutraceuticals, and other chemical manufacturing often use chiral liquid chromatography to identify and purify important compounds for commercial and biological benefits. The latest developments in the past 5 years emphasize the use of ultra-high-efficiency supports (such as fully porous silica particles with a narrow particle size distribution (NPSD-FPP) or surface porous particles (SPPs) with a nominal smaller than 2μm) to increase the number of plates (Patel et al. 2015; Ismail et al. 2016). The surface porous particles have a solid core on which a thin layer of porous shell is formed, while the fully porous particles consist of the entire porous framework. These advances in chiral chromatography mean that throughput, robustness, selectivity, and efficiency can be significantly improved in many industrial applications.

3.11 Summary

According to the mechanism of chromatographic retention, existing liquid chromatography methods can be divided into two categories: single mode and multimode. Single-mode chromatography involves a single type of interaction between solute and ligand, such as polar interaction in normal phase chromatography, hydrophobic interaction in reversed-phase chromatography, and electrostatic interaction in ion exchange chromatography. In contrast to this, multimode chromatography involves a variety of interaction types between solutes and ligands, such as the recognition of biomolecules in affinity chromatography and the recognition of chiral molecules in chiral chromatography. These molecular recognitions are all based on the binding site matching, geometric fit, or complexation between the interaction pairs, and the superposition of multiple types of interactions. Therefore, in fact, it uses the multivalent effect in the biological system for efficient separation and purification. Correspondingly, the stationary phase of single-mode chromatography is composed of a single type of functional group, and the binding of these functional groups with the solute determines the retention of the solute. The residual groups or link arms on the surface of the matrix have a secondary effect on the retention of solutes and can usually be eliminated by adjusting the composition of the mobile phase. For example, the peak tailing effect of residual silanol groups on basic compounds in reversed-phase chromatography can be eliminated by adjusting the pH of the mobile phase or adding amino modulators. In ion exchange chromatography, the peak tailing caused by the hydrophobic groups of the matrix or spacer can be eliminated by adding organic solvents to the mobile phase.

Multimode chromatography can provide higher selectivity than single-mode chromatography. Many such examples are provided in affinity chromatography and chiral chromatography. For example, protein A chromatography can specifically adsorb the Fc fragment of the antibody and directly separate it from the cell culture medium. One-step chromatography can achieve a purity of more than 99%. Analysis of the crystal structure of protein A and Fc fragments provided information about the binding interface, confirming the characteristics of multipoint binding of amino acid residues between the two. The disadvantage of using natural biomolecules as multimodal ligands is that they are easily degraded in the presence of proteases, and the leached ligands enter the product, causing contamination. In addition, not all compounds can find matching bioselective ligands. Therefore, it is a challenging task in the current chromatographic field to find artificial ligands similar to the mechanism of action of biomolecular ligands to achieve multimodal separation through synthetic pathways.

References

Abbott SR, Berg JR, Achener P et al (1973) Chromatographic reproducibility in high-performance liquid chromatographic gradient elution. J Chromatogr 126:421–437

Ares AM, Bernal J (2012) Hydrophilic interaction chromatography in drug analysis. Cent Eur J Chem 10:534–553

Batista-Viera F, Jan-Christer Janson J-C, Carlsson J (2011) Affinity chromatography. In: Janson J-C (ed) Protein purification: principles, high resolution methods, and applications, 3rd edn. Wiley, Hoboken, NJ

Bernal J, Ares AM, Pól J et al (2011) Hydrophilic interaction liquid chromatography in food analysis. J Chromatogr A 1218:7438–7452

Berthod A (2006) Chiral recognition mechanisms. Anal Chem 78:2093–2099

Bo H, Wang J, Chen Q et al (2013) Using a single hydrophobic interaction chromatography to purify pharmaceutical-grade supercoiled plasmid DNA from other isoforms. Pharm Biol 51:42–48

Cavazzini A, Catani M, Felinger A (2017) Chapter 6 Hydrophilic interaction liquid chromatography. In: Fanali S, Haddad PR, Poole CF, Riekkola M-L (eds) Liquid chromatography: fundamentals and instrumentation, 2nd edn. Elsevier, Amsterdam

Chothia C, Janin J (1975) Principles of protein-protein recognition. Nature 256:705–708
Cope L, McNeely T (2009) Bacteria immunoglobulin-binding proteins: biology and practical applications. In: An Z (ed) Therapeutic monoclonal antibodies: from bench to clinic. Wiley, Hoboken, NJ
Cuatrecasas P, Wilchek M, Anfinsen CB (1968) Selective enzyme purification by affinity chromatography. Proc Natl Acad Sci U S A 68:636–643
Dalgliesh CE (1952) The optical resolution of aromatic amino-acids on paper chromatograms. J Chem Soc 137:3940–3942
Davankov VA, Navratil JD, Walton HF (1988) Ligand exchange chromatography. CRC Press, Roca Raton, FL
Easson EH, Stedman E (1933) Studies on the relationship between chemical constitution and physiological action: molecular dissymmetry and physiological activity. Biochem J 27:1257–1266
Eriksson K-O, Belew M (2011) Hydrophobic interaction chromatography. In: Janson J-C (ed) Protein purification: principles, high resolution methods, and applications, 3rd edn. Wiley, Hoboken, NJ
Fernandes C, Palmerira A, Santos A et al (2013) Enantioresolution of chiral derivatives of xanthones on (S,S)-Whelk-O1 and L-phenylglycine stationary phases and chiral recognition: mechanism by docking approach for (S,S)-Whelk-O1. Chirality 25:89–100
Hage DS, Anguizola JA, Li R et al (2017) Chapter 12 Affinity chromatography. In: Fanali S, Haddad PR, Poole CF, Riekkola M-L (eds) Liquid chromatography: fundamentals and instrumentation, 2nd edn. Elsevier, Amsterdam
Haverick M, Mengisen S, Shameem M et al (2014) Separation of mAbs molecular variants by analytical hydrophobic interaction chromatography HPLC: overview and applications. MAbs 6: 852–858
Hong P, Koza S, Bouvier ESP (2012) Size-exclusion chromatography for the analysis of protein biotherapeutics and their aggregates. J Liq Chromatogr Relat Technol 35:2923–2950
Horvath C, Melander W (1977) Liquid chromatography with hydrocarbonaceous bonded phases: theory and practice of reversed phase chromatography. J Chromatogr Sci 15:393–404
Horvath C, Melander W, Molnar I (1976) Solvophobic interactions in liquid chromatography with nonpolar stationary. J Chromatogr 125:129–156
Hyun MH, Cho YJ (2005) Preparation and application of a chiral stationary phase based on (+)-(18-crown-6)-2,3,11,12-tetracarboxylic acid without extra free aminopropyl groups on silica surface. J Sep Sci 28:31–38
Ismail OH, Ciogli A, Villani C et al (2016) Ultra-fast high-efficiency enantioseparations by means of a teicoplanin-based chiral stationary phase made on sub-2μm totally porous silica particles of narrow size distribution. J Chromatogr A 1427:55–68
Janin J (1999) Wet and dry interfaces: the role of solvent in protein-protein and protein-DNA recognition. Structure 7(12):R277–R279
Jennissen HP (1976) Evidence for negative cooperativity in the adsorption of phosphorylase b on hydrophobic agaroses. Biochemistry 15:5683–5692
Jennissen HP (1986) Protein binding to two-dimensional hydrophobic binding the lattices: sorption kinetics of phosphorylase b on immobilized butyl residues. J Colloid Interface Sci 111:570–586
Jennissen HP, Heilmeyer LMG (1975) General aspects of hydrophobic chromatography. Adsorption and elution characteristics of some skeletal muscle enzymes. Biochemistry 14:754–759
Jones S, Thornton JM (1996) Principles of protein-protein interactions. Proc Natl Acad Sci U S A 93:13–20
Kagedal L (2011) Immobilized metal ion affinity chromatography. In: Janson J-C (ed) Protein purification: principles, high resolution methods, and applications, 3rd edn. Wiley, Hoboken, NJ
Kahsay G, Song H, Van Schepdael A et al (2014) Hydrophilic interaction chromatography (HILIC) in the analysis of antibiotics. J Pharm Biomed Anal 87:142–154
Karger BL, Gant JR, Hartkopf A et al (1976) Hydrophobic effects in reversed-phase liquid chromatography. J Chromatogr 128:65–78

Karlsson E, Hirsh I (2011) Ion exchange chromatography. In: Janson J-C (ed) Protein purification: principles, high resolution methods, and applications, 3rd edn. Wiley, Hoboken, NJ

Kastritis PL, Bonvin AMJJ (2013) On the binding affinity of macromolecular interactions: daring to ask why proteins interact. J R Soc Interface 10:20120835

Kauzmann W (1959) Some factors in the interpretation of protein denaturation. Adv Protein Chem 14:1–63

Li Z, Han J, Sun SA et al (2016) Hydrophilic interaction liquid chromatography tandem mass spectrometry: an attractive and prospective method for quantitative bioanalysis in drug metabolism. Curr Drug Metab 17:386–400

Lu Y, Williamson B, Gillespie R (2009) Recent advancement in application of hydrophobic interaction chromatography for aggregate removal in industrial purification process. Curr Pharm Biotechnol 10(4):427–433

Mandl A, Nicoletti L, Lämmerhofer M et al (1999) Quinine- versus carbamoylated quinine-based chiral anion exchangers: a comparison regarding enantioselectivity for N-protected amino acids and other chiral acids. J Chromatogr A 858:1–11

Martell AE, Smith RM (2004) NIST critically selected stability constants of metal complexes. NIST Standard Reference Database 46, Version 8.0

Melander W, Horvath C (1977) Salt effects on hydrophobic interaction in precipitation and chromatography of proteins: an interpretation of lyotropic series. Arch Biochem Biophys 183: 200–215

Melander W, Corradini D, Horvath C (1984) Salt mediated retention of proteins in hydrophobic interaction chromatography. Application of solvophobic theory. J Chromatogr 317:67–85

Nesterenko PN, Jones P, Paull B (2011) High performance chelation ion chromatography. RSC Publishing, Cambridge

Noga S, Felinger A, Buszewski B (2012) Hydrophilic interaction liquid chromatography and per aqueous liquid chromatography in fungicides analysis. J AOAC Int 95:1362–1370

Ouyang J (2013) Drug-to-antibody ratio (DAR) and drug load distribution by hydrophobic interaction chromatography and reversed phase high-performance liquid chromatography. Methods Mol Biol 1045:275–283

Patel DC, Breitbach ZS, Wahab MF et al (2015) Gone in seconds: praxis, performance, and peculiarities of ultrafast chiral liquid chromatography with superficially porous particles. Anal Chem 87:9137–9148

Pirkle WH, Pochapsky TC (1989) Considerations of chiral recognition relevant to the liquid chromatography separation of enantiomers. Chem Rev 89:347–362

Pyrzynska K, Sentkowska A (2015) Recent developments in the HPLC separation of phenolic food compounds. Crit Rev Anal Chem 45:41–51

Queiroz JA, Tomaz CT, Cabral JMS (2001) Hydrophobic interaction chromatography of proteins. J Biotechnol 87:143–159

Ruhaa LR, Huhn C, Waterreus WJ (2008) Hydrophilic interaction chromatography-based high-throughput sample preparation method for n-glycan analysis from total human plasma glycoproteins. Anal Chem 80:6119–6126

Schlichtherle-Cerny H, Affolter M, Cerny C (2003) Hydrophilic interaction liquid chromatography coupled to electrospray mass spectrometry of small polar compounds in food analysis. Anal Chem 75:2349–2354

Sharon N, Lis H (2004) History of lectins: from hemagglutinins to biological recognition molecules. Glycobiology 14:53R–62R

Shukla AA, Thömmes J (2010) Recent advances in large scale production of monoclonal antibodies and related proteins. Trends Biotechnol 28:253–256

Sinanoglu O, Abdulnur S (1965) Effect of water and other solvents on the structure of biopolymers. Fed Proc 24:12–23

Tomaz CT (2017) Chapter 7 Hydrophobic interaction chromatography. In: Fanali S, Haddad PR, Poole CF, Riekkola M-L (eds) Liquid chromatography: fundamentals and instrumentation, 2nd edn. Elsevier, Amsterdam

Treadway JW, Philibert GS, Olesik SV (2011) Enhanced fluidity liquid chromatography for hydrophilic interaction separation of nucleosides. J Chromatogr A 1218:5897–5902
Vadinská M, Bosáková Z, Tesařová E et al (2008) LC with a teicoplanin aglycone chiral sorbent for the separation of the enantiomers of non-steroidal anti-inflammatory drugs: an evaluation of chiral capillary columns. Chromagraphia 67:33–40
van Nuijs ALN, Tarcomnicu I, Covaci A (2011) Application of hydrophilic interaction chromatography for the analysis of polar contaminants in food and environmental samples. J Chromatogr A 1218:5964–5974
Wahab MF, Weatherly CA, Patil RA et al (2018) Chapter 12 Chiral liquid chromatography. In: Polavarapu PL (ed) Chiral analysis advances in spectroscopy, chromatography and emerging methods, 2nd edn. Elsevier, Amsterdam
Wang F, Li S, Zheng Y (2007) Studies on the activation condition and stability of pork pancreas carboxypeptidase B in the process of extraction and purification. Chin J Biochem Pharm 28: 250–253
Wells JA, McClendon CL (2007) Reaching for high hanging fruit in drug discovery at protein–protein interfaces. Nature 450:1001–1009
Wiggins B, Shin LL, Yamaguchi H et al (2015) Characterization of cysteine-linked conjugation profiles of immunoglobin G1 and immunoglobin G2 antibody-drug conjugates. J Pharm Sci 104:1362–1372
Zhou T, Lucy CA (2008) Hydrophilic interaction chromatography of nucleotides and their pathway intermediates on titania. J Chromatogr A 1187:87–93
Zywicki B, Catchpole G, Draper J et al (2005) Comparison of rapid liquid chromatography-electrospray ionization-tandem mass spectrometry methods for determination of glycoalkaloids in transgenic field-grown potatoes. Anal Biochem 336:178–186

Chapter 4
Chromatographic Matrices

Abstract Porous materials are used as the chromatographic matrix to give the column packing a certain geometric shape and mechanical strength. The supporting matrix can be roughly classified into inorganic materials, synthetic polymers, and polysaccharides. Inorganic materials, such as silica and porous glass, provide high mechanical strength for high-pressure operation, while polysaccharides, such as agarose and cellulose, provide very hydrophilic materials for protein purification. In this chapter, we will review their preparation, characterization, and functionalization.

4.1 Introduction

The matrix is the solid support part of the chromatographic packing, which gives the packing a certain geometric shape and mechanical strength. At the same time, it forms a certain selective adsorbent through the ligands bound on the surface. The choice of matrix is essential to maximize chromatographic separation performance. Ideally, the matrix should have the following properties:

- Suitable shape in order to obtain a uniformly packed column bed structure.
- Suitable particle size and distribution to facilitate the flow and mass transfer in the column bed structure.
- Larger specific surface area to bond more ligand molecules.
- Appropriate reactivity to facilitate the derivatization reaction.
- Appropriate mechanical strength for use under high flow rate or pressure.
- Chemically inert. It will not dissolve, swell, react, or produce nonspecific adsorption under operating conditions.

Among the ideal properties of the abovementioned substrates, some interfere with each other, for example, the specific surface area and the pore size are opposite to each other, and increasing the pore size results in a decrease in the specific surface area. Surface reactivity and chemical inertness are also contradictory to each other. Therefore, so far, there is no material that can fully meet the requirements of the ideal medium. In fact, the medium used has compromised in some aspects.

Q.-H. Wan, *Mixed-Mode Chromatography*,
https://doi.org/10.1007/978-981-16-5485-5_4

The matrix used as a chromatographic carrier can be roughly divided into inorganic materials, synthetic organic polymers and polysaccharides. All three types of materials can be used for atmospheric or low-pressure chromatography, but only some of them can be used for high-performance liquid chromatography. The matrix materials to be covered in this chapter are as follows:

Inorganic materials

- Porous silica gel
- Hydroxyapatite
- Controlled pore glass

Synthetic organic polymers

- Polystyrene
- Polymethacrylate

Polysaccharides

- Agarose
- Dextran
- Cellulose

4.2 Silica

4.2.1 Structure

According to the concept of ideal chromatography packing described above, silica gel is almost an ideal matrix material. This is because in addition to good mechanical strength, easy-to-control pore structure and specific surface area, good chemical stability, and thermal stability, silica gel has a distinct advantage that its surface is rich in silanol, which can be used for various surface chemical modifications and for the introduction of required functional groups. It is not surprised that silica gel is the earliest developed and most widely used packing material in liquid chromatography.

The chemical composition of porous silica gel is $SiO_2 \times H_2O$, which belongs to an amorphous structure. The basic unit is the accumulation of silica nanoparticles of different sizes in the space to form the framework of silica gel. The gaps between the stacked particles are the pores of the silica gel. The water in silica gel can be classified as adsorbed water and structured water. The latter is bonded to silicon atoms in the form of hydroxyl groups and covers the surface of silica gel.

Silica gel is formed by silicic acid sol through the polycondensation reaction of orthosilicic acid. Submicroscopic primary particles are formed, which retain the micelles of the initial acid and are bound by Si–O–Si siloxane bonds in their interior. Each primary particle touches the surface of several neighboring particles, thus forming an aggregate containing capillary spaces (cavities with a diameter of 30–300 Å). After drying, they form a large specific inner surface of several hundred m^2/g. Due to the presence of closely adjacent Si–OH groups (in addition to other types of OH groups) on the surface, siloxane bonds are formed during the activation of the silica. Under elevated temperature (such as 200–400 °C), the various types of hydroxyl groups on the surface of silica gel particles condense and eliminated. This will have some impact on the adsorption performance of silica gel. The presence of free hydroxyl groups on the surface is essential for optimal separation activity of macroporous silica gels with an average pore size (e.g., 100 Å) and a medium specific surface area (e.g., 300 m^2/g).

4.2.2 Preparation

Emulsion polymerization Inverse emulsion polymerization is a classic method for preparing spherical silica gel. In this method, the silica gel precursor, usually silica sol, is dispersed in an organic solvent that is immiscible with water. Under the action of an acidic or alkaline catalyst, the precursor molecules undergo a hydrolysis and condensation reaction in the emulsion droplets to form a silica gel with a network structure. After aging, washing, drying, calcination, and sieving, the porous spherical silica gel with the targeted particle size can be obtained. The main consideration of this method is to select suitable organic solvents and emulsion stabilizers in order to obtain a stable emulsion system under heating conditions. The size of the silica gel particles is determined by the size of the emulsion droplets, and the size of the droplets is determined by the stirring speed. The pore structure of silica gel particles is affected by many factors in the process of synthesis and posttreatment such as precursor concentration, catalyst type, and pH value. According to the type of precursor, the method can be subdivided into silicate method and organosilane hydrolysis method.

The silicate acidification method is to hydrolyze soluble silicate such as water glass under acidic conditions to generate monosilicic acid. Monosilicic acid molecules are prone to polycondensation to form polysilicic acid. By controlling the experimental conditions, such as pH, silicate concentration, salt concentration, and temperature, different types of silica sols can be prepared. Among them, pH has a great influence on the polycondensation reaction of polysilicic acid. Under acidic conditions of pH 2.0–4.0, the rate of hydrolysis of silicate is greater than the rate of polycondensation, resulting in polysilicic acid. Under acidic conditions, polysilicic acid has not yet reached equilibrium and is unstable. As the pH increases, the rate of polycondensation increases. When under alkaline conditions, polysilicic acids cross-link, condense, and form a stable silica sol. On the other hand, since the isoelectric point of silica sol is about pH 3.0, the surface of the colloid under alkaline conditions is negatively charged, and the sol is stabilized by formation of an electric double layer on the surface of the particles. The stability of the electric double layer is not only related to pH but also affected by the electrolyte concentration in the solution. Increasing the electrolyte concentration can reduce the thickness of the diffusion layer, leading to sol gelation.

The organosilane hydrolysis method is a hydrolysis reaction in water under acidic conditions, and the hydrolyzed product polycondenses to form a sol. There are many types of hydrolyzable organosilanes, but tetraethoxysilane (TEOS) is commonly used in the preparation of silica gel. The operation of the hydrolysis reaction is simple. Generally, an appropriate amount of HCl is added to the ethanol solution of TEOS and stirred for a certain period of time to obtain a silica sol. Under acidic conditions, TEOS first generates ethoxysilanol groups, then dehydrates and condenses to form dimers between molecules, and further polycondenses into chain-like polyethoxysilane (PES). In this reaction, the amount of water and ethanol is the key to the polycondensation product. One mole of TEOS needs two moles of water to be completely hydrolyzed. When the amount of water is less than the stoichiometric ratio, TEOS cannot be completely hydrolyzed. The silanol group that is initially generated needs to be dehydrated and condensed to produce water to continue the hydrolysis. Water becomes a limiting factor for the hydrolysis and polycondensation reaction. Therefore, precise control of the amount of water is the key to preparing silica sol. From the perspective of reaction equilibrium, ethanol plays a dual role in the hydrolysis and polycondensation reaction; on the one hand, ethanol is a good solvent for TEOS and PES and acts as a dispersant during the reaction, diluting the reactants and reducing the reaction rate; on the other hand, ethanol is also a product of the hydrolysis and polycondensation reaction, which to some extent inhibits the hydrolysis and polycondensation. Therefore, the dual roles of ethanol work together to affect the size of colloidal silica sol particles.

Sol–gel method The sol–gel process is the most commonly used method for preparing high-purity silica gel microspheres. Usually, the silicon source is hydrolyzed and polycondensed under acidic or alkaline conditions to obtain silica gel microspheres. Under acidic conditions, the hydrolysis rate of the silicon source is greater than its polycondensation rate, and it is easy to form a microporous structure; while under the weakly alkaline conditions of pH 7–9, the polycondensation rate of the silicon source is greater than its hydrolysis rate, and finally a gel-like structure is formed. The microspheres prepared by the traditional sol–gel method are usually micron or submicron nonporous silica gel microspheres. Unger and Scharf (1976) used the two-step catalyzed sol–gel method to prepare large-particle porous silica for the first time. The method is carried out in two steps. The first step is to catalyze and hydrolyze ethyl orthosilicate (TEOS) under acidic conditions to obtain a polyethoxysilane (PES) sol with a controlled degree of polymerization; the second step is to perform hydrolysis under alkaline conditions. The PES is completely hydrolyzed and allowed to stand for aging to form silica gel microspheres.

In the first step, in a system containing certain amounts of TEOS, ethanol sol, and hydrochloric acid catalyst, water is added dropwise under constant stirring, and stirring is continued for a certain period of time to allow TEOS to react and hydrolyze, and a PES sol with a certain molecular weight is obtained, which is the aforementioned preparation of silica sol. The properties of the PES sol, such as viscosity, colloidal particle size, and concentration, have a great influence on the particle size of the silica gel microspheres prepared in the second step.

In the second step, the ethanol and hydrochloric acid in the PES sol are removed by rotary evaporation. Water, isopropanol, and template compound are added to the PES and stirred at high speed to form an oil-in-water (O/W) emulsion, and then ammonia solution is added to initiate the secondary hydrolysis and polycondensation reaction. The ammonia catalyst and water molecules penetrate into the dispersed-phase PES droplets, and the PES continues to be hydrolyzed into silica sol and slowly turns into silica gel. Obviously, the size of the dispersed-phase droplets in the emulsion preliminarily determines the particle size of the silica gel microspheres. As the stirring speed increases, the emulsion is dispersed into small droplets under the action of mechanical force, and after the subsequent hydrolysis and polycondensation reaction, silica gel microspheres of small particle size are finally obtained. The particles are collected, washed, dried, and calcined to obtain porous silica gel microspheres.

In the sol–gel method, the type of silicon source, concentration of reactants, pH, temperature, stirring speed, and other conditions have an important influence on the particle size and structure of silica gel. In addition, adding trace electrolytes to the emulsion can also adjust the particle size of the silane microspheres through the formation of the emulsion and the influence of the dispersed droplet size. During the second step catalytic hydrolysis process, DMF, cyclohexane, polyethylene oxide, and polypropylene oxide can be used as porogens to control the pore structure of silica gel. By optimizing various parameters in the preparation process, it is possible to prepare silica gel microspheres with targeted particle size and with uniform pore structure.

Because the particle size of the microspheres prepared by the emulsion polymerization method is controlled by controlling the stirring speed, the uneven shearing force in the solution causes the particle size of the resulting microspheres to be uneven. Generally, a sieving step after synthesis is required to obtain microsphere products with particle size and particle size distribution that meet the requirements. The most commonly used particle screening technology in the industry is airflow classification. The principle of this technique is briefly described as follows. First, the sample of particles to be classified is added to the high-speed rotating classification disc. The small particles are taken away by the updraft and enter the small particle collection device, while the large particles are driven by the centrifugal force into the edge of the disc and enter the large particle collection device. By controlling the rotation speed, the particle sample can be divided into two fractions, large and small. By repeating this process, the components with uniform particle size distribution can be obtained. From the perspective of particle classification, the biggest disadvantage of the sol–gel method is that the product yield is low. A considerable part of the microspheres is removed as waste due to the large or small particle size, resulting in excessive consumption of raw materials and additional resources.

Spray drying The principle of this method is to form a mist of the reaction mixture solution, and each droplet becomes a microreactor. When the reactants in the droplets are heated, they polymerize and solidify to yield microspheres proportional to the droplet size (Wang and Friess 2017). If the template reagent is added to the precursor solution, the composite microspheres will be obtained. After the template reagent is removed, the porous microspheres will remain. The main instrument used in this method is a spray dryer. When preparing silica gel microspheres, a certain concentration of inorganic salt and silica sol mixture is added to the spray dryer system. After the solution is dispersed by the atomizer, it is sprayed into the heating vessel in the form of a mist by the high-speed airflow, and the solvent is taken away by the airflow, leaving behind porous microspheres. In this method, silica sol characteristics (viscosity, colloidal particle size), reaction mixture concentration, spray drying parameters, etc. determine the size of silica gel particles, whereas the type of salt is the key factor to control the pore structure of the microspheres. Macroporous structure can be formed with low melting point salts as template.

If the precursor solution does not contain inorganic salts, after the mist droplets enter the heating furnace, the surface is heated first, and the water at the interface of the mist droplets is first evaporated. Then, the heat is transferred to the droplet core, and as the heat transfer proceeds, the water diffuses from the core to the droplet surface and continues to evaporate. Finally, the silica sol colloidal particles are dehydrated to form compact spherical particles, and nonporous silica gel microspheres are obtained. When the high melting point salt exists, as the surface water evaporates, the water and dissolved salt in the droplets diffuse to the surface, resulting in an increase in the salt concentration at the interface between the droplets and the gas and finally phase separation due to salt precipitation, forming a salt packet-covered microspheres. After cleaning, the resulting microspheres have a porous structure. Mixed salt can be used as low melting point salt, because the melting point of mixed salt is lower than that of any constituent salt; or low melting point salt such as $NaNO_3$ (308 °C), $LiNO_3$ (255 °C) can be used alone.

Polymerization-induced assembly The polymerization-induced assembly method uses the adsorption of silica sol nanoparticles by polymer chains generated by monomer polymerization during the polymerization reaction to assemble the nanoparticles into polymer microspheres. The formed hybrid microspheres of polymer/silica nanoparticles are then heated and calcined to remove the polymer to obtain porous silica gel microspheres (Hao et al. 2011). One of the main chemical reactions on which this method is based is the urea-formaldehyde, or melamine-formaldehyde polycondensation reaction. The specific steps are as follows: Urea and formaldehyde are added to a silica sol containing silica nanoparticles with a certain particle size. Under acid catalysis, urea and formaldehyde polymerize, and the resulting low polymer is adsorbed on the surface of the colloidal particles and further condenses to form urea-formaldehyde resin/silica gel nanoparticle composite microspheres. In the reaction system, polymer molecules act as a bridge connecting colloidal particles, and the colloidal particles are connected and assembled through hydrogen bonding and electrostatic interaction to form composite microspheres of resin and silica gel particles. After the composite microspheres are collected, washed, dried, and calcined, the organic resin is decomposed and escaped, leaving behind porous silica microspheres with uniform particle size.

Taking the generation of urea-formaldehyde resin as an example, when the molar ratio of urea and formaldehyde is close to 1:1, linear urea-formaldehyde resin is formed; when the ratio of formaldehyde is higher, cross-linked or branched urea-formaldehyde resin will be formed. Both acids and bases can catalyze the reaction but the former tends to produce high-molecular-weight polymers. The silica sol, formaldehyde, and urea are mixed in an aqueous solution in a certain ratio, wherein the ratio of silica sol (based on the content of SiO_2) to urea is roughly maintained at 1:1. With adjustment of the pH of the solution to be acidic and stirring evenly, the polymerization reaction starts. In the initial stage of the polymerization reaction, the low molecular weight polymer is formed and the polymerization system is homogeneous. As the reaction progresses, the molecular weight of the polymer gradually increases, the polymer begins to precipitate, leading to a phase separation. During the formation and precipitation of the urea-formaldehyde resin, the colloidal silica sol particles in the system are dispersed in the urea-formaldehyde resin composite microspheres. After the reaction, the composite microspheres are collected, washed, and dried, and the composite microspheres are calcined according to a certain heating program. The organic components in the microspheres are thermally cracked and volatilized and removed. The silica particles in the composite microspheres accumulate to form porous silica gel microspheres. The commonly used calcination temperature is set at about 600 °C. The calcination temperature should not be too high; otherwise, the sintering phenomenon will occur and cause the destruction of the pore structure. This method can prepare 3–6-μm microspheres with narrow particle size distribution, which may be used directly without sieving. The pore structure in the microspheres is formed by the packing of colloidal particles. Therefore, the size of the colloidal particles is directly related to the pore size of the microspheres. Using commercially available silica sol of about 20 nm, a pore diameter of about 8–10 nm can be obtained. By repeating the above operation in the same system, the size of the microspheres can be increased.

Polymer microsphere template method The porous polymer microspheres with uniform particle size are used as the microreaction vessel, and the silica gel precursor is impregnated therein. Under the action of an acidic or alkaline catalyst, the hydrolysis and condensation reaction of the silica gel precursor is initiated to form a gel. After drying, the polymer microsphere template is removed by calcination, leaving behind the porous silica gel microspheres with monodisperse particle size, which can be directly used as a chromatographic packing without particle size classification. This method represents an energy-saving, high-efficiency, and environment-friendly preparation method, which has generated great interest. Meyer et al. (2002) first reported the preparation of silica gel microspheres by this method. They used commercially available monodisperse polystyrene matrix microspheres, including SOURCE 15 RPC, SOURCE 15 ETH, SOURCE 30 RPC, and SOURCE 30 Q, as templates, tetramethoxysilane as the precursor, hydrochloric acid

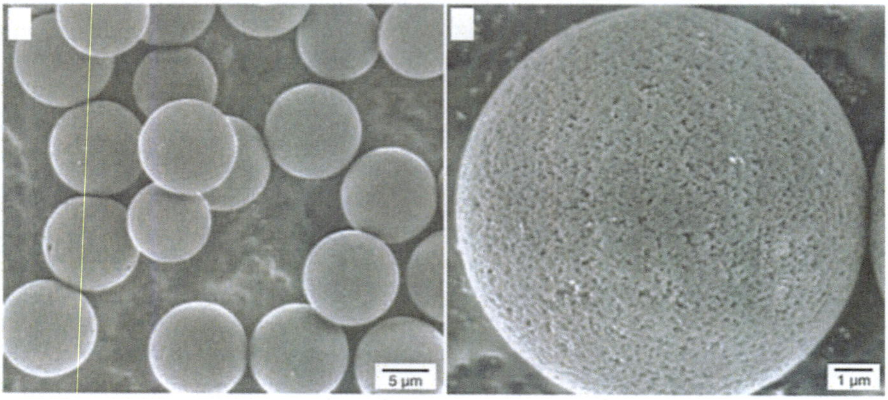

Fig. 4.1 Silica gel microspheres prepared with SOURCE 15 ETH as a template. The left picture shows the uniformity of the particle size of the microspheres; the right picture shows the porous structure on the surface of the microspheres. (Reproduced from Meyer et al. 2002)

as the catalyst, and aqueous solution as media for the reaction. The mixture was heated at 70 °C for 72 h, allowing the hydrolysis and condensation of the precursor to complete, forming organic/inorganic composite microspheres. The composite microspheres are collected, dried, and calcinated at 550 °C for 10 h to remove the organic template and produce inorganic silica gel (Fig. 4.1). Among the four tested templates, only the microspheres with hydrophilic surface treatment, namely, SOURCE 15 ETH and SOURCE 30 Q, can form porous microspheres with uniform structure, which means that the functional groups on the template surface attract the precursor through hydrogen bonding or electrostatic interaction. The hydrolysis and condensation products of the polymer aggregate in the pores of the polymer to form a gel structure. After heat treatment, the structure of silica gel microspheres is more compact, and its particle size is 10–20% smaller than that of template microspheres.

He et al. (2012) reported a novel method for preparing monodisperse porous silica microspheres with controllable morphology and structure. First, a two-step swelling method was used to prepare monodisperse macroporous polyglycidyl methacrylate/ ethylene glycol dimethacrylate (GMA/EGDMA) microspheres. The particles were then amine functionalized by the reaction of alkylene oxide on the surface with ethylene diamine (EDA). In the presence of tetra-*n*-butylammonium bromide (TBAB) or tetramethylammonium hydroxide (TMAH), a 10% tetraethoxysilane (TEOS) isopropanol solution was added dropwise to the suspension of the particles. After heating at 30 °C for 24 h, the silica nanoparticles produced by the hydrolysis and condensation of TEOS were deposited in porous polymer microspheres to form polymer/silica hybrid microspheres. The hybrid microspheres were calcined at 600 °C for 10 h to remove the polymer template, and porous silica microspheres were obtained (Fig. 4.2). The morphology, internal structure, and properties of porous silica microspheres were studied by field emission scanning electron microscope, Fourier transform infrared spectroscopy, thermogravimetric analysis, and

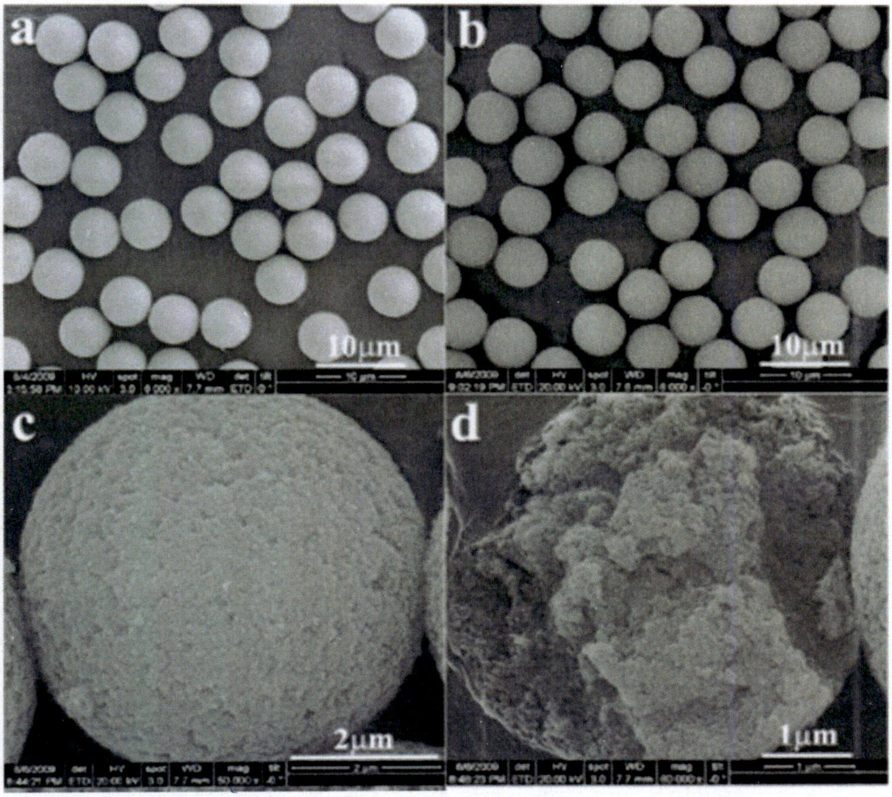

Fig. 4.2 (**a**) is the SEM image of polymer/silica hybrid microspheres obtained from the improved sol–gel method in the presence of TBAB; (**b**) the SEM image of silica microspheres; (**c**) the SEM image of the silica microspheres, showing the structure of its outer surface; (**d**) the SEM image of the silica microspheres, showing the internal porous structure. (Reproduced from He et al. 2012)

mercury intrusion method. The results showed that during this process, the presence of TBAB or TMAH not only prevented the agglomeration of hybrid microspheres but also controlled the controllable morphology from the porous internal structure to the hollow cage structure. The obtained porous silica microspheres did not show shrinkage with polymer microspheres, and the yield was about 98%.

Grama and Horak (2015) reported the use of monodisperse macroporous polyglycidyl methacrylate (PGMA) microspheres as a template for preparing porous silica particles. By using a modified Ugelstad method for multistep swelling polymerization, the starting polymer microspheres with a size of 9.3 μm were synthesized (Fig. 4.3). Then, the template microspheres were immersed in a mixed solution of tetramethoxysilane (TMOS) or methyltrimethoxysilane (MTMS) and hydrochloric acid aqueous solution, heated at 60 °C for 24 h. The resulting silica sol was deposited on the internal surface of the PGMA microspheres, forming poly (glycidyl methacrylate)–silica hybrid particles (PGMA–SiO_2) or poly(glycidyl

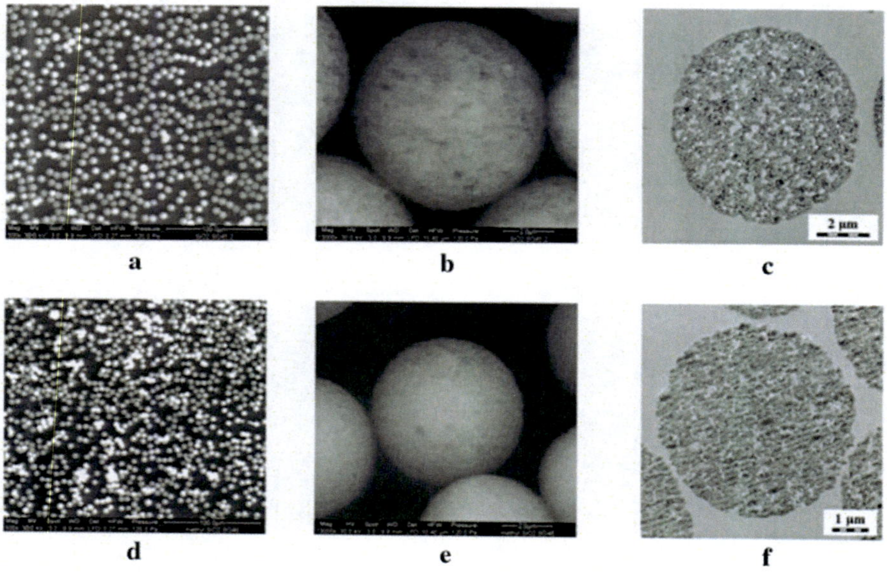

Fig. 4.3 SEM (**a**, **b**) and TEM (**c**) of SiO_2 prepared from precursor TMOS and SEM (**d**, **e**) and TEM (**f**) of SiO_2 prepared from precursor MTMS. TEM (**c**, **f**) showing particle cross-sections. (Reproduced from Grama and Horak 2015)

methacrylate)–methyl–silica hybrid particles (PGMA–M–SiO_2). The particle size of the former is slightly decreased to 8.6 μm, while the particle size of the latter is slightly increased to 9.7 μm, which may reflect the different formation mechanisms of the two. The silica gel nanoparticles formed by the hydrolysis condensation reaction tend to deposit in the hydrophilic pores of the polymer, while the hydrophobic methyl silica gel tends to phase separate and coat the surface of the template, thereby increasing the particle size of the hybrid particles.

The dried hybrid particles were calcined at 650 °C for 6 h. After removing the polymer template, porous silica microspheres were obtained. During this process, the silica gel microspheres showed partial shrinkage, and the final size of the silica gel microspheres obtained from the SiO_2 and Me–SiO_2 hybrid particles were 7.9 μm and 6.0 μm, respectively. This may be because TMOS mainly penetrates into the PGMA pores to form silica with a relatively uniform structure. MTMS tends to cover the surface of the particles. During the calcination process, the structure collapses, causing serious shrinkage of the microspheres. This result confirms that the compatibility of the surface of the template microspheres with the silica sol is an important factor for preparing silica gel microspheres with uniform pore structure.

Xia et al. (2016) developed a new method to synthesize micron-sized monodisperse macroporous silica microspheres with polymer microspheres as templates. The monodisperse macroporous polyglycidyl methacrylate–vinyl dimethacrylate (PGMA–EDMA) microspheres prepared by the two-step swelling method were reacted with tetraethylenepentamine (TEPA) at 80 °C for 24 h to obtain TEPA

Fig. 4.4 SEM images of PGMA/EDMA microspheres (**a–c**), PGMA/EDMA/silica hybrid microspheres (**d–f**), and monodisperse macroporous silica microspheres (**g–l**). (Reproduced from Xia et al. 2016)

functionalized template microspheres. In a 1000-mL three-neck round-bottom flask, 5 g of TEAP-functionalized PGMA–EDMA microspheres were mixed with 200 mL of isopropanol and 40 mL of water and sonicated for 5 min. One milliliter of ammonium hydroxide was added to the suspension, followed by 125 mL of TEOS (10% isopropanol solution) at a rate of 0.2 mL/min. The reaction mixture was stirred at 30 °C for 24 h. The obtained microspheres were washed with ethanol and water and dried at 60 °C for 12 h. The dried product was then calcined at 600 °C at a temperature increase rate of 10 °C/min for 10 h. The obtained silica microspheres with porous internal structure maintained the same size as the initial polymer microspheres, and 95.7% of monodisperse macroporous silica microspheres were obtained under a high concentration of polymer template (Fig. 4.4). Compared with the method of He et al. (2012), the yield of silica microspheres is significantly increased. The synthesized monodisperse macroporous silica microspheres with controllable size and structure can be used as a matrix for the separation and analysis of biological macromolecules and have good performance.

Silicone microsphere template method Polysilsesquioxane (PSQ), the empirical formula is $(RSiO_{1.5})_n$, as a new organic/inorganic hybrid material, has unique chemical, electrical, and mechanical properties, so it has attracted great interest.

The surface of polymethylsilsesquioxane (PMSQ) is rich in methyl groups, which may be used as alternative to traditional silica-based reversed-phase chromatography packing materials. In the past 10 years, the monolithic PMSQ stationary phase has been successfully synthesized and fully characterized. Kanamori et al. (2004) and Kanamori and Nakanishi (2011) used nonionic surfactants to control phase separation in the sol–gel process and successfully prepared mesoporous materials for the separation of polar and nonpolar molecules. Brennan et al. (Dong et al. 2005; Dong and Brennan 2006a, b) introduced the method of preparing macroporous PMSQ monolithic columns by a two-step acid-base reaction. The separation of the hydrolysis and condensation steps facilitates precise control of the gel process, making it easier to obtain a monolithic column with a hierarchical pore structure. Based on the formation of porous PMSQ gels initiated by surfactants, Zhu et al. (2015) introduced an alkaline-heating posttreatment method to further enlarge the pores and reduce the mass transfer resistance of the material.

Huo et al. (2018) reported a new reversed-phase packing based on porous PMSQ microspheres and evaluated its application in pharmaceutical analysis. PMSQ microspheres were synthesized through a two-step hydrolysis and condensation process, and their pore structure was further modified by post-synthesis aging treatment. The typical procedure is illustrated as follows: in a three-necked reaction flask equipped with a Teflon-coated stirrer, methyltrimethoxysilane (MTMS) (20 g) was dispersed in a large amount of water (100 mL). The dispersion was stirred at a moderate rate for about 30 min at room temperature to obtain a homogeneous solution. Under stirring, the catalyst ammonium hydroxide was added dropwise to the solution to catalyze the condensation reaction of the organosilane. The initial translucent solution became turbid within 30 s and quickly became completely milky white due to the formation of micron-sized particles. The suspension was allowed to react for 4 h under standing. The particles obtained were collected by filtration and washed thoroughly with ethanol and water. Then, the particles were suspended in a mixture of 80% ethanol (70 mL) containing 0.1-M ammonium hydroxide, and heated to reflux for 10 h. The suspension was cooled to room temperature; the particles were collected, washed, and dried under vacuum at 80 °C overnight to obtain free-flowing monodisperse porous PMSQ microspheres. The microspheres were characterized by a variety of physical and chemical methods, including scanning electron microscopy, Coulter counter, nitrogen adsorption-desorption measurement, elemental analysis, and solid-state 13C and 29Si NMR. Figure 4.5 shows the morphology and particle size distribution of the resulting PMSQ microspheres. Figure 4.6 shows the nitrogen adsorption curve and pore size distribution of microspheres.

Huo et al. (2020) developed a new method of monodisperse porous silica gel microspheres based on the preparation of monodisperse porous polymethylsilsesquioxane microspheres. The methyl silica gel microspheres were calcined in the air at 600 °C for 6 h to remove methyl groups and produce silica microspheres. By changing the amount of basic catalyst ammonium hydroxide during the condensation process, silica microspheres with particle sizes of 7, 5, and 3 μm were obtained (Fig. 4.7). Compared with the polymer microsphere

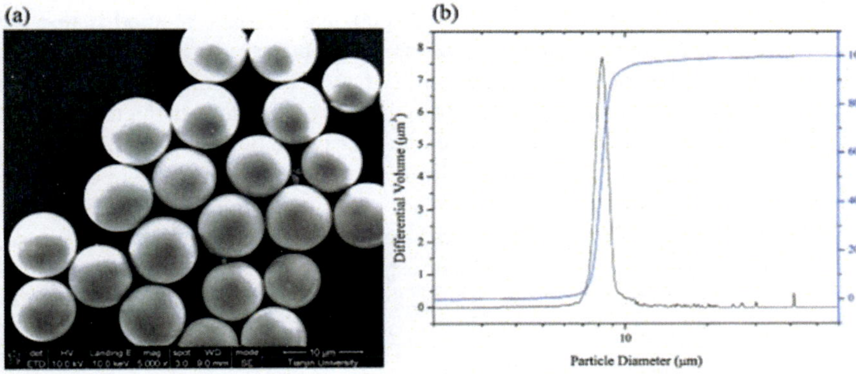

Fig. 4.5 SEM image (**a**) and particle size distribution (**b**) of PMSQ particles. (Reproduced from Huo et al. 2018)

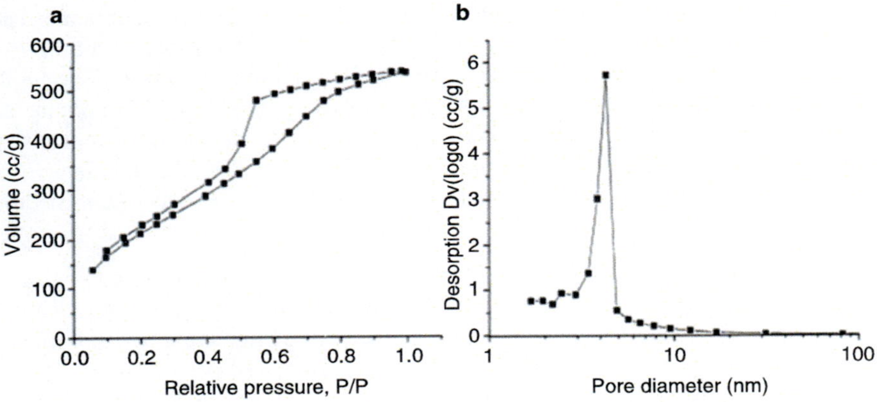

Fig. 4.6 (**a**) N2 adsorption-desorption isotherm and (**b**) pore size distribution of PMSQ particles. (Reproduced from Huo et al. 2018)

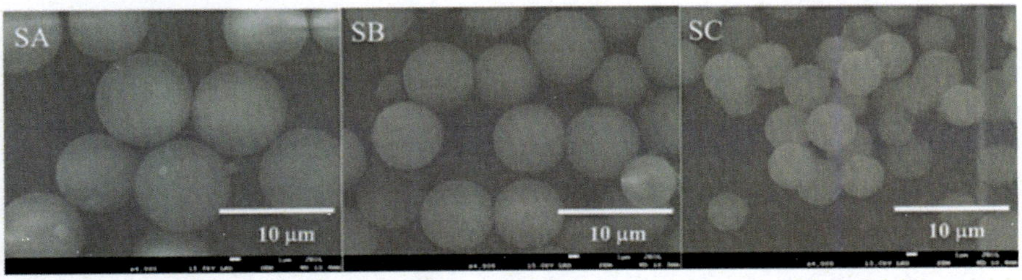

Fig. 4.7 SEM micrograph of the synthesized SiO_2 microspheres. (Reproduced from Huo et al. 2020)

template method, the difference of the silsesquioxane template method is that the high-temperature calcination process removes less than 20% of the organic matter, and about 80% by weight of the inorganic silica part of the organic/inorganic hybrid microsphere template is all retained, so the structure of the obtained silica gel microspheres is more dense and uniform. In the synthetic polymer microsphere template method, the organic part of the organic/inorganic hybrid microsphere accounts for more than 60%. After treatment by high-temperature calcination, only less than 40% of the silica gel remains (Grama and Horak 2015). In the latter case, there are two types of pores in silica gel microspheres: one is the macropores after the polymer skeleton is removed, and the other is the mesopores of the silica gel skeleton itself.

Membrane emulsification Membrane emulsification (ME) is a relatively new technology that was introduced by Nakajima and Shimizu in the late 1980s to produce monodisperse polymer microspheres (Vladisavljević 2015). In the process of membrane emulsification, pressure is applied to the dispersed phase and to force it to enter the continuous phase through the pores of the microporous membrane. The emulsified droplets are formed and separated at the exit of the pores by a drop-by-drop mechanism. Compared with the conventional emulsification process, the emulsion prepared with lower shear stress has a lower energy input per unit volume, so that a very fine emulsion with controlled droplet size and narrow droplet size distribution can be obtained. The main limiting factor of ME is the low dispersed-phase flux. In order to expand industrial applications, the productivity of this method must be increased.

Dragosavac et al. (2012) reported the use of membrane emulsification to produce nearly monodisperse spherical silica particles with controllable porosity. Using a hydrophobic metal membrane with a pore diameter of 15 μm and a pore spacing of 200 μm, nearly monodisperse droplets are produced, the average size of which can be controlled between 65 and 240 μm. The acidified sodium silicate solution (SiO_2 content of 4 and 6 wt%, pH 3.5) that enters the stirred dispersion tank through the metal membrane forms a uniform emulsion in the kerosene, and the sodium silicate gels and ages in the droplets. After drying, the average size of the final silica particles is between 30 and 70 μm (Fig. 4.8). The coefficient of variation of both droplets and particles does not exceed 35%. The average diameter of the most uniform particles is 40 μm, and the coefficient of variation is 17%. By changing the pH value of the sodium silicate solution and aging the gel particles in water or acetone, the internal structure of the silica particles was successfully modified, and almost monodispersed spherical particles with micropores and mesopores were obtained. The pore size is 1 and 6 nm, and the average surface area is between 360 and 750 m^2/g.

Core-shell silica The preparation process is generally divided into two steps. The first step is to prepare a solid core (nonporous spherical silica gel); the second step is to assemble a certain thickness of shell on the surface of the solid core (Hayes et al. 2014).

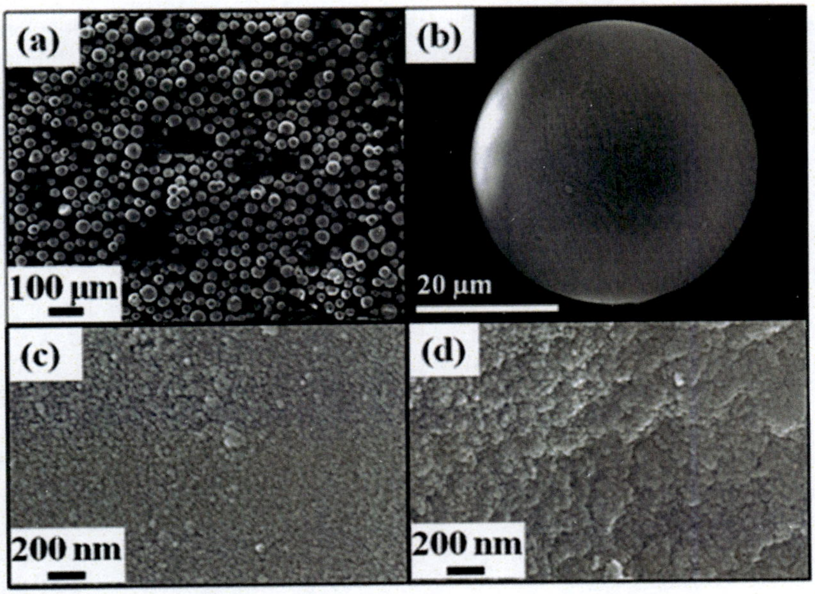

Fig. 4.8 (**a**) Scanning electron micrograph (SEM) of silica particles with an average particle size of 40 μm. (**b**) SEM of a single silica sphere. (**c**) Field emission gun (FEG) SEM of the outer surface structure of the silica sphere. (**d**) FEG SEM of broken silica balls. (Reprinted from Dragosavac MM, Vladisavljević GT, Holdich RG et al. (2012) Production of porous silica microparticles by membrane emulsification. Langmuir 28:134–143. Copyright {2012} American Chemical Society)

In the first step, the diameter of the nonporous silica gel microspheres is selected as required. For the core-shell-type silica gel of about 2 μm, the Stoeber method is generally used to prepare monodisperse nonporous silica gel (Stoeber et al. 1968). In the Stoeber method, TEOS or TMOS is mixed with 3-aminopropyltriethoxysilane in a mixture of water/alcohol/ammonia solution and hydrolyzed and condensed. Monodisperse silica particles with a diameter of about 1 μm can be obtained. At present, by adjusting the ratio of each component and adding electrolyte to the solution, it is possible to prepare nonporous silica gel microspheres of about 3 μm. For the preparation of core-shell silica gel with a diameter of 3 μm or more, commercial porous silica gel is generally used as the starting material and sintered at high temperature (above 1000 °C) to prepare nonporous silica gel.

The second step is to assemble the core-shell layer on the surface of the nonporous silica gel. At present, the commonly used assembly methods include the layer-by-layer assembly method, precipitation method, and spray drying method.

Layer-by-layer assembly has been successfully used to prepare nano-coatings on the surface of colloidal particles, and the prepared materials are called core-shell particles or core-shell composites. Based on this technology, a core-shell chromatographic packing with silica gel as the core and other inert inorganic materials as the shell can be prepared. At present, the layer-by-layer assembly method has become a common preparation method for core-shell silica gel. The basic principle is based on electrostatic attraction. Negatively charged silica core can adsorb positively charged polyelectrolytes, and the polyelectrolyte layer electrostatically adsorbs the negatively charged silica sol, so that an alternating structure of electrolyte layer and silica sol layer is formed on the surface of the silica core. The specific procedures are illustrated as follows: First, the activated nonporous silica gel microspheres are suspended into a solution containing positively charged electrolytes or surfactants, such as cetyltrimethylammonium chloride (CTMAB), polydiallyldimethylammonium chloride, polyallylamine salt or polyethyleneimine. After a layer of organic monomolecular film is formed on the surface of silica core microspheres, the core particles are collected, rinsed, and dried to produce organic film/silica microspheres. Secondly, the silica gel microspheres coated with the organic film are suspended in the silica sol, the negatively charged sol is assembled on the positively charged surface of the silica core. The above steps are cycled several times until the desired thickness is reached. Finally, the prepared composite microspheres are calcined at a certain temperature to obtain silica/silica core-shell silica gel. The size and uniformity of the core-shell silica gel, the thickness and pore diameter of the core-shell layer, etc. are related to the type and molecular weight of the polyelectrolyte, the size of the colloidal particles of the silica sol and the pH of the solution, the ratio of the silica sol to the nonporous silica gel, the assembly times, the drying temperature and other parameters. The method is simple to operate, and the obtained silica gel can be used directly without size classification. However, repeated assemblying takes a long time. Using the layer-by-layer assembly method, zirconium oxide and titanium oxide layers can also be assembled on the surface of the silica gel to form zirconia/silicon oxide or titanium oxide/silicon oxide core-shell microspheres.

The principle of the deposition method is basically the same as that of the polymerization induction method used to prepare porous silica gel. The specific process is as follows: first, silica gel microspheres are mixed with silica sol, and then a certain amount of urea and formaldehyde are added. The polymerization is initiated, by adjusting the pH of the solution, and proceeds for a certain amount of time to obtain composite microspheres. In the reaction system, oligomers formed by the polymerization reaction of urea and formaldehyde are adsorbed on the surface of colloidal particles and deposited on the surface of silica core microspheres to form a film. The thickness of the film can be adjusted by controlling the concentration of silica sol and the ratio of silica sol to urea-formaldehyde. After the composite microspheres are collected, washed, and dried, they are heated to decompose the organic resin to produce porous core-shell microspheres.

In the layer-by-layer assembly method and the deposition method, the pore size in the outer core-shell depends on the size of the silica sol particles. The larger the colloidal silica sol particles, the larger the pore size. Generally speaking, the pore diameter of the core-shell layer is about one-half of the diameter of the colloidal sol. In the comparative experiment, the particles prepared by the deposition method and the layer-by-layer assembly method with the same thickness and similar diameter have very similar chromatographic properties. From the viewpoint of pore size, the deposition method is suitable for the preparation of small pore packings, while the layer-by-layer assembly method is suitable for the preparation of large pore packings.

Ordered mesoporous silica Ordered mesoporous material is a kind of porous material with a pore size between 2 and 50 nm (Galarneau et al. 2006). It has a large pore volume, a high specific surface area, and an orderly and adjustable pore structure. Since Beck et al. (1992) reported the ordered mesoporous silica material based on aluminosilicate—M41S, mesoporous silica gel has become a hot topic in materials science research. The study of ordered mesoporous silica gel as liquid chromatography stationary phases has generated considerable interest (Unger et al. 2000).

The classical synthesis of MCM-41 is carried out using silicate solution and long-chain *n*-alkylammonium salt at a high temperature of about 100 °C (Beck et al. 1992). Detergent molecules interact with each other to form micelles, which then attract silicate anions to the their outer surfaces (Monnier et al. 1993). In the case of MCM-41, these micelles are arranged into hexagonal packed rods to form a so-called meso-structure phase. The micelles are inside the rods and the walls are composed of amorphous silica. Therefore, the detergent constructs a supramolecular structure, in which the inorganic silica source and the supramolecular template interact through ionic bonds, hydrogen bonds, and covalent bonds to form organic–inorganic composites. In this structure, the average pore diameter can be adjusted by the length of the *n*-alkyl chain of the surfactant used. The calcination removes the detergent molecular template, leaving behind a porous material with an amorphous silica wall about 1-nm thick. The material is agglomerated silica of primary particles with a diameter of about 50 nm, with a specific surface area greater than 1000 m^2/g and a specific pore volume between 0.8 and 2 mL/g. Due to its high porosity and agglomerated structure, the material cannot withstand high pressures and therefore cannot be used as a packing material for HPLC.

Improved Stoeber method The Stoeber method uses TEOS as a silicon source, ammonia as a catalyst, and ethanol as a solvent to prepare nano- or submicron monodisperse nonporous silica gel (Stoeber et al. 1968). In 1997, Unger and his colleagues introduced the template cetyltrimethoxyammonium bromide (CTAB) or cetylpyridinium chloride into the Stoeber reaction system and established a method for preparing mesoporous silica gel microspheres, that is, the improved Stoeber method (Grün et al. 1997). However, the silica gel particles prepared by this method have a small particle size and are not suitable for the stationary phase of high-performance liquid chromatography. In order to increase the particle size of silica gel particles, Unger and colleagues (Büchel et al. 1998) adopted the synthetic route shown in Fig. 4.9 to prepare mesoporous silica particles with an average particle diameter in the range of 0.2–2 μm and an average pore diameter of 2–5 nm (Fig. 4.10). In this method, a certain amount of hexadecylamine is dissolved in an isopropanol aqueous solution at room temperature, ultrasonicated to ensure that the alkylamine is completely dissolved. An appropriate amount of ammonia solution is added, followed by TEOS, and the reaction proceeds at room temperature for a

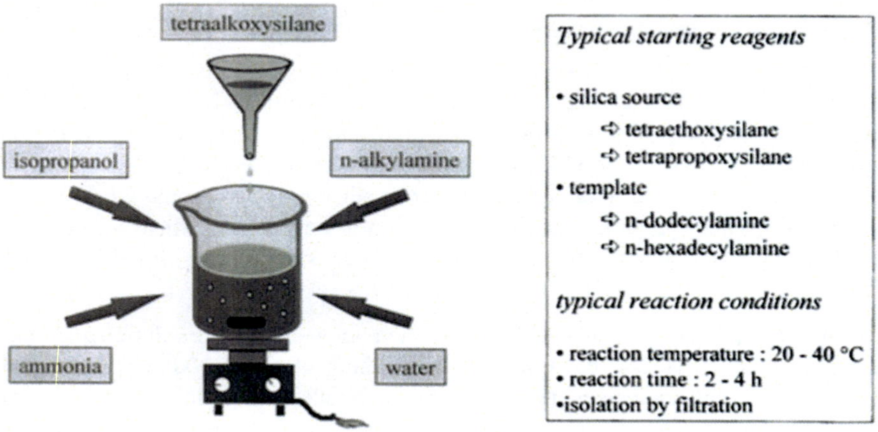

Fig. 4.9 Synthesis of spherical mesoporous silica. (Reproduced from Unger et al. 2000)

certain time. The resulting particles are collected and calcined to remove the template agent, and finally monodisperse mesoporous silica gel is obtained. Conditions such as the water/TEOS molar ratio in the reaction mixture, the reaction temperature, and the carbon chain length of the template have important effects on the particle size, while the reaction time, the type and quantity of ammonia, and the template affect the pore size and pore structure of the particles. The distinct advantage of this method is the strict control of particle morphology and particle size, which makes classification unnecessary. This means that there will be no material loss due to undersize and oversize.

In this improved Stoeber method, in addition to cationic surfactants and long-chain amines as templates, block copolymers and cation surfactants can also be used as co-templating agents. Zhao and colleagues (Zhao et al. 2002) used polyoxyethylene–polyoxypropylene–polyoxyethylene triblock copolymer P123 ($EO_{20}PO_{70}EO_{20}$) as the template, CTAB as the auxiliary template, TEOS as the silica source. A spherical mesoporous silica gel SBA-15 with a particle size of 2 μm, a pore diameter of 8 nm and a specific surface area of 800 m^2/g was prepared under acidic conditions. It was used as a stationary phase for reversed-phase chromatography to successfully separate protein and peptide mixtures. If ethanol is added to the reaction system as a cosolvent and an autoclave is used for the reaction, a spherical SBA-15 with a particle size of 3–6 μm, a pore diameter of 6.6–9.8 nm, and a specific surface area of 660–713 m^2/g can be obtained. In addition, the triblock copolymer F127 ($EO_{105}PO_{70}EO_{106}$) can also be used as the template. The advantage of this method is that it can prepare large particles with large pores. When the mesoporous silica gel prepared by the co-template method is used as the stationary phase of high-performance liquid chromatography, the column back pressure is low, the column capacity is high, and the mass transfer rate is fast.

Two-step synthesis method The two-step synthesis method includes an assembly stage and a condensation stage. The first stage is to emulsify TEOS and surfactant

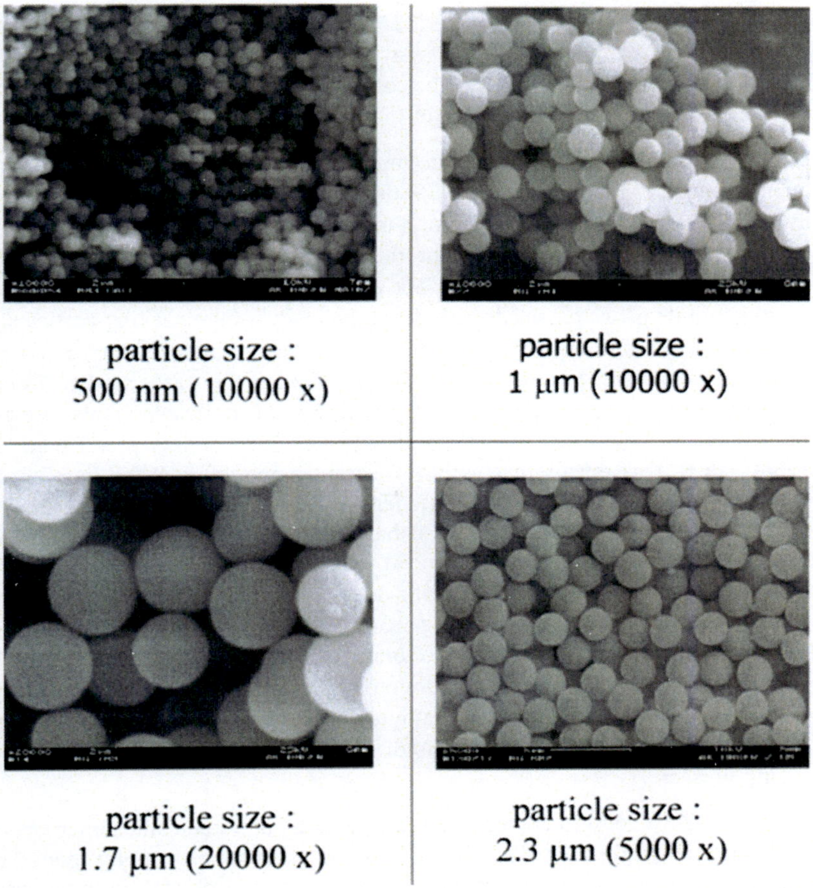

Fig. 4.10 Scanning electron micrographs of various sizes of silica beads produced by the amine-templated synthetic route. (Reproduced from Unger et al. 2000)

solutions, that is, mix TEOS with a molar ratio of (6–10):1 with nonionic surfactants, and then acidify, so that the stable microemulsion is formed quickly. In the second stage, a small amount of sodium fluoride solution is added to the obtained emulsion, and the emulsion is shaken at a constant temperature for an appropriate time to induce the hydrolysis and condensation of TEOS. The resulting particles are collected, washed, dried, and calcined to obtain MSU-X mesoporous silica gel (Boissière et al. 2000). The particle size of the mesoporous silica gel obtained by this method is usually 3–8 μm, the pore diameter is 2–4 nm, the specific surface area is 700–1000 m^2/g, the surface is smooth, the size is uniform, and the pore structure is orderly worm-shaped. If sodium silicate and colloidal silica are used as the silicon source in the first stage, MSU-X mesoporous silica can also be obtained. The pore size and pore structure of the mesoporous silica gel can be modulated by adding the swelling agent 1,3,5-trimethylbenzene (TMB) or isopropanol, increasing the length of the hydrophobic chain of the surfactant, changing the amount of NaF, etc. Since the reaction is divided into two steps, the controllability of the operation is increased, thereby increasing the reproducibility of the results.

Perfusion silica Perfusion chromatography is a separation technology for biomacromolecules developed in the 1980s (García et al. 2000). The characteristic of this technology is that in the separation medium, there are ultra-large pores with a pore size of 600–800 nm throughout the entire particle, as well as the usual 50–150 nm mesopores or macropores. The through pores provide a channel through which the mobile phase passes directly from the inside of the packing particles, and the mass transfer process of solute on the stationary phase has changed from diffusion mass transfer to convective mass transfer. In this case, the van Deemter equation, describing the band broadening process, is not applicable to this case as the theoretical plate height has not changed with the linear velocity and tends to a constant value. This means that within a certain range, even if the mobile phase flow rate is increased, the separation efficiency of the chromatographic column will not be reduced. Since the packing also has a usual porous structure, its specific surface area has not been greatly reduced with the appearance of through pores. Therefore, the sample load on this packing does not decrease with the increase of the flow rate. Furthermore, the through pores on the packing greatly increase the permeability of the chromatographic column, so that even at a very high flow rate, the operating pressure of this chromatographic column does not need to be very high. Therefore, this perfusion chromatographic column can simultaneously have the characteristics of high flow rate, high efficiency, high sample load, and low operating pressure.

Wei et al. (2009) described a method for preparing perfusion silica microspheres. This method uses high-molecular-weight polymers as the pore-forming agent for the through pores. The specific steps are as follows: 90-mL TEOS, 150-mL hydrochloric acid (0.01 mol/L), and 15.0-g polyethylene oxide (PEO) were mixed at 60 °C for 2 h to obtain a transparent solution. Then, it was added to a 2-L three-necked round-bottom flask containing 800 mL of paraffin oil. The whole solution was stirred vigorously for 10 min to form a water-in-oil emulsion, and then it was incubated in a water bath at 50 °C to undergo sol–gel transformation to form particles. The emulsion was stirred at 2000 revolutions per minute. Twenty hours later, a white precipitate appeared. The product thus obtained was rinsed sequentially with petroleum ether and ethanol. After that, it was hydrothermally treated at 200 °C for 10 h and calcined at 600 °C for 2 h to obtain free flowing silica gel microspheres. Since the particles are formed by the sol-gelation reaction of the water-phase droplets, the particle size is related to the droplet size, which depends on the stirring speed of the emulsion. Using elutriation technology to classify the microspheres, relatively uniform microspheres can be obtained. The scanning electron microscope (SEM) image (Fig. 4.11) shows that the average particle size of the microspheres is 5 μm. The surface of the microspheres is interwoven with the framework and macropores to form a network structure. The size of the framework is about 0.8 μm, while the

Fig. 4.11 Scanning electron micrograph of silica microspheres. (Reproduced from Wei et al. 2009)

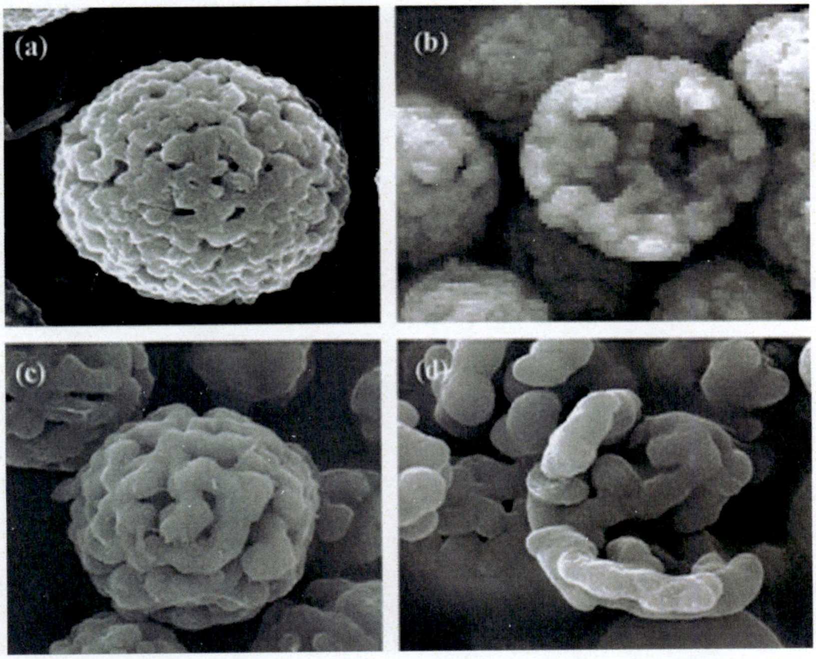

Fig. 4.12 Scanning electron micrographs of silica gel microspheres (**a**) and fragments (**b**) and carbon replicas (**c**) and fragments (**d**). (Reproduced from Wei et al. 2009)

size of the macropores is 0.6–1.3 μm range. To verify the presence of through pores, the microspheres were used as sacrificial templates to prepare carbon replicas. Figure 4.12 shows the SEM images of through-pore silica gels ((a) and (b)) and their carbon replicas ((c) and (d)). It is not difficult to find that the carbon replica is similar to its template, with spherical shape and obvious large pores on the surface. The images of fragmentation products of broken silica gel microspheres and their carbon replicas provide information about the internal structure of the perfusion silica gel. Spherical carbon can only be obtained when the macropores are permeable and interconnected within the silica spheres. The successful preparation of carbon replicas provides strong evidence for the existence of through pores in silica gel microspheres. This packing can be used for rapid separation. At the same flow rate, the column pressure is lower than the regular porous packing of the same size.

The formation of through-pore silica microspheres is attributed to the synergistic effect of phase separation and sol–gel transition (or gelation) in the closed emulsion droplets. In the presence of PEO, the sol–gel transition process is usually accompanied by phase separation. The relative speed that determines the gelation and phase separation of the final material structure is highly related to synthesis parameters (such as reactant ratio, solvent, temperature, etc.). Therefore, the structure control of the material can be achieved by selecting appropriate synthesis conditions. In some cases, when gelation and phase separation are performed in a balanced manner, a material with interlaced phases of skeleton and solvent can be formed. After necessary treatments (such as drying and calcination), the solvent phase will evolve into macropores. As a result, the prepared silica microspheres showed a macroporous structure. In addition to macropores, mesopores also exist in the framework of the material. Their formation may be attributed to hydrothermal treatment.

4.2.3 Structural Characterization and Evaluation

From the perspective of separation kinetics and thermodynamics, the particle size and distribution of the packing, the pore size and its distribution, and the specific surface area have an important influence on the efficiency of the chromatographic separation column. Therefore, for the prepared packings, it is necessary to characterize its structure, clarify its structure, and further study the relationship between structure and performance, so as to provide theoretical guidance for preparing packings that meet the design requirements.

The particle size and particle size distribution of the packing are usually measured with a Coulter particle size analyzer, also known as Coulter counter or electrical zone sensor. For particles smaller than 2 μm, transmission electron microscopy can be used for measurement. The principle of the Coulter counter is based on the change in electrolyte conductivity caused by microspheres passing through the orifice of a test tube. The closer the particle size of the microspheres is to the diameter of the orifice, the greater the resistance change caused, and thus the particle size of the packing can be determined. The particle size distribution of the packing can be measured by counting the microspheres passing through the orifice and statistical analysis of the resulting data.

Gas adsorption method and mercury intrusion method are commonly used for the measurement of pore size and its distribution in porous materials. The principle of the gas adsorption method for measuring the pore size and its distribution is based on the adsorption characteristics of the gas on the solid surface. Under a certain pressure, the surface of the measured sample particle (adsorbent) undergoes a reversible physical adsorption of gas molecules (adsorbent) at ultralow-temperature. The equilibrium adsorption capacity is determined, and the theoretical model is used to calculate the physical quantities related to gas adsorption, such as the specific surface area, pore size, and pore size distribution of the sample.

Mercury intrusion method, also known as mercury porosity method, is a method to determine the pore size distribution of some mesopores and macropores. Mercury does not wet general solids, and a certain external pressure needs to be applied to make mercury enter the pores. The greater the external pressure, the smaller the radius of the pore that mercury can enter. By measuring the amount of mercury entering the pores under different external pressures, the pore volume of the corresponding pore size can be calculated. The mercury porosimeter currently used has a maximum pressure of 200 MPa, and the measurable pore radius range is 3.75–750 nm. In recent years, inverse size exclusion chromatography has become more and more popular for pore size determination. The so-called inverse chromatography refers to the use of a compound of a certain molecular weight as a probe and the tested porous particles as a chromatographic stationary phase. By measuring the retention value of the probe under specific chromatographic conditions, the pore size and its distribution of the packing can be determined.

The mechanical strength of the packing is usually measured from the plots of back pressure against flow rate by which the fluid passes through the packed column. The method is based on the relationship between the pressure difference of the packed bed and the particle diameter of the packing. If the packing particles are rigid and do not deform under pressure, there is a linear relationship between pressure and flow rate. However, if the packing deforms under pressure, lower flow rate will be observed under the same pressure. In this case, the relationship between the pressure difference and the flow rate deviates from the linear relationship and deflects upward.

The morphology of the packing is usually observed by scanning electron microscope. Transmission electron microscopy is used for particles smaller than 2 μm.

4.2.4 Surface Modification and Functionalization

After more than half a century of development, liquid chromatography has become an indispensable tool for research and production, is widely used in the chemical industry, drug development, life sciences, food safety, environmental protection, and other related fields. Since liquid chromatography separation is mainly achieved through the interaction between analyte molecules and the chromatographic stationary phase, the availability of various chromatographic column packings has always been a key factor in promoting the development and application of liquid chromatography. The matrix of these packings is mainly silica gel and polymer. Spherical silica gel is the most important matrix material in liquid chromatography.

Traditionally, high-performance liquid chromatography is performed on the column packed with 3–10-μm (especially 5 μm) silica particles. Over the last decade, however, ultrahigh pressure liquid chromatography on the column packed with sub-2-μm small particles has become increasingly popular as more efficient, fast, accurate, and sensitive analysis can be achieved with smaller packings. The advantages of silica microspheres over other materials are reflected in their rigidity, structural stability (specific surface area and pore size, etc.), and easy surface modification. The surface modification and functionalization of silica microspheres have laid a solid foundation for the application and development of liquid chromatography.

Although the silica matrix itself can also be used directly as a stationary phase for normal-phase chromatography, some polar bonded stationary phases (such as amino, cyano, glycol group, etc.) provide better chromatographic performance and separation options than silica itself. In fact, the most widely used chromatographic mode in liquid chromatography is reversed-phase chromatography, in which the column packing is mainly silica bonded phase. For the separation and analysis of complex samples, especially biological samples, on the one hand, we need to overcome the defects of the silica bonded stationary phase to meet the needs of solving practical problems. On the other hand, continuous innovations and breakthroughs in the preparation of silica stationary phases often become the most effective way to some difficult problems.

Over the last decades, considerable progresses have been made in developing high-efficiency silica-based chemically bonded stationary phases. The most noticeable developments are summarized as follows: (1) Silica-based packings are prepared by ordinary silane bonding, but efforts have been made to reduce the residual silanol groups in the packing and improve the stability of the bonded stationary phase under acidic or alkaline conditions. (2) The versatility of silica-based packings is demonstrated by various functional groups and functional structures introduced, and their ability to separate complex samples and special isomers is greatly improved. (3) Advanced organic synthesis methods and polymer chemistry methods have been used to develop new chromatographic stationary phases. So far, silica-based bonded stationary phases have played an important role in different chromatographic modes such as normal-phase chromatography, reversed-phase chromatography, chiral separation, ion exchange chromatography, size exclusion chromatography, affinity chromatography, and hydrophilic interaction chromatography. The surface modification of silica gel microspheres has also evolved from simple silanization reactions on the surface of silica gel to the current advanced polymer graft polymerization reactions, cutting-edge organic synthesis reactions, and increasingly precise biomolecular grafting schemes.

4.2.4.1 Liquid-Phase Reaction

The bonded stationary phase of silica microspheres is generally prepared by the condensation reaction of a silanizing agent with the silanol on the surface of the silica gel. This method forms a thin layer of stationary phase on the surface of the silica microspheres through the Si–O-Si–C bond, which has the advantages of fast mass transfer, solvent resistance, chemical stability, and repeatability. Commonly used silylation reagents include alkylchlorosilanes or alkyltrialkoxysilanes. A typical silanization reaction is shown in the following formula.

$$\mathrm{Si-OH+RSi(CH_3)_2Cl=Si-O-Si(CH_3)_2R} \quad (4.1)$$

where *R* can be octadecyl, octyl, phenyl, aminopropyl, and cyanoethyl. When chlorosilane is bonded, it is necessary to add a basic compound as an acid scavenging agent to remove the hydrochloric acid generated by the reaction. The standard operating procedures for the preparation of commonly used C18 bonded silica stationary phases are described as follows (du Fresne von Hohenesche et al. 2004):

Hundred gram of silica microspheres were weighted out and transferred to a crucible in an oven. After dried at 120 °C for more than 12 h, they were put in a 1000-mL three-neck round-bottom flask. The flask was fitted in an oil bath on heating pot, connected with the condenser tube containing the desiccant, stirrer, and a 500-mL dropping funnel. Under stirring, 400 mL of dry toluene in the 500-mL dropping funnel was added; the oil bath was heated to about 120 °C to reflux the toluene, and then 16.4 g of imidazole (or equimolar pyridine or triethylamine or other basic compounds) was added. The mixture was stirred for 10 min. A mixed solution of 86.4-g octadecyldimethylchlorosilane and 200-mL toluene was added through the dropping funnel within 15 min; the suspension was stirred under reflux for 8 h and then cooled to room temperature. The solvent was removed by filtration and then washed with 200 mL of toluene, 300 mL of methanol, 300 mL of methanol/water (1:1, volume ratio) mixed solvent, and 250 mL of methanol. Next, the above C18 bonded silica gel was put into a 1000-mL three-neck round-bottom flask. The flask was placed in an oil bath heating pot, connected with the condenser tube containing the desiccant, stirrer, and a 500-mL dropping funnel. Under stirring, 400 mL of tetrahydrofuran was added through the dropping funnel. The oil bath was heated to about 80 °C and at reflux for 1 h; then, the particles were collected on filtration while hot, rinsed with 300 mL of tetrahydrofuran and 300 mL of methanol, and finally dried under vacuum at 60 °C for 12 h to obtain a C18 bonded silica phase.

Due to the steric hindrance effect, it is impossible for all the silanols on the surface of the silica gel to react with the silylation reagent, and the unreacted residual silanols can adsorb basic compounds, resulting in peak tailing. The most common solution is the endcapping of the bonded silica stationary phase with trimethylchlorosilane or hexamethyldisilazane. Because the ligands in these endcapping agents are much smaller than those of alkylsilanes, they can overcome the steric hindrance to react with residual silanol groups. However, these methyl groups bonded on silica are unstable under acidic conditions, with a pH of less than 2, and are easily degraded and lost. Substituting bulky substituents, such as isopropyl or isobutyl, to replace the methyl group in octadecyldimethylsilane can produce a steric shielding effect on the silanol group. Another solution is to insert a polar amide bond at the junction of the hydrophobic alkyl chain and the silica gel to prepare a polar embedded reversed-phase stationary phase. The embedded polar groups can form hydrogen bond with the residual silanol groups on the surface of the silica gel, thereby reducing their adverse effects.

4.2.4.2 Vapor-Phase Reaction

Because the traditional solution reaction method consumes a large amount of organic solvents, the disposal cost of waste organic solvents is very high. In order to reduce the consumption of organic solvents and reduce the environmental burden caused by the preparation of stationary phases, considerable efforts have been made to the development of new preparation methods that are green, environmentally friendly, energy-efficient, and consumption-reducing. The vapor-phase reaction method is one of the many methods proposed. This method is based on the vapor deposition method widely used in the semiconductor industry. The principle of the method is as follows: silanization reagents of low boiling points such as aminopropyl silane vaporize at a temperature near its boiling point, and the generated vapor reacts with the silanol groups on the surface of the silica gel to directly generate a silane bonded stationary phase (Li and Wan 2018). The typical operation is described as follows:

Ten gram of dry silica gel was placed in a 100-mL tetrafluoroethylene-lined hydrothermal synthesis reactor, an appropriate amount of silane reagent and triethylamine or other basic catalysts were added, and a stream of nitrogen gas was blown to remove oxygen in the reactor. After being sealed tightly, the reactor was placed in an oven and heated to 150 °C for 15 h. After the reaction, the particles were collected on filtration, washed with methanol, methanol/water mixed solvents, and methanol in sequence. The above procedure was repeated for the endcapping of the bonded silica phase.

The high-boiling silane reagent can be uniformly coated on the surface of the silica gel microspheres by the dipping method, then transferred to the hydrothermal synthesis reactor, and heated at 150 °C for 15 h. After the endcapping treatment, a reversed-phase silica-based stationary phase is obtained, which shows symmetric peaks for basic compounds.

4.2.4.3 Surface Graft Polymerization

Surface graft copolymerization is an effective method to form a uniform and controllable organic polymer film on the surface of silica gel (Mallik et al. 2008). It can be carried out in two different ways: grafting to and grafting from. The former is to chemically bond a polymer with reactive terminal functional groups to the surface of silica gel; the latter is achieved by polymerization reaction initiated by an initiator bonded on the surface of silica gel or by copolymerization of unsaturated bonds bonded to the surface. It is believed that the continuous extension of polymer chains through the reaction of surface groups is more advantageous because it can produce dense polymer coatings with high graft density, uniform distribution, and high surface coverage.

Surface-initiated living radical polymerization mainly includes surface-initiated atom transfer radical (ATRP) polymerization and reversible addition /fragmentation chain transfer (RAFT) polymerization. In the ATRP process, an initiator containing chlorine or bromine atoms is bonded to the surface of the silica gel. Under the catalysis of the combination of copper and cuprous ions and organic amines, the monomers are removed from the silica gel without being affected by steric hindrance. The surface starts to polymerize, and a brush-type polymer stationary phase with high molecular weight and uniform chain length is obtained. This technology has been used to prepare polyacrylamide-coated silica packings as stationary phases for size exclusion chromatography and hydrophilic interaction chromatography. In the RAFT process, the chain transfer agents, such as dithioester, trithioester, or dithioamide ester, are bonded on the surface of silica gel, and then the surface polymerization reaction of the monomer is initiated under light or heating conditions. The advantages of the RAFT method include a wide range of monomers available for polymerization and no need to use metal catalysts; the brush-shaped polymer chains obtained by the reaction have a very narrow molecular weight distribution.

4.2.4.4 Click Chemical Reaction

Click chemistry is a synthetic concept introduced by chemist Barry Sharpless in 2001. The main purpose is to quickly and reliably complete the chemical synthesis of various molecules through the splicing of small units (Kolb et al. 2001; Evans 2007). It particularly emphasizes the development of new combinatorial chemistry methods, based on the synthesis of carbon-heteroatom bonds (C-X-C), and the use of these reactions (click reactions) to obtain molecular diversity simply and efficiently. The concept of click chemistry has made a great contribution to the field of chemical synthesis. In many fields of drug development and molecular biology, it has become one of the most useful and attractive synthesis concepts.

Among them, copper-catalyzed azide–alkyne cycloaddition and the click chemistry reaction between thiol–ene have been used to synthesize silica gel bonded stationary phases. Zhao et al. (2012) coupled three alkyne-carboxylic acids to azide–silica to obtain three new types of weak cation exchange chromatographic stationary phases. Six standard proteins including myoglobin, RNase A, RNase B, cytochrome C, α-chymotrypsin A, and lysozyme were completely separated on these weak cation exchange chromatographic stationary phases. Compared with commercial weak cation exchange chromatography columns, the three new weak cation exchange chromatographic packing materials prepared by click chemistry have better resolution and selectivity. The mass recovery yields of the tested proteins were all above 97%, and the biological activity recovery yield of lysozyme on the preparation column was 96%. In addition, a new type of weak cation exchange chromatographic column was used to successfully purify lysozyme from egg white in one step. The purity is greater than 97%, and the specific activity is as high as 81,435 U/mg.

Click chemistry based on "thiol-ene" was used to prepare cysteine bonded zwitterionic hydrophilic interaction chromatography (HILIC) stationary phase (Shen et al. 2012). The effects of water content, salt concentration, and pH value of buffer solution on the retention of model compounds were studied. The results show that the retention of polar analytes is mainly dominated by the partition mechanism, while the contribution of electrostatic interaction is small. The separation of protein trypsin digests using hydrophilic interaction chromatography–electrospray ionization mass spectrometry (HILIC-ESI-MS) shows the great potential of zwitterionic stationary phases in the separation of complex samples and its application in two-dimensional liquid chromatography (2D-LC).

4.3 Hydroxyapatite

Hydroxyapatite (HA) is a biologically active material with a chemical formula of $Ca_{10}(PO_4)_6(OH)_2$. It has the same composition as the inorganic structure in human hard tissues. It has good biocompatibility. When it is implanted in the human body, it can be closely integrated with the soft and hard tissues of the human body in a short time. It is an ideal bone tissue replacement material and can be used for surface modification of dental implants. It has been widely used in clinical medicine and dentistry. Due to its unique crystal structure and surface charge distribution, HA can also be used as a chromatographic separation medium for separation of proteins, antibodies, and other biological macromolecules. HA is unique in the way it interacts with biopolymers and thus occupies a very special position in biological processing.

4.3.1 Development

In 1956, Tiselius et al. (1956) first introduced hydroxyapatite crystal particles prepared by precipitation method into chromatography for the separation of proteins, enzymes, antibodies, and other biological macromolecules. From 1959 to the 1970s, Bernardi (1973) conducted a series of studies on the behavior of biomolecules on HA and proposed a retention mechanism on HA (Bernardi and Kawasaki 1968; Bernardi et al. 1972). In 1972, Moss and Rosenblum (1972) reported the HA separation of proteins in the presence of detergent. In 1973, Martinson (1973a, b, c) systematically studied the mechanism of interaction between nucleic acid and HA. In 1978, Spencer and his colleagues used different methods to study the chromatographic properties of several different HAs. They used electron microscopes to observe the surface structure of HA crystals and infer the size of the crystals that make up HA particles from X crystal diffraction data (Spencer and

Grynpas 1978; Spencer 1978; Spencer et al. 1978). In 1984, Gorbunoff (1984a, b) systematically studied the retention mechanism of biomolecules on HA. In 1986, Kawasaki and Kadoya almost simultaneously reported the preparation of spherical HA aggregated from HA crystals with an average particle size of 5–20 μm and its application in high-performance liquid chromatography (Kawasaki 1991). With the commercialization of HA microspheres, the scope of HA continues to expand, and it has gradually become one of the important separation and purification tools in the preparation of monoclonal antibody drugs (Kawasaki 1991; Gagnon 2009).

4.3.2 Crystal Structure and Properties

The most common form of HA is the hexagonal crystal system with $P6_3/m$ space group symmetry. The unit cell parameters are $a = b = 9.432$ Å, $c = 6.881$ Å, and $\gamma = 120$. The structure of the unit cell is characterized by the reverse arrangement of two axes composed of an array of PO_4 tetrahedrons (Kawasaki 1991). These arrays are held together by Ca ions dispersed in them, forming an asymmetric biaxial structure. Ca ions appear in two distinct positions, in the form of an array (CaI) precisely aligned with the spiral axis around the center of the unit cell and an equilateral triangle (CaII) array centered on the edge of the unit cell (6_3 axis). There are two ways of existence of CaI atoms, one is surrounded by six phosphate oxygen atoms, and the other is surrounded by three phosphate oxygen atoms. The former CaI site (usually called the P-site) basically functions as a cation exchanger, while the latter CaI site may constitute another weak adsorption site. The P-sites are arranged on the (001) plane in the form of a hexagon with a pitch of 9.42 Å, that is, on the c plane. In addition, CaII atoms exist on the (100) and (010) planes of the crystal, namely, the a plane and the b plane. It forms a triangle with the Ca ions outside the unit cell, and the hydroxyl is located in the center of the triangle. In an aqueous solution, the hydroxyl group in the crystal is dissociated, and the two CaII ions on the a-face and b-face constitute a positively charged adsorption site, which is called the C-site. The zeta-potential measurement of the hydroxyapatite matrix shows that in the pH range of 5–11, the matrix has a negative net charge, indicating that the C-site may be saturated with hydroxyl or other anions, so that the matrix mainly shows the negative charge of the P-site. In addition to having an anion exchange function through electrostatic interaction, C-sites can also interact with molecules containing phosphoric acid or carboxylic acid through metal coordination. This principle is used to explain the extremely strong retention observed for negatively charged nucleic acids on the same negatively charged HA matrix.

From the late 1970s, Gorbunoff and Timasheff (1984) conducted a systematic study on the retention behavior of proteins on hydroxyapatite. By investigating the effects of the type of charged groups of the protein, the pH value, the composition of the eluent, and the ionic strength on protein retention, they concluded that there are two binding sites on hydroxyapatite: the negatively charged P-site points and positively charged C-sites. The protein can be bound by the interaction of the positive charge group on the surface with the P-position or by the coordination of the acidic group (carboxylic acid or phosphoric acid) of the protein with the metal at the C-position. Therefore, for different binding mechanisms, corresponding elution strategies can be adopted. The P-site interaction is mainly electrostatic interaction, and it can be eluted by increasing the ionic strength in the eluent. The C-site interaction is a metal coordination effect, and a specific displacer is needed for elution. If a gradient of increasing phosphate concentration is used for elution, the proteins adsorbed by the two different mechanisms will be eluted. Kawasaki (1991) proposed the use of a double gradient method to improve the selectivity of hydroxy-apatite chromatography. First, a common salt, such as potassium chloride gradient, is used to elute the protein adsorbed at the P-site, and then a phosphate gradient is switched on to elute the protein adsorbed at the C-site. So far, the mixed mode retention mechanism on the hydroxyapatite matrix has been widely recognized.

4.3.3 Preparation

Hydroxyapatite can be prepared by a variety of methods. The prepared HA presents a variety of morphologies, including needle-like, rod-like, flake-like, and spherical shapes. Among them, spherical particles have the characteristics of large specific surface area, low flow resistance, and high mechanical strength, which have attracted attention in the field of chromatography. At present, the methods for preparing spherical HA include spray drying, hydrothermal synthesis, emulsion polymerization, and template method.

Spray drying The spray drying method uses the principle of spontaneous formation of droplets under the action of gravity and surface tension. The solution of precursor substances is dispersed into single droplets by high-pressure jet, and the droplets are dried into microspheres by rapid dehydration. Luo and Nieh (1996) reported the preparation of 1–8 μm spherical HA by chemical precipitation combined with spray drying. X-ray crystal diffraction and transmission electron microscopy measured the average crystal diameter of HA synthesized by precipitation method to be 10 nm. The spray-dried microspheres are aggregates of nanocrystals. The morphology, particle size, and its distribution of the microspheres vary with the spray conditions. Using high-concentration crystal homogenate, solid microspheres can be obtained; otherwise, hollow microspheres can be obtained. Increasing the atomization pressure can reduce the particle size of the resulting microspheres. Bastan et al. (2017) further studied the influence of precipitation pH, solid content, slurry feed rate, atomization pressure, hot air temperature, and hot airflow rate on the particle size distribution of HA microspheres prepared by spray drying. The results show that the increase of precipitation pH value leads to a slight increase in particle size and high-temperature stability; the increase of heat treatment temperature reduces the specific surface area, pore volume, and particle size of HA microspheres. At the same time, the crystallinity and powder density are increased. The spray drying method for preparing HA microspheres is a simple process with low cost, but the resulting microspheres have rough surfaces and irregular spherical shapes.

Hydrothermal synthesis The hydrothermal method uses the temperature difference between the upper and lower parts of the reactor to generate strong convection under high-temperature and high-pressure conditions to transport ions to the low-temperature zone to form a supersaturated solution and finally crystallize to form the target product. The morphology of HA produced by calcium and phosphate ions under hydrothermal conditions has a strong relationship with the pH of the solution. When the pH is low, rod-shaped HA with higher crystallinity is usually produced. When the pH reaches 14, nearly spherical nano-HA crystals are formed. In order to obtain spherical HA, it is generally necessary to add a morphology modifier to the reaction solution. Ma and Zhu (2009) used *N,N*-dimethylformamide as a regulator and sodium tartrate as a Ca^{2+} chelating agent to prepare HA hollow microspheres with a diameter of 3.6 μm under hydrothermal conditions at 200 °C for 4 h. The electron microscope showed that the microspheres were self-assembled from HA nanorods with a size of 100 nm. Jiang et al. (2012) used polyaspartic acid to prepare hollow HA microspheres with a size of about 800 nm under hydrothermal conditions at 180 °C and 24 h. The microspheres are self-assembled by nanorods. Polyaspartic acid acts as a chelating agent and a surface capping agent, mediating the self-assembly of hollow HA microspheres. Zhang et al. (2012) synthesized HA hollow microspheres with a diameter of 500 nm and a shell thickness of 100 nm in a mixed solvent of glycerol and water under hydrothermal synthesis conditions. The microspheres are assembled from nano flakes with a thickness of 10 nm. By changing the ratio of glycerol/water, hollow microspheres of 6 and 12 μm can be obtained. Wang et al. (2013) used ethylenediaminetetraacetic acid (EDTA) as a morphology modifier to synthesize FHA microspheres formed by self-assembly of nanorods. They found that the pH of the reaction solution can affect the release rate of Ca^{2+} in the EDTA–Ca complex, thereby affecting the morphology of the reaction product. Microspheres with spherical morphology are formed at high pH, and rod-like structures are formed at low pH.

Emulsion method In the microemulsion method, the surface tension of the two immiscible solvents is different to prepare spherical particles. Teng et al. (2007) used a water-in-oil emulsification method with gelatin droplets as a microreactor to coat needle-shaped HA crystals to form HA/gelatin composite microspheres with an average particle size of 7.5 μm. The needle-shaped nano HA crystals are clustered. The clusters are uniformly and orderly precipitated in the gelatin matrix. Lee et al. (2008) dispersed HA nanoparticles in a dimethyl sulfoxide solution of polyvinyl butyral, volatilized to remove dimethyl sulfoxide, and then subjected to further heat treatment to obtain HA microparticles with a size of 200–600 nm.

Template method The template method to prepare HA microspheres is based on the rigid spherical structure of the precursor template or the chemical bonding between the functional groups in the template molecule and calcium phosphate ions. In the reaction, HA ions are used to replace the original ions in the template or regulate the growth rate of the HA crystals in different crystal orientations. Descamps et al. (2009) used porous PMMA microspheres as templates to prepare $CaCO_3$ framework materials with porous spherical cavities. Then, through the steps of solution impregnation, drying, and burning, HA beads with diameters ranging from 300 μm to 3 mm were obtained. The pore structure of HA beads can be densified by further high-temperature treatment. Wang et al. (2006) reported a method of preparing hollow HA microspheres in situ using lithium calcium borate glass beads as a template. The components are melted at 1100 °C for 1 h to produce lithium calcium borate ($10Li_2O$–$15CaO$–$75B_2O_3$) (mol%) glass, which is ground into powder, and the powder is spheroidized in a flame at 1400 °C. The resulting glass microspheres (diameter 106–125 μm) were reacted in a 0.25 M K_2HPO_4 solution at 37 °C and pH 10–12 for 5 days, resulting in porous hollow calcium phosphate microspheres. The outer diameter is comparable to the size of the original glass particles. Heat treatment at 600 °C for 4 h can convert the CaP material into HA and increase the strength of the microspheres.

4.4 Porous Glass

Porous glass is a silica-based glass with a purity greater than 96% and a microporous glass with a pore size distribution of tens to thousands Å. It is a kind of porous material with connected pore structure prepared by using a glass system of suitable components and using the principle of phase separation. It has the advantages of controllable pore size, uniform pores, large specific surface area, high mechanical strength, and stable chemical properties. It has a wide range of applications in virus filtration, optical fiber communication, chromatographic purification (Schnabel and Langer 1991), nucleic acid synthesis, etc.

4.4.1 Development

In 1938, Hood and Nordberg (1938) published their invention of porous glass. Through heat treatment and leaching of the glass components of the RO–B_2O_3–SiO_2 ternary system (RO refers to alkali or heavy metal oxides), porous glass with a silica content of 96% was prepared. As an intermediate of the final high-temperature-treated high-silica glass, the typical pore structure data of porous glass is a specific surface area of 121 m^2/g, a pore volume of 0.18 mL/g, and a pore size of 60 Å. The pore volume accounts for about 28% of the total volume. MacDonell and colleagues first reported the application of porous glass in liquid chromatography, gas chromatography, and electrophoresis. The porous glass plate can be used as a porous medium for chromatography and electrophoresis to separate dyes in ballpoint pen refills. The crushed porous glass particles can also be used as a chromatographic carrier, and after loading an appropriate stationary phase, gas chromatographic separation is performed. However, Haller is accredited with the discovery of unique properties of porous glass for chromatographic applications. In 1965 Haller (1965) reported the synthesis of a series of porous glasses with different pore sizes, which were then crushed and sieved for size exclusion chromatographic separation of virus and cell components. Compared with chromatographic media based on polysaccharides and silica gel, porous glass has many advantages. First of all, porous glass has higher mechanical strength and can be used under higher operating pressures. Secondly, the pore size of the porous glass is adjustable, the pore channels are uniform, the difference between the pore size at 95% of the pore volume and the average pore size is less than 20%, and for silica gel this ratio is greater than 35%. Therefore, porous glass has better chromatographic separation performance. As a chromatographic medium and solid-phase synthesis carrier, porous glass has been widely used in the separation and purification of biological macromolecules and the solid-phase synthesis of nucleic acids.

4.4.2 Preparation

The most commonly used method for preparing porous glass is thermally induced phase separation. First, the molten soda borosilicate glass is subjected to a heat treatment for phase separaion, and then acid leaching is carried out to remove the soluble boron component, and the porous glass is obtained. The specific steps are to mix certain components of high-purity boric acid, sodium carbonate, quartz sand, and other additives uniformly, melt them at a suitable temperature, and keep them for a period of time. After molding and annealing, the bulk glass is obtained, and is treated at a specific temperature. Internal particle migration leads to segregation of certain components, forming high boron phase and silicon phase. The separated glass is immersed in an acid solution to dissolve the sodium and boron phase, leaving behind mainly silica-based three-dimensional porous skeleton structure.

The main factors affecting the pore structure of porous glass include the composition of the starting glass raw material, acid leaching treatment conditions, and heat treatment conditions during the phase separation process. The heat treatment of the initial glass in the temperature range of 580 to 700 °C is used to prepare macroporous glass.

There are two models for the phase separation mechanism in the preparation of porous glass. According to Haller (1965), the phase separation in alkali metal borosilicate glasses proceeds through nucleation and growth mechanisms. The interconnect structure is produced by the growth of microphases. It depends on the volume fraction of the phase obtained by the nucleation. Cahn and Hilliard (1958) proposed another phase transition model, that is, spinodal decomposition or metastable separation causes the phase separation of alkali metal borosilicate glass. The original glass components are located in the unstable spinodal region of the miscible gap, and the interwoven structure already exists in the original glass. Phase separation occurs with the increase of statistically unstable concentration fluctuations, making the texture coarser.

This can be illustrated by the phase diagram of the miscible gap. As long as the material transitions to the unstable region of the phase diagram, phase separation occurs. By performing the common tangent construction of the free energy diagram, the boundary of the unstable region (sometimes called a binodal or coexistence curve) can be found. Inside the binodal curve is an area called spinodal. The spinodal area can be determined by determining the area where the curvature of the free energy curve becomes negative. The binodal line and the spinodal line meet at the critical point. Spinodal decomposition occurs when the material moves into the spinodal region of the phase diagram.

Haller (1965) prepared porous media from a ternary system containing RO–B_2O_3–SiO_2. RO refers to alkali metal or heavy metal oxides, such as Li_2O, Na_2O, K_2O, CaO, BaO, MgO, PbO, etc. The selected oxide mixture should show a non-existence of immiscible gaps, that is, after reaching a preset temperature, the resulting molten mixture of oxides is a substantially homogeneous liquid. When the temperature drops to the set temperature, the melt separates into two immiscible liquids. In addition, the volume ratio of the two phases should be between 1:2 and 2:1. The chemical stability difference between the two phases is large enough to allow selective leaching. The composition of the oxide is usually 50–83% by mass of the basic glass silica gel, and RO means soda, potash, lithia, etc., accounting for 2–10%, while boron oxide accounts for between 38% and 48%.

4.4.3 *Application*

Porous glass has high mechanical strength, high chemical inertness, and low flow resistance. It is used to separate tobacco mosaic virus and other components of cells. In the solid-phase synthesis chemistry of oligonucleotides, porous glass particles with pore diameters of 500, 1000, 1500, 2000, and 3000 Å are used as solid-phase supports to prepare oligomers of about 50, 80, 100, 150, and 200 nucleotides, respectively. MilliporeSigma's ProSep-vA is an affinity chromatography material with porous glass as the matrix and protein A as the ligand, used for the separation and purification of monoclonal antibodies and their derivatives. Due to the rigidity of porous glass, the material can withstand a certain pressure. Therefore, compared with other polysaccharide gel media, it has a higher dynamic binding capacity and is widely used in the biopharmaceutical industry.

4.5 Synthetic Polymer

As mentioned earlier, porous silica gel microspheres are widely used as a chromatographic matrix material. However, silica-based packings also have two difficult problems: (1) The usable pH range is narrow. When the pH of the solution is greater than 8, the silica itself is unstable, and the silica skeleton will be hydrolyzed. In the presence of phosphate and carbonate, this hydrolysis reaction is greatly accelerated; when the pH is <2, the siloxane bond connecting the functional group is unstable, causing the loss of the bonded phase, and the solute retention characteristics and separation efficiency are changed. (2) Residual silanol groups on the surface of silica gel can interact electrostatically with nitrogen-containing basic compounds and cause irreversible adsorption of biological macromolecules, especially peptides, proteins, etc., which greatly limits their use in the separation and analysis of biologically interested compounds.

Polystyrene–divinylbenzene (PS–DVB) copolymer particles have a long history of application in the field of chromatography. When the separation requires the use of alkaline conditions, researchers often seek packings based on synthetic polymer matrix, such as PS–DVB resin. These resins have a broad range of useable pH as they stable from pH 1 to 14, good mechanical strength, and excellent surface chemical inertness; there are no active groups such as silanol groups. Therefore, in terms of separation of ionic compounds, polymer materials often show greater advantages than silica gel. Ion exchange chromatography for separating inorganic and organic ions mostly uses a polymer matrix. In gel permeation chromatography, the polymer column has become the gold standard for determining the molecular weight of polymers. Since the advent of perfusion chromatography, polymer packings have also been used more extensively in the separation of biological macromolecules. However, polymer matrix materials also have some shortcomings. The most prominent problem is that the column efficiency of polymer microspheres is generally lower than that of silica gel microspheres of the same particle size due to the large mass transfer resistance. Secondly, polymer materials with a low degree of cross-linking are easy to swell in organic solvents, which will cause compression of the column bed and increase in operating pressure (Lubomirsky et al. 2021).

Most polymer matrices used in liquid chromatography are based on cross-linked polystyrene-divinylbenzene. These resins can be used directly as packings for reversed-phase chromatography or size exclusion chromatography, or they can be formed into ion exchangers through sulfonation or amination reactions. In addition, polyglycidyl methacrylate microspheres are more commonly used today. Because its surface is rich in epoxy groups, this type of resin has good chemical reactivity and can be easily derivatized and prepared into various chromatographic separation materials under mild conditions. For example, under acidic conditions, the epoxy group can be ring-opened to obtain a hydrophilic hydroxyl-functionalized resin. They can react with different types of ionizing reagents to obtain corresponding ion exchangers and chelating resins or react with an affinity ligand to obtain an affinity resin, which can be used for the separation and purification of biological macromolecules.

4.5.1 *Synthesis*

Generally, macroporous polymer microspheres are prepared by conventional free radical copolymerization, in which phase separation or templating methods are used to customize the pore structure. Specific methods mainly include suspension polymerization and seeded emulsion polymerization of monomer mixtures in the presence of an inert diluent, as well as membrane emulsion polymerization technology (Liu et al. 2007).

Suspension polymerization A typical suspension polymerization system consists of an oil phase, containing monomers, cross-linkers, inert diluents, and free radical initiators, and a continuous aqueous phase containing additives (Coutinho et al. 2000). During the polymerization, the oil phase is dispersed in the continuous water phase in the form of droplets by stirring. It should be noted that the inert diluent must generally be soluble in the monomer mixture but insoluble in the continuous phase of the suspension polymerization. At an appropriate reaction temperature, the initiator decomposes into free radicals, thereby initiating the cross-linking copolymerization of the monomer mixture in the monomer-diluent droplets. After polymerization, the diluent is removed by extraction with another organic solvent in a Soxhlet extractor and then vacuum dried to obtain highly cross-linked macroporous polymer microspheres (Garcia-Diego and Cuellar 2006). The photomicrograph of typical St-DVB microspheres prepared by suspension polymerization is shown in Fig. 4.13.

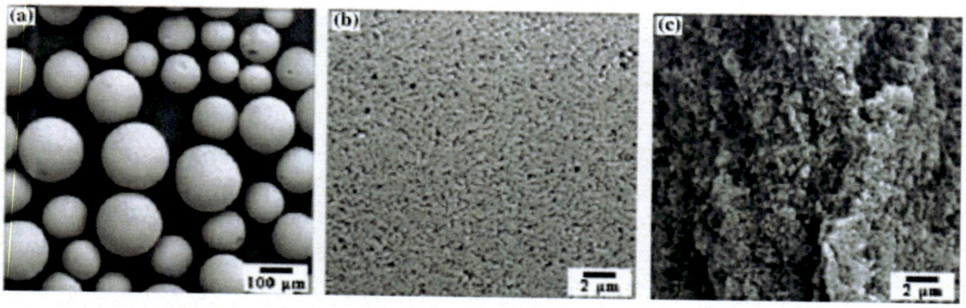

Fig. 4.13 SEM micrograph of macroporous St-DVB microspheres prepared by suspension polymerization. (**a**) Spherical morphology, (**b**) outer surface, and (**c**) internal structure. (Reprinted from Garcia-Diego and Cuellar (2006) Determination of the quantitative relationships between the synthesis conditions of macroporous poly(styrene-co-divinylbenzene) microparticles and the characteristics of their behavior as adsorbents using bovine serum albumin as a model macromolecule. Ind Eng Chem Res 45:3624–3632. Copyright {2006} American Chemical Society)

Therefore, in order to produce macroporous St-DVB copolymer microspheres, one of the key factors is to select an inert diluent, namely, a pore former, which can be a single organic solvent, such as toluene, *n*-heptane, bis(2-ethylhexyl)phosphonic acid, cyclohexanol, and dibutyl phthalate (DBP), or a mixture of two organic solvents, for example, *n*-heptane/toluene, bis(2-ethylhexyl)phosphonic acid/isooctane, and cyclohexanol/toluene. Solvating reagents lead to micropores, while non-solvating reagents lead to void structures. The control of the microsphere pore structure requires a balance between the two limits. The second is the amount of cross-linking agent DVB. A low cross-linking agent dosage results in a macroporous structure while a high cross-linking agent dosage results in a microporous structure.

Different polymerization parameters will affect the porous properties of the polymer beads (specific surface area, pore volume, and pore size distribution). Among them, it can include the polymerization time and temperature, the type and concentration of the pore-forming agent, the concentration of the cross-linking agent, the ratio of the dispersed/continuous phase, the stirring speed, etc.

Seeded emulsion polymerization The macroporous polymer microspheres prepared by the suspension polymerization have a relatively wide particle size distribution, and size classification is required to obtain microspheres within the target particle size range. In contrast, preparation of microspheres with uniform particle size has great cost advantages. Seeded emulsion polymerization has been developed for the preparation of monodisperse macroporous St-DVB microspheres. Generally, this technology can be divided into four stages: (1) synthesis of latex seeds; (2) latex seeds swelling, absorbing monomers, cross-linking agents, etc.; (3) copolymerization of styrene and divinylbenzene; and (4) removing the diluent and porogen. Latex seeds are monodisperse particles of polystyrene, which can be prepared by direct emulsion polymerization of styrene or by dispersion polymerization using an emulsifier (Cheng et al. 1992). In the swelling stage, the latex seeds are swollen by styrene, divinylbenzene, and diluent, and the swollen microspheres appear as individual polymerization droplets. During the copolymerization process, phase separation occurs between the copolymer and the diluent, and after removing the inert diluent, the porosity of the St-DVB microspheres can be observed. In this case, the pore former may be linear polystyrene or a mixture of linear polystyrene and solvated or non-solvated diluents, such as dibutyl phthalate, *n*-hexane, and toluene. Through this technology, monodisperse macroporous St-DVB microspheres with a diameter of 10 μm were produced using polystyrene latex seeds with a diameter of 8.7 μm. The picture of typical St-DVB microspheres produced by seeded emulsion polymerization is shown in Fig. 4.14.

Seeded emulsion polymerization has been used to prepare poly(St-co-DVB) particles, poly(glycidyl methacrylate-ethylene dimethacrylate) [(GMA-co-EDMA)], poly(methacrylic acid)-co-EDMA [poly(MA-co-EDMA)], and poly (HEMA-co-EDMA).

Membrane emulsion polymerization The key step of membrane emulsion polymerization technology is to use a porous membrane with a uniform pore size, to

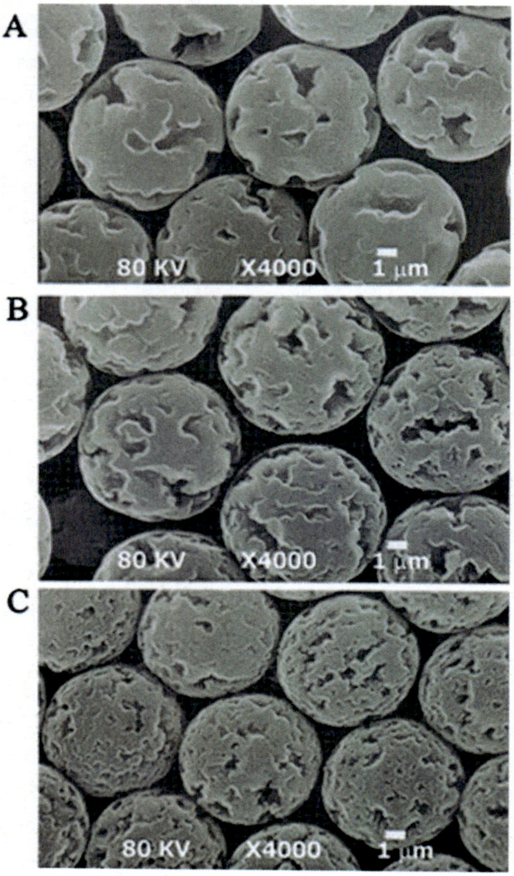

Fig. 4.14 SEM micrograph of macroporous St-DVB microspheres prepared by seed emulsion polymerization. DBP/seed latex ratio: (**a**) 1.0 mL/g, (**b**) 1.5 mL/g, (**c**) 2.0 mL/g. (Reproduced from Camli et al. 2004)

prepare emulsion droplets with a uniform particle size, and use the droplets as a microreactor for the polymerization reaction. Omi (1996) reported a specific emulsification process as shown in Fig. 4.15. Microporous glass membranes (MPG; NA-1, Ise Chemical Co., Ltd.) with average pore diameters of 0.5, 0.9, 1.4, 1.8, and 5.25 μm were used to prepare monomer emulsion. The glass membrane is a cylindrical tube (outer diameter = 10 mm, $L = 150$ mm, surface area = 50 cm^2), as shown in the detailed schematic diagram of Fig. 4.15b, installed in a stainless steel cylinder. The dispersed phase is a mixture of monomers, solvents, initiators, and water-insoluble substances. It is stored in a glass storage tank and penetrates the membrane into the circulating flow under appropriate pressure. The droplets are suspended in the continuous phase and transferred to the reactor. Under nitrogen gas and stirring, the reactor is heated to reaction temperature, usually 70 or 75 °C, and polymerization proceeds under nitrogen atmosphere for 24 h.

An example of porous microspheres synthesized by membrane emulsion polymerization is shown in Fig. 4.16. The ratio of styrene to divinylbenzene is 1:1. The

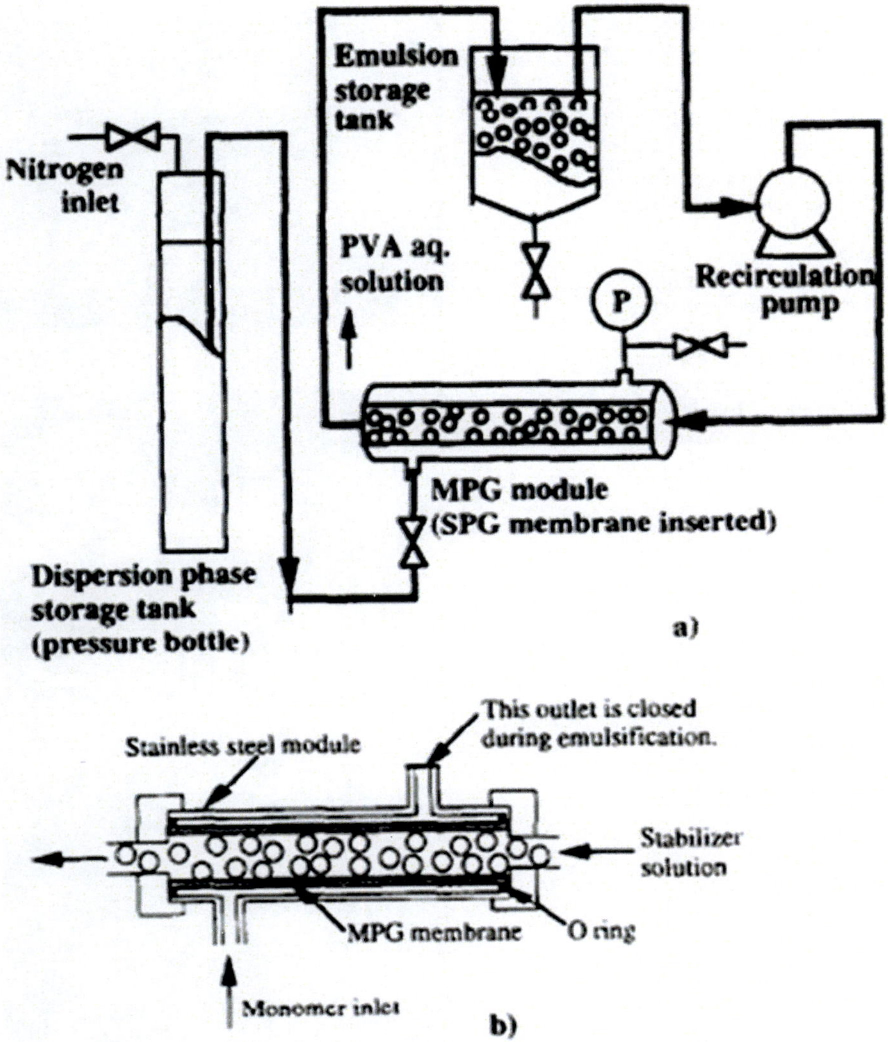

Fig. 4.15 Schematic diagram of SPG emulsification: (**a**) the entire system; (**b**) the MPG module. (Reproduced from Omi 1996)

average pore diameter of the microspheres (a) obtained by using heptane as the diluent is 80 nm, and the specific surface area is 140 m^2/g. They can be used as a packing for GPC columns. The microspheres (b) obtained by using lauryl alcohol as the diluent have an average pore diameter of 80 nm and a specific surface area of 142 m^2/g. They can be used as enzyme carriers. The polymethyl methacrylate–ethylene glycol dimethacrylate (PMMA–EGDMA) microspheres (c) prepared by the membrane emulsion-two-step swelling method also showed good porous structure.

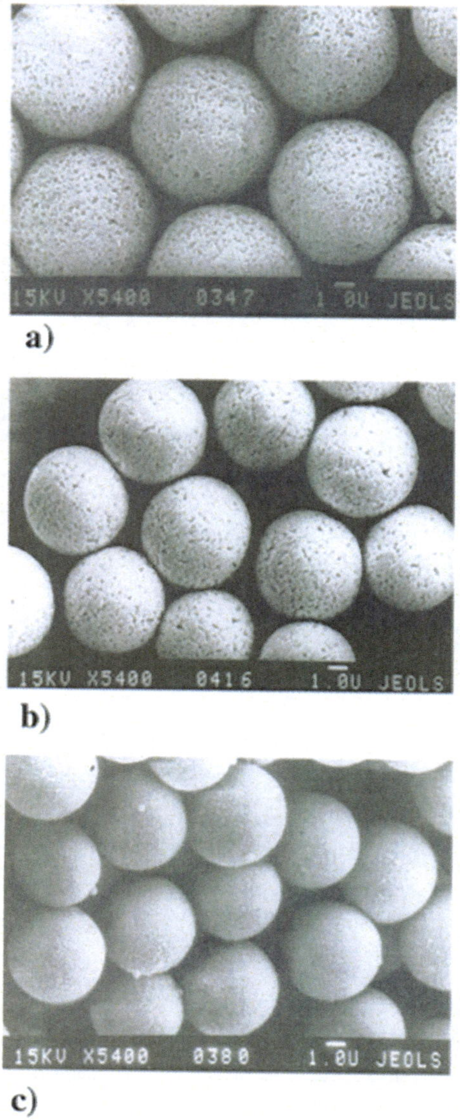

Fig. 4.16 SEM image of cross-linked porous spheres: (**a**) PS–DVB spheres, $d_p = 7.71$ μm, RSD = 9.99%, specific surface area = 140 m^2/g, used for GPC packing; (**b**) PS–DVB spheres, $d_p = 5.20$ μm, RSD = 10.9%, specific surface area = 142 m^2/g, used for enzyme carrier; (**c**) PMMA–EGDMA microspheres, $d_p = 4.99$ μm, RSD = 10.6%, specific surface area = 138 m^2/g, used for enzyme carrier. (Reproduced from Omi 1996)

As the ratio of heptane to hexanol in the diluent increases, a rough structure is formed, which is used as a carrier for enzyme immobilization, for example, immobilized glucoamylase is used to produce glucose from oligosaccharides (Omi et al. 1995a, b).

The membrane emulsion polymerization method has been used to prepare PS–DVB and PMMA–EGDMA microspheres.

4.5.2 *Characterization*

The properties of macroporous St-DVB copolymer microspheres are characterized by specific surface area, total pore volume, average pore size, pore size distribution, apparent and true density, total porosity, and swelling rate. The specific surface area is usually determined using nitrogen adsorption and desorption isotherms. In order to calculate the specific surface area of macroporous microspheres, the BET equation is usually used to analyze the relatively low-pressure region of the isotherm.

The pore size and its distribution can be measured by the following techniques: nitrogen adsorption, mercury intrusion method, electron microscope, and size exclusion chromatography. Various techniques are applicable to microspheres with different pore diameters. For example, applying the Kelvin equation to a nitrogen desorption isotherm can measure pore diameters with a maximum diameter of about 200 nm. A common method to characterize the pore size distribution is the mercury intrusion method, which can be used to determine a wide range of pore sizes. The technology is based on the penetration of mercury into the pores of dry microspheres under a certain pressure. However, the determination of pores with a diameter of less than 10 nm by mercury intrusion method is questionable, because the high pressure necessary to push mercury into pores of less than 10 nm will cause the deformation of the macroporous microspheres rather than the penetration of mercury. In addition, size exclusion chromatography is an alternative method for measuring pore diameters in the 200–500 nm range.

4.5.3 *Surface Functionalization*

Sulfonation of cross-linked polymer microspheres Strongly acidic cation exchange resin is made by sulfonation of polystyrene microspheres and introducing sulfonic acid groups on the benzene ring of cross-linked polystyrene. Generally, 93% industrial sulfuric acid is used at 70–80 °C. In the presence of a swelling agent, sulfonation is a very effective reaction. In polymers with a low to moderate degree of cross-linking, almost all aromatic rings may be substituted by sulfonic acid groups. The main limitation of this strongly acidic cation exchange stationary phase is that the selectivity to H^+ is not high and the selectivity to monovalent (alkali metal and NH_4^+) and divalent (alkaline earth metal) cations is different. In order to elute divalent cations in a short time, multivalent eluent ions, such as diaminopropionic acid, are needed. With this strong eluent, the peaks of monovalent cations, such as sodium and potassium ions, will be close to each other, or even co-elute.

Carboxylation of cross-linked polymer microspheres The use of weakly acidic ion exchange groups such as carboxylic acid can improve the selectivity to H^+. Carboxylic acid cation exchange resins generate carbonyl groups, through the reaction of polystyrene microspheres and diacetic anhydride, and then oxidize the carbonyl groups with potassium permanganate to convert them into carboxylic acids.

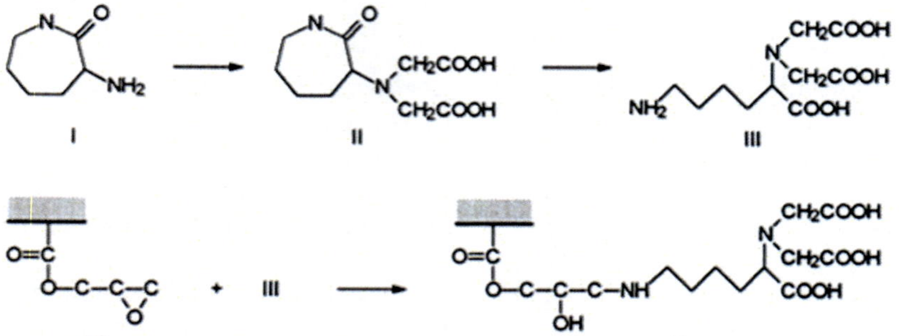

Fig. 4.17 Synthesis of chelating ion exchangers. (Reprinted from Inoue Y, Kumagai H, Shimomura Y et al. (1996) Ion chromatographic separation of rare-earth elements using a nitrilotriacetate-type chelating resin as the stationary phase. Anal Chem 68:1517–1520. Copyright {1996} American Chemical Society)

This type of stationary phase has high selectivity to H^+. A single acid (such as sulfuric acid, methanesulfonic acid) can be used as the eluent to separate monovalent and divalent cations at the same time with one injection.

Crown ether is a kind of macrocyclic complex with hydrophilic inner pores and hydrophobic outer shells. It combines with metal ions to form a stable complex. The selectivity of this cyclic crown ether to cations mainly depends on the pore size of the crown ether and the radius of the metal ion. The radius of the inner pore of 18-crown-6-ether is 1.38 nm, and the ionic radius of K^+ is 1.38 nm, so it shows strong retention of K^+. The new cation exchanger contains three functional groups: carboxylic acid, phosphonic acid, and 18-crown-6-ether, so that the monovalent K^+ elutes last.

Amination of cross-linked polymer microspheres The anion exchangers used in ion chromatography are mainly strongly basic anion exchange resins with quaternary amine groups, including alkyl quaternary amine groups and alkanol quaternary amine groups. The direct modification of polymer microspheres by chemical derivatization is one of the main preparation methods. Quaternary amine groups and tertiary amine groups and secondary amine groups can be introduced through the two-step reaction of chloromethylation and amination of polystyrene microspheres.

The surface of cross-linked polyglycidyl methacrylate microspheres (PGMA) is rich in epoxy groups, which can react with hydroxyl, amino, or sulfhydryl groups under fairly mild conditions. Suzuki and colleagues reported the preparation of several chelating ion exchangers, bonded with lysine-*N*,*N*-diacetic acid (Fig. 4.17) (Inoue et al. 1996) and 1-aminobenzyl1,2-diaminopropane-*N*,*N*,*N*, on 10 μm PGMA microspheres *N*-tetraacetic acid (Kumagai et al. 1999).

4.5.4 Applications

Size exclusion chromatography Since SEC is a separation method based on the size of the hydrodynamic diameter of the molecule, any interaction with the surface of the stationary phase will cause the elution volume to deviate from the theoretical prediction. Therefore, nonspecific interaction between the stationary phase and the separated polymer should be avoided. The stationary phase and mobile phase that match the polarity of the sample molecules are usually selected to suppress the secondary interaction effects of the chromatographic system. Cross-linked polystyrene microspheres prepared by methods, including suspension polymerization, seeded swelling polymerization, and membrane emulsion polymerization, have been used to characterize the molecular weight of synthetic polymers. After the poly(GMA-co-EDMA) microspheres are hydrolyzed, they are made into hydrophilic microspheres, which have been successfully used for the separation of protein and polystyrene standards. It is worth noting that the same column can be used for both hydrophobic and hydrophilic analytes, because the hydrolysis of poly(GMA-co-EDMA) will increase the polarity of the stationary phase.

Ion exchange chromatography Polymer matrix is the preferred material in ion exchange chromatography and is widely used in the separation and analysis of inorganic and organic ionic compounds, amino acids, nucleic acids, and other biological molecules, and metabolites in body fluids.

Chelation ion exchange chromatography Synthetic polymer microspheres with metal chelating groups on the surface are the most widely used stationary phase in chelation ion exchange chromatography. Chelex-100 produced by the Dow Chemical Company, a PS–DVB resin functionalized with iminodiacetic acid, is used in large-scale extraction and concentration of trace metals. Figure 4.18 shows a cross-linked PGMA resin with lysine-*N*,*N*-diacetic acid covalently bonded to the surface for the separation of rare earth elements (REE) (Inoue et al. 1996).

Chiral chromatography The polymer matrix provides an example of the earliest successful application in chiral chromatography. In 1968, Rogozhin and Davankov (1970) first bonded L-proline to polystyrene resin, complexed with divalent metal ions to form a ligand-exchange chiral stationary phase, by which the enantiomeric amino acids were separated by ligand exchange chromatography for the first time. Since then, a large number of chiral ligand exchange separation systems have been reported. L-Proline, L-hydroxyproline, L-isoleucine, etc. were bonded to monodisperse cross-linked polyglycidyl methacrylate-ethylene glycol diester (GMA-EDMA) microspheres, and the obtained chiral ligand exchange stationary phases were used for the direct optical resolution of a variety of derivatized and nonderivatized amino acids, α-hydroxy acids, and amino alcohol drugs (Ma et al. 2008).

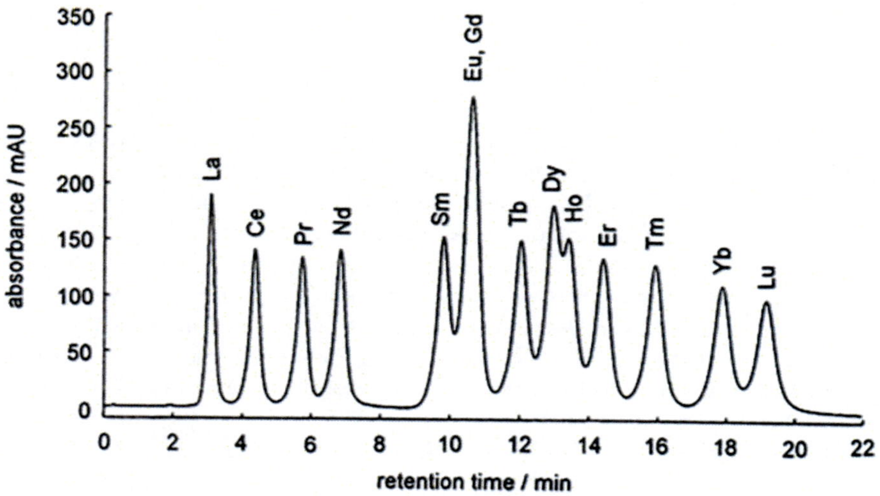

Fig. 4.18 REE chromatogram using the post-column detection method. At a flow rate of 1.0 mL/min, the concentration of nitric acid as the mobile phase increased from 20 mM at 0 min to 80 mM at 25 min. Column temperature, 40 °C; sample concentration, 5 mg/L each; injection volume, 0.05 mL. (Reprinted from Inoue Y, Kumagai H, Shimomura Y et al. (1996) Ion chromatographic separation of rare-earth elements using a nitrilotriacetate-type chelating resin as the stationary phase. Anal Chem 68:1517–1520. Copyright {1996} American Chemical Society)

4.6 Agarose

Agar is a natural polysaccharide extracted from red seaweed, composed of agar and agar gum. Agarose is a linear polymer composed of repeating units of agarobiose. Agarobiose is a disaccharide composed of D-galactose and 3,6-anhydro-L-galactose pyranose. Agar gum is a mixture of sulfated galactose, containing different ratios of sulfate esters, D-glucuronic acid, and a small amount of pyruvate. Agarose is purified by removing the agar gum in the agar. Since agarose contains almost no charged groups, it has better performance than agar in separation techniques such as chromatography and electrophoresis. Therefore, it has been widely used in molecular biology and biochemistry. As a separation medium for slab gel electrophoresis, agarose gel is used for the separation of biological macromolecules, especially DNA (Serwer 1983). As a chromatographic medium, agarose gel is the most important matrix material in biological processing. Through the functionalization of surface hydroxyl groups, agarose gels are made into various chromatographic packings for the separation and purification of recombinant proteins, monoclonal antibodies, and other biological agents. In addition, agarose gel can also be used as a solid medium for cell culture.

4.6.1 Structure and Performance

Agarose is a linear polymer with a molecular weight of approximately 120,000 Da. It consists of alternating D-galactose and 3,6-anhydro-L-galactose pyranose linked by α-(1,3) and β-(1,4) glycosidic bonds. Each agarose chain contains about 800 galactose molecules. The polymer chains form a double helix structure with 0.95 nm axial periodicity. Each chain in the double helix forms a left-handed triple fold with a spacing of 1.9 nm and is axially translated by half this distance relative to its partner (Choji 1956; Abraham et al. 1975).

Agarose is a white powder, soluble in boiling water, and forms a gel after cooling. The gelation of agarose involves the transition from the random coils in the solution to the double helix at the initial stage of gelation and the transformation into the double helix bundle at the final stage (Lahoussine et al. 2001). Agarose gel is a porous network structure, and the size of the pores depends on the concentration of agarose in the gel. The higher the agarose concentration, the smaller the pores, and vice versa.

4.6.2 Preparation

In the literature, several preparation methods have been described of agarose gel microspheres, including emulsion coagulation, spray drying, and membrane emulsification.

Emulsion coagulation Hjerten (1964) first used the emulsion coagulation method to prepare agarose gel microspheres. The basic steps of this method are as follows: dissolve the agarose powder in boiling water, add it to an organic solvent containing an emulsifier, and heat it to 80–90 °C under mechanical stirring to form a water-in-oil emulsion. Then, reduce the reaction temperature to the solidification temperature of the agarose to obtain agarose gel microspheres or beaded agarose gel. In industrial production, high-shear mixers can be used to prepare emulsions to increase production capacity. The particle size and its distribution are related to the preparation conditions.

The advantage of the emulsion coagulation method is that the equipment is simple and the operation is convenient and easy to operate. The disadvantage is that the particle size distribution of the obtained product is wide, and size classification is required to obtain the microspheres of the target particle size.

The spray coagulation method is proposed by Bengtsson and Philipson (1964) in 1964. The hot agarose solution is atomized through a nozzle under pressure, and the droplets formed are suspended in water–ether mixture and continuously stirred to induce gelation of the agarose beads. The particle size is controlled by the aperture of the atomizer and the applied pressure. The resulting agarose beads have a relatively wide particle size and need to be classified (Egorov et al. 1970).

Membrane emulsification The common disadvantage of the abovementioned emulsion coagulation method and spray coagulation method for preparing agarose gel microspheres is that the obtained microspheres have a wide particle size distribution and need to be classified. In the membrane emulsification, the particle size distribution can be better controlled and uniform size microspheres are produced. Zhou et al. (2007, 2009) reported the preparation of agarose gel microspheres by membrane emulsification. First, agarose powder is dissolved in boiling water to prepare a dispersed phase. Then, under a certain pressure, the dispersed phase is allowed to pass through a porous membrane with a uniform pore size and enter immiscible continuous phase to obtain an agarose emulsion with uniform droplet size distribution. Then, under conditions similar to those in the emulsification coagulation method, the agarose emulsion is cooled to induce gelation of the beads. The coefficient of variation of particle size distribution of agarose gel microspheres prepared by membrane emulsification is usually within 20%, which is particularly suitable for chromatographic applications. Since there is no need for size classification, it not only simplifies the production process but also saves raw materials and reduces production costs. Detailed procedure is illustrated as follows:

Agarose gel microspheres were prepared by membrane emulsification combined with cooling and solidification process (Zhou et al. 2007). The water phase is an aqueous solution containing 4 wt% agarose (0.9 wt% NaCl), and the oil phase is liquid paraffin/petroleum ether containing 4 wt% PO-500, 7:5 (v/v), and the reaction temperature is 55 °C. The agarose powder was dissolved in 0.9 wt% NaCl boiling water. Liquid paraffin was mixed with 4 wt% PO-500 in a ratio of 7:5 (v/v) to obtain an oil phase containing 4 wt% emulsifier. The oil phase was heated to 55 °C. Then, the water phase was poured into the storage tank of the membrane emulsification device, forced through the membrane pores under nitrogen pressure, and entered the oil phase to form a uniform water-in-oil emulsion. The emulsion was then cooled to room temperature under gentle stirring to induce the solidification of the agarose emulsion droplets. The microspheres were collected by filtration, washed with petroleum ether, ethanol, and water in sequence to obtain monodisperse agarose microspheres.

The optical micrographs of the agarose gel microspheres prepared by the membrane emulsification method and the conventional emulsification method are shown in Fig. 4.19. The diameter of the microspheres prepared by the membrane emulsification method is 30 μm, the coefficient of variation is 15%, and they are mainly distributed between 28 and 32 μm. The average diameter of the microspheres prepared by the mechanical stirring method is 34 μm, the coefficient of variation is 71%, and the particle size distribution is between 10 and 60 μm.

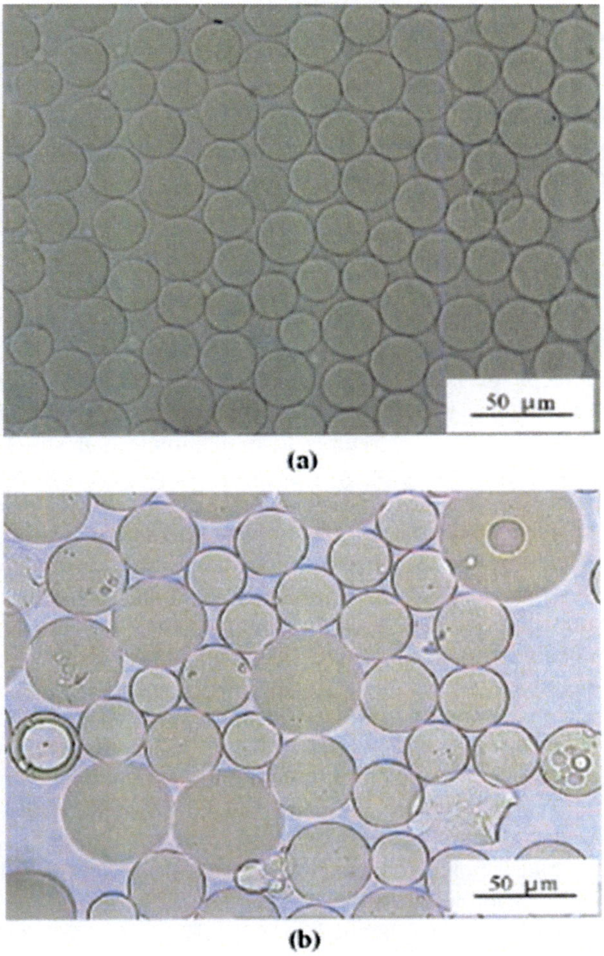

Fig. 4.19 Optical micrograph of agarose beads: (**a**) Prepared by standard formula membrane emulsification method (membrane pore size is 10.2 μm); (**b**) Prepared by mechanical stirring emulsification method (average diameter is 34 μm). (Reproduced from Zhou et al. 2007)

4.7 Cellulose

Cellulose is a renewable biopolymer that exists widely in nature. It is usually combined with lignin and hemicellulose to form plant cell walls. This homopolysaccharide is formed by the linear connection of D-glucose units condensed with β-(1–4)-glycosidic bonds (Gericke et al. 2013). Depending on the source, the degree of polymerization (DP) of cellulose can vary from 100 to 20,000. From a single anhydroglucose pyranose unit (AGU) to microfibrils and macrofibrils, cellulose is organized in quite complex ways. Among them, intramolecular and intermolecular hydrogen bonds play a key role in constructing cellulose supramolecular structure. Each AGU has three hydroxyl groups, which can form strong intra- and intermolecular hydrogen bonds. Intramolecular hydrogen bonds can provide chain rigidity, and intermolecular hydrogen bonds can assemble linear polymer molecules into a sheet structure. In addition, the hydrocarbon groups on the plane of the glucopyranose ring constitute the hydrophobic center. This synergistic effect of hydrogen bonds and hydrophobic bonds can further promote the formation of sheet-like structures, making cellulose neither meltable nor soluble in most common organic solvents.

4.7.1 Dissolution and Regeneration

Except for a small part of cellulose directly used, most of it is regenerated after dissolution. The so-called regeneration means that after cellulose is dissolved in a suitable solvent, it returns to a supramolecular structure composed of hydrogen bonds and hydrophobic bonds in a non-solvent coagulation bath. Efficiently dissolving cellulose is a challenge. Cellulose solvents are classified into derivatized solvents and nonderivatized solvents according to whether a derivatization reaction occurs during the dissolution process. The most important ones are CS_2/NaOH system, carbamic acid system, ammonium oxide system, lithium chloride/dimethylacetamide system, ionic liquid, copper ammonia solution, and alkali/urea aqueous solution system. In the past two decades, due to its high efficiency, nontoxicity, low pollution, and low cost, alkali/urea aqueous solutions have received great attention. Alkali/urea aqueous solution can quickly dissolve cellulose in a few minutes, when it is precooled to -12 to -8 °C, and can be used to prepare cellulose functional materials for different purposes (Qi et al. 2008).

The study of its mechanism of action shows that at low temperatures, NaOH, urea, and water associate in a solvent to form an inclusion compound. First, NaOH hydrate is bound to the cellulose molecular chain through the formation of hydrogen bonds with cellulose and further captures urea to form a wormlike pipe inclusion compound containing cellulose-NaOH guest as the backbone. It is highly stable at low temperatures. Therefore, the dissolution of cellulose at low temperatures is caused by dynamic self-assembly driven by hydrogen bonds between solvent small molecules (NaOH, urea, water) and cellulose macromolecules (Qin et al. 2013).

4.7.2 Preparation

Cellulose microspheres can be prepared by different methods, such as emulsion solidification and spray drying. First, the cellulose or cellulose derivative solution is prepared and then dispersed into a specific immiscible continuous phase to form microspheres or directly dispersed into hot air or the solidification liquid through an atomizer.

Emulsion solidification The emulsion solidification method, as described by Stamberg (1988), is based on the use of viscose to pass through the hot sol–gel transition process. This method can be implemented by the following steps: (1) preparing a cellulose solution, (2) preparing an emulsion of a cellulose solution, and (3) curing the droplets formed in emulsion, and (4) regeneration of the cellulose beads. For cellulose aqueous solutions, the continuous phase is an organic solvent containing emulsifiers, such as liquid paraffin, edible vegetable oil, etc. The curing process is the phase transition process of cellulose in the emulsion system, through heating, acid or salt precipitation, cross-linking, etc. to achieve the purpose of curing. The regeneration process is the process of returning the cellulose solution state to solid cellulose. There is no clear boundary between curing and regeneration. In most cases, they are done at the same time.

Membrane emulsification O'Brien et al. (2017) reported a membrane emulsification method of cellulose microbeads as a sustainable alternative to plastic microparticles. It is the first demonstration that the cross-flow membrane emulsification-phase inversion process can be used to produce stable cellulose droplets. The emulsification system consists of cellulose dissolved in an organic electrolyte solution (1-ethyl-3-methylimidazole acetate: DMSO) as the dispersed phase and sunflower oil–Span 80 as the continuous phase. The emulsion is then coagulated with antisolvent, resulting in the formation of solid, spherical, and biodegradable cellulose microbeads. The systematic analysis of the process parameters (continuous- and dispersed-phase flow rates, viscosity, and applied pressure) indicated that it is possible to produce uniform cellulose microspheres using porous glass membranes with a pore size of 10 μm.

Alkali/urea/water solution method Luo and Zhang (2010) reported that the NaOH/urea/water solution system was used as a cellulose solution to quickly dissolve the cellulose at low temperature, and then the cellulose solution was dispersed in liquid paraffin and stirred to form an emulsion. The droplets in the emulsion were solidified to form regenerated cellulose microspheres by adjusting the pH value of the solution to neutral. Specific steps are as follows:

The preparation of the cellulose solution: the NaOH/urea/water solution with a mass ratio of 7/12/81 was precooled to −12.3 °C. Immediately 8 g of cellulose were added to the 200 mL solvent system at room temperature and stirred quickly for 3 min to obtain a transparent solution. After degassing, the solution was added to the liquid paraffin containing Span 80, stirred, and emulsified for 3 h. Then, dilute hydrochloric acid was added dropwise to adjust the pH to 7.0. The droplets solidify to form regenerated cellulose microspheres. After standing the solution is divided into two layers. The upper liquid paraffin organic phase was removed and the lower aqueous phase filtered to recover cellulose microspheres. The particles were washed with distilled water and acetone and stored in a 20% ethanol solution.

***N*-Methylmorpholine-*N*-oxide (NMMO) method** Cellulose is mixed with NMMO aqueous solution (60% by mass), swollen at 80 °C for 2 h, and 0.5% by mass of cellulose antioxidant (pyrogallic acid) is added. At 80–120 °C and the vacuum greater than 0.095 mPa, the cellulose is dissolved under vigorous stirring, and the excess water is evaporated to obtain a light brown transparent solution, which is solid when cooled to room temperature.

In a 250-mL three-necked flask, a certain proportion of organic phase (liquid paraffin, kerosene, or transformer oil) is added, which is followed by the addition of cellulose solution (actually solid) and an appropriate amount of emulsifier. The mixture is heated to dissolve the solid, stirred to disperse the cellulose phase, and then cooled until microspheres solidified. Finally, after soaking in water for 24–48 h to dissolve NMMO, white cellulose microspheres are obtained.

Viscose method Technical grade viscose (100 g, containing 82% cellulose, 6% NaOH, $\gamma = 40$) is dispersed in a 400-mL transformer oil (viscosity 26 cSt). At room temperature, the mixture is stirred at 460 rpm and heated to 90 °C for 1.5 h. The droplets solidify (the viscose solidifies when the temperature is increased), and the solids are collected by filtration. The particles are dispersed in a mixed solution of acetic acid (20 mL)/ethanol (80 mL) and stirred for 1 h to decompose cellulose xanthogen acid, filtered, washed, and vacuum dried (Peska et al. 1977).

4.8 Dextran

Dextran is a product secreted by bacteria and consists of a D-glucose main chain connected by α-(1,6)-glycosidic bonds and different proportions of branched chains located at different branch points. Natural dextran has a relatively high molecular weight, from 9×10^6 to 5×10^8 g/mol. The dispersibility of the molecular weight increases with the increase of the molecular weight. This is due to the increase in branch density. Generally speaking, α-(1,6)-glycosidic bonds provide higher chain mobility, resulting in high solubility in different solvents. These solvents include water, DMSO, DMA/LiCl, formamide, ethylene glycol, glycerol, 6 M urea, 2 M glycine, and 4-methylmorpholine 4-oxide (Heinze et al. 2006).

4.8.1 Preparation

The dextran gel microspheres are prepared by the process of emulsion-cross-linking and solidification. The dextran aqueous solution containing the cross-linking agent is dispersed in the organic phase to form a water-in-oil emulsion. Heating initiates a cross-linking reaction and solidifies the droplets. Commonly used cross-linking agent is epichlorohydrin. The pore size of dextran gel microspheres is affected by both the concentration of polysaccharide and the concentration of cross-linker, which decreases with the increase of the polysaccharide concentration and the amount of cross-linking agent.

The two end groups of epichlorohydrin react with the free hydroxyl groups on the two glucose residues, respectively. The epoxy group opens the ring to form an ether bond. The chlorine atom reacts with other sugar hydroxyl groups, to remove a molecule of HCl, and also forms an ether bond between the molecules, so that the dextran molecules are cross-linked to form a network structure. The amount of cross-linking agent and reaction conditions when synthesizing cross-linked dextran determine the degree of cross-linking and mesh size of the gel, which also determines the compressive strength and exclusion limit of the medium.

4.9 Surface Functionalization of Polysaccharide Matrix

The surface of the polysaccharide matrix is rich in hydroxyl groups, which provides a large number of active sites for the derivatization reaction. The functionalization of the polysaccharide matrix can be roughly divided into three types: one is that the polysaccharide matrix reacts directly with the active groups of the ligand to bond the ligand to the surface of the matrix. This method is called the direct coupling method; the other is that after the polysaccharide matrix is activated, it reacts with the active group of the ligand to obtain the chromatographic stationary phase. This method is called the activation coupling method; the third is that after activation of the polysaccharide matrix, it reacts with one of the two functional groups of the spacer and then reacts with the active group of the ligand to obtain the chromatographic stationary phase. This method is called the spacer coupling method.

4.9.1 Direct Coupling Method

The most commonly used weak anion exchange packing in protein ion exchange chromatography is Diethylaminoethyl Sepharose, also called DEAE Sepharose. It is prepared by the etherification reaction between the hydroxyl groups on the surface of agarose gel microspheres and diethylaminochloroethane hydrochloride under alkaline catalysis. In this reaction, NaOH is both a catalyst and an acid scavenger, which neutralizes the hydrogen chloride released by the reaction.

$$R-OH+(CH_3CH_2)_2NCH_2CH_2Cl=R-OCH_2CH_2N(CH_2CH_3)_2 \tag{4.2}$$

The most commonly used weak cation exchange packing is Carboxymethyl Sepharose, also called CM Sepharose. It is prepared by the etherification reaction between the hydroxyl groups on the surface of agarose gel microspheres and 2,3-epoxypropyltrimethylammonium chloride under alkaline catalysis.

$$R-OH+ClCH_2COOH=R-O-CH_2COOH \tag{4.3}$$

The most commonly used strong anion exchange packing is Trimethylamino Sepharose, also called Q Sepharose. It is prepared by the etherification reaction between the hydroxyl groups on the surface of agarose gel microspheres and 2,3-epoxypropyltrimethylammonium chloride under alkaline catalysis.

$$R - OH + CH_2OCHCH_2N(CH_3)_3{}^{+} = R - OCH_2CH(OH)CH_2N(CH_3)_3{}^{+} \quad (4.4)$$

The most commonly used strong cation exchange packing is Sulfopropyl Sepharose, also called SP Sepharose. It is obtained through a two-step reaction. The agarose gel microspheres are firstly reacted with allyl glycidyl ether under alkaline catalysis to obtain allyl agarose gel microspheres and then reacted in a sodium metabisulfite solution of pH 6.5 at room temperature to obtain SP agarose gel.

4.9.2 *Activation Coupling Method*

The activation of the polysaccharide matrix is to use a bifunctional reagent to react with sugar hydroxyl groups to obtain an activated matrix with reactive groups and then react with the ligand to obtain a functionalized chromatographic packing.

Cyanogen bromide (BrCN) activation Cyanogen bromide activation method is a classic method for preparing polysaccharide matrix chromatographic stationary phases. It is carried out in two steps: the activation of the agarose microspheres by cyanogen bromide and coupling of the functional ligand to the activated matrix. The reaction of activating and coupling should be carried out at a lower temperature (4 °C to room temperature). The pH value of the activation reaction medium is controlled at pH 10.8–11.0, which can be controlled by NaOH titration or a high-concentration buffer solution, such as a 2-M carbonate solution with pH 10.9. The pH of the reaction to couple the ligand is between pH 4.5 and 10 (depending on the reactants).

The reaction results in a mixture of three different structures. The amount of each structure depends on the type and conditions of the polysaccharide of the matrix. The agarose matrix mostly undergoes cyanate ester active intermediates and then forms isourea derivatives with ligands. Isourea derivatives tend to be positively charged.

The advantage of this method is that the reaction conditions are mild, and the reaction temperature is from 4 °C to room temperature, which is particularly useful for the preparation of affinity chromatography packings by coupling proteins. The disadvantage is that cyanogen bromide is a highly toxic substance, which is harmful to human health; the ligand needs to contain amino groups for the coupling reaction to occur; the generated isourea derivative is not stable, and the ligand is easy to leak and lose; isourea derivatives introduce positive charges and thus added sites for interaction with anionic groups, which complicates solute retention behavior.

Trichlorotriazine or dichlorotriazine activation Trichlorotriazine is also called cyanuric chloride. Dichlorotriazine is a derivative of trichlorotriazine with one of the chlorines being substituted. A water-soluble triazine can be obtained by substituting a carboxyethyloxy group or carboxyethylamino group for a chlorine in trichlorotriazine. Under the conditions of pH 9–11 and 20 °C, it can quickly react with the substrate to form a monochlorotriazinyl–polysaccharide complex. Subsequently, under the conditions of pH 7–9 and 0–20 °C, the triazinyl-activated matrix reacts slowly with strong nucleophilic molecules such as primary amino groups (16–20 h). Similarly, trichlorotriazine can also react with hydroxyl-containing substrates to form dichlorotriazine derivatives. At pH 7 or so, it is very easy to couple with proteins.

Epoxy activation The epoxy activation method introduces epoxy groups through the reaction with sugar hydroxyl groups. The epoxy groups can react with amino groups, hydroxyl groups, and sulfhydryl groups for the next step of ligand coupling. Unlike the BrCN activation method, the epoxy activation method can react with nonamino groups without introducing positively charged groups. The epoxy compound used for activation is a bifunctional reagent. Commonly used reagents include epichlorohydrin and diepoxy butane (or ethylene) glycol bisglycidyl ether.

Epoxy activation is equivalent to attaching a spacer arm with an epoxy end group to the substrate. Epoxy-activated products can easily react with ligands with amino and sulfhydryl groups and can also react with ligands with hydroxyl groups under certain conditions.

The epoxy activation reaction is carried out under alkaline conditions. In addition to the activation and attachment of epoxy groups, there are three side reactions that may occur: firstly, the epoxy group attached to the substrate after activation will undergo ring-opening reaction to yield diol groups; secondly, epoxy groups attached can react with another hydroxyl group of the polysaccharide matrix to form a cross-linked product; thirdly, the epoxy compound itself self-polymerizes. Due to the existence of side reactions, the excess amount of epoxy compound is generally used in activation.

Divinyl sulfone activation Divinyl sulfone is an activating reagent with stronger reactivity than epoxy compounds. It contains two active alkenyl groups: one can activate the polysaccharide matrix in alkaline solution to generate active intermediates, and the other alkenyl group in the active intermediates can be coupled with the amino, hydroxyl, and sulfhydryl groups of the ligand.

Benzoquinone activation Quinone reacts with hydroxyl-containing substrates more easily than divinyl sulfone to connect to the substrate to obtain intermediates and then couples with amino-containing ligands to obtain chromatography packing. This method has high coupling efficiency and can be carried out in a wide pH range.

Periodate activation The periodate activation method uses an oxidant with moderate oxidizing ability, such as periodate, which can oxidize the *o*-dihydroxy group of the polysaccharide matrix into an aldehyde group and react with the primary amine group in the ligand under the condition of pH 4–6. A Schiff base is generated, and then the Schiff base is reduced with sodium cyanoborohydride at pH 6–6.5 to obtain a chromatographic packing.

***N,N′*-Disuccinimidyl carbonate (DSC) activation** *N,N′*-disuccinimidyl carbonates activate polysaccharide matrix in a nonaqueous organic solvent without active hydrogen. The reaction is carried out under the catalysis of base (triethylamine, pyridine). Because *N,N′*-disuccinimidyl carbonate reacts with the hydroxyl groups of polysaccharides, the resulting intermediates with succinimidyl carbonate groups will be rapidly hydrolyzed in water. The activated intermediate reacts with the amino-containing ligand to form a stable, uncharged carbamate. The coupling reaction can be carried out in a neutral (pH 6–8) aqueous solution or in an organic solvent.

***N,N′*-Carbonyl diimidazole (CDI) activation** The hydroxyl-containing substrate reacts with carbonyl diimidazole (CDI) in an anhydrous organic solvent without active hydrogen to form the active intermediate imidazolyl carboxylate. In an alkaline solution, the active intermediate can react with amino-containing ligands to form carbamates to obtain functional chromatographic packing.

Organic sulfonyl chloride activation The organic sulfonyl chloride reacts with the hydroxyl group of the substrate to form an easily leaving sulfonyl ester group. Generally, *p*-toluenesulfonyl chloride or trifluoroethanesulfonyl chloride is used in an anhydrous organic phase without active hydrogen, preferably in acetone, to activate the matrix. When the ligand is coupled, the nucleophilic group of the ligand, such as amino, sulfhydryl, etc., replaces the sulfonyl group. The coupling reaction can be carried out in water or organic solvents.

4.9.3 Spacer Coupling Method

In the direct coupling and activation coupling methods introduced in the previous section, some ligands have no spacer function when they are coupled to the matrix, and the ligand has a relatively large steric hindrance effect, which is not conducive to the interaction with solute molecules. In this case, it is usually considered to connect a spacer arm between the substrate and the ligand to extend the distance between the ligand and the substrate and expand the movement space of the ligand. This is the spacer coupling method. The spacer method is usually carried out in three steps: matrix activation, spacer arm coupling, and ligand coupling to the spacer arm.

The commonly used spacers are bifunctional reagents, such as aminocaproic acid, hexamethylene diamine, diaminopropyl imine, and so on. The amino groups of these spacers can react with the activated groups of the matrix, and another reactive functional group can react with the ligand under the action of the coupling agent to prepare functional chromatography packing.

For example, cyanogen bromide activates the polysaccharide matrix, connects the spacer arm, and then connects the ligand to synthesize affinity chromatography packing material, which is a common method. The activated polysaccharide matrix reacts with a bifunctional reagent as a spacer arm, and a spacer arm is attached to extend a certain length of molecular arm outward. At the same time, its end group has a group that can be coupled with a ligand. This group is usually carboxy, or amino. In the presence of the coupling agent, the carboxyl or amino group on the substrate reacts with the amino or carboxyl group on the ligand to form a functional chromatographic packing with spacer arms.

Commonly used coupling agents can be divided into two categories: reagents that react in the aqueous phase and reagents that react in the organic phase or the aqueous organic phase. The most commonly used coupling agent for aqueous reaction is water-soluble carbodiimide, such as *N*-ethyl-*N*′-(3-dimethylaminopropyl)-carbodiimide (EDC). Glutaraldehyde was also commonly used as a coupling agent in the past. The coupling agent for the aqueous organic phase reaction is dicyclohexylcarbodiimide and the like. Generally, the coupling agent for aqueous phase reaction is preferred, and the coupling agent for aqueous organic phase reaction is secondly selected.

Carbodiimide coupling There are two types of carbodiimide coupling agents: water-soluble and non-water-soluble. Dicyclohexylcarbodiimide (DCC) is insoluble in water but can be used in 80% (volume fraction) pyridine or acetonitrile aqueous solution or in dioxane and dimethyl sulfoxide to make ligands with amino groups. The polysaccharide matrix with ω-carboxyl group is coupled to each other. However, the produced by-product urea is insoluble in water and must be thoroughly washed with organic solvents such as ethanol or butanol to remove it. If water-soluble carbodiimide is used, such as *N*-ethyl-*N*′-(3-dimethylaminopropyl)-carbodiimide (EDC) or *N*-cyclohexyl-*N*′-(2-morpholineethyl)-carbodiimide (CMC) *p*-toluenesulfonate can overcome this difficulty.

Glutaraldehyde coupling The two aldehyde groups at both ends of the glutaraldehyde molecule can form Schiff bases with the amino groups on the carrier and the ligand, respectively, and connect them through a five-carbon bridge. The reaction conditions are mild, can be carried out at 4–40 °C and pH 6–8, and the operation is simple, so it is widely used.

***N*,*N*′-Carbonyl diimidazole (CDI) coupling** CDI can activate both hydroxyl and carboxyl groups. CDI is used to activate the substrate with carboxyl spacer. The carboxyl group and imidazole form an intermediate amide. In the presence of the primary amine compound, the imidazole is substituted, and a stable amide bond is formed between the substrate and the ligand.

***N*-Hydroxysuccinimide (NHS) coupling** *N*-Hydroxysuccinimide (NHS) can activate substrates with carboxyl groups. Under the action of NHS, the carboxyl group is converted into active ester, and the reaction is carried out in the presence of carbodiimide (such as dicyclohexylcarbodiimide (DCC)) in anhydrous organic medium. The resulting *N*-hydroxysuccinimide ester reacts with the amino group of the ligand at pH 5–8 to form an amide and release *N*-hydroxysuccinimide.

4.10 Summary

The stationary phase in liquid chromatography usually consists of a porous solid-phase carrier and functional groups covalently bound to the surface. The porous solid-phase carrier, also called the matrix, plays a very important role in the chromatographic process. The morphology and size distribution of its particles determine the uniformity of the packed column bed and the band dispersion caused by flow; its pore structure determines the mass transfer resistance and binding capacity of the solute; its mechanical strength determines the pressure-bearing capacity of the packed column. Existing chromatographic matrices can be divided into three categories: inorganic materials, synthetic polymers, and polysaccharides. Inorganic materials include silica gel, porous glass, and hydroxyapatite. Among them, silica gel is the most widely used. Especially in high-performance liquid chromatography, silica gel has excellent mechanical strength and can withstand pressures up to 1000 bar. However, the pH range of silica gel is relatively narrow and cannot withstand the clean-in-place treatment by sodium hydroxide, which limits its application in biological processing. The synthetic polymer is represented by cross-linked polystyrene–divinylbenzene polymer, which has good mechanical strength and a wide pH range. After surface functionalization, it can be used for ion exchange chromatography. With the introduction of hydrophilic macroporous materials with through pores, the application of polymer materials in the separation and purification of biomolecules is increasing, showing great potential. Polysaccharides are natural polymers formed by the condensation of glucose and other units, including agarose, cellulose, and dextran. Due to the balance between hydrophilicity and mechanical strength, polysaccharide materials have become the preferred matrix in the chromatographic purification of biological macromolecules. Among them, agarose materials are the most widely used. The surface functionalization of the silica matrix is usually through the reaction of silane coupling reagents with silanol groups to covalently bond functional groups to the silica gel surface. The surface functionalization of the polysaccharide matrix can be conveniently combined with ligands containing amino, hydroxyl, or sulfhydryl groups through the activation reaction of surface hydroxyl groups.

References

Abraham A, Er-el Z, Shaltiel S (1975) Ultrastructure of beaded agarose. Arch Biochem Biophys 171:673–677

Bastan FE, Erdogan G, Moskalewicz T et al (2017) Spray drying of hydroxyapatite powders: the effect of spray drying parameters and heat treatment on the particle size and morphology. J Alloys Compounds 724:586–596
Beck JS, Vartuli JC, Roth WJ et al (1992) A new family of mesoporous molecular sieves prepared with liquid crystal templates. J Am Chem Soc 114:10834–10843
Bengtsson S, Philipson L (1964) Chromatography of animal viruses on pearl-condensed agar. Biochim Biophys Acta 79:399–406
Bernardi G (1973) Chromatography of proteins on hydroxyapatite. Methods Enzymol 27:471–479
Bernardi G, Kawasaki T (1968) Chromatography of polypeptides and proteins on hydroxyapatite columns. Biochim Biophys Acta 160:301–310
Bernardi G, Giro M-G, Gaillard C (1972) Chromatography of polypeptides and proteins on hydroxyapatite columns: some new developments. Biochim Biophys Acta 278:409–420
Boissière C, Larbot A, Prouzet E (2000) Preparation of spherical micrometric MSU-X mesoporous silica particles for chromatography applications. Stud Surf Sci Catal 129:31–36
Büchel G, Grün M, Unger KK et al (1998) Tailored syntheses of nanostructured silicas: control of particle morphology, particle size and pore size. J Supramol Sci 5:253–259
Cahn JW, Hilliard JE (1958) Free energy of a nonuniform system. I. Interfacial free energy. J Chem Phys 28:258–267
Camli ST, Unsal E, Senel S et al (2004) Chromatographic performance of monodisperse macroporous particles produced by modified seeded polymerization. II. The effect of the diluent/seed-latex ratio. J Appl Polym Sci 92:3685–3696
Cheng CM, Micale FJ, Vanderhoff JW et al (1992) Synthesis and characterization of monodisperse porous polymer particles. J Polym Sci A Polym Chem 30:235–244
Choji A (1956) Structure of the agarose constituent of agar–agar. Bull Chem Soc Jpn 29:543–544
Coutinho FMB, Alves GM, Neves AFS et al (2000) Modified suspension polymerization of styrene-divinylbenzene for application in size exclusion chromatography. Macromol Symp 151:399–405
Descamps M, Hornez JC, Leriche A (2009) Manufacture of hydroxyapatite beads for medical applications. J Eur Ceramic Soc 29:369–375
Dong H, Brennan JD (2006a) Controlling the morphology of methylsilsesquioxane monoliths using a two-step processing method. Chem Mater 18:541–546
Dong H, Brennan JD (2006b) Macroporous monolithic methylsilsesquioxanes prepared by a two-step acid/acid processing method. Chem Mater 18:4176–4182
Dong H, Brook MA, Brennan JD (2005) A new route to monolithic methylsilsesquioxanes: gelation behavior of methyltrimethoxysilane and morphology of resulting methylsilsesquioxanes under one-step and two-step processing. Chem Mater 17:2807–2816
Dragosavac MM, Vladisavljević GT, Holdich RG et al (2012) Production of porous silica microparticles by membrane emulsification. Langmuir 28:134–143
du Fresne von Hohenesche C, Ehwald V, Unger KK (2004) Development of standard operation procedures for the manufacture of n-octadecyl bonded silicas as packing material in certified reference columns for reversed-phase liquid chromatography. J Chromatogr A 1025:177–187
Egorov AM, Vakhabov AK, Chernyak VY (1970) Isolation of agarose and granulation of agar and agarose gel. J Chromatogr 46:143–148
Evans RA (2007) The rise of azide–alkyne 1,3-dipolar 'click' cycloaddition and its application to polymer science and surface modification. Aust J Chem 60:384–395
Gagnon P (2009) Monoclonal antibody purification with hydroxyapatite. New Biotechnol 25:287–293
Galarneau A, Iapichella J, Brunel D et al (2006) Spherical ordered mesoporous silicas and silica monoliths as stationary phases for liquid chromatography. J Sep Sci 29:844–855
García MC, Marina ML, Torre M (2000) Perfusion chromatography: an emergent technique for the analysis of food proteins. J Chromatogr A 880:169–187
Garcia-Diego C, Cuellar J (2006) Determination of the quantitative relationships between the synthesis conditions of macroporous poly(styrene-co-divinylbenzene) microparticles and the

characteristics of their behavior as adsorbents using bovine serum albumin as a model macromolecule. Ind Eng Chem Res 45:3624–3632
Gericke M, Trygg J, Fardim P (2013) Functional cellulose beads: preparation, characterization, and applications. Chem Rev 113:4812–4836
Gorbunoff MJ (1984a) The interaction of proteins with hydroxyapatite. I. Role of protein charge and structure. Anal Biochem 136:425–432
Gorbunoff MJ (1984b) The interaction of proteins with hydroxyapatite. II. Role of acidic and basic groups. Anal Biochem 136:433–439
Gorbunoff MJ, Timasheff SN (1984) The interaction of proteins with hydroxyapatite. III. Mechanism. Anal Biochem 136:440–445
Grama S, Horak D (2015) Preparation of monodisperse porous silica particles using poly(glycidyl methacrylate) microspheres as a template. Physiol Res 64:S11–S17
Grün M, Unger KK, Matsumoto A et al (1997) In: McEnaney B, Mays TJ, Rouquerol J et al (eds), Proceedings of the COPS-IV symposium, London, p 81
Haller W (1965) Chromatography on glass of controlled pore size. Nature 206:693–696
Hao Z-X, Luo H, Cheng Y-Y et al (2011) Synthesis of silica microspheres under hybridization of urea-formaldehyde resin in the presence of surfactants. Acta Chim Sin 69:2531–2538
Hayes R, Ahmed A, Edge T et al (2014) Core-shell particles: preparation, fundamentals and applications in high performance liquid chromatography. J Chromatogr A 1357:36–52
He J, Yang C, Xiong X et al (2012) Preparation and characterization of monodisperse porous silica microspheres with controllable morphology and structure. J Polym Sci A Polym Chem 50: 2889–2897
Heinze T, Liebert T, Heublein B et al (2006) Functional polymers based on dextran. Adv Polym Sci 205:199–291
Hjerten S (1964) The preparation of agarose spheres for chromatography of molecules and particles. Biochim Biophys Acta 79:393–398
Hood HP, Nordberg ME (1938) Treated borosilicate glass. U.S. patent 2106744
Huo Z, Wan Q, Chen L (2018) Synthesis and evaluation of porous polymethylsilsesquioxane microspheres as low silanol activity chromatographic stationary phase for basic compound separation. J Chromatogr A 1553:90–100
Huo Z, Wan Q-H, Chen L (2020) Energy-efficient and environment-friendly method to prepare monodispersed silica stationary phases for simultaneous separation of compound drugs. J Chromatogr A 1618:460866
Inoue Y, Kumagai H, Shimomura Y et al (1996) Ion chromatographic separation of rare-earth elements using a nitrilotriacetate-type chelating resin as the stationary phase. Anal Chem 68: 1517–1520
Jiang S-D, Yao Q-Z, Zhou G-T et al (2012) Fabrication of hydroxyapatite hierarchical hollow microspheres and potential application in water treatment. J Phys Chem C 116:4484–4492
Kanamori K, Nakanishi K (2011) Controlled pore formation in organotrialkoxysilane-derived hybrids: from aerogels to hierarchically porous monoliths. Chem Soc Rev 40:754–770
Kanamori K, Yonezawa H, Nakanishi K et al (2004) Structural formation of hybrid siloxane-based polymer monolith in confined spaces. J Sep Sci 27:874–886
Kawasaki T (1991) Hydroxyapatite as a liquid chromatographic packing. J Chromatogr 544:147–184
Kolb HC, Finn MG, Sharpless KB (2001) Click chemistry: diverse chemical function from a few good reactions. Angew Chem Int Ed 40:2004–2021
Kumagai H, Yokoyama T, Suzuki TM et al (1999) Liquid chromatographic selectivity and retention behavior of rare-earth elements on a chelating resin having a propylenediaminetetraacetate type functional group. Analyst 124:1595–1597
Lahoussine Q, Joachim Kohlbrecher J, Konstantin Startchev K et al (2001) Fractal structure of agarose gel: effects of the ionic strength of the solvent and cooling rate during gel formation, SINQ Proposal Number II/00 S–22

Lee H-H, Hong S-J, Kim C-H et al (2008) Preparation of hydroxyapatite spheres with an internal cavity as a scaffold for hard tissue regeneration. J Mater Sci Mater Med 19:3029–3034

Li R, Wan Q-H (2018) A thioether-embedded mixed-mode cyano-bonded chromatographic stationary phase: preparation, characterization and retention mechanism. Chromatographia 81: 1623–1630

Liu Q, Wang L, Xiao A (2007) Research progress in macroporous styrene-divinylbenzene co-polymer microspheres. Design Monom Polym 10:405–423

Lubomirsky E, Khodabandeh A, Preis J et al (2021) Polymeric stationary phases for size exclusion chromatography: a review. Anal Chim Acta 1151:338244

Luo L, Nieh TG (1996) Preparing hydroxyapatite powders with controlled morphology. Biomaterials 17:1959–1964

Luo X, Zhang L (2010) Creation of regenerated cellulose microspheres with diameter ranging from micron to millimeter for chromatography applications. J Chromatogr A 1217:5922–5929

Ma M-G, Zhu J-F (2009) Solvothermal synthesis and characterization of hierarchically nanostructured hydroxyapatite hollow spheres. Eur J Inorg Chem 2009:5522–5526

Ma GJ, Gong B-L, Yan C (2008) Preparation of polymer-bonded chiral ligand exchange chromatographic stationary phase and resolution of racemates. Chin J Anal Chem 36:275–279

Mallik AK, Rahman MM, Czaun M et al (2008) Facile synthesis of high-density poly(octadecyl acrylate)-grafted silica for reversed-phase high-performance liquid chromatography by surface-initiated atom transfer radical polymerization. J Chromatogr A 1187:119–127

Martinson HG (1973a) Nucleic acid-hydroxylapatite interaction. I. Stabilization of native double-stranded deoxyribonucleic acid by hydroxylapatite. Biochemistry 12:139–145

Martinson HG (1973b) Nucleic acid-hydroxylapatite interaction. II. Phase transitions in the deoxyribonucleic acid-hydroxylapatite system. Biochemistry 12:145–150

Martinson HG (1973c) Basis of fractionation of single-stranded nucleic acids on hydroxylapatite. Biochemistry 12:2731–2736

Meyer U, Larsson A, Hentze HP et al (2002) Templating of porous polymeric beads to form porous silica and titania spheres. Adv Mater 14:1768–1772

Monnier S, Schüth G, Huo Q et al (1993) Cooperative formation of inorganic-organic interfaces in the synthesis of silicate mesostructures. Science 261:1299–1303

Moss B, Rosenblum EN (1972) Hydroxylapatite chromatography of protein-sodium dodecyl sulfate complexes. J Biol Chem 247:5194–5198

O'Brien JC, Torrente-Murciano L, Mattia D et al (2017) Continuous production of cellulose microbeads via membrane emulsification. ACS Sustain Chem Eng 5:5931–5939

Omi S (1996) Preparation of monodisperse microspheres using the Shirasu porous glass emulsification technique. Colloids Surf A Physicochem Eng Asp 109:97–107

Omi S, Katami K, Taguchi T et al (1995a) Synthesis and applications of porous spg (shirasu porous glass) microspheres. Makromol Symp 92:309–320

Omi S, Katami K, Taguchi T et al (1995b) Synthesis of uniform PMMA microspheres employing modified SPG (shirasu porous glass) emulsification technique. J Appl Polym Sci 57:1013–1024

Peska J, Stamberg J, Blace Z (1977) Method for manufacturing of spherical cellulose particles. US patent 4055510

Qi H, Chang C, Zhang L (2008) Effects of temperature and molecular weight on dissolution of cellulose in NaOH/urea aqueous solution. Cellulose 15:779–787

Qin X, Lu A, Cai J, Zhang L (2013) Stability of inclusion complex formed by cellulose in NaOH/urea aqueous solution at low temperature. Carbohydr Polym 92(2):1315–1320

Rogozhin SV, Davankov VA (1970) A chromatographic method of resolving racemates of optically active compounds. German patent 1932190

Schnabel R, Langer P (1991) Controlled-pore glass as a stationary phase in chromatography. J Chromatogr 544:137–146

Serwer P (1983) Agarose gels: properties and use for electrophoresis. Electrophoresis 4:375–382

Shen A, Guo Z, Cai X et al (2012) Preparation and chromatographic evaluation of a cysteine-bonded zwitterionic hydrophilic interaction liquid chromatography stationary phase. J Chromatogr A 1228:175–182
Spencer M (1978) Hydroxyapatite for chromatography: II. Sources of variability and improved methods of preparation. J Chromatogr 166:435–446
Spencer M, Grynpas M (1978) Hydroxyapatite for chromatography: I. Physical and chemical properties of different preparations. J Chromatogr 166:423–434
Spencer M, Neave EJ, Webb NL (1978) Hydroxyapatite for chromatography: III. Cation and pH effects on fractionation of tRNA for crystallization. J Chromatogr 166:447–454
Stamberg J (1988) Bead cellulose. Sep Purif Methods 17:155–183
Stoeber W, Fink A, Bohn E (1968) Controlled growth of monodispersed silica spheres in the micron range. J Colloid Interface Sci 26:62–69
Teng S, Chen L, Guo Y et al (2007) Formation of nano-hydroxyapatite in gelatin droplets and the resulting porous composite microspheres. J Inorg Biochem 101:686–691
Tiselius A, Hjerten S, Levin O (1956) Protein chromatography on calcium phosphate columns. Arch Biochem Biophys 65:132–155
Unger K, Scharf B (1976) Controlled porosity silica supports from hydrolytic polycondensation reaction of poly(ethoxysiloxane). J Colloid Interface Sci 55:377–380
Unger KK, Kumar D, Grün M et al (2000) Synthesis of spherical porous silicas in the micron and submicron size range: challenges and opportunities for miniaturized high-resolution chromatographic and electrokinetic separations. J Chromatogr A 892:47–55
Vladisavljević GT (2015) Structured microparticles with tailored properties produced by membrane emulsification. Adv Colloid Interf Sci 225:53–87
Wang B, Friess W (2017) Spray drying of silica microparticles for sustained release application with a new sol-gel precursor. Int J Pharm 532:281–288
Wang Q, Huang W, Wang D et al (2006) Preparation of hollow hydroxyapatite microspheres. J Mater Sci Mater Med 17:641–646
Wang Y, Wu C, Lin K et al (2013) Facile fabrication of nanorod assembled fluorine substituted hydroxyapatite (FHA) microspheres. Chemistry 8:990–996
Wei J-X, Shi Z-G, Chen F et al (2009) Synthesis of penetrable macroporous silica spheres for high-performance liquid chromatography. J Chromatogr A 1216:7388–7393
Xia H, Wan G, Zhao J et al (2016) Preparation and characterization of monodisperse large-porous silica microspheres as the matrix for protein separation. J Chromatogr A 1471:138–144
Zhang X, Zhang W, Yang Z et al (2012) Nanostructured hollow spheres of hydroxyapatite: preparation and potential application in drug delivery. Front Chem Sci Eng 6:246–252
Zhao J, Gao F, Fu Y et al (2002) Biomolecule separation using large pore mesoporous SBA-15 as a substrate in high performance liquid chromatography. Chem Commun:752–753
Zhao K, Bai Q, Song C et al (2012) Preparation of weak cation exchange packings for chromatographic separation of proteins using click chemistry. J Sep Sci 35:907–914
Zhou Q-Z, Ma G-H, Su Z-G (2009) Effect of membrane parameters on the size and uniformity in preparing agarose beads by premix membrane emulsification. J Membr Sci 326:694–700
Zhou Q-Z, Wang L-Y, Ma G-H et al (2007) Preparation of uniform-sized agarose beads by microporous membrane emulsification technique. J Colloid Interface Sci 311:118–127
Zhu Y, Morimoto Y, Shimizu T et al (2015) Synthesis of hierarchically porous polymethylsilsesquioxane monoliths with controlled mesopores for HPLC separation. J Ceram Soc Jpn 123:770–778

Chapter 5
Ligand Design

Abstract With the development of molecular biology, structural biology, and computing technology and the improvement of computing power, the structure of biological macromolecules and small molecule compound libraries have been updated more and more rapidly. Structure-based drug design methods have been widely used for the discovery of new drugs and accelerated the approval of a series of new drugs on the market. In view of the structural similarity between small molecule ligands and drugs, the application of structure-based drug discovery methods in the design of chromatographic ligands has also attracted great interest and has become an important research direction in the field of chromatography. In this chapter, we will review the background and techniques of ligand design used in chromatography.

5.1 Introduction

With the exception of a few adsorption chromatography based on inorganic minerals, such as hydroxyapatite, silica gel, or graphitized carbon, most chromatography involves the use of stationary phases with ligands immobilized on a suitable matrix to obtain the selectivity required for separation. Depending on the solutes to be separated, the structure of the ligand is diverse, from simple alkyl chains, such as octadecyl or ionic groups, to complex proteins and antibodies. In single-mode chromatography, the ligand contains a single functional group, and its chemical structure is relatively simple. In multimode chromatography, the ligand contains multiple functional groups. The type, arrangement, and presentation of functional groups have an important influence on chromatographic selectivity.

Traditionally, the design of ligands is based on the principle of complementation with the target in space and energy. According to the chemical structure of the target, the corresponding ligand is found. The most typical example is the design of chiral chromatography ligands. The ligands involved in chiral chromatography are based on the "three-point interaction" model (Berthod 2006). The concept of "three-point contact" was proposed in 1933 (Easson and Stedman 1933). The stereochemical difference used to explain the activity of the drug is due to the differential binding of

Q.-H. Wan, *Mixed-Mode Chromatography*,
https://doi.org/10.1007/978-981-16-5485-5_5

the two optical isomers at the common binding site on the receptor surface. Fifteen years later, Ogston (1948) revisited the "three-point interaction model," but it was Dalgliesh (1952) who first used the "three-point interaction" model to explain the chiral separation of amino acids in paper chromatography. He pointed out that the condition for a chiral selector to recognize an optical isomer is that at least three configuration-dependent active sites of the selector molecule interact with the complementary and configuration-dependent active sites of the optical isomer molecule. Based on the "three-point interaction" model, research groups, such as Pirkle (1997) and Davankov (1997), have successfully developed donor–acceptor chiral chromatography and ligand exchange chiral chromatography, respectively. It confirms the importance of the "three-point interaction" model in understanding the phenomenon of chiral recognition.

An important source of multimodal chromatography ligands are natural active molecules that can bind to the target with affinity, such as enzymes and substrates, receptors and ligands, nucleic acids and proteins, and antibodies and antigens. Traditionally, affinity chromatography ligands are mostly bioactive molecules, such as RNA and DNA fragments, nucleotides, coenzymes, vitamins, lectins, antibodies, receptor proteins, and oligosaccharides from organisms. Small molecule substrates or coenzymes are used as ligands to purify macromolecular enzymes, or IgG-binding proteins are used to purify antibodies. The limitation of this method is that there must be a natural molecule that binds to the target with affinity.

In recent years, with the development of molecular biology, structural biology, and computing technology and the improvement of computing power, the structure of biological macromolecules and small molecule compound libraries have been updated more and more rapidly. Structure-based drug design methods have been widely used for the discovery of new drugs and accelerated the approval of a series of new drugs on the market. In view of the structural similarity between small molecule ligands and drugs, the application of structure-based drug discovery methods in the design of chromatographic ligands has also attracted great interest and has become an important research direction in the field of chromatography. In this chapter, the background and development status of this technology will be introduced.

5.2 Fragment Growth

The binding site of a protein is usually composed of multiple different sub-binding sites. That is to say, the binding pocket can be divided into several regions. Different regions are suitable for different fragments. The selected fragments are divided into different binding sites. The complementarity of the two is expanded or connected to form a ligand with stronger binding energy and better selectivity (Clonis et al. 2000; Li 2020). Like traditional drug design, it must go through steps such as compound screening and structural optimization. Structural optimization includes the growth and connection of fragments. The fragment growth method is basically similar to the traditional method of optimizing lead compounds in medicinal chemistry. According to the interaction information provided by NMR, X-ray crystallography, and other methods, combined with the structural characteristics of the protein itself, the fragment molecules are selected and spliced to turn them into larger and more complex molecules, so as to fit in the binding pocket of the protein with enhanced affinity. When two different fragment molecules bind to different sites on the target protein, if the two binding sites are close enough in space, the two fragment molecules can be connected by a suitable linking arm. Because each fragment can be combined with the receptor, when the fragments are connected, they can exert additive and synergistic effects to produce lead compounds with higher activity. Another situation is that if the two fragments are very close in space, there will be overlaps between the fragments. The two fragments are merged in an appropriate way to become a more active ligand.

According to the three-dimensional structure of the binding pocket of the target protein, a suitable ligand is selected to complement the active site of the protein in space and energy. Malate dehydrogenase (EC 1.1.1.37) (MDH) is an enzyme that reversibly catalyzes the oxidation of malate to oxaloacetate by reducing NAD^+ to NADH. This reaction is part of many metabolic pathways, including the citric acid cycle. In most organisms, malate dehydrogenase (MDH) exists in the form of homodimer molecules and is structurally closely related to lactate dehydrogenase (LDH). It is a large protein molecule with subunit masses between 30 and 35 kDa (Banaszak and Bradshaw 1975).

Each subunit of the malate dehydrogenase dimer has two structural and functional domains. The parallel β-sheet structure constitutes the NAD^+-binding domain, while four β-sheets and one α-helix constitute the central NAD^+-binding site. The subunits are held together by extensive hydrogen bonding and hydrophobic interactions (Hall et al. 1992).

Malate dehydrogenase has also been shown to have a movable loop region, which plays a vital role in the catalytic activity of the enzyme. Studies have shown that after binding to the substrate, the conformational change of the loop region from the open conformation to the closed conformation enhances the catalysis of MDH by shielding the substrate and the catalytic amino acid from the solvent. Studies have also shown that this loop region is highly conserved in malate dehydrogenase (Goward and Nicholls 1994).

Labrou et al. (1996) synthesized a series of homologues containing carboxyl or keto acid structure through the substitution reaction on the parent anthraquinone dichlorotriazine dye Vilmafix Blue AR (VBAR) to simulate L-malate dehydrogenase (MDH) cofactor, substrate, or inhibitor. Figure 5.1 shows the structure of the synthesized biomimetic ligand (BM) and the dissociation constant K_d value with bovine heart MDH. Each biomimetic ligand is a chimeric molecule consisting of two enzyme recognition parts. All ligands contain an anthraquinone moiety connected to the triazine ring, which is the same as the parent dichlorotriazine dye VBAR and binds to the nucleotide-binding site of MDH. The carboxyl or keto acid structure (R–) attached to the other side of the triazine ring binds to the binding site of MDH substrate or inhibitor. Molecular modeling, kinetic inhibition, differential

Fig. 5.1 The binding characteristics of biomimetic dye BM (1–7), parent dyes Vilmafix Blue AR (VBAR) (10) and Cibacron Blue 3GA (CB3GA) (8) and control dyes choline-VBAR (VBARCh) (9) and Procion Blue MX-3G (BMX-3G) (10) with bovine heart mitochondrial L-malate dehydrogenase. For all dyes, X = H, except for BMX3G (11), where one X = SO_3^- and the other X = H. (Reproduced from Labrou et al. 1996)

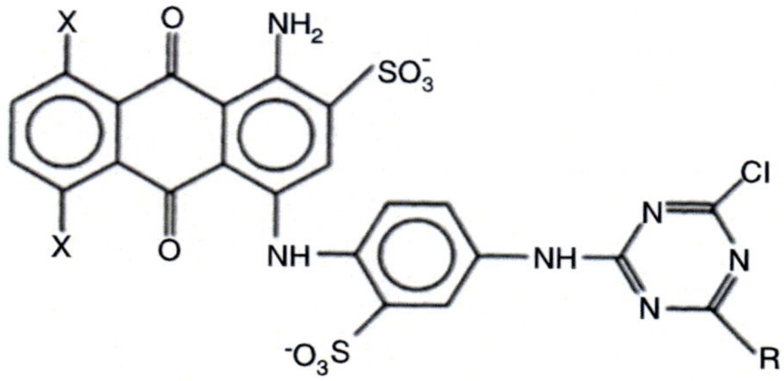

Dye-ligand (-R)	K_D (μM) by inactivation (pH 8.5)	K_D (μM) by difference spectra (pH 7.5)
(1) -*m*-HN-C_6H_4-COO^-	6.6 ± 0.5	7.1 ± 0.2
(2) -*m*-HN-C_6H_4-CH_2COO^-	3.7 ± 0.04	5.5 ± 0.4
(3) -$HNCH_2COO^-$	20.3 ± 1.60	17.1 ± 0.8
(4) -$HN(CH_2)_2COO^-$	24.9 ± 1.80	25.7 ± 1.9
(5) -*p*-HN-C_6H_4-$NHCOCOO^-$	2.0 ± 0.05	1.3 ± 0.03
(6) -SCH_2COCOO^-	6.9 0.70	18.3 ± 1.5
(7) -$HN(CH_2)_2NHCOCOO^-$	3.0 ± 0.2	1.8 ± 0.03
(8) -o-HN-C_6H_4-SO_3^- (CB3GA)	4.5 ± 0.30	5.9 ± 0.5
(9) -$O(CH_2)_2N^+(CH_3)_3$ (VBARCh)	-	285 ± 15
(10) -Cl (VBAR)	14.4 ± 1.3	-
(11) -OH (BMX3G)	-	245 ± 16

spectroscopy, and enzyme inactivation studies all show that, compared with the parent VBAR (10), Cibacron Blue 3GA (8), and two control non-biomimetic dyes (9 and 11), the chimeric ligands (BM5 and BM7) exhibit lower experimental K_d values and therefore have a higher affinity for MDH. The partial chimeric ligand (BM5) with *p*-aminoaniline oxalate mimetic enzyme substrate exhibits the highest affinity (K_d 1.3 μM) and shows a competitive inhibitory effect on both NADH and oxaloacetic acid.

Bioinformatics-based molecular modeling shows a three-dimensional view of the coenzyme binding site of chimeric ligand BM5 in porcine mitochondrial MDH (Fig. 5.2).

The anthraquinone part is located at the position of NAD^+ adenine in the hydrophobic gap of MDH. The bottom of the crevice is blocked by Tyr-32 and Leu-100, so anthraquinone cannot enter the deeper cracks. This causes the sulfonate of the ring and one of the amino group and the carbonyl group to be exposed to the solvent, while the carbonyl group on the other side of the anthraquinone forms a hydrogen bond with Asp-33. The triazine ring binds in the region where the coenzyme pyrophosphate binds, one side is close to the MDH surface, and the other side faces the inside of the binding site. The substrate mimic part is connected to the triazine ring through an amine group, which can be connected to the carbonyl group of Gly-77 to form a hydrogen bond. The benzene ring is placed in the

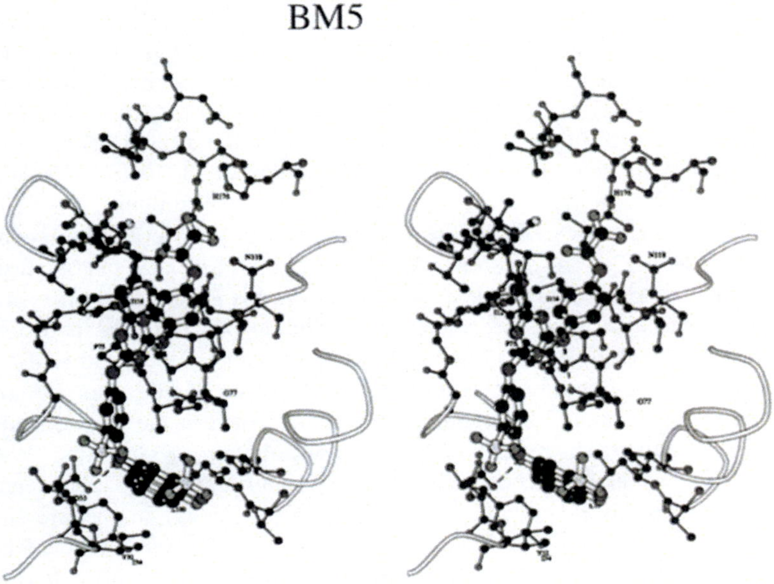

Fig. 5.2 A three-dimensional view of the keto-carboxyl chimeric ligand BM5 on the MDH coenzyme binding site of porcine mitochondria. (Reproduced from Labrou et al. 1996)

hydrophobic cavity formed by residues Ile-12, Ile-116, Asn-118, Pro-75, and MDH backbone fragments; this space is occupied by the second ribose ring of NAD^+ (connected to nicotinamide). The use of a benzene ring can maintain the orientation of the ligand toward the catalytic site and provide a gap without introducing additional degrees of freedom. Ketocarboxylic acid esters tend to overlap with the amide group of the NAD^+ niacinamide moiety, while other analogs are shorter.

The chimeric ligand was immobilized on a cross-linked agarose Ultrogel A6R via a triazine group for one-step purification of mitochondrial MDH (mMDH) from unpretreated bovine heart extract (Labrou and Clonis 1996). Using a mixture containing 0.1-mM β-nicotinamide adenine dinucleotide (NAD^+) and 1.5-mM sulfite for stepwise elution, this procedure provides a specific activity of approximately 1300 U/mg (25 °C), gel electrophoresis shows a single band of mMDH, and a total yield is 54%. The specific activity of commercial analytical grade bovine heart mitochondrial MDH does not exceed 950 U/mg when measured under the same conditions. The well-known adsorbent Cibacron Blue 3GA-agarose reduces the recovery rate of mMDH by 8% and the purification by 25%.

Lactate dehydrogenase (LDH or LD) is an enzyme found in almost all living cells. LDH catalyzes the mutual conversion of pyruvate and lactic acid and is accompanied by the mutual conversion of NADH and NAD+. When hypoxia or insufficient oxygen supply, it converts pyruvate, the final product of glycolysis, into lactic acid and reverses it in the Cori cycle of the liver. At high concentrations of lactic acid, the enzyme exhibits feedback inhibition, and the rate of conversion of pyruvate to lactic acid decreases. Lactate dehydrogenase is composed of four subunits (tetramers). The two most common subunits are LDH-M and LDH-H proteins. These two subunits can form five possible tetramers (isozymes): 4H, 4M, and three mixed tetramers (3H1M, 2H2M, 1H3M). These five isoforms are similar in enzymology but show different tissue distributions: M4, the main isoenzyme of the skeletal muscle and liver, has four muscle (M) subunits, and H4, the main isozyme of the myocardium in most species, contains four heart (H) subunits.

LDH in humans uses His-193 as a proton donor and works with coenzyme (Arg-99 and Asn-138) and substrate (Arg-106, Arg-169, Thr-248) binding residues. His-193 active site not only exists in the human form of LDH but also exists in many different animals, showing the convergent evolution of LDH. The two different subunits of LDH (LDHA, also known as the M subunit of LDH, and LDHB, also known as the H subunit of LDH) retain the same active site and the same amino acids involved in the reaction. The significant difference between the two subunits that make up the tertiary structure of LDH is the substitution of glutamine (in the H chain) for alanine (in the M chain). This small but significant change is believed to be the reason why the H subunit can bind NAD faster, and in the presence of acetylpyridine adenine dinucleotide, the catalytic activity of the M subunit is unchanged, while the activity of the H subunit is reduced by a factor of 5.

LDH is widely expressed in body tissues, such as blood cells and heart muscle. Because it is released during tissue damage, it is a sign of common injuries and diseases such as heart failure. Labrou et al. (1999) used the three-dimensional structure of pig heart L-lactate dehydrogenase (LDH)–NAD^+ complex to design a ligand for L-lactate dehydrogenase based on the triazine backbone. The ligand is an analog of the textile dye Cibacron Blue 3GA, and its end is modified with 2-(4-aminophenyl)-ethyloxamic acid to match the polar and hydrophobic regions of the enzyme (Fig. 5.3). The length of the keto acid moiety is determined to approach the enzyme catalytic site and form a salt bridge with Arg-171. The designed ligand shows a higher purification capacity for LDH in bovine heart. When integrated into a two-step purification scheme, the pure enzyme produced has a specific activity of 600 U/mg and a recovery of 50%. The purity of the enzyme is twice that of the reference commercial dye Cibacron Blue 3GA.

Similar to the chimeric ligand used for MDH purification, the new chimeric ligand used for LDH contains binding sites that mimic coenzymes and substrates (Fig. 5.4). The anthraquinone part mimics the adenine part of NAD^+ and is contained in the hydrophobic cavity formed by the five chain regions of LDH. The bottom of the crevice is blocked by Tyr-85 (not shown) and Ile-123, so anthraquinone cannot enter the deeper cavity. Like MDH, this causes the sulfonate, the amino group, and one carbonyl group of the anthraquinone to point to the solvent, while the carbonyl group on the other side of the anthraquinone is located within the hydrogen bond distance between Asp-53 and Lys-58 (not shown) (Fig. 5.4). The sulfonate group of the bridged diaminobenzene methanesulfonate is located in a polar cavity, where it can form hydrogen bonds with the backbone nitrogen atoms of Val-29 and Gly-30 (Fig. 5.4). The bridged amino group on the benzene ring forms a stable hydrogen bond with Asp-53. The triazine ring of the chimeric ligand is bound in the

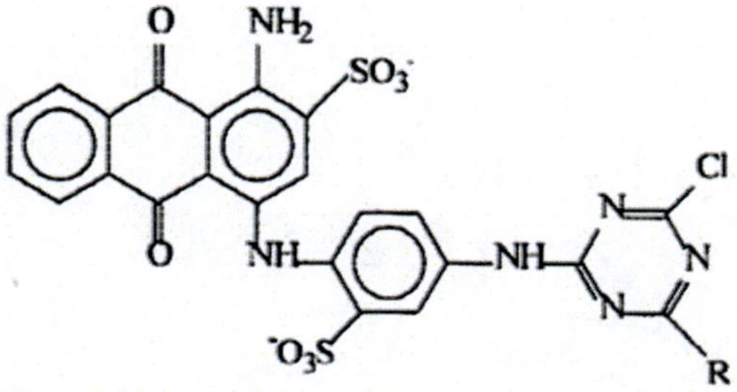

Immobilized dye (-R)	Capacity (U/μmol dye)	SA (U/mg) KCl	SA (U/mg) NADH	Purification (-fold) KCl	Purification (-fold) NADH
-p-NHBenzCH_2CH_2NHCOCOO$^-$	130	119	196	8.8	14.5
-o-NHBenzSO_3^- (CB3GA)	122	62	103	4.6	7.6

Fig. 5.3 The structure and chromatographic evaluation of the chimeric ligand purified from the bovine heart extract for LDH activity. (Reproduced from Labrou et al. 1999)

pyrophosphate binding region of NAD^+, one side close to the surface of the enzyme. On the other side of the triazine ring, toward the inside of the binding site, it is connected to the biomimetic terminal part via the -NH- group within the hydrogen bond distance from the Thr-97 hydroxyl group (Fig. 5.4).

The biomimetic part of the chimeric ligand has a benzene ring connected to the triazine ring through an amino group. The charged carboxyl group at the end is selected to mimic the natural substrate of LDH and can adapt to its interaction with Arg-171. The benzene ring is located in the space occupied by NAD^+ nicotinamide linked ribose, which is the hydrophobic cavity formed by the Val-32, Val-138, Thr-246, Ile-250, Gln-102, and Arg-101 side groups (Fig. 5.4), and the LDH backbone. The terminal carboxyl group is connected to the benzene ring through the ethylene spacer -CONH$(CH)_2$-. They are located near the catalytic site, where the cofactor nicotinamide is located. Carboxylate is surrounded by several polar amino acids, namely, Gln-102, Arg-109, Arg-171, Asn-166, His-195, and Thr-246. In the case of the reference ligand CB3GA, the charged bulky sulfonate group of the terminal aminobenzene methanesulfonate abuts the hydrophobic Val-32 and Gln-102 with a fairly limited volume to accommodate the sulfonate group. The position of the terminal keto acid of the biomimetic part at the catalytic site was modeled to simulate the position of the terminal amide of the natural substrate (*S*-lactic acid) and NAD^+. This is achieved by introducing aliphatic and amide groups at the para position of the benzene ring. From the molecular model (Fig. 5.4), it is obvious that if the spacer does not have the ability to form an H bond after the phenyl group, the keto acid end group cannot reach the polar catalytic site of LDH. Therefore, the introduction of an ethyl spacer removes the keto acid from the

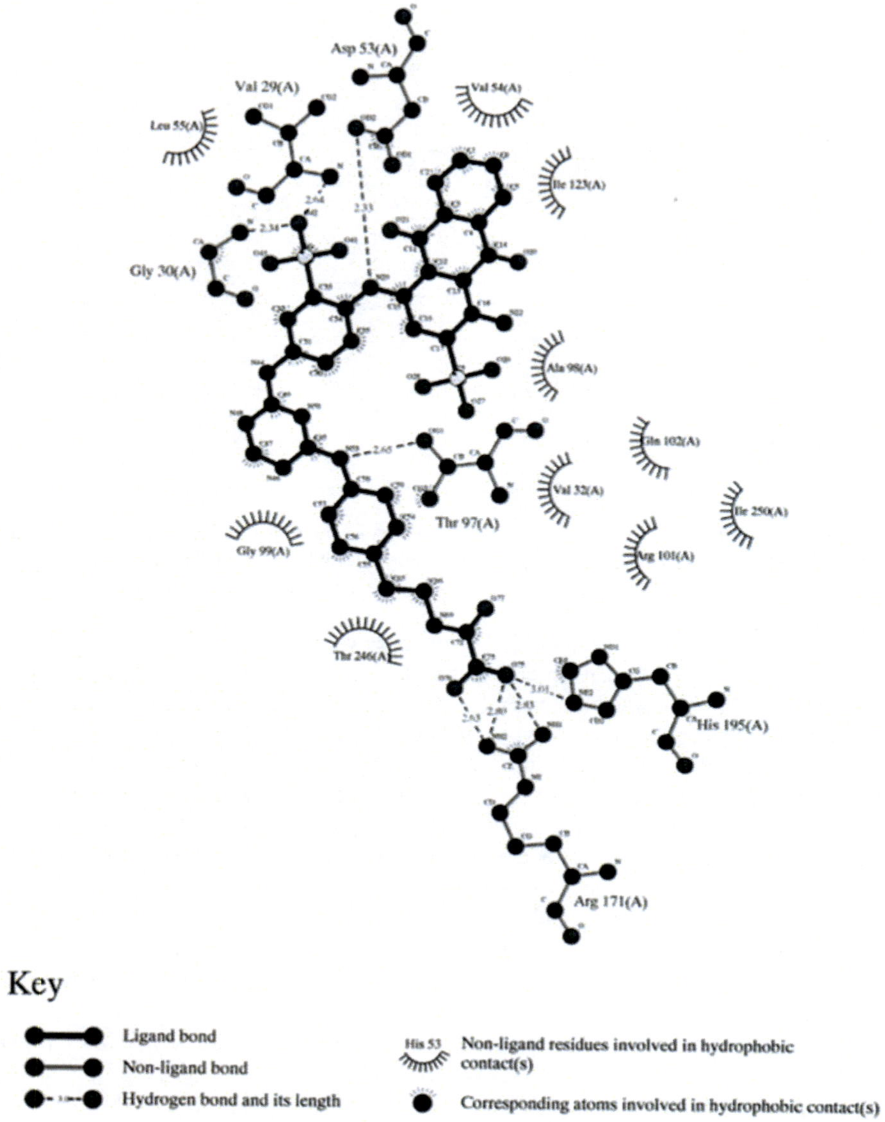

Fig. 5.4 Chimeric ligand in the porcine heart LDH coenzyme binding site. (Reproduced from Labrou et al. 1999)

hydrophobic portion of the binding site, while the introduction of additional amide allows the keto acid to enter the catalytic site. Therefore, the use of the benzene ring and the ethylene group preserves the direction of the biomimetic chain and introduces a maximum length. The latter increases the possibility of additional hydrogen bonding with amine donors and carbonyl acceptors.

Melissis et al. (2001) reported the design of chimeric ligands for the purification of glutathione recognition enzymes. These ligands contain two enzyme recognition moieties: the anthraquinone chromophore and the terminal glutathione moiety attached to the triazine ring. Three anthraquinone glutathione chimeric ligands, including glutathione analogs (glutathione sulfonic acid, *S*-methyl glutathione, and glutathione) as terminal biomimetic parts, were synthesized and characterized. After the chimeric ligand is immobilized on the agarose aggregation matrix, these stationary phases are used to purify several glutathione recognition enzymes, such as glutathione *S*-transferase, $NAD(P)^+$-dependent glutathione reductase (glutathione reductase, GSHR), and NAD^+-dependent formaldehyde dehydrogenase (FaDH). Compared with the single-function reference ligand Cibacron Blue 3GA, all chimeric adsorbents show higher purification ability for the target enzyme. In particular, the adsorbent containing glutathione sulfonic acid (BM1) as a terminal biomimetic part exhibits the highest purification ability for NAD^+-dependent formaldehyde dehydrogenase (FaDH) and recombinant corn glutathione *S*-transferase I (GSTI), and the adsorbent containing methyl glutathione (BM2) as the terminal biomimetic part shows the highest purification ability for GSHR. Molecular modeling was used to visualize the binding of BM1 with FaDH, indicating the favorable positioning of key structural features of biomimetic dyes. The anthraquinone part provides the driving force for the correct positioning of the glutathione biomimetic part at the binding site. It is located deep in the active site crevice and forms many favorable hydrophobic contacts with the hydrophobic residues of the enzyme. The positioning of the glutathione-like biomimetic part is mainly achieved by the strong ionic interaction with the Zn^{2+} ions of FaDH and Arg-114 and the hydrophobic contact with Tyr-92 and Met-140. A molecular model was also generated for binding BM1 and BM3 (glutathione substitution) to GSTI. In both cases, the biomimetic dye forms multiple hydrophobic interactions with enzymes by binding to surface pockets.

5.3 Combinatorial Peptide Library

Combinatorial chemistry is an organic synthesis technique that produces a library of diverse compounds. It uses different components to synthesize a large number of different compounds at the same time. According to the combination method used, it can be divided into two categories: parallel synthesis and combinatorial synthesis. Parallel synthesis means that multiple building block molecules with the same function and a single building block simultaneously undergo multiple bonding reactions. Combinatorial synthesis refers to the simultaneous reaction of multiple building blocks with the same function to generate multiple products in one step.

The basic methods of combinatorial chemical synthesis are divided into solid-phase synthesis and liquid-phase synthesis. Solid-phase synthesis refers to a synthesis reaction in which an insoluble solid material is used as a carrier, and different solvents are used as a medium, and the component molecule is connected to the carrier through a linking group. In each step of the reaction process, the reaction reagents react with the product molecules bound to the carrier, and the excess reaction reagents can be removed by filtration or washing. Therefore, solid-phase synthesis simplifies post-synthesis processing and improves efficiency. Through the multistep cycle, the structure of the compound continues to grow. Finally, through a specific reaction, the product is cut from the carrier to obtain the target molecule. Liquid-phase synthesis can be illustrated by multicomponent Ugi condensation reaction (Ugi et al. 2003). The reaction components including aldehyde, isocyanide, carboxylic acid, and amine are mixed in an equimolar number and reacted in polar solvents, such as methanol, to obtain a four-component condensation reaction product, a dipeptide. The reaction is very active and can usually be completed in a few seconds or minutes at room temperature. Due to the relatively low solubility of the product, it forms a precipitate in the polar solvent, which can be recovered by filtration, and the post-processing is convenient. The compound library can be constructed by changing the substituents of the substrate.

A synthetic peptide library is a collection of random peptide molecules, covering all possible sequences of a given peptide length. Through the experimental method of random screening, potential ligands that bind to the target protein can be screened from the synthetic peptide library (Fang et al. 2018). In 1986, Geysen et al. (1986) proposed the concept of peptide ligands in affinity chromatography based on the study of synthetic peptide libraries. The main points of Geysen's proposal include the following: (1) short peptide molecules containing key amino acid residues can mimic the determinants on the protein surface, and (2) the non-covalent interaction force between the key residues and the target protein is the main force for the binding of the ligand to the protein. The Geysen's proposal laid the theoretical foundation for the study of affinity peptide ligands. First, affinity peptide ligands are usually short peptide sequences composed of only a few amino acids. Compared with protein ligands, the production cost is lower, and it does not cause immunogenic reactions. Second, the short peptide ligand has a small molecular weight and will not be degraded by the protease in the raw material solution to lose activity; even if it falls off the stationary phase and is incorporated into the product, it can be easily removed from the final product using dialysis or size exclusion chromatography. Third, the peptide ligands are stable in physical and chemical properties and can withstand the strong acid–base elution or regeneration environment in the separation operation. Fourth, the affinity peptide ligand has intermediate affinity to target protein and does not require harsh elution conditions, which avoids denaturation of the target protein. Finally, peptide ligands can be easily modified by chemical methods and can be immobilized to a solid support in high-density and selective orientation.

The combinatorial synthetic peptide library contains all possible random peptide molecules under a given peptide sequence length. This method first uses amino acids as the basic building blocks and uses solid-phase peptide synthesis methods to react amino acids with solid-phase medium (resin) multiple times in a permutation and combination manner to obtain polypeptides with the same length that contain all possible sequences. The peptides are cleaved from the solid-phase medium to form a soluble peptide library. Then, the peptide library mixture is incubated with the immobilized target protein, and the peptide that can bind to the target protein is the ligand that has affinity for it (Denholt et al. 2009). The combinatorial synthetic peptide library screening includes three steps:

1. Peptide library synthesis. First, the sequence length of the peptide molecule is set, and then, the desired amino acid types are selected, and quality control and optimization methods for the automated synthesis of the peptide library are developed. At present, the commonly used synthesis methods include "tea bag method," "mixed-splitting method," "light-controlled synthesis method," and "multicenter synthesis method."
2. Cluster screening. Suitable indicator methods, such as radio labeling, fluorescent labeling, and dye staining, are chosen for automated high-throughput screening. First, the target protein is immobilized on the chromatographic medium, the medium is loaded on the column, and then, affinity chromatography is used to screen out peptide molecules with high affinity to the target protein from the peptide library (Tozzi et al. 2003).
3. Analysis of peptide library. A variety of methods are used to determine the structure of the peptides screened in the previous step, mainly using mass spectrometry methods, such as MS/MS, capillary electrophoresis/MS, and HPLC/MS (Huang and Carbonell 1999).

Yang and colleagues (2005, 2009) used the "couple-divide-recombine" strategy to create solid-phase synthesized "one-bead–one-peptide (OBOP)" combinatorial libraries and then used three rounds of screening to obtain three hexapeptide ligands for human IgG, which are characterized by the "histidine + aromatic group amino acid + positively charged amino acid" combination pattern. These three peptides (HWRGWV, HYFKFD, and HFRRHL) can specifically bind to the Fc fragment.

Kaufman et al. (2002) screened a hexapeptide library synthesized by combinatorial chemistry with human fibrinogen as the target protein and developed a hexapeptide ligand with the sequence FLLVPL for purification of fibrinogen. The immobilized ligand and corresponding adsorbent are used to purify human fibrinogen. In a low-ionic-strength solution of pH 4, the adsorbed fibrinogen was successfully eluted from the chromatographic column, with high purity and high biological activity. The peptide ligand is expected to be useful for large-scale purification.

Wang et al. (2004) used combinatorial synthetic peptide library to obtain the hexapeptide affinity ligand YYWLHH for staphylococcal enterotoxin B. This ligand can replace the antibody-like ligand to specifically purify enterotoxin B from the fermentation broth of *Staphylococcus aureus* in order to reduce production costs.

Synthetic peptide libraries have been used for the screening of ligands for immunoglobulin purification (Fassina et al. 1996). The first method uses three screening cycles to generate a protein A mimetic peptide. In the first screening cycle, a sub-library with arginine at the N-terminus was identified (60% inhibition). The N-terminal arginine was fixed to synthesize the second multimer sub-library. In the second screening cycle, it was found that the second amino acid residue, threonine, had the best inhibitory activity. The third sub-library consists of 18 tripeptide tetramers with the general formula $(R\text{-}T\text{-}X)_4\text{-}K_2\text{-}K\text{-}G$, from which the most active polymer $(RTY)_4$-K_2-K-G (TG19318) was identified. The antibody adsorption on the TG9318 chromatographic column is carried out in a low-ionic-strength neutral buffer. The conditions are completely compatible with the use of crude raw materials derived from cell culture fluid. The adsorbed immunoglobulin can be eluted by adjusting the pH of the buffer to acidic or weakly alkaline conditions (pH 3.3 with acetic acid or pH 9.0 with sodium bicarbonate). The new ligand not only can purify IgG from different sources but also can purify IgY from egg yolk; purify IgM from ascites, serum, or cell culture supernatant; and purify IgA and IgE from cell culture supernatant.

The synthetic peptide library strategy is also used for the identification and evaluation of cyclic peptide ligands. Kang et al. (2016) reported the development and utilization of small cyclic peptide ligands that specifically bind to the Fc domain of antibodies (i.e., IgG Fc domain binding peptides containing 13 amino acid residues, FcBP). FcBP immobilized on agarose gel resin showed good binding ability to humanized IgG (trastuzumab) and chimeric IgG (cetuximab). Antibodies with a purity of 95% were isolated from pig serum.

5.4 Ugi Multicomponent Reaction

Ugi multicomponent reaction is an efficient synthesis method. In a typical four-component reaction, the carbonyl component (aldehyde or ketone), primary or secondary amine, carboxylic acid, and isonitrile groups are condensed, and a single diamide scaffold product is formed in a one-pot process (Fig. 5.5a). Lowe and colleagues (Haigh et al. 2009) used the derivatized chromatographic matrix carrier as one of the four components and used the Ugi reaction to directly generate selective ligands with diamide as the scaffold on the surface of the chromatographic matrix (Fig. 5.5b). The structure of the ligand can be optimized by adjusting the functional groups and combinations of component amine (R2), isonitrile (R3), and

Fig. 5.5 Ugi reaction product and ligand formation on solid-phase chromatography support. (**a**) In the "one-pot method" reaction, the Ugi reactant composed of carbonyl-containing components (aldehydes or ketones) (R1), primary or secondary amines (R2), isonitrile groups (R3), and carboxylic acids (R4) is condensed to produce a single product (1). (**b**) The solid-phase Ugi reaction contains an aldehyde-functionalized matrix and three other reaction components to produce multifunctional ligands immobilized on the chromatographic medium (2). (Reproduced from Haigh et al. 2009)

carboxylic acid (R4) in the Ugi reaction to obtain the specificity required for interaction with the target protein.

Compared with the conventional "two-component" methods for synthesizing ligands such as triazine chemistry, the Ugi four-component reaction has many obvious advantages. First of all, by incorporating three or four reactants, the generated ligands have a greater diversity, and each reactant can be systematically changed to generate a ligand library with variable structures and explore a larger chemical space. Second, the "one-pot" nature of the Ugi reaction can save a lot of time and reagent costs. Complex ligands that usually require multistep reactions can be completed in a single synthesis step, which can meet chemical diversity while delivering quickly. The basic steps for developing a target protein-specific ligand based on Ugi reaction include: (1) analyze the crystal structure of the target protein and its natural binding partner and determine the key amino acid residues of the binding site in the binding partner; (2) select the appropriate Ugi reaction component, simulate the key amino acid in the binding site, and synthesize Ugi ligand; and (3) use pure target protein to evaluate the static binding capacity and binding constant of the synthesized ligand library and select the lead ligand from the cell culture medium to purify the target protein to evaluate the selectivity of synthetic ligands, recovery, and enrichment factor.

Haigh et al. (2009) first reported the application of Ugi four-component reaction in the synthesis of chromatographic ligands for the purification of immunoglobulins and their fragments. As mentioned above, in the four-component reaction, the aldehyde-functionalized agarose gel microspheres are used as a fixed component, while the other three components (carboxylic acid, primary or secondary amine, and isonitrile) change in combination to generate a trisubstituted peptide scaffold ligand, which has a certain degree of rigidity and functionality, and can bind to immunoglobulins with high affinity. First, the structural data of the main binding interface between the immunoglobulin-binding protein, protein L, and the human Fab fragment (PDB:1HEZ) was studied, and it was determined that there are seven key residues that are conserved in different domains of protein L. They are buried in the complex formed between protein L and IgG light chain. According to the functions of these residues, the corresponding Ugi library analogs are selected and incorporated into the Ugi combinatorial library. For example, the Ugi counterparts of Gln35 are 4-aminobenzamide (A2) and Boc-glutamine (C4); the counterpart of Thr36 is 1-amino-2-propanol (A4); the counterpart of Ala37 is acetic acid (C6); the counterparts of Glu38 are glutaric acid (C1), 2,4-pyridinedicarboxylic acid (C2), and isophthalic acid (C3); the counterpart of Phe39 is benzoic acid (C5); and for Lys40 and Tyr53, the corresponding counterparts are tyramine (A1), 4-aminophenol (A5), 3-aminophenol (A6), and 4-hydroxybenzylamine (A7). A ligand library composed of 8 (A) $\times$ 6 (C) = 48 ligands was synthesized through a one-pot reaction.

The Ugi ligand library was chromatographically screened against intact human IgG and its fragments (Fc and Fab), and four Fab-specific lead ligands (A3C1, A3C2, A3C3, and A3C4) were selected (Fig. 5.6). The ability of the four candidate ligands to separate IgG from human serum was evaluated by preparative

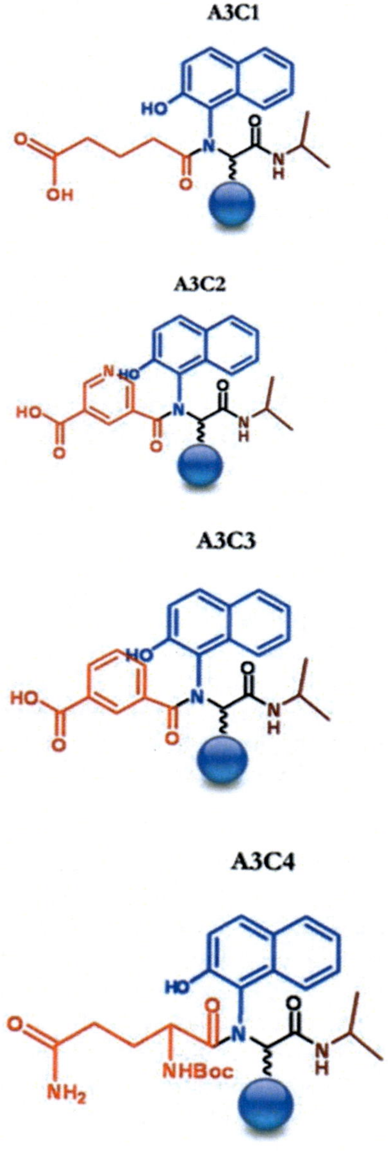

Fig. 5.6 The structure of the Fab-specific lead ligand. (Reproduced from Haigh et al. 2009)

chromatography. A3C1 shows specificity for Fab fragments at low ionic strength, the highest dynamic binding capacity, and the highest purity (81%) of IgG isolated from human serum. High-purity Fab and IgG separation from yeast and *E. coli* lysates has been achieved on A3C1 ligand. The adsorbent shows a dynamic binding capacity of 73.06-mg IgG/mL wet gel for full-length IgG, a static binding capacity of 16.1-mg Fab/mL wet resin for Fab, and an affinity constant of

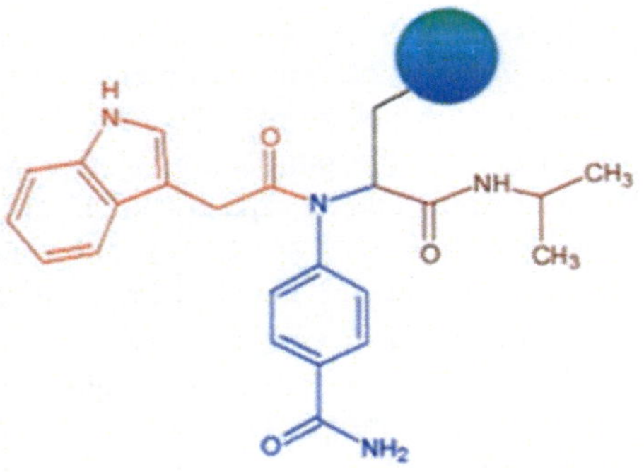

Fig. 5.7 Ugi ligand A2C11I1 mimics the active site of protein G. (Reproduced from Qian et al. 2012)

$K_d = (2.6 \pm 0.3) \times 10^{-6}$ M. These results clearly indicate the potential of Ugi scaffolds in future ligand development.

Qian et al. (2012) used a solid-phase multicomponent Ugi reaction to generate a multifunctional ligand that mimics protein G to adsorb mammalian immunoglobulins, including only heavy chain camelid IgG, and elute them effectively under neutral pH conditions. From the crystal structure data of protein G and IgG complex, the binding site information of protein G is determined. The C2 fragment of protein G includes an α-helix diagonally spanning four-strand β-sheets. The amino acid residues that bind to IgG are located at the C-terminus of the α-helix (Asn35), the N-terminus of the third β-chain (Glu42 and Trp43), and the ring connecting these two structures (Asp40). In addition, the following amino acid residues of protein G have been determined: Glu27, Lys28, Lys31, Gln32, Asn35, Asp40, Glu42, and Trp43 to interact with the Fc domain of human IgG, mainly involving three residues in the CH2 domain (Ile253-Ser254 and Gln311) and two regions in the CH3 domain (Glu380, Glu382 and residues His433-Gln438). His433 and Asn434 on the Fc domain play a key role in binding to protein G. Asn434 on the Fc domain forms a hydrogen bond (H bond) with Asn35 and Trp43 on the C2 fragment of the protein G and may also form an H bond with the backbone nitrogen atom of Gly41 and the carboxyl group of Asp40. The involvement of Asn434 in many interactions indicates that it is essential for protein–protein interactions; therefore, Asn35 and Trp43 in protein G, which form an H bond with Asn434 in Fc, were selected as templates to design Ugi ligands. 4-Aminobenzamide (A2) and indole-3-acetic acid (C11) were used to simulate amide (Asn35) and indole (Trp43) functional groups, respectively. They were mixed with the third component isopropyl isocyanate (I1) and aldehyde-functionalized agarose gel to synthesize the dipeptide scaffold ligand A2C11I1 that mimics the active site of protein G as shown in Fig. 5.7.

The ligand A2C11I1 is docked into the Fc domain by computer, showing a key binding interface very similar to the natural protein G. The candidate ligand showed affinity for chicken IgY, and recombinant camelid Fc domain, and IgG derived from human, bovine, goat, mouse, sheep, pig, rabbit, and rat serum, among which bovine and sheep IgG show 100% binding under selected conditions. When Ugi ligand is used to separate IgG from human serum, the purity of IgG eluted with standard buffer is about 65%, which is better than 62%, the purity of protein G separation, which is about 62%. This adsorbent shows the highest IgG elution at neutral pH (0.1-M sodium phosphate pH 7.0, 30% (v/v) ethylene glycol), which is a significant advantage for the purification of extremely pH-sensitive antibodies. The ligand shows a static binding capacity of 24.6-mg IgG/mL resin and a dissociation constant of $K_d = 4.78 \times 10^{-6}$ M. It has application prospects in the purification of immunoglobulins and other potentially high-value biological therapeutic proteins.

El Khoury and coworkers used a small library generated by the Ugi four-component reaction to screen for IgG selective ligands that mimic the third domain (domain III) of protein G (El Khoury and Lowe 2013). Protein G is a cell surface protein from *Streptococcus* strains. It contains about 600 residues and forms three different binding domains that bind to the Fab and Fc parts of IgG molecules from various species. Protein G can recognize a wide range of IgG and is usually the ligand of choice for purifying immunoglobulins that bind weakly to protein A (such as sheep IgG and IgG3). The IgG-binding domain of protein G has been identified as three homologous sequences, each with approximately 55 amino acid residues (Sjobring et al. 1991). Nuclear magnetic resonance (NMR) studies of protein G and IgG identified two Fc binding domains on B1 and B2 of protein G. The third IgG-binding domain of protein G (called domain III) participates in the binding of Fab fragments through the highly conserved amino acid residues of the CH1 domain.

The analysis of the main interaction sites between these two proteins showed that protein G binds to Fab through hydrogen bonds. The hydrogen bonds mainly involve the backbone atoms of the Pro126, Val128, Ser209, Lys212, Asp214, and Lys216 residues in the CH1 domain (8 of the 12 bonds), which helps expand the range of IgG γ-chains recognized by protein G. The remaining four interactions use side chain atoms from residues Tyr129, Ser210, Thr211, and Lys215; these residues are highly conserved in the amino acid sequence of the IgG γ-heavy chain of several species (i.e., human, mouse, rat, rabbit) and subclasses (Derrick and Wigley 1994). Among all these key amino acid residues, Pro126 and Val128 located on the first β-chain of CH1 are the most conserved residues and are in contact with Tyr38 and Asn42 at the C-terminus of the α-helix of domain III. Therefore, the affinity ligand was designed to mimic the Tyr38 and Asn42 residues of protein G (domain III). 4-Aminobenzamide (A2) and 4-hydroxyphenylacetic acid (C7) were used to simulate the functional groups of Asn42 and Tyr38, respectively. They were mixed with the isonitrile component, isopropyl isocyanate (I1), and aldehyde groups on the functionalized agarose gel to synthesize the dipeptide ligand A2C7I1, as shown in Fig. 5.8. The small isonitrile (I1) is selected to minimize the steric hindrance of the interaction between the 3D structure of the ligand and the target molecule, because the bulky isonitrile structure will interfere with the orientation of the ligand on the substrate and the accessibility of the key functional groups of the A2 and C7 components.

Chromatographic adsorption experiments showed that under neutral pH conditions, the ligand A2C7I1 showed adsorption rates of 98%, 46%, and 1.5% (w/w) to intact human IgG, Fab, and Fc, respectively. Under optimized chromatographic conditions for pure human IgG and its Fab fragments, the binding behavior of

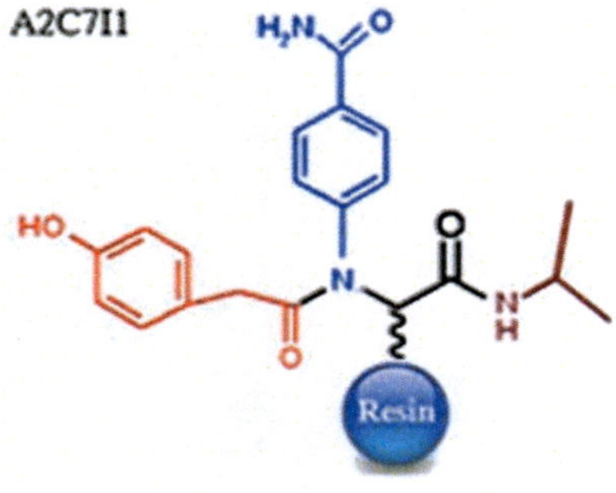

Fig. 5.8 Ugi ligand A2C7I1 mimicking the active site of the domain III of protein G. (Reproduced from El Khoury and Lowe 2013)

pure IgG samples derived from cows, goats, mice, pigs, rabbits, rats, and sheep and pure IgY from chickens on the ligand A2C7I1 was studied and compared with that on protein G. The results showed that similar binding strengths to these immunoglobulins were observed on Ugi ligand and protein G. Except for rat IgG and chicken IgY, all IgG from other species and human Fab showed significant binding (80% and 100% (w/w)) to each adsorbent. This confirms the ability of the ligand A2C7I1 to mimic the domain III of protein G and shows its potential to purify immunoglobulins derived from different sources.

The Ugi four-component reaction is also used to synthesize specific ligands that mimic erythropoietin (EPO) receptors (El Khoury et al. 2013). Detailed analysis of the binding site of erythropoietin and receptor revealed that there is a specific region on the erythropoietin receptor, which is composed of Phe93, His114, and Glu117 residues, among which Phe93 is the main contributor to hydrophobic interactions and interacts with Thr44 and Asn147 of EPO through the formation of H bonds through the oxygen atom on the backbone. And His114 and Glu117 interact with Asn147 and Arg150 of EPO through H bond and electrostatic interaction, respectively. Then, four Ugi ligands (A9C1I8, A9C10I8, A9C1I10, and A9C10I10) were synthesized on aldehyde-functionalized agarose beads (Fig. 5.9). They all contain the same amine component histamine (A9), which is used to simulate His114. In addition, glutaric acid (C1) and succinic acid (C10), which are used to simulate Glu117; and benzyl isocyanate (I8) containing benzene ring and isocyanate (I10) containing ester groups are used to simulate the hydrophobic and hydrogen bonding interaction shown by Phe93. Recombinant human EPO protein was used to test the binding ability of the four ligands under physiological pH conditions, it was found that the ligand A9C10I8 has the highest binding strength, reaching 98% of the load. Finally, affinity chromatography was used to test the ability of the immobilized ligand A9C10I8 to separate rHuEPO from the spiked CHO cell extract, and the EPO protein with a purity of 80% in the eluted fractions was obtained.

Chen et al. (2014) described the rational design and combinatorial synthesis of carbohydrate-binding ligands based on solid-phase multicomponent Ugi reaction. Aldehyde-functionalized Sepharose™ solid support constitutes one component (aldehyde) in a four-component reaction, while the other three components (primary/secondary amine, carboxylic acid, and isocyanide) are changed in combination

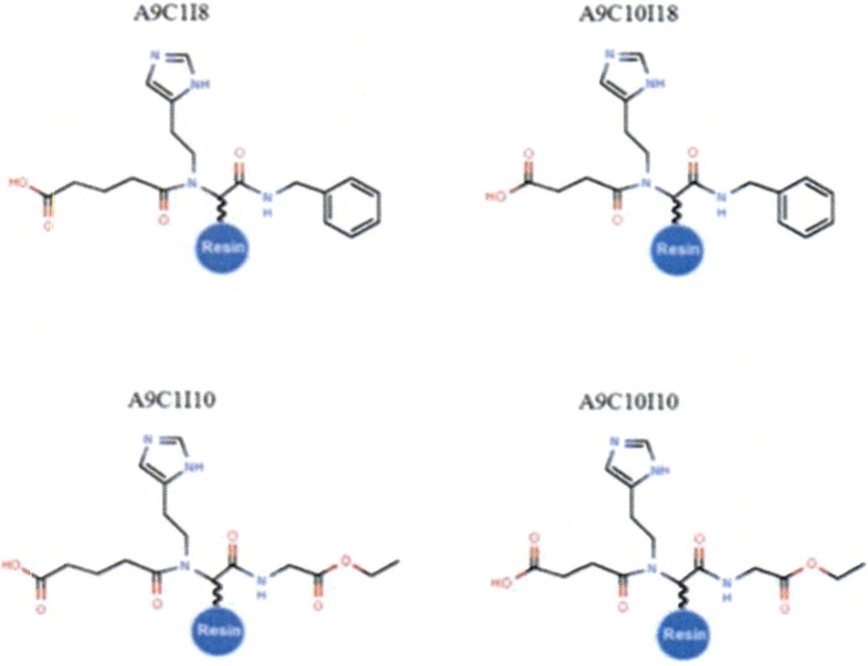

Fig. 5.9 Ugi ligand immobilized on an aldehyde-activated matrix (blue ball: Sepharose™ CL-6B resin), mimicking Phe93, His114, and Glu117 on the EPO receptor. Each ligand includes primary amine (A9: histamine), carboxylic acid (C1: glutaric acid or C10: succinic acid), and isonitrile component (I8: benzyl isocyanate or I10: ethyl isocyanate). (Reproduced from El Khoury et al. 2013)

to generate a trisubstituted Ugi scaffold, which provides a certain degree of rigidity and is functionally suitable for interacting with the glycan portion of glycoproteins. A Ugi library containing 48 ligands was screened against glucose oxidase (GOx) as a model glycoprotein, and the candidate ligand A13C24I8 was identified, which showed affinity for GOx through its carbohydrate portion. The immobilized ligand A13C24I8 (Fig. 5.10a) showed a static binding capacity of 16.7-mg GOx/mL resin and an apparent dissociation constant (K_d) of 1.45×10^{-6} M at pH 7.4. The adsorbent can also bind to 8.1-mg AGP/mL resin and display an apparent affinity constant $K_d = 1.44 \times 10^{-5}$ M. The ligand has the sugar specificity of the following sequence: sorbitol > fructose > mannitol > ribose > arabinose > xylose > galactose > mannose > glucose; however, it did not show any specificity for sialic acids or methyl α-D-glycosides. The control ligand produced by substituting C7 (4-hydroxyphenylacetic acid) for C24 (3-carboxyphenylboronic acid) (Fig. 5.10b) failed to show affinity for carbohydrate moieties, supporting the important role of boronic acid groups in sugar binding. The GOx spiked *E. coli* sample was loaded onto the immobilized ligand A13C24I8, 3-aminophenylboronic acid (APBA), and concanavalin A, and the fractions collected at peak elution were analyzed by

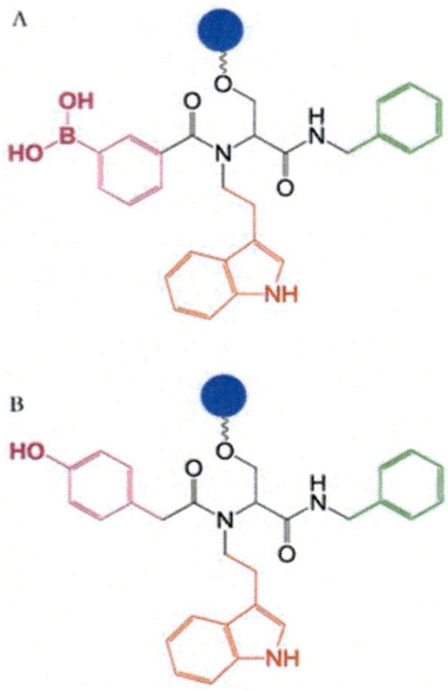

Fig. 5.10 The structure of glycoprotein binding Ugi ligand. (**a**) Lead ligand A13C24I8. (**b**) The control ligand A13C7I8 prepared by substituting C7 for C24. (Reproduced from Chen et al. 2014)

SDS-PAGE and gel densitometry, which indicate that the ligand A13C24I8 was able to purify GOx to 92% (w/w). Under the current chromatographic conditions, the protein-specific activity is 69%. The solid-phase Ugi scaffold provides a strategy for the development of cost-effective pseudobiological-specific ligands for glycoprotein purification.

5.5 Phage Display

Phage display is an important biotechnology method developed in the 1990s (Smith 1985; Rahbarnia et al 2017). An exogenous gene is inserted into the filamentous phage display vector, and connect it to the gene encoding the phage coat protein, so that the polypeptide or protein encoded by the exogenous gene and the coat protein are displayed on the surface of the phage in the form of a fusion protein. The displayed polypeptide or protein can maintain relatively independent spatial structure and biological activity. Screening ligands using phage display usually involves the following steps:

1. Construct a phage display library. Often, a set of random polypeptide coding sequence gene groups are inserted into phage display vectors to form a display library. Each phage particle in the library only displays a foreign peptide chain of one sequence, and different phage particles display peptide chains of different sequences.

2. Screen for a target phage. Recombinant phage with foreign genes is efficiently selected from the phage display library. Usually, the affinity selection method is used to select the recombinant phage that displays the exogenous polypeptide from the library. With a specific antibody against the foreign peptide as an affinity ligand, the phage with the foreign peptide will bind to the antibody immobilized on a solid matrix, and all the other phages will pass through without retention. The bound phage with the foreign peptide is then eluted.
3. Repeat the construction and selection cycles to obtain a foreign peptide with high affinity. The sequence of the gene encoding the foreign peptide can be randomized to increase the diversity and thus the chance of finding the peptide of higher affinity. It is also possible to cut out foreign genes from recombinant phage DNA and clone them into other host cells. A large amount of foreign gene products can be obtained after expression. Phage display has found widespread applications. It can be used in the discovery of fully humanized antibodies, directed evolution of enzymes, and study of protein recognition.

Following the establishment of phage display technology, yeast surface and ribosome display technologies have emerged one after another. Unlike prokaryotic bacteriophages, yeast is a eukaryotic display system that fuses foreign proteins to a lectin and displays them on the surface of yeast cells. Because the protein folding and secretion mechanism of yeast is very similar to that of mammalian cells, it can express and display human proteins more correctly than prokaryotes. In addition, due to the relatively large yeast cells, flow cytometry can be used for separation and screening. Yeast surface display technology is widely used to study the mutual recognition of protein molecules, new vaccines, and directed evolution. The ribosome display technology uses a DNA library to encode a target peptide fused to the carboxy-terminal spacer, allowing the peptide product to be clearly displayed on the ribosome. A strong transcription promoter is introduced upstream of the gene, and transcription and translation are performed in vitro. The deletion of the stop codon prevents the release of mRNA and peptide from the ribosome and forms a stable ternary complex under appropriate conditions. The complex is then exposed to a molecular target. The molecular target is usually immobilized on the surface. Peptides with sufficient affinity bind to the target and are retained, while peptides with low affinity show no retention and are washed away. The complex is usually dissociated by EDTA, thereby recovering the mRNA bound in the complex. Using mRNA as a template, cDNA is produced by reverse transcription and amplified by polymerase chain reaction.

Phage display has been employed as a tool for the discovery of high-affinity chromatography ligands. Figure 5.11 depicts the basic principle of phage display. The affinity selection of phage-displayed peptides can be performed in two different ways. First, the library is incubated with the immobilized target. Like affinity chromatography, a fraction of the library members are captured on the solid-phase carrier, and a fraction of the library members flow out with the mobile phase. By washing, the weakly interacting part is removed, and then, the high-affinity adsorbed phages are eluted by specific or nonspecific elution. The screened phage can be

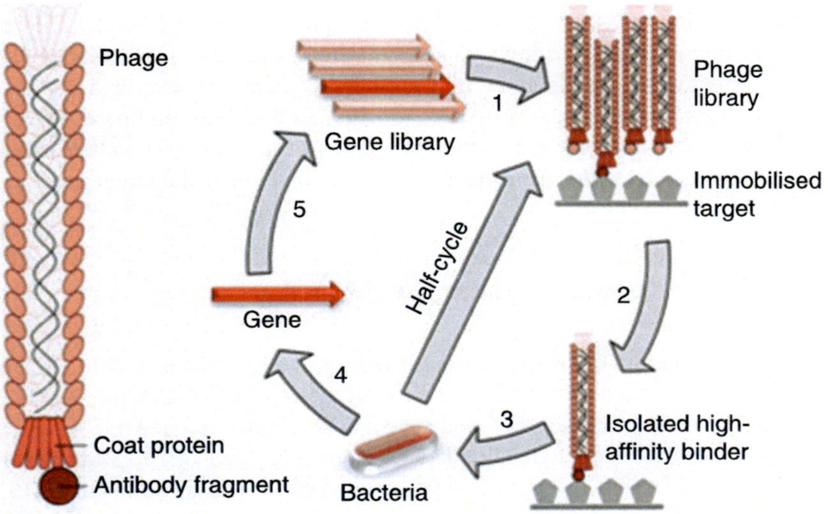

Fig. 5.11 Phage display cycle. (1) Fusion proteins for a viral coat protein + the gene to be evolved (typically an antibody fragment) are expressed in bacteriophage. (2) The library of phage is washed over an immobilized target. (3) The remaining high-affinity binders are used to infect bacteria. (4) The genes encoding the high-affinity binders are isolated. (5) Those genes may have random mutations introduced and used to perform another round of evolution. The selection and amplification steps can be performed multiple times at greater stringency to isolate higher-affinity binders. (Reproduced from Wikipedia at https://en.wikipedia.org/wiki/Phage_display)

amplified by bacterial infection to increase its copy number. The screening and amplification process can be repeated as needed to obtain higher-affinity phage-displayed peptides. The sequence of the target peptide is obtained by sequencing the isolated phage DNA.

The successful application of this technique in the selection of peptide ligands in chromatography has been reported (Nord et al. 1997). Nord et al. (2000) used a phage display system to develop two small protein-binding ligands that specifically bind to apolipoprotein and Tag polymerase, respectively. These two ligands were selected from a library constructed based on the 7-kDa domain of staphylococcal protein A. They bind to their respective targets—apolipoprotein A-1m and Tag DNA polymerase with dissociation constants in the micromolar range—and are used in their immobilized form to purify their respective target proteins from *E. coli* cell lysates. In another example, phage peptide library screening resulted in a lead peptide (RVRSFY) that interacts with von Willebrand Factor (vWF) (Nord et al. 2001). The conservative substitution of the terminal residues of the lead peptide resulted in the second peptide RVRSFV, which was more efficient in purification. Using immobilized *N*-acetyl-RVRSFYK to adsorb vWF from human plasma and eluting with divalent salts, such as calcium chloride or magnesium chloride, the purified target protein can be obtained with a recovery of 82.5%.

Gaskin et al. (2001) described the use of a heptapeptide library displayed on the surface of filamentous bacteriophage M13 to screen and purify *Rhizomucor miehei* lipase. Experimental results show that the heptapeptide sequence and part of the phage coat protein work together to produce the best binding with lipase. The entire bacteriophage was immobilized on the surface of the magnetic microsphere and used as an affinity adsorbent for the separation and purification of the target protein.

5.6 Oligonucleotide Aptamer by SELEX

Systematic evolution of ligands through exponential enrichment (SELEX), also known as in vitro selection or in vitro evolution, is a combinatorial chemical technique in molecular biology used to produce single-stranded DNA or RNA oligonucleotides. These single-stranded DNA or RNA are often called aptamers (Tuerk and Gold 1990; Ellington and Szostak 1990). The basic idea of SELEX technology is when the target interacts with a large-capacity single-stranded random oligonucleotide library, there may be sequences in the library that can specifically bind to the target molecule to form a nucleic acid–target complex. A certain method is used to remove unbound sequences to obtain sequences that bind to the target, and then, these sequences are amplified by PCR for the next round of screening. Through repeated screening, aptamer sequences with high specificity and high affinity to the target can be obtained. These sequences can activate or inhibit the function of the target, or they may only play a role in recognition. Compared with other combinatorial chemical libraries, such as random peptide libraries, antibody libraries, and phage display libraries, it has the characteristics of simplicity and speed, with great potential in molecular diagnosis and drug development.

At present, there are three types of random oligonucleotide libraries: random single-stranded DNA libraries, random RNA libraries, and random modified RNA libraries (Perret and Boschetti 2018). Generally speaking, compared with single-stranded DNA, RNA of the same molecular weight will form a wider variety of structures. In the base pairing form of RNA, there are not only AU and GG but also GU variant base pair form. Therefore, RNA will form a richer secondary structure, thus increasing the diversity of the library and expanding the scope of target screening. However, RNA is unstable in nature and easily degraded by nucleases. Therefore, it is necessary to modify the 3′ end or 5′ end of the RNA in its natural state, which increases the production cost and operational difficulty of such libraries. And single-stranded DNA can also form a rich spatial structure, has the ability to bind with high affinity and high specificity to the target, and is simple to synthesize and has stable properties. Therefore, it is more widely used in molecular biomedicine and other fields.

In the random oligonucleotide library synthesized in vitro, there is a certain length of random sequence in the middle of each nucleotide chain. It contains four freely arranged nucleotides: A, G, C, and T (U). Assuming that the length of the random region nucleic acid is n, then there will be at least 4^n kinds of nucleotides in the

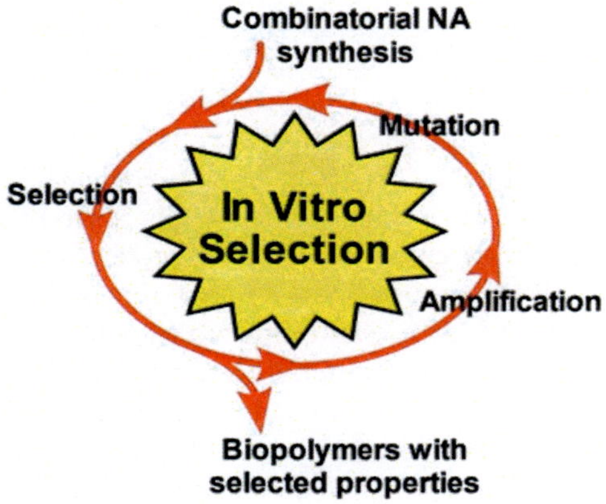

Fig. 5.12 A general overview of in vitro selection protocol. NA stands for nucleic acids (DNA, RNA, PNA), which start as a random pool and are enriched through the selection process. (Reproduced from Wikipedia at https://en.wikipedia.org/wiki/Systematic_evolution_of_ligands_by_exponential_enrichment)

random sequence. If the length of the random sequence is 35–60 bases, it can easily meet the screening requirement of the nucleotide library capacity of 10^{15}. Regardless of whether the nucleic acid in the screening library is single-stranded DNA or RNA, it will form different types of spatial structures through the pairing and packing of its own bases, hydrogen bonds, and hydrophobic interactions, such as stem loops, pseudoknots, hairpins, and G-quadruplexes. In addition, some very subtle differences, such as base addition or deletion, base exchange, and reversal in the order, may lead to a big change in the spatial structure formed, so that it has a rich three-dimensional conformation of nucleotides to bind to the target. Therefore, almost all substances in nature, including metal ions, peptides, proteins, polysaccharides, nucleic acids, organic dyes, cell fragments, and even whole living cells, can find aptamers that bind to them.

The principle of SELEX is shown in Fig. 5.12. First, a DNA library with constant sequences on both sides of the random mutation sequence region is synthesized. The T5 RNA polymerase promoter is located at the 5′end. The DNA fragment is amplified by several PCR cycles and transcribed in vitro to form an RNA library. Then, RNA molecules are selected based on binding of the synthesized RNA molecules to the target protein. For example, affinity chromatography is used to select RNA molecules that bind to the target. cDNA is obtained by reverse transcription, cDNA is amplified by PCR, and RNA is transcribed. The whole cycle can be repeated to obtain a tight and stringent binder. The aptamer selection technique in vitro has been used to identify RNA aptamers, whose targets include simple ions, peptides, proteins, organelles, viruses, and even whole cells.

Romig et al. (1999) reported the successful application of immobilized DNA aptamer to purify recombinant human L-selectin-immunoglobulin fusion protein from Chinese hamster ovary (CHO) cells, achieving a 1500-fold purification and 83% recovery. The DNA aptamer specific for L-selectin is composed of 36 nucleotides and has a dissociation constant of 2×10^{-9} M with L-selectin. The aptamer is tagged with a biotin on its 5′ end and bound to avidin immobilized on the solid support (Kuehne et al. 2017).

Miyakawa et al. (2008) reported an attempt to use RNA-based aptamers for the purification of human IgG. IgG is a class of antibodies mainly used in therapeutic agents. They can be derived from polyclonal libraries from human plasma, or they can be derived from cell cultures such as monoclonal antibodies. At present, polyclonal IgG is purified by conventional separation methods, such as precipitation and ion exchange chromatography from plasma, while monoclonal IgG is purified by affinity chromatography on protein A as a specific ligand.

The authors described for the first time the selection of aptamers specific for IgG and used them as affinity chromatography ligands. The RNA aptamer ligand has a C18 spacer with a terminal amino group, which is chemically linked to the solid support through NHS activation or Tresyl activation. The density of the ligand thus obtained is about 1 mmol/mL adsorbent. The human serum was directly injected into the chromatographic column, and the adsorbed IgG was eluted with 10-mM Tris-HCl buffer containing 200-mM KCl and 10-mM EDTA, pH 7.6. As assessed by SDS polyacrylamide gel electrophoresis, the purity of the IgG isolated from RNA aptamer is very similar to the purity of the antibody isolated from protein A. The binding capacity of the column is close to 30-mg/mL affinity resin. The aptamer resin can be reused and can be regenerated many times. It is washed with 6-M urea or heated in pure water at 85 °C for 5 min without significant loss of activity. Further study emphasizes the necessity of GGUGCU sequence to achieve high affinity for hIgG-Fc. The aptamer ligand offers a distinct advantage over protein A in that the antibody can be eluted without the use of acidic solutions.

Attempts have also been made to use DNA aptamers as affinity chromatography ligands to purify lectins (Ahirwar and Nahar 2015). Lectins are carbohydrate-binding proteins that are highly specific for sugar groups and so cause agglutination of particular cells or precipitation of glycoconjugates and polysaccharides. Lectins play numerous roles in biological recognition phenomena involving cells, carbohydrates, and proteins. Lectins also mediate attachment and binding of bacteria, viruses, and fungi to their intended targets. Recently, an aptamer ligand with ~172 nM affinity was selected, which has the ability to distinguish Con A from other lectins. A 41-nt ssDNA aptamer, which has a hexylamino spacer at its end, was immobilized on NHS-activated agarose beads. After quenching the unreacted active ester groups with a pH 9 Tris-HCl buffer, the affinity adsorbent for lectins was obtained. The final ligand density on the adsorbent is close to 5 mM, and the coupling yield is 80%. The crude protein extract was loaded on the aptamer phase, and the unbound proteins were washed; the adsorbed Con A was eluted by increasing the sodium chloride concentration to 1.5 M. Under these conditions, the recovery of lectin was 66%, and the enrichment was about 366-folds. The purity was assessed by SDS-PAGE, which showed a single protein band when stained with classic Coomassie Blue. After seven separation cycles on a 19-mL column at a constant flow rate, the Con A recovery dropped by only 12.9% while maintaining a constant purity. These results demonstrate the potential of aptamer affinity chromatography in one-step protein purification. Immobilized sugars are usually used for the purification of lectins. However, this method is not very specific as other lectins or other proteins that have affinity for sugar ligands are retained too, resulting in cross contamination. When using aptamers, this risk is completely eliminated.

Forier et al. (2017) demonstrated the use of DNA aptamers in one-step chromatographic purification of three human therapeutic proteins from various sources: Factor VII, Factor H, and Factor IX. The DNA aptamers were chemically bonded to the NHS-activated agarose through a six-carbon spacer with one end attached to the 5′end of the aptamer and the other end terminated by a primary amine.

DNA aptamers were selected using the SELEX technology against coagulation Factor VII, complement Factor H, coagulation Factor IX, and GLA domain containing proteins. The sequences of anti-FVII aptamer Mapt2.2CS is as follows: 5′-CCGCACGCTACGCGCATGAACCCGCGCACACGACTTGAAGTAGC-3′. The sequence of anti-blood coagulation Factor IX aptamer (Anti-FIX) called Nonapta5.1 is as follows: 5-′-ACCTTTAACCCACTAGTCTAAGCGGGTCTAGCGTCGTCC-3′. The sequence of anti-Factor H aptamer called MaptH1.1CSO is as follows: 5-′-GGTCTCGGGCACGGGTCAGGCGGTTATACGGTGCCC-3′. The sequence of anti-GLA domain aptamer Mapt1 is as follows: 5′--GGAGATAGCCACGACCTCGCACATGACTTGAAGTTAAACGCGAATTAC-AAACCCAGCCC CCTCCAGGCTGTGCGAAAGC-3′. This sequence has also been shortened to a smaller size (Mapt1.2CS) with a similar specificity, corresponding to the following modified sequence: 5-′-CCACGACCTCGCACATGACTTGAAGTTAAACGCGAATTAC-3′.

The coagulation FVII is a 50-kDa single-chain glycoprotein present in blood plasma at a normal concentration of 0.5 μg/mL. This is a very low-abundance protein representing less than 0.001% of the total plasma proteins, involved in the blood coagulation process, and is mediated by a tissue factor. It is used in clinical applications of acquired hemophilia and also when alloantibodies to coagulation factors are developed. It is classically purified by a combination of several chromatographic separation steps, involving ion exchange and immunoaffinity chromatography either from human cryoprecipitate-poor plasma or from recombinant preparations. Coagulation Factor VII purification trials were performed using a DNA aptamer named Mapt2.2CS and a pre-purified preparation of human origin containing about 10% of protein impurities. The adsorption condition involved the presence of calcium chloride at 10-mM concentration buffered in 50-mM Tris-HCl pH 7.4 containing 150-mM sodium chloride. The adsorbed FVII was eluted with an aqueous buffered solution of 50-mM EDTA. The analysis of the flowthrough and the eluted fraction demonstrated a high degree of purity reached by the present affinity process, with a purification over 98% and recovery in excess of 80%. Since Factor VII is posttranslationally modified at the level of glutamate residues through a γ-carboxylation process, the selected aptamers have been successfully tested for their ability to distinguish isoforms.

The second example of an aptamer-based purification process deals with the separation of human complement Factor H (FH) either from a semi-purified human plasma preparation or directly from a crude supernatant of recombinant cells, secreting the humanized rFH protein. FH, a member of the regulators of complement activation family, is a large plasma glycoprotein (155 kDa) found in normal blood at a concentration of 200–400 μg/mL. It is a low-abundance protein, representing about 0.4% of the global plasma protein amount, but actively involved in the protection action on self cells and surfaces. It is generally purified by combining precipitations and classical chromatography steps. Both native and recombinant Factor H was purified using the anti-Factor H aptamer as single band in gel electrophoresis. After loading the biological material and washing with 1 M NaCl to remove noninteracting proteins, FH was eluted with 50-mM Tris-HCl buffer containing 100-mM EDTA, pH 8.

The third example deals with the purification of human coagulation Factor IX. It was discovered in 1952 after a young boy named Stephen Christmas was found to be lacking this exact factor, leading to hemophilia. This protein is a member of the vitamin K-dependent coagulation system. It is a glycosylated polypeptide chain having a molecular mass of 56 kDa. It is constituted of four main domains: (1) the interaction domain of EGF-1, (2) the activation domain of EGF-2, (3) the GLA domain interacting with calcium ions, and (4) the serine protease domain. Its normal concentration in blood plasma is 4.4 μg/mL. Its deficiency engenders the so-called hemophilia B, which is treated by injections of purified Factor IX obtained either from blood plasma or from recombinant technology. Purification of this protein from human plasma was performed on an anti-Factor IX aptamer. The biological material was loaded on the column and washed successively with washing buffer 1 (50-mM Tris-HCl pH 7.4 containing 150-mM NaCl, 2-mM $CaCl_2$, and 1-mM $MgCl_2$) and washing buffer 2 (2-M sodium chloride). The bound Factor IX was eluted using 200-mM EDTA solution pH 8. The purity of the protein increased from 0.5% to 98%. The biological activity of aptamer-purified Factor IX, determined as a ratio between the amidolytic activity and the antigen rate [Am]/[Ag], was perfectly preserved with a value of 1.1.

The DNA aptamers were also used to distinguish native proteins from their variants. As many proteins, coagulation Factor VII and Factor IX are posttranslationally modified to acquire specific binding properties or their full biological activity. In this case and also for a number of other vitamin K-dependent plasma proteins involved in the coagulation cascade, a γ-carboxylation process takes place at glutamic acid residue levels. In the so-called GLA domain, the carboxylation of the γ-carbons is catalyzed by the γ-glutamyl carboxylase. This posttranslational modification is required to confer to GLA domain binding properties to calcium and phospholipids and to acquire the necessary protease activity for the coagulation effect. While this protein modification is present during its maturation in human cells, recombinant proteins contain a certain amount of incompletely modified proteins with the absence or partial presence of GLA domain (des-GLA forms). These protein variants with largely reduced biological activity must be eliminated during the purification process. The aptamer ligands were tested for their ability to discriminate between active proteins and des-GLA forms. As a model, food cow milk was spiked at 0.21 mg/mL of pure human plasma coagulation Factor VII containing a significant part of des-GLA forms. This sample was directly submitted to affinity separation on anti FVII-GLA Mapt2.2CS aptamer ligand. During the affinity chromatography process, the des-GLA forms, very similar to the active protein, are completely eliminated. The lower molecular mass of des-GLA variant result from two distinct phenomena: (1) the absence of glutamic acid carboxylation and (2) the partial cleavage of the GLA domain.

The results presented by Forier and coworkers confirmed the potential of aptamer-based affinity ligands, which can be used not only for the purification of plasma-derived proteins but also for the purification of any other general proteins due to their excellent molecular recognition capabilities (Murphy et al. 2003; Kokpinar et al. 2011).

5.7 Rational Ligand Design

Traditional drug design is to conduct quantitative structure-activity relationship studies on a large number of small molecule active compounds to obtain lead compounds and then perform molecular optimization to finally obtain candidate drug molecules and enter the next stage of clinical evaluation. This method requires chemical synthesis, activity determination, screening, and optimization of each active compound, which requires a lot of time, material, and labor costs. Therefore, the traditional drug development process is a costly, time-consuming, and laborious process. However, the rational drug design method developed in the last two decades takes a completely different approach. Computer modeling technology is used to identify active sites in the three-dimensional structure of drug targets, active compounds are designed based on the principle of space and energy complementarity, and then the candidate compounds are screened and optimized through molecular techniques, such as molecular docking and molecular dynamics simulation, and hits are selected from them for further synthesis and evaluation. Since most of the work in this process is done in a computer system and does not require experimental synthesis and activity determination, it not only saves costs and accelerates the development process but also helps to better understand the drug–target interactions.

5.7.1 Molecular Docking

Molecular docking is a structure-based drug design method. The binding site information in the protein–ligand binding pocket is mainly based on the three-dimensional structure information of the target. According to the shape and physical and chemical properties of the binding site, small molecule design or virtual screening by database search is used to predict the binding mode and affinity of the compound molecule and the receptor. The compounds obtained by screening were synthesized experimentally, and their physicochemical properties and biological activities were determined. Drug design based on molecular docking usually proceeds by the following steps:

Obtaining the receptor structure: The three-dimensional structure of the receptor is generally downloaded from the protein structure database, and its structure is mainly obtained through X-ray diffraction techniques, nuclear magnetic resonance spectroscopy, and other means. For proteins that do not exist in the database, they can also be obtained by homology modeling. Homologous modeling is generally based on amino acid sequences, and sequence alignment is performed by searching for homologous sequences in a sequence library. Generally, the pros and cons of a sequence are judged by the similarity between the sequences. It is generally accepted that sequences with a similarity greater than 50% can be used for modeling. After the three-dimensional structure model is established, it is necessary to evaluate the obtained model through computation methods. The commonly used homology modeling tools include Modeller, SwissModel, etc. After obtaining the three-dimensional structure of the receptor from database, it is generally necessary to process it to complete the missing atoms, residues, and chain information, determine the protonation status of the residues in the structure, and correct the valence bond information in the structure.

Determining the binding site: If the downloaded crystal structure contains ligand molecules, the binding site can be determined based on the three-dimensional structure of the small molecule. If there is no small molecule in the download crystal structure or the small molecule is not in its binding site, some other tools can be used to determine its binding site. The commonly used tools are Fpacket, PocketDepth, CASTp, etc. GRID is one of the most commonly used tools to determine the binding site of a target. Generally, the probe atoms/groups are placed in the grid near the target cavity, and the interaction mode between the receptor and the probe is used to determine the interaction type of each grid, thereby determining the binding site of the target. However, some empirical rules can also be used to define the binding site, such as the HSITE module in the LUDI software, which mainly determines the hydrogen bonding site by analyzing the three-dimensional structure of small molecules to obtain hydrogen bond parameters.

Generation of ligand molecules: The generation of ligand molecules is mainly divided into three categories:

1. Database search method. Mainly obtain the initial small molecule structure from existing compound databases, such as ChemBL, DrugBank, PubChem, STITCH, BindingDB, ZINC, SIDER, etc. If the initial molecule is a two-dimensional structure, first transform it into a three-dimensional structure (such as the CONCORD program), and then optimize the structure and generate a conformation. Programs based on database search include CAVEAT, ALLADIN, etc. The disadvantage of this type of method is that new molecules cannot be discovered.
2. Fragment growth method. First construct a library of molecular fragments, and then construct a new compound library in a freely combined manner, or select appropriate fragments according to the site information at the binding pocket

to splice into a complete molecule to construct a specific molecular library. The commonly used fragment growth tools are LUDI, CLIX, SPLICE, SPROUT, etc. This method ignores the internal flexibility of the fragment.

3. Atomic growth method. Starting from atoms, the enumeration method is used to generate molecules. There are differences in the types of bonds formed between atoms, spatial positions, and atom types, and the atoms that grow each time have greater randomness, making the number of molecules generated exponentially increase. The commonly used tools are LEGEND and GenStar.

Implementation of molecular docking: Molecular docking mainly realizes molecular recognition through intermolecular interaction matching. The basic principle is to place the ligand molecule in the receptor binding pocket, and predict the binding mode and binding constant of the protein–ligand based on the principle of geometric shape and energy complementarity. In the molecular docking process, the three-dimensional space coordinates of the macromolecular receptor and the small molecule ligand are input into the docking program, and the predicted geometric coordinates of the complex are output from the docking program. According to whether the receptor and ligand molecules undergo conformational changes during the docking process, molecular docking can be divided into rigid docking, semiflexible docking, and flexible docking. In rigid docking, the conformational changes of ligand and acceptor molecules are not considered. This docking mode has fast calculation speed, but low accuracy, and is suitable for processing molecules with larger structures, such as protein–protein and protein–nucleic acid interactions. After the molecular model is established, this method is generally used for preliminary analysis to determine whether the molecules can be bound. During the semiflexible docking process, the conformation of the receptor is fixed, and only the conformation of the ligand is allowed to change to a certain extent to adapt to the environment of the binding pocket. This docking method improves its prediction energy, while taking into account the calculation efficiency, and is widely used in drug design, involving the docking of large molecules and small molecules. In fact, both receptor macromolecules and ligand small molecules have a certain degree of flexibility, and conformational changes may occur during the docking process. The inductive effect of the ligand will cause the conformation of the receptor to change. At the same time, the conformation of the ligand will also change due to the inductive effect of the receptor. The two adapt to each other to achieve the best binding mode. This flexible docking method is more in line with the real physiological environment. At the same time, considering the flexibility of receptors and ligands can improve the accuracy of the prediction results, but it also increases the amount of calculation and requires a lot of technical resources. Therefore, this docking technology is generally only used in the process of precise identification between molecules.

Score evaluation: The scoring function is mainly used to evaluate the binding ability of ligand molecules and receptors. The free energy perturbation method is the most rigorous method combining affinity calculation, but it will take a lot of computation time, and the simpler scoring function does not require too much computing resources and can improve the screening speed. Most scoring functions are not based on a complete physical model, so they often do not strictly follow the many-body expansion theory, conservation laws, symmetry invariance, etc., and even the expressions of some knowledge-based scoring functions have no physical meaning at all. In fact, as a tool used in the high-throughput screening of drugs, most of the scoring functions are centered on efficiency, and the balance of accuracy and efficiency is pursued through approximate means. According to the literature (Su et al. 2019), scoring functions can be roughly divided into the following four types: scoring functions based on physics (force field), scoring functions based on experience (regression), scoring functions based on mean force potential (knowledge), and scoring functions based on descriptors (machine learning). The ideal scoring function should be able to correctly evaluate the interaction between the compound and the potential target. However, so far, there is no scoring function that can be well adapted to different systems. Therefore, the consistency scoring evaluation method is usually used, that is, multiple scoring functions are used to simultaneously evaluate the results of a round of docking.

5.7.2 Molecular Dynamics Simulation

Molecular dynamics simulation is a method to describe the evolution of molecular motion over time on the basis of molecular mechanics. It is assumed that the motion of all particles in the system follows the Newton's second law of motion, that is, $F = ma$, where F is the external force on the system, m is the mass of the particle, and a is the acceleration. Different from molecular mechanics, molecular dynamics solves the state, behavior, and process of molecules that change over time, which can make the molecular conformation cross a larger energy barrier. It regards the atoms in the molecule as a series of elastic balls. At any moment of motion, the force and acceleration on each atom are calculated through the Newtonian equation of motion to determine their coordinate position and speed of motion, and from this motion, various macro properties are calculated in the trajectory. Molecular dynamics simulation is used to understand some dynamic processes in biological systems, such as ligand-receptor molecular recognition mechanism, conformational changes, solvent effects, etc. The main steps in molecular dynamics simulation are as follows:

Determine the starting configuration: The starting configuration of biological macromolecules mainly comes from experimental data, such as the molecular structure determined by X-ray crystal diffraction or nuclear magnetic resonance spectroscopy or the structure obtained by homology modeling based on the known molecular structure. The configuration of small organic molecules in the system can also be calculated by quantum chemistry.

Choose the appropriate force field and simulation software: Different force fields have different scopes and limitations, which need to be selected appropriately according to the system and problems studied. When multiple force fields are used at the same time, attention should be paid to the compatibility of the force fields used. The choice of force field is closely related to the accuracy of the final simulation result. The choice of software is often related to the force field used, and the required algorithms, software operating speed, and parallel computing capabilities should be considered.

System construction and energy minimization: Once the initial configuration of the target and interacting molecule has been established, a simulation system must be constructed according to the environment in which the target and the ligand is located, such as an aqueous solution or a transmembrane environment. Add enough solvent molecules around the biological macromolecules. The size of the system and the shape of the simulated cell depend on the specific system, and both rationality and feasibility should be considered. Local irrationality often exists in the initially established system (such as the distance between adjacent atoms is too close), and the system energy needs to be minimized first. The most commonly used energy minimization methods are the steepest descent method and the conjugate gradient method. The steepest descent method optimizes fast, but the convergence is poor. The conjugate gradient method has a slow search speed but good convergence. Therefore, the two are often used in combination. Generally, the steepest descent method is used to quickly optimize the molecular structure, and then the conjugate gradient method is used to continue the optimization until convergence.

Equilibrate process: After the system is constructed, each atom must be given an initial velocity. This speed is randomly generated based on the Boltzmann distribution at a certain temperature. Then, the movement speed of each atom is adjusted so that the sum of the kinetic energy of the system in all directions is zero, that is, to ensure that the system has no translational displacement. The initial velocity is usually generated at low temperature (to avoid excessively high initial temperature, causing atomic collisions, system instability, etc.), and then restrain the solute under NVT conditions and gradually increase the temperature to prevent damage to the reasonable configuration of the solute during the heating process. After rising to the required temperature, simulation is performed under NPT conditions to adjust the pressure and density of the system. In this process, it is necessary to monitor the energy, temperature, pressure, density, etc. of the system to see if it converges until the system reaches equilibrium.

Data collection: After the system reaches equilibrium, a long-term simulation can be carried out, and samples can be collected for analysis from this process. Sampling should be performed after the system is equilibrated, and three trajectories can be recorded, that is, the changes in the coordinates, velocity, and energy of the particles in the system over time. Energy includes different energy items, as well as changes in temperature and pressure of the system over time, such as electrostatic energy, van der Waals energy, kinetic energy, potential energy, and total energy of the system. The frequency of recording the trajectory can be selected according to the specific simulation system, the simulation step length, and the phenomenon and properties studied, and the length of the simulation time and the storage requirements of the output file are taken into consideration. If the trajectory is not frequently recorded, it is easy to miss information, and it is not possible to observe some related short-time phenomena. Recording trajectories too frequently will generate huge trajectory files, causing unnecessary trouble for post-processing and data backup, and may exceed the available storage space, causing simulation interruption. The simulation time should be as long as possible to ensure that the phenomenon or property under study can be observed and to ensure the repeatability of the phenomenon.

Data analysis: Obtain macroscopic physical quantities that can be compared with experimental results through ensemble averaging. Common data analysis methods include the average energy and its fluctuation with the simulation time; the system's average atomic coordinates and its fluctuation with the simulation time; the calculation of the binding free energy of the system; the analysis of intra- and intermolecular hydrogen based on geometric criteria, bond structure, and its hydrogen bond occupancy; protein secondary structure and its changes over time; and calculation of the dynamic properties of the system.

5.7.3 Rational Design of Ligands

Liu et al. (2007a) first reported the rational design of protein-specific peptide ligands using the flexible docking simulation in the Sybyl package. They chose α-amylase as the model protein. Based on the X-ray crystal structure of α-amylase with a resolution of 2.21 Å, they determined the binding pocket of α-amylase, which consists of 40 residues including phenylalanine and tryptophan. The presence of these hydrophobic amino acid residues indicates that ligands containing hydrophobic groups can be selected to bind to the target protein through π-π stacking. Therefore, phenylalanine and tryptophan were chosen as the two building blocks. In addition, histidine, glutamic acid, and asparagine, which are easy to provide protons or accept protons, were selected to construct a pentapeptide with hydrogen bonding function. There are 120 combinations of pentapeptides ($5 \times 4 \times 3 \times 2 \times 1 = 120$), so a peptide library containing 120 pentapeptides was constructed for molecular docking simulation. The scoring function is used to rank the candidate pentapeptides, and it is found that FHENW has the highest score. However, visual inspection with the MolCAD program revealed that the C-terminus of the pentapeptide was embedded in the protein binding pocket. If the C-terminus is used as an anchor to connect to a solid carrier, it may weaken its binding force to α-amylase. Therefore, serine was selected as the anchor group to increase the length of the ligand, and the hexapeptide of FHENWS was constructed. Molecular modeling shows that the hexapeptide is docked into the binding pocket of α-amylase, the C-terminal serine is exposed to the outside of the binding pocket, the N-terminal phenylalanine interacts electrostatically with the negatively charged residues in the binding pocket, and the ligand interacts with the negatively charged residues in the binding pocket. The protein forms ten hydrogen bonds. In addition, the side chain atoms of the ligand can form many van der Waals contacts with the protein, and there is a π-π stacking interaction between the phenylalanine in the ligand and the Tyr-62 in the binding pocket. Finally, the hexapeptide obtained by solid-phase synthesis was connected to EAH-Sepharose microspheres, and the protein adsorption isotherm on the stationary phase was tested. The apparent binding constant was $2.5 \times 10^5\ M^{-1}$. The synthesized peptide resin was used to purify α-amylase from *Bacillus subtilis* culture solution. α-Amylase with a purity of 80.4% was obtained, which confirmed the high affinity and selectivity of the designed hexapeptide ligand.

Liu et al. (2007b) also reported the rational design of tissue-type plasminogen agonist (t-PA)-specific binding ligands by combining molecular docking and molecular dynamics simulation. It is known that the binding pocket of t-PA can be divided into three regions: inner P1, middle P2, and outer P3. Twenty natural amino acids were placed in the binding pocket of t-PA one by one, and arginine, asparagine, glutamine, histidine, phenylalanine, tryptophan, and tyrosine were found to have good interactions with site P1, glycine, and aspartic acid with site P2 and glutamic acid with site P3. Therefore, ten amino acids were selected and linked sequentially to form a library of 14 tripeptides ($7 \times 2 \times 1 = 14$). Visual inspection showed that the C-terminus of the tripeptide was embedded with a binding pocket, which was not conducive to the immobilization of the ligand. Therefore, a serine at the C-terminus was chosen as the spacer arm for the immobilization of the ligand. Therefore, a library of 14 tetrapeptides with serine as the C-terminus was constructed. The tetrapeptides were docked to the t-PA binding pockets one by one, the scores were calculated and ranked, and the six tetrapeptides with the highest scores were selected and connected to the EAH-Sepharose microspheres, and then the molecular docking simulation was performed. The results show that electrostatic interaction and hydrogen bonding are the most important stabilizing factors. Further analysis showed that among the six tetrapeptides, only the complex of QDES and the target was relatively stable, and the binding conformation of the other five tetrapeptides more or less changed. Therefore, QDES was chosen for affinity analysis and molecular dynamics (MD) simulation. A molecular dynamics simulation of 8 ns was performed on the QDES–t-PA complex. In the simulation process, the conformation of the tetrapeptide and t-PA binding pocket was basically unchanged, while the conformation of the whole protein changed a little bit, indicating that the tetrapeptide forms a stable complex with the target protein and reduces their flexibility. At the same time, MD simulations showed that the electrostatic interaction between the N-terminal glutamine of the tetrapeptide and the negatively charged residue of the t-PA binding site played an important role in the association. The six hydrogen bonds formed between the tetrapeptide and t-PA are very stable. Hydrogen bonds mediated by water molecules play an important role in stabilizing protein–ligand complexes. Finally, the QDES tetrapeptide was connected to agarose gel to purify t-PA from the crude pig heart extract, which confirmed the affinity and specificity of the tetrapeptide ligand.

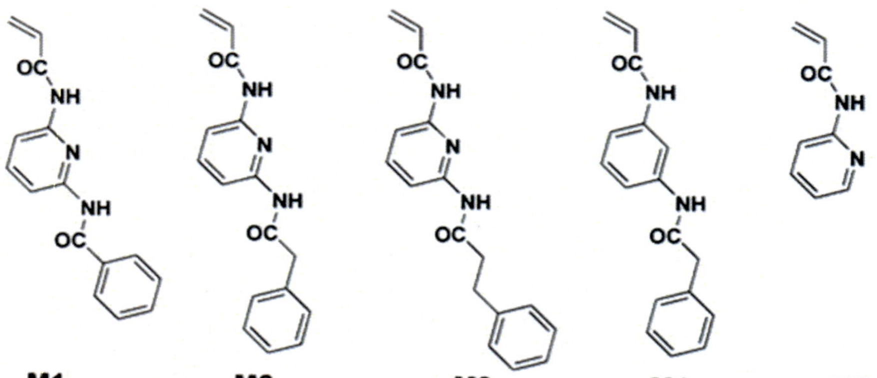

Fig. 5.13 Chemical structures of the functional monomers. (Reprinted with permission from Qiao Y, Zhao J, Li P et al. (2010a) Adsorbents with high selectivity for uremic middle molecular peptides containing the Asp-Phe-Leu-Ala-Glu sequence. Langmuir 26:7181–7187. Copyright {2010} American Chemical Society)

Qiao et al. (2010a) reported the design and simulation of selective adsorbents used to remove mid-molecular-weight peptides from the serum of patients with uremia. Taking the common sequence Asp-Phe-Leu-Ala-Glu (DE5) of medium-molecular-weight peptides as the target oligopeptide, according to the structural characteristics of hydrophobic residues in the middle and negatively charged residues on both sides, three bifunctional monomers (M1, M2, and M3) with both coordination and hydrophobic groups were designed, as shown in Fig. 5.13. They are based on 2,6-diaminopyridine as the skeleton, one side is connected with a hydrophobic group, and the other side is connected with a vinyl group to obtain functional monomers M1, M2, and M3. The chemical composition of the three monomers is basically the same, except for the spacing between the terminal benzene group and the pyridyl group increases sequentially. In addition, two reference monomers, M4 and M5, were synthesized. M4 was based on 2,6-diaminobenzene, while M5 lacked the hydrophobic end group. Through the copolymerization of functional monomer with acrylamide and cross-linking agent, three corresponding bifunctional adsorbents, CP1, CP2, and CP3, were synthesized. Their adsorption capacity for DE5 was evaluated by static adsorption experiments. All three adsorbents show good adsorption. Among them, CP2 has the best adsorption capacity due to the best geometric matching between them. By comparing the adsorption capacity of CP2 to six peptides of different sequences, the specificity of CP2 to DE5 sequence was confirmed. CP2 shows good adsorption capacity for peptides containing DE5 sequence but has much lower affinity for peptides without DE5 sequence.

Isothermal titration calorimetry (ITC) was used to quantify the association behavior of functional monomers and linear polymers of acrylamide P1, P2, P3, and DE5, and their binding constants were determined to be 8.68, 10.4, and $7.8 \times 10^4\ M^{-1}$,

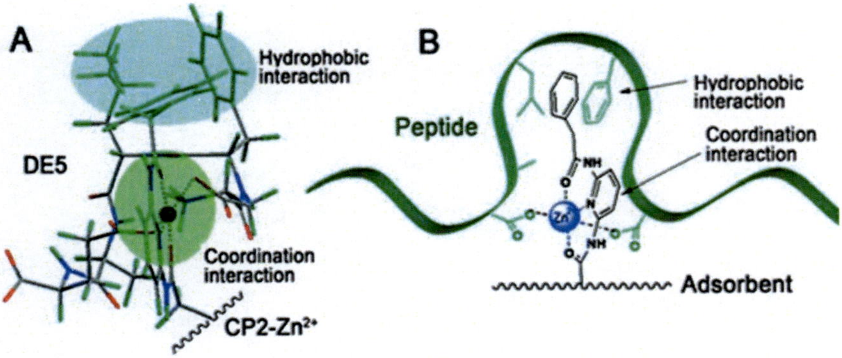

Fig. 5.14 Interaction conformation of CP2–Zn^{2+} with DE5 by molecular docking (**a**) and schematic diagram for the adsorption of DE5-containing peptides onto CP2–Zn^{2+} (**b**). (Reprinted with permission from Qiao Y, Zhao J, Li P et al. (2010a) Adsorbents with high selectivity for uremic middle molecular peptides containing the Asp-Phe-Leu-Ala-Glu sequence. Langmuir 26:7181–7187. Copyright {2010} American Chemical Society)

respectively, consistent with the results from batch adsorption experiments. Next, the Sybyl program package was used to simulate the molecular docking between the ligand and DE5. The binding energies were −24.71, −32.92, and −19.31 kcal/mol, respectively, for the complexes of CP1, CP2, and CP3 with DE5. The interaction conformation of CP2 and DE5 is shown in Fig. 5.14. When the DE5-containing peptide is adsorbed on the ligand, the hydrophobic residue and the phenyl group in the ligand undergo π-π stacking, and the two carboxyl moieties are bound to the ligand through Zn^{2+}-mediated coordination bonding. The synergy of these interactions and the geometric matching of DE5 and CP2 result in the ligands having good selectivity and adsorption capacity for peptides containing DE5 sequences.

Qiao et al. (2010b) reported the design, optimization, and evaluation of oligopeptide selective ligands. They used a two-step strategy to design and optimize the structure of the ligand. The medium-molecular-weight oligopeptide DFLAE (DE5) is one of the main toxins in the serum of uremic patients, and it is usually removed by selective adsorption in hemoperfusion. The target pentapeptide DFLAE is composed of a hydrophobic side chain as the core and negatively charged residues on both sides (Fig. 5.15a). According to the structural characteristics of DE5 and the principle of space and energy complementarity, a cyclodextrin derivative ligand with both hydrophobic inclusion and electrostatic interaction sites was designed. Cyclodextrin is connected to the carboxyl group of histidine through diamine or triamine to obtain chimeric ligands, CDenHis (Fig. 5.15b) and CDdnHis. The hydrophobic cavity of cyclodextrin can be complexed with the phenyl group in DE5, and the amino group of histidine can form an ion pair with the carboxyl groups at both ends of DE5. Therefore, the ligand is expected to have a strong affinity for DE5. Through isothermal titration microcalorimetry and batch adsorption experiments, the binding free energy and adsorption isotherm of the ligand and the target oligopeptide were determined. Then, with the help of molecular modeling, the binding energy of ligand

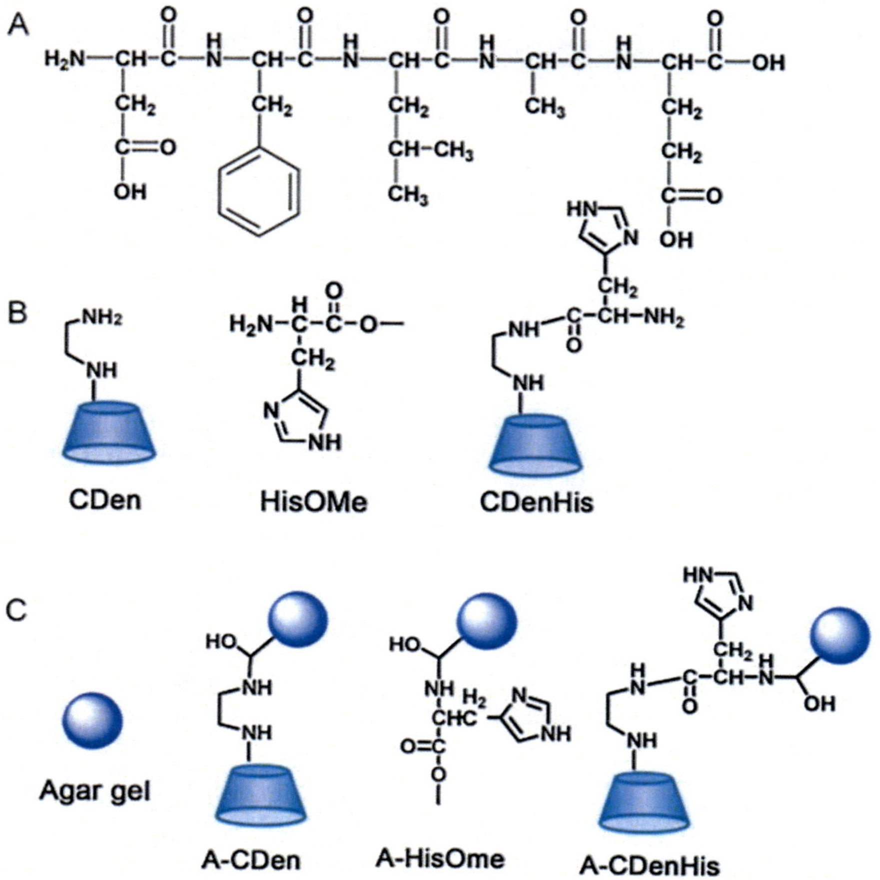

Fig. 5.15 Chemical structures of DE5 (**a**), affinity ligands (**b**), and corresponding affinity adsorbents (**c**). (Reproduced from Qiao et al. 2010b)

and target was calculated through molecular docking and combined with visual inspection of the docking pose of oligopeptide with the ligand. It was found that the distance between CD and His on CDenHis, in which the diethylamine is as the linker, is slightly shorter than the distance between Phe and Glu in DE5. The ligand CDdnHis with diethylenetriamine as the linker shows a closer match. The lowest binding energies of the two ligands to DE5 are −64.71 and −79.67 kJ/mol, respectively. ITC experiments show that their binding constants are 6.23 and 10.2×10^4 M^{-1}, respectively. Surface plasmon resonance (SPR) experiments show that the binding of DE5 to CDdnHis is fast and stable, and the flow rate of DE5 solution in the sample cell has almost no effect on the binding of ligand.

In order to verify the adsorption selectivity of CDdnHis to DE5, the adsorption behaviors of six oligopeptides with different sequences were compared, with DE5 exhibiting the highest binding capacity. Finally, ^{1}H NMR analysis was used to characterize the chemical shift caused by the formation of a complex between DE5 and CDdnHis. It was observed that the proton of the phenyl group of DE5 produced a high field shift of 0.09 ppm, which was attributed to the inclusion of the benzene ring in the hydrophobic cavity of CD. The three hydrogens near the carboxyl group of the N-terminal and C-terminal side chains of DE5 display chemical shifts of 0.008, 0.03, and 0.05 ppm, respectively, which are attributed to the electrostatic interaction with the histidine side chain on the ligand. These NMR data provide clear evidence for the proposed mode of binding between the chimeric ligand and DE5.

Aghaee et al. (2014) introduced a method for designing human serum albumin (HSA)-specific ligands based on molecular docking and molecular dynamics simulation. First, a dipeptide of Ala-X (X is any one of the 20 natural amino acids) and Y-Ala (Y is any one of the 19 natural amino acids except Ala) was synthesized on the computer, resulting in a ligand library of 39 dipeptides. Then, these dipeptides were docked on to the warfarin binding pocket of HAS one by one, and it was found that four dipeptides containing tryptophan (Trp), tyrosine (Tyr), histidine (His), and phenylalanine (Phe) bind to the internal site P1, and the three dipeptides containing tryptophan (Trp), tyrosine (Tyr), and phenylalanine (Phe) bind favorably to the external site P2. One of the four amino acids in the first group was coupled with one of the three amino acids in the second group to form a ligand library of 12 dipeptides ($4 \times 3 = 12$). After docking simulation, the Trp-Trp dipeptide with the highest score was selected as the lead ligand. Visual inspection revealed that the dipeptide Trp-Trp is shorter than the binding pocket, which means that the N-terminal Trp is embedded in the binding pocket, which is not conducive to coupling with the solid-phase medium. Therefore, lysine was selected as the spacer arm to be connected to the N-terminus of the dipeptide, and thus the Lys-Trp-Trp tripeptide was synthesized. A 20-ns molecular dynamics simulation was performed on the binding of the tripeptide and HAS. It was found that water molecules bridge the ligand and proteins through hydrogen bonds and play an important role in stabilizing protein–ligand complexes. In addition, the distance between the hydrophobic residues on the protein and the Trp-Trp in the ligand did not change significantly, indicating the key role of the hydrophobic interaction in stabilizing the complex. Finally, the competitive binding with warfarin and the difference spectrum confirmed that the ligand can specifically bind to the warfarin binding site of HAS.

5.8 Summary

With the development of molecular biology, structural biology, and computational technology, drug discovery has evolved from traditional experimental screening to computer-aided design and virtual screening, making structure-based drug design methods widely used in the pharmaceutical industry. In view of the structural similarity between chromatographic ligands and drugs, the application of structure-based approaches to the design of chromatographic ligands will speed up the discovery of new ligands. Rational design of chromatographic ligands has become an important research direction in the field of chromatography. Among them, the fragment growth method is the most commonly used method. By bonding different functional groups on a single-mode ligand, the pluripotency of the ligand's mode of action is increased to form a highly selective multimode ligand. The second is the synthesis and screening of combinatorial peptide libraries. Through parallel or random synthesis, a large number of peptide libraries with different structures have been prepared. According to the binding ability to the target, the peptide library is screened for ligands with high affinity and selectivity. The Ugi four-component reaction has also been used in the synthesis and screening of multimodal ligands, resulting in a series of affinity ligands that are highly selective for biomolecules.

Phage display represents the most promising approach for peptide based highly specific ligands. By inserting exogenous genes into filamentous phage vectors and connecting them to the gene encoding phage coat protein, the polypeptide or protein encoded by the exogenous gene and coat protein are displayed on the surface of the phage as a fusion protein. The displayed polypeptide or protein can maintain a relatively independent spatial structure and biological activity and can form a high-affinity complex with the target molecule. Through multiple rounds of screening procedures, a series of multimodal chromatography ligands with high selectivity can be obtained. Similar to the phage display, the SELEX technology based on the screening of single-stranded oligonucleotide libraries has been used for the screening of highly selective affinity ligands. Due to the richer diversity of compound libraries produced by biological methods, the probability of obtaining selective ligands for target molecules is higher. The disadvantage of these methods is that the experimental process is complicated and there are many interference factors.

In recent years, the technology of rational ligand design has developed rapidly, showing great potential. This approach consists of three steps: (1) use computer simulation technology to identify the active sites in the three-dimensional structure of the target molecule and design an active compound library based on the principle of space and energy complementarity; (2) then, screen and optimize the candidate compound library through molecular docking and molecular dynamics simulation techniques; and (3) select the hits for further synthesis and evaluation. Since most of the work in this process is done in a computer system, the experimental synthesis and activity determination are significantly reduced. Therefore, it not only saves costs and accelerates the development process but also helps to better understand interactions between ligands and targets. Computer-aided ligand design represents a dynamic development direction in the field of chromatography and is worthy of attention.

References

Aghaee E, Ghasemi JB, Manouchehri F et al (2014) Combined docking, molecular dynamics simulations and spectroscopic studies for the rational design of a dipeptide ligand for affinity chromatography separation of human serum albumin. J Mol Model 20:2446

Ahirwar R, Nahar P (2015) Development of an aptamer-affinity chromatography for efficient single step purification of Concanavalin A from Canavalia ensiformis. J Chromatogr B Biomed Sci Appl 997:105–109

Banaszak LJ, Bradshaw RA (1975) Malate dehydrogenase. In: Boyer PD (ed) The enzymes, vol 11, 3rd edn. Academic Press, New York, pp 369–396

Berthod A (2006) Chiral recognition mechanisms. Anal Chem 78:2093–2099

Chen C, El Khoury G, Lowe CR (2014) Affinity ligands for glycoprotein purification based on the multi-component Ugi reaction. J Chromatogr B Biomed Sci Appl 969:171–180

Clonis YD, Labrou NE, Kotsira VP et al (2000) Biomimetic dyes as affinity chromatography tools in enzyme purification. J Chromatogr A 891:33–44

Dalgliesh CE (1952) The optical resolution of aromatic amino-acids on paper chromatograms. J Chem Soc 137:3940–3942

Davankov VA (1997) The nature of chiral recognition: is it a three-point interaction? Chirality 9: 99–102

Denholt CL, Hansen PR, Pedersen N et al (2009) Identification of novel peptide ligands for the cancer-specific receptor mutation EFGRvIII using a mixture-based synthetic combinatorial library. Biopolymers 91:201–206

Derrick JP, Wigley DB (1994) The third IgG-binding domain from streptococcal protein G. J Mol Biol 243:906–918

Easson EH, Stedman E (1933) Studies on the relationship between chemical constitution and physiological action: molecular dissymmetry and physiological activity. Biochem J 27:1257–1266

El Khoury G, Lowe CR (2013) A biomimetic Protein G affinity adsorbent: an Ugi ligand for immunoglobulins and Fab fragments based on the third IgG-binding domain of Protein G. J Mol Recognit 26:190–200

El Khoury G, Wang Y, Wang D et al (2013) Design, synthesis, and assessment of a de novo affinity adsorbent for the purification of recombinant human erythropoietin. Biotechnol Bioeng 110: 3063–3069

Ellington AD, Szostak JW (1990) In vitro selection of RNA molecules that bind specific ligands. Nature 346:818–822

Fang Y-M, Lin D-Q, Yao S-J (2018) Review on biomimetic affinity chromatography with short peptide ligands and its application to protein purification. J Chromatogr A 1571:1–15

Fassina G, Verdoliva A, Odierna MR et al (1996) Protein A mimetic peptide ligand for affinity purification of antibodies. J Mol Recognit 9:564–569

Forier C, Boschetti E, Ouhammouch M et al (2017) DNA aptamer affinity ligands for highly selective purification of human plasma-related proteins from multiple sources. J Chromatogr 1489:39–50

Gaskin DJH, Starck K, Turner NA et al (2001) Phage display combinatorial libraries of short peptides: ligand selection for protein purification. Enzym Microb Technol 28:766–772

Geysen HM, Rodda SJ, Mason TJ (1986) A priori delineation of a peptide which mimics a discontinuous antigenic determinant. Mol Immunol 23:709–715

Goward CR, Nicholls DJ (1994) Malate dehydrogenase: a model for structure, evolution, and catalysis. Protein Sci 3:1883–1888

Haigh JM, Hussain A, Mimmack ML et al (2009) Affinity ligands for immunoglobulins based on the multicomponent Ugi reaction. J Chromatogr B Biomed Sci Appl 877:1440–1452

Hall MD, Levitt DG, Banaszak LJ (1992) Crystal structure of Escherichia coli malate dehydrogenase. A complex of the apoenzyme and citrate at 1.87 A resolution. J Mol Biol 226:867–882

Huang PY, Carbonell RG (1999) Affinity chromatographic screening of soluble combinatorial peptide libraries. Biotechnol Bioeng 63:633–641

Kang HJ, Choe W, Min J-K et al (2016) Cyclic peptide ligand with high binding capacity for affinity purification of immunoglobulin G. J Chromatogr A 1466:105–112

Kaufman DB, Hentsch ME, Baumbach et al (2002) Affinity purification of fibrinogen using a ligand from a peptide library. Biotechnol Bioeng 77:278–289

Kokpinar O, Walter JG, Shoham Y et al (2011) Aptamer-based downstream processing of his-tagged proteins utilizing magnetic beads. Biotechnol Bioeng 108:2371–2379

Kuehne C, Wedepohl S, Dernedde J (2017) Single-step purification of monomeric L-selectin via aptamer affinity chromatography. Sensors 17:226–232

Labrou NE, Clonis YD (1996) Biomimetic-dye affinity chromatography for the purification of mitochondrial l-malate dehydrogenase from bovine heart. J Biotechnol 45:185–194

Labrou NE, Eliopoulos E, Clonis YD (1996) Molecular modelling for the design of chimaeric biomimetic dye—ligands and their interaction with bovine heart mitochondrial malate dehydrogenase. Biochem J 315:695–703

Labrou NE, Eliopoulos E, Clonis YD (1999) Molecular modeling for the design of a biomimetic chimeric ligand: application to the purification of bovine heart L-lactate dehydrogenase. Biotechnol Bioeng 63:322–332

Li Q (2020) Application of fragment-based drug discovery to versatile targets. Front Mol Biosci 7: 180

Liu FF, Wang T, Dong XY et al (2007a) Rational design of affinity peptide ligand by flexible docking simulation. J Chromatogr A 1146:41–50

Liu FF, Dong XY, Wang T et al (2007b) Rational design of peptide ligand for affinity chromatography of tissue-type plasminogen activator by the combination of docking and molecular dynamics simulations. J Chromatogr A 1175:249–258

Melissis SC, Rigden DJ, Clonis YD (2001) New family of glutathionyl-biomimetic ligands for affinity chromatography of glutathione-recognising enzymes. J Chromatogr A 917:29–42

Miyakawa S, Nomura Y, Sakamoto T et al (2008) Structural and molecular basis for hyperspecificity of RNA aptamer to human immunoglobulin G. RNA 14:1154–1163

Murphy MB, Fuller ST, Richardson PM et al (2003) An improved method for the in vitro evolution of aptamers and applications in protein detection and purification. Nucleic Acids Res 31:e110

Nord K, Gunneriusson E, Ringdahl J et al (1997) Binding proteins selected from combinatorial libraries of an alpha-helical bacterial receptor domain. Nat Biotechnol 15:772–777

Nord K, Gunneriusson E, Uhlen M et al (2000) Ligands selected from combinatorial libraries of protein A for use in affinity capture of apolipoprotein A-1M and taq DNA polymerase. J Biotechnol 80:45–54

Nord K, Nord O, Uhlen M et al (2001) Recombinant human factor VIII-specific affinity ligands selected from phage-displayed combinatorial libraries of protein A. Eur J Biochem 268:4269–4277

Ogston AG (1948) Interpretation of experiments on metabolic processes, using isotopic tracer elements. Nature 162:963–963

Perret G, Boschetti E (2018) Aptamer affinity ligands in protein chromatography. Biochimie 145: 98–112

Pirkle WH (1997) On the minimum requirements for chiral recognition. Chirality 9:103

Qian J, El Khoury G, Issa H et al (2012) A synthetic Protein G adsorbent based on the multi-component Ugi reaction for the purification of mammalian immunoglobulins. J Chromatogr B Biomed Sci Appl 898:15–23

Qiao Y, Zhao J, Li P et al (2010a) Adsorbents with high selectivity for uremic middle molecular peptides containing the Asp-Phe-Leu-Ala-Glu sequence. Langmuir 26:7181–7187

Qiao Y, Li P, Chen Y et al (2010b) Design, optimization and evaluation of specific affinity adsorbent for oligopeptides. J Chromatogr A 1217:7539–7546

Rahbarnia L, Farajnia S, Babaei H et al (2017) Evolution of phage display technology: from discovery to application. J Drug Target 25:216–224

Romig TS, Bell C, Drolet DW (1999) Aptamer affinity chromatography: combinatorial chemistry applied to protein purification. J Chromatogr B Biomed Sci Appl 731:275–284

Sjobring U, Bjorck L, Kastern W et al (1991) Streptococcal protein G. Gene structure and protein binding properties. J Biol Chem 266:399–405

Smith GP (1985) Filamentous fusion phage-novel expression vectors that display cloned antigens on the virion surface. Science 228:1315–1317

Su M, Yang Q, Du Y et al (2019) Comparative assessment of scoring functions: the CASF-2016 update. J Chem Inf Model 59:895–913

Tozzi C, Anfossi L, Giraudi G (2003) Affinity chromatography techniques based on the immobilisation of peptides exhibiting specific binding activity. J Chromatogr B Analyt Technol Biomed Life Sci 797:289–304

Tuerk C, Gold L (1990) Systematic evolution of ligands by exponential enrichment: RNA ligands to bacteriophage T4 DNA polymerase. Science 249:505–510

Ugi I, Werner B, Dömling A (2003) The chemistry of isocyanides, their multicomponent reactions and their libraries. Molecules 8:53–66

Wang G, De J, Schoeniger JS et al (2004) A hexamer peptide ligand that binds selectively to staphylococcal enterotoxin B: isolation from a solid phase combinatorial library. J Pept Res 64: 51–64

Yang H, Gurgel PV, Carbonell RG (2005) Hexamer peptide affinity resins that bind the Fc region of human immunoglobulin G. J Pept Res 66:120–137

Yang H, Gurgel PV, Carbonell RG (2009) Purification of human immunoglobulin G via Fc-specific small peptide ligand affinity chromatography. J Chromatogr A 1216:910–918

Chapter 6
Retention Mechanisms

Abstract Mixed-mode chromatography based on coupled hydrophobic and ionic interaction has long been used for the purification of proteins and nucleic acids. However, the solute retention mechanism in mixed mode chromatography is not well understood. Cox and Stout were the first to systematically study the effect of electrostatic interactions on the retention of basic solutes in reversed-phase chromatography. Horvath and colleagues proposed a quantitative model to explain the U-shaped retention curve of proteins in ion exchange chromatography. Carr and colleagues introduced the concept of hydrophobic-assisted electrostatic interaction to explain the excessive retention of basic compounds on the zirconia based reverse phase packing. Wan and colleagues developed a retention model based on the concept of multivalency to explain the retention behavior of amphiphilic compounds on bifunctional stationary phases. In this chapter, we will review the progress made in uncovering retention mechanisms in mixed-mode chromatography.

6.1 Introduction

Since its inception, high-performance liquid chromatography has become an indispensable tool for clinical drug analysis and drug quality control. Most of the drugs used in clinical applications are ionic compounds, with the majority carrying basic groups. Basic compounds account for more than half of all prescription drugs. The separation of basic drugs on the silica-based reversed-phase packing is still facing challenges, mainly because the basic functional groups of such compounds can interact with the remaining silanol groups on the silica matrix, causing peak tailing and poor reproducibility. Ion suppression chromatography and ion pair chromatography have been proposed to mitigate the detrimental effects of the residual silanol groups. Ion suppression chromatography refers to the deprotonation of basic compounds by adjusting the pH of the mobile phase to higher than the dissociation constant of basic compounds. However, this method is not applicable to strongly basic drugs as their charge statues are not affected by the pH of the solution. In this case, ion pair chromatography may be used as alternative to regular reversed-phase chromatography. A surfactant agent such as alkyl sulfuric acid or sulfonic acid is

Q.-H. Wan, *Mixed-Mode Chromatography*,
https://doi.org/10.1007/978-981-16-5485-5_6

added in the mobile phase and bound to the reversed-phase ligand to form dynamic cation exchanger. The basic compounds will be retained and separated by dual electrostatic and hydrophobic interactions. However, the ion-pairing agents are usually nonvolatile, excluding their use with detectors that require evaporation of solvents, such as mass spectrometer and evaporative light scattering detector.

Efforts have been made to understand the retention behavior of basic compounds in reversed-phase chromatography. As early as 1987, Cox and Stout (1987) reported a systematic study on the retention characteristics for basic compounds on silica and octyl (C8) silane-bonded silica reversed-phase packings. Experimental results show that under the conditions of elution for reversed-phase chromatography, the neutral solutes were retained by hydrophobic interactions alone; however, the ionized solutes were retained by both hydrophobic and electrostatic interactions. On bare silica, the solutes engage in hydrophobic interaction with the siloxane bridges but not the silanol groups and electrostatic interaction with silanol groups only. On C8 bonded silica, the solutes engage in hydrophobic interaction with the C8 chains and electrostatic interaction with residual silanol groups. The relative contributions of these interactions to the solute retention depend upon the bonding phase coverage.

Both ion exchange and hydrophobic interaction chromatography have been extensively used in purification of biopolymers, in which the type and concentration of salt in the eluent is the basic retention modulator. Horvath and colleagues found that the effect of the salt on the retention behavior of proteins in ion exchange and hydrophobic interaction chromatography can be described by a three-parameter equation. In the low salt concentration region, the retention is dominated by electrostatic interaction and decreases with the salt concentration whereas in the high salt concentration region, the retention is dominated by hydrophobic interaction and increases with the salt concentration.

Although the influence of hydrophobic and electrostatic interactions on the retention of basic solutes in reversed-phase chromatography is qualitatively well recognized, the quantitative relationship between hydrophobic and electrostatic interactions remains unclear. Carr and colleagues investigated the retention mechanisms on both octadecyl (C18) silane-bonded silica and polybutadiene-coated zirconia. The retention of a homologous series of *p*-alkylbenzylamines was studied as a function of the ammonium concentration in the eluent, and the relative contributions of the hydrophobic interaction and electrostatic interaction were assessed, allowing unambiguous classification of packings in terms of mixed mode retention characteristics.

In mixed mode chromatography, the solute retention is not only determined by the type and concentration of the functional groups in the stationary phase but also influenced by the presentation of various functional groups on the surface of the stationary phase. Wan and colleagues introduced the effective molarity and cooperativity factor into the characterization of solute retention in mixed mode chromatography. The values of effective molarity and cooperativity factor reflect the geometric and energetic complementarity between the solute and the mixed mode ligand.

In this chapter, we will review the progresses made in understanding the retention mechanisms in mixed mode chromatography involving hydrophobic and ionic interactions as well as hydrophilic and ionic interactions.

6.2 Retention Mechanism I: Cox Model

Let us first consider the chemical composition of the silica gel surface. As we all know, the surface of silica gel is composed of two types of functional groups: silanol and siloxane. According to reports in the literature, the silanol group is a weak acid, usually dissociates in the pH range of 4–7, and is the center of ion action on the surface of silica gel. Since silanol groups can exist in the form of free, gemini, hydrogen-bonded silanol groups, the acid dissociation constant of silanol groups is distributed in a wide range. Deprotonated silanol groups are observed even under the condition of pH 3. Siloxane is the product of silanol group condensation and dehydration and is the hydrophobic center on the silica surface. Siloxane can also be hydrolyzed to form silanol groups under suitable conditions. Since silica is a typical "living polymer," it seems reasonable to assume that the silanol and siloxane groups exist in a slow equilibrium. In this context, it is not difficult to understand that the chromatographic behavior on the silica matrix may change with time and environmental factors, especially when the compounds to be separated undergo simultaneous interactions with the chemically bonded ligand and the silanol groups on the silica surface.

6.2.1 *Ion Exchange Equilibrium of Silanol Groups*

Under the conditions of elution for reversed-phase chromatography, the solute may have hydrophobic interactions with the siloxane on the silica gel and ionic interactions with the silanol groups. Therefore, we first consider the various aspects of the ion exchange equilibrium involving the silanol groups. The silanol group dissociates to form a negatively charged cation exchange active center SiO^-, and the ion exchange process with the basic compound NH^+ can be expressed as

$$SiO^-M^+ + NH^+ \rightleftharpoons SiO^-NH^+ + M^+ \tag{6.1}$$

The ion exchange equilibrium constant K_{ix} is given by

$$K_{ix} = [SiO^-NH^+][M^+]/[SiO^-M^+][NH^+] \tag{6.2}$$

If the solute is dissociable, the pH of the system will control the concentration of cations through its ionization constant K_a. If it is assumed that the solute in the charged form is bound with the stationary phase and retained, and the neutral solute has no effect on the stationary phase, it can be concluded that the distribution coefficient of the solute between the stationary phase and the mobile phase, D_{ix}, is expressed as:

$$D_{ix} = \frac{[SiO^-NH^+]}{[NH^+] + [N]} = \frac{K_{ix}[SiO^-M^+]}{\left(1 + \frac{K_a}{[H^+]}\right)[M^+]} \tag{6.3}$$

It is not difficult to see that the solute distribution coefficient is proportional to the ion exchange equilibrium constant K_{ix} and the number of accessible ionized silanol groups $[SiO^-M^+]$ (also dependent on the pH and pK of the silica surface). It is also related to the reciprocal of the counterion concentration $[M^+]$ in the eluent. The relationship between the solute retention factor k and the distribution coefficient of the solute in the two phases is as follows:

$$k = \phi D_{ix} \tag{6.4}$$

The retention factor k is a dimensionless parameter. Under normal circumstances, when the distribution coefficient D is a dimensionless parameter, ϕ in the above equation is the ratio of the volume of the stationary phase to the mobile phase. However, when the distribution coefficient D has a unit of length, ϕ is the ratio of the surface area of the stationary phase to the volume of the mobile phase. Therefore, if the solute retention only involves the ion exchange process, the retention factor (k) directly related to the distribution coefficient is plotted against the reciprocal of the counterion concentration, and a straight line passing through the origin should be obtained, the slope of which is proportional to the ion exchange equilibrium constant and the number of ionized silanol groups on the silica surface. In addition, if solute retention involves hydrophobic interactions and other retention mechanisms that are not related to ionic processes, an intercept will be generated on the k-axis. The size of the intercept corresponds to the contribution of nonionic processes to retention. This is because when the concentration of competing ions is "infinite," the ionic interaction is cancelled. Therefore, by examining the relationship between the retention factor and the salt concentration in the mobile phase, we can obtain quantitative information about the impact of the ion exchange process on solute retention.

6.2.2 Coupled Ionic and Hydrophobic Interaction

Normally, chromatographic separation experiments are conducted under conditions dominated by a single retention mechanism, such as reversed-phase chromatography, ion exchange chromatography, and hydrophilic interaction chromatography. In this case, the solute retention mechanisms are well understood, and the operating parameters can be selected to optimize the retention and separation. In addition, the retention factor and selectivity can often be predicted based on the structure of the solute. However, the retention mechanism in a single-mode separation can be complicated by the secondary equilibrium effects. Furthermore, in order to improve the selectivity of chromatographic separation, it is necessary to introduce additional retention mechanism in a single chromatographic system. Thus, a mixed mode or multimodal chromatography is introduced. One way to achieve multimodal chromatography is to bond different types of functional groups on the chromatographic matrix in order to interact with solute molecules at multiple sites. For example, chiral separation requires three-point contact between different functional groups on the chiral carbon and the surface of the stationary phase, and it is impossible to achieve chiral separation with only one or two interactions.

When several retention mechanisms work at the same time, the situation becomes extremely complicated. For the sake of simplicity, we only consider the adsorption of amphiphilic solute molecules on the surface of silica gel. Amphiphilic molecules mean that the molecular structure contains both a hydrophobic part and a hydrophilic part. Silica gel can simultaneously interact with different parts of the molecule through the reversed-phase interaction of siloxane and the electrostatic interaction of silanol groups to retain the solute. If it is assumed that each individual interaction only occurs on a specific part of the molecule, and has nothing to do with the existence (or non-existence) of other interactions, the total energy of the interaction between silica gel and the solute will be the sum of the energy component of the interaction of each part. In other words, under the assumption that different interactions are independent of each other, we can apply the principle of additivity of the free energy of each interaction.

According to the principle of thermodynamics, the Gibbs free energy ΔG of the chromatographic system can be related to the equilibrium constant K of the solute between the two phases through the following relationship:

$$\Delta G = -RT \ln K \tag{6.5}$$

where R is the molar gas constant and T is the absolute temperature. The total distribution coefficient resulting from the retention mechanism coupled by different interactions will be equal to the product of the individual coefficients. For the retention equilibrium with the partition coefficient K_i, there is a related phase ratio (ϕ_i), which depends on the volumes of the packing and the mobile phase. In the case of a single retention mechanism, the phase ratio is the ratio of the volume of the stationary phase to that of the mobile phase. If there are several mechanisms, the stationary phase may be divided into several phases and thus there are several phase ratios. If two or more retention mechanisms are coupled, there is a related phase ratio, which reflects the number of groups on the surface where this coupling interaction occurs in the appropriate configuration. This ratio defines the degree of interaction. Since the volume of the mobile phase in the column is not a function of the retention mechanism, the key parameter in the phase ratio is the volume or surface area of the stationary phase for each interaction mechanism. Alternatively, this phase ratio may be considered as a parameter related to the frequency of occurrence of interacting groups on the surface of the packing, to which the solute molecules should be sufficiently close for interaction to occur.

In the case of uniform and random distribution of interaction sites on the chromatographic packing, when the concentrations of different interaction sites are equal, the maximum distribution coefficient due to the coupling interaction mechanism should be observed. If these sites are unevenly distributed on the surface, for example, as assumed by reversed-phase chromatography, the highly hydrophobic alkane chain on the bonded phase can aggregate in an aqueous environment to form a so-called “alkane chain stack.” Therefore, it can be assumed that there are hydrophobic “alkyl chain stacks” and hydrophilic residual silanol groups on the surface of the bonded phase. Then the mechanism of coupling interaction will be more likely to occur in local areas, such as near the edge of the “stack,” where solute molecules can access the bonded phase and the silica surface at the same time. For the retention mechanism of these coupling interactions, the relative stationary phase volume must be a function of the product of the respective phase volumes.

Even in the case of a single retention mechanism, it is difficult to specify the volume of these stationary phases, especially when there is uncertainty in the actual mechanism of the retention process. Some suggestions have been made to solve this problem. Perhaps the most useful of these is to think of the phase ratio as the ratio of the molecules participating in the retention of the stationary phase to the total number of molecules in the mobile phase. This avoids the difficulty of having to deal with the combination of surface area and volume and therefore avoids the derivation of the distribution coefficient in units of the reciprocal of the length.

The previous argument implies that if the chromatographic process involves multiple retention mechanisms, the observed solute retention factor depends not only on the product of the individual partition coefficients and the sum of their phase ratios but also on the sum of the products of the partition coefficients (K_i) and phase ratios (ϕ_i). Taking the coupling of hydrophobic (1) and ionic (2) interaction as an example, we can give the total retention factor k as composed of the contributions of the hydrophobic term, the ionic term, and the coupling term:

$$k = K_1\phi_1 + K_2\phi_2 + K_1K_2\phi_1\phi_2 \tag{6.6}$$

One way to study these processes is to select appropriate probe molecules, investigate their retention behavior in a system coupled with multiple retention mechanisms, and through comparison, qualitatively reveal the different interactions involved in the retention process. Another method is to systematically change the composition of the mobile phase, such as the organic phase content and salt concentration, examine the trend of solute molecules changing with the change of the mobile phase composition, and compare them with the behavior in single-mode chromatography to reveal the nature of coupled retention mechanism qualitatively or quantitatively. A more complicated way is to systematically change the concentration and ratio of the surface functional groups of the bonded phase packing and investigate the trend of the retention of solute molecules with change of the ligand coverage in the packing. In this case, the phase ratios of the various retention processes will be related to the surface coverage of the bonded phase: the reverse phase process directly related to the alkane chain will obviously show a positive relationship, and the mechanism involving the surface of the silica gel will be related to the residual silica surface area on the bonded phase, that is, the function of the difference between the surface area of the original silica gel and the surface area covered by the bonded phase. Therefore, in this case, the retention factor of amphiphilic solute molecules may not exhibit a simple linear relationship with the surface coverage as expected by the "pure" reverse phase mechanism but a more complex nonlinear relationship. The relationship needs to be described by a quadratic or higher power function of the surface coverage (depending on the number of participating coupling interaction mechanisms).

6.2.3 Experimental Verification

Cox and Stout (1987) selected several basic compounds as probe molecules to reveal the characteristics and nature of the interaction with the pure silica surface during chromatography. As shown in Fig. 6.1, these compounds are thiamine (quaternary ammonium compound), morphine (basic tertiary amine), and caffeine (not basic and are generally considered neutral compounds that perform well in reversed-phase systems).

Thiamine, also known as thiamine or vitamin B1, is a vitamin found in food, and it can also be synthesized artificially as a dietary supplement and medicine. Thiamine can be used to treat and prevent thiamine deficiency and the diseases it causes, including beriberi and Wernicke's encephalopathy (McCandless 2010). Other uses include the treatment of maple syrup urine disease and Lee syndrome. Its structure consists of an aminopyrimidine and a thiazolium ring connected by a methylene bridge. Thiazole is substituted with methyl and hydroxyethyl side chains. Thiamine is soluble in water, methanol and glycerin and is almost insoluble in organic solvents with less polarity. It is stable at acidic pH, but unstable in alkaline solutions.

Morphine is an analgesic in the opiate family, and it occurs naturally in many plants and animals, including humans (Jonsson et al. 1988). It acts directly on the

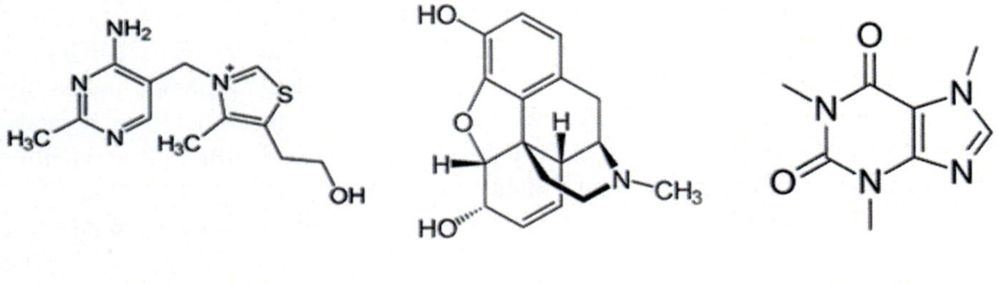

A B C

Fig. 6.1 The structure of a basic compound. (**a**) Thiamine; (**b**) morphine; and (**c**) caffeine central nervous system (CNS) to relieve pain. It can be used for both acute and chronic pain and is often used for myocardial infarction, kidney stones, and pain during childbirth. Potential serious side effects of morphine include reduced breathing effort, vomiting, nausea, and low blood pressure. Morphine is addictive and easily abused. If the dose is reduced after long-term use, symptoms of opioid withdrawal may occur. Common side effects of morphine include drowsiness, vomiting, and constipation. Morphine is a benzyl isoquinoline alkaloid with two additional ring closures.

Caffeine is a central nervous system (CNS) stimulant of methylxanthines (Wright et al. 2013). It is the most widely consumed psychotropic drug in the world. Unlike many other psychoactive substances, it reversibly prevents the effect of adenosine on its receptors, thereby preventing the drowsiness caused by adenosine. Caffeine can also stimulate certain parts of the autonomic nervous system. Caffeine is a bitter white crystalline purine that has positive and negative effects on health. It can treat and prevent respiratory diseases of premature infants, including bronchopulmonary dysplasia and apnea of premature infants. Caffeine citrate has been included in the WHO standard list of essential medicines. It may have a moderate protective effect on certain diseases, including Parkinson's disease. Caffeine produces a mild drug dependence and is accompanied by withdrawal symptoms such as drowsiness, headaches, and restlessness. Long-term use can lead to tolerance to the autonomic effects of increased blood pressure and heart rate and increased urine output (i.e., these symptoms become less obvious or no longer appear after continuous use).

In a reversed-phase chromatography system using silica gel as the stationary phase and a 15% methanol-buffer solution as the mobile phase, the effect of pH in the mobile phase on solute retention was investigated. In the studied pH range (pH 2.1–7.0), the retention factor k of both thiamine and morphine increases exponentially with the increase of pH, which corresponds to the expected increase in the number of ionized silanol groups on the surface of silica, which shows that ionic interaction plays a very important role in solute retention. The retention factor of caffeine decreases slightly with the increase of pH, reflecting that the retention of neutral solutes on silica gel is mainly through the hydrophobic interaction with siloxane, and the hydrophobicity of the silica decreases with the increasing dissociation of silanol groups. An increase of pH in the eluent leads to a decline in caffeine retention.

By changing the content of organic solvent in the mobile phase, the hydrophobic interaction between the stationary phase and the solute can be changed. Organic solvent molecules can compete with solute molecules for hydrophobic interaction sites on the stationary phase, thereby reducing solute retention. Therefore, it can be expected that the contribution of the hydrophobic interaction to the solute retention will decrease as the organic solvent content in the mobile phase increases. Figure 6.2 shows the retention factors of the three solutes on the silica stationary phase as a function of the percentage of methanol in a 0.05 M phosphate buffer at pH 4.6. When the methanol content increased from 15% to 50%, the retention factors of the three solutes all showed a downward trend, with thiamine showing the largest decline, and the decline of morphine and caffeine was smaller. From the methanol content

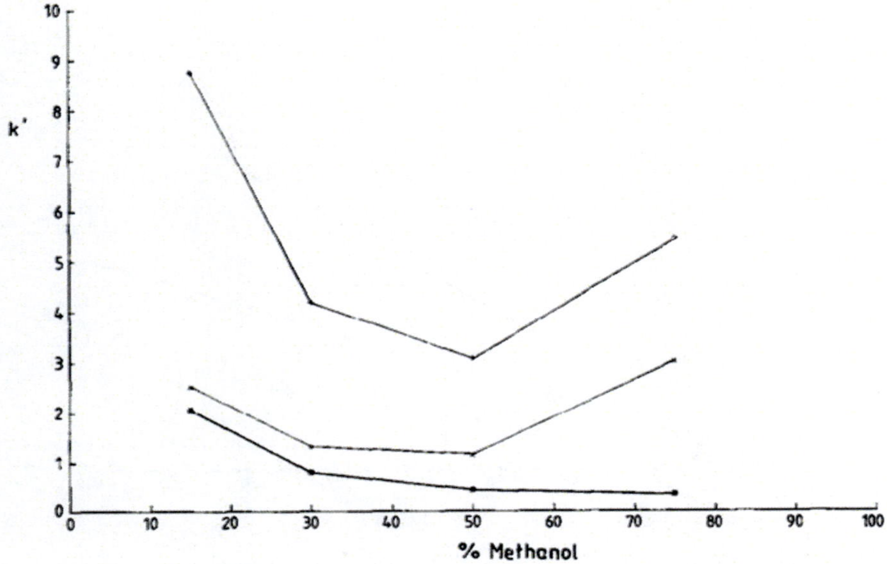

Fig. 6.2 The relationship between retention factor and methanol concentration. S-type silica gel. The mobile phase is made of a mixture of methanol and 0.05-M sodium phosphate buffer (pH 4.6). + = thiamine; * = caffeine; x = morphine. (Reproduced from Cox and Stout 1987)

increased from 50% to 75%, the retention factors of thiamine and morphine showed an upward trend, while the retention factors of caffeine continued to decrease slightly. The U-shaped graphs of thiamine and morphine are typical for the separation of many polar molecules on silica gel, such as crown ethers and basic amino compounds. This shape is generally attributed to hydrophobic interactions at low organic solvent concentrations and polar interactions at high organic solvent concentrations. For basic compounds, the ionic interaction with the silanol group on the silica gel can be superimposed on the conventional hydrophobic and polar interactions, further enhancing the retention of amphiphilic solutes.

As mentioned earlier, if the solute retention is dominated by pure ion exchange mechanism, then the plot of retention factor k against the reciprocal of the concentration of competing ions or counter ions in the mobile phase should produce a straight line passing through the origin of the plot. The slope is proportional to the ion exchange equilibrium constant and the ion exchange capacity of the packing. If there are other retention mechanisms in addition to the ion exchange mechanism, the linear relationship between the retention factor and the reciprocal of the competing ion concentration will produce a straight line crossing the vertical axis, resulting in a positive intercept. Figure 6.3 shows the relationship between the retention factor and the reciprocal of the sodium ion concentration for thiamine and morphine at 15% and 30% methanol, respectively. When the sodium ion concentration increased from 0.01 to 0.2 M, the solute retention factor showed a downward trend. The curves obtained by plotting the retention factor and the reciprocal of the sodium ion

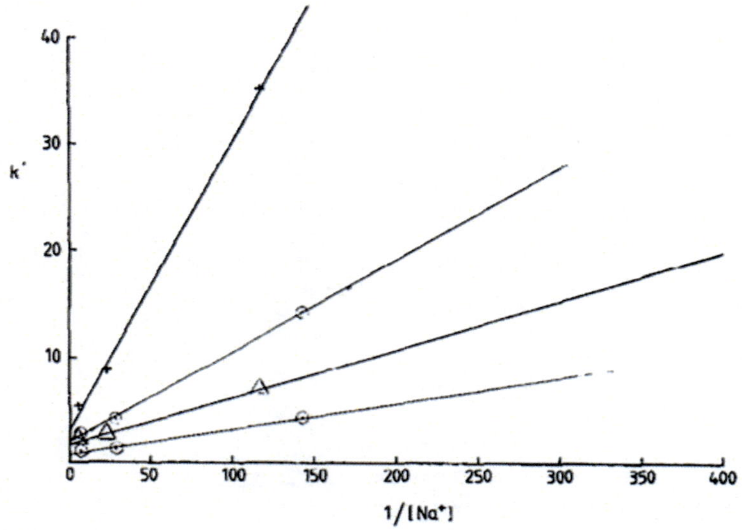

Fig. 6.3 The relationship between retention factor and the reciprocal of competing ion concentration. S-type silica gel, mobile phase, 15% and 30% methanol phosphate buffer (pH 4.6). + = thiamine, 15% methanol; ○ = thiamine, 30% methanol; Δ = morphine, 15% methanol; O = morphine, 30% methanol. (Reproduced from Cox and Stout 1987)

concentration are all straight lines, but the straight line does not pass through the origin but crosses the vertical axis and produces a positive intercept. The inverse relationship between the retention factor and the concentration of competing ions proves the contribution of ionic interactions to retention, while the intercept on the linear relationship shows the existence of nonionic interactions besides ionic interactions. In the low organic solvent content range, this nonionic interaction is mainly hydrophobic interaction, while in the high organic solvent content range, the polar interaction dominates.

It is worth noting that the slope and intercept of the straight line both change with the change of methanol concentration. When the methanol concentration increased from the initial 15% to 30%, both the slope and intercept showed a trend of sharp decline. According to the ion exchange theory, the slope of the straight line is related to the characteristics of the ion exchanger, and the decrease in the slope reflects the change in the equilibrium constant of the ion exchanger or the change in the ion exchange capacity. In addition, the change of the slope of the straight line with the concentration of methanol also implies that this parameter may be related to the characteristics of hydrophobic interaction. The positive intercept is obtained under the condition of "infinite" competing ion concentration. It is not affected by the ion concentration in the solution, so it is essentially nonionic. This nonionic retention contribution decreases with the increase of methanol concentration, indicating that the hydrophobic interaction between the solute and the stationary phase is weakened due to the solvation of methanol, in agreement with what observed in reversed-phase

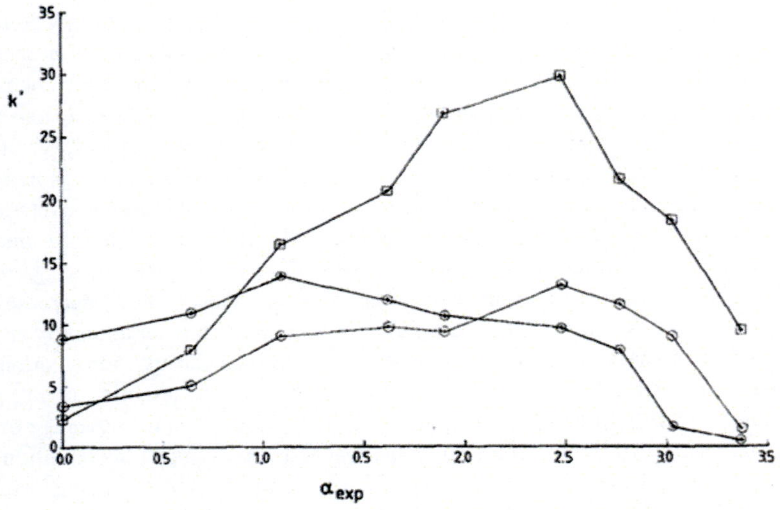

Fig. 6.4 The relationship between retention factor k and octyl surface coverage (α_{exp}). Matrix: spherical silica gel; mobile phase containing 15% methanol in 0.05% phosphate buffer (pH 4.6). ○ = thiamine; □ = caffeine; ◊ = morphine. (Reproduced from Cox and Stout 1987)

chromatography. Therefore, in the current system, the observed intercept value actually reflects the contribution of hydrophobic interaction in the mechanism that couples hydrophobic and ionic interactions.

The silica-based reversed-phase bonded stationary phase is obtained by the condensation reaction of alkylsilane reagents with silanol groups on the surface of silica gel. Due to the large molecular volume of the alkyl silane reagent, there is a steric hindrance effect, which leads to incomplete silanization on the surface of the silica gel, and a considerable proportion of silanol groups fail to participate in the reaction. Cox and Stout (1987) prepared a series of octyl silane-bonded silica gel stationary phases with different surface coverages and investigated the behavior of solutes on these stationary phases. The octyl group on the stationary phase serves as the hydrophobic interaction site, and the unreacted silanol group serves as the ionic interaction site, which can adsorb amphiphilic solutes through the coupling mechanism of hydrophobic and ionic interaction. Figure 6.4 shows that the retention factors of thiamine, morphine, and caffeine vary with the octyl coverage of the stationary phase in the mobile phase of 15% methanol-0.05 M phosphate buffer (pH 4.6). The ligand coverage varies from 0 to 3.36 μmol/m^2. Obviously, the retention of the three compounds increased with the increase of the octane surface coverage, and after reaching the highest point, the retention began to decrease. However, the range in which the retention varies and the surface coverage corresponding to the highest point for each solute are not the same. The retention of thiamine increased the least, and the retention of caffeine increased the most. The retention of thiamine reached the apex at the earliest and the retention of morphine reached the apex at the latest. These differences in behavior can be explained by the changes in the surface properties of the silica-based bonded phase. At low coverage, there are a large number of silanol groups on the surface of silica, ion exchange dominates the retention of thiamine and morphine, and hydrophobic interaction dominates the retention of caffeine. As the coverage of octyl groups increases, the hydrophobic contribution of the stationary phase increases accordingly, resulting in an increase in solute retention. However, as the coverage of octyl groups increases, the concentration of silanol groups decrease proportionally, resulting in a decrease in the contribution of ionic interactions. Therefore, for compounds such as thiamine, their strong hydrophilicity determines that their affinity for octane is not great, and therefore their retention is dominated by the ionic interaction, which decreases with increase of octyl coverage. The hydrophobicity of morphine is significantly stronger than that of thiamine, and the retention can be enhanced through ionic and hydrophobic coupling mechanisms in a wide coverage range. The retention of caffeine is controlled by the hydrophobic effect and therefore increases as the coverage of octyl group increases. After the surface coverage reached 2.5 $\mu mol/m^2$, the retention of all solutes showed a downward trend, which may be due to the "stacking" effect of octane, which buried some nonpolar alkyl groups, and polar solute molecules were excluded. The higher the polarity of the molecule, the earlier the tendency of its retention to decrease with the increase in coverage, such as thiamine $<$ morphine and caffeine.

6.3 Retention Mechanism II: Horvath Model

Horvath and colleagues (Melander et al. 1989) proposed a simplified model to describe the effect of salt concentration on the retention behavior of proteins in ion exchange chromatography. The surface of the chromatographic stationary phase used has both fixed charge and hydrophobic binding sites. In addition, they also assumed (1) the size of the internal pores of the rigid carrier is much larger than that of the solute molecules, so the pores can be approximated by a cylinder with an infinite radius, and there is no size exclusion effect. (2) In the hydrated stationary phase layer at the pore wall, the fixed charge and hydrophobic binding sites are evenly spaced and equal. (3) Protein solutes are spherical molecules with uniformly distributed fixed charges and hydrophobic spots on the surface that are easy to contact. (4) When transferring between the two phases and interacting with the functional groups in the stationary phase region, the conformation of the protein molecule is preserved. (5) A single valued phase ratio is used to express the ratio of the volume occupied by the fixed charge and hydrophobic binding site to the volume of the mobile phase in the column. (6) The conditions of linear elution chromatography are dominant, that is, the solute only occupies a small part of the binding domain. (7) The specific interaction between the elution salt and the protein is negligible.

When the retention of protein molecules involves both electrostatic and hydrophobic interactions, the magnitude of the retention depends on the sum of free energy for the equilibrium distribution of the solute molecules between the mobile phase and the stationary phase. The corresponding equilibrium constant K can be written in form

$$\Delta G_{\mathrm{total}} = \Delta G_{\mathrm{es}} + \Delta G_{\mathrm{h}} = -RT \ln K \tag{6.7}$$

where ΔG_{es} and ΔG_{h} are Gibbs free energy retained by electrostatic and hydrophobic interactions, respectively, R is the universal gas constant, and T is the absolute temperature.

The solute retention factor k (which can be measured directly from the chromatogram) can be approximately related to K by phase ratio (strictly speaking, the retention factor is related to the distribution coefficient of the solute between the two phases, rather than directly related to the equilibrium constant. In most cases, there is little difference between the two):

$$k = \phi K \tag{6.8}$$

where ϕ is the phase ratio, that is, the ratio of the volume of the stationary phase to that of the mobile phase.

6.3.1 Electrostatic Interaction Energy

Manning's counterion condensation theory (Manning 1969) was used for the thermodynamic analysis of electrostatic interactions involved in chromatography. Counterion condensation theory deals with the distribution of counterions around a linear polyelectrolyte chain with discrete charges according to the linear charge density parameter ξ. ξ is defined as

$$\xi = \lambda_{\mathrm{B}}/b \tag{6.9}$$

where λ_{B} is the Bjerrum length and b is the average distance between fixed charges on the polyelectrolyte. The Bjerrum length is given by $\lambda_{\mathrm{B}} = e^2/\varepsilon k_{\mathrm{B}}T$. e is the unit charge, ε is the dielectric constant of the bulk solution, k_{B} is the Boltzmann constant, and T is the absolute temperature. In water at room temperature (300 K), $\lambda_{\mathrm{B}} = 0.7$ nm. The Bjerrum length is the distance when the electrostatic interaction energy between two-point charges is equivalent to the energy of thermal motion. When the distance between two ions is less than λ_{B}, an ion pair can be formed. Manning proposed that for polyelectrolytes with $\xi > 1$ and counter ions with a valence of Z_{s}, if $\xi > 1/Z_{\mathrm{s}}$, the counter ion atmosphere around the polyelectrolyte is unstable, and some of the counter ions condense or adsorb to the polyelectrolyte. In this case, the ξ of the polyelectrolyte is reduced to $1/Z_{\mathrm{s}}$.

It is assumed that the protein with a characteristic charge of Z_{p} is spatially bound in a "condensation" layer with a thickness of δ_{p} on the surface of the stationary phase, where the area occupied by each fixed charge is b^2. In this case, the thickness of the protein condensation layer is

$$\delta_\mathrm{p} = b\exp\left(\left(1 + Z_\mathrm{p}\right)\left(\xi - Z_\mathrm{p}^{-1}\right)\right) \tag{6.10}$$

The corresponding layer thickness of the salt counterion with the valence of Z_s is

$$\delta_\mathrm{s} = b\exp\left(\left(1 + Z_\mathrm{s}\right)\left(\xi - Z_\mathrm{s}^{-1}\right)\right) \tag{6.11}$$

It can be seen from the above equation that the thickness of the counterion layer decreases as the distance between the fixed ions increases or the charge density decreases, and it increases with the increase of the counterion valence (increased ion action causes the adsorption layer to become thicker). When $\xi = 1/Z$, $\delta = b$.

The protein binds to the oppositely charged surface on the stationary phase through electrostatic interaction and quantitatively discharges the condensed salt counterion from the surface area. The free energy of this process is

$$\Delta G_\mathrm{es}^\mathrm{o} = -RT\ln\left(\frac{N_\mathrm{A}b^2\delta_\mathrm{p}}{1000}\right) - \left(Z_\mathrm{p}/Z_\mathrm{s}\right)RT\ln\left[\frac{1000}{\left(N_\mathrm{A}b^2\delta_\mathrm{s}m_\mathrm{s}\right)\left(1 - Z_\mathrm{s}\xi\right)}\right] \tag{6.12}$$

where m_s is the salt molar concentration and N_A is the Avogadro constant. When $\xi = 1$ and $Z_\mathrm{s} = 1$, $(1 - Z_\mathrm{s}\xi) = 0$, there are no adsorbed ions on the surface of the stationary phase, and protein adsorption does not produce discharged ions; when $\xi > 1$, it causes counterion adsorption, and protein adsorption produces discharged ions, thus favorable entropy change.

6.3.2 *Hydrophobic Interaction Energy*

Sinanoglu's solvophobic theory (Sinanoğlu and Fernández 1985) was used to account for the contribution of hydrophobic interactions involved in chromatography to retention. The free energy change $\Delta G_\mathrm{h}^\mathrm{o}$ of the hydrophobic interaction between the hydrophobic spots on the protein surface and the stationary phase can be expressed as

$$\Delta G_\mathrm{h}^\mathrm{o} = \Delta G_\mathrm{es}^\mathrm{o} + \Delta G_\mathrm{vdw}^\mathrm{o} + \Delta G_\mathrm{assoc}^\mathrm{o} + \Delta G_\mathrm{red}^\mathrm{o} + N_\mathrm{A}\left(\gamma_\mathrm{S}' A_\mathrm{S} - \gamma_\mathrm{M}' A_\mathrm{M}\right) + RT\ln\left(\frac{RT}{P^\mathrm{o}V}\right) \tag{6.13}$$

where $\Delta G_\mathrm{es}^\mathrm{o}$ is the change of free energy related to electrostatic effect during binding; $\Delta G_\mathrm{vdw}^\mathrm{o}$ is the change of free energy due to van der Waals interaction; $\Delta G_\mathrm{assoc}^\mathrm{o}$ is the free energy change of the solute-ligand association in the gas phase; $\Delta G_\mathrm{red}^\mathrm{o}$ is the free energy reduced due to the interactions between solvent-ligand and solvent-solute, which has not been accounted for in the previous items. In the last item concerning free volume; V and P^o are the average molar volume and standard pressure of the solvent, respectively. The microthermodynamic surface tensions associated with the mobile phase and stationary phase domains are γ'_M and γ'_S, respectively, while the surface areas of free and bound solute molecules exposed to the aqueous mobile phase are A_M and A_S, respectively.

The electrostatic effect represented by the ΔG_{es} term in Eq. (6.13) is evaluated by combining the Debye-Huckel treatment with the solvophobic theory. In the absence of a fixed charge on the surface, the electrostatic contribution will be completely determined by this term, which vanishes at sufficiently high salt concentrations. When there is a fixed charge on the surface of the stationary phase, Eq. (6.12) can be used to determine the value of ΔG_{es}.

Examination of Eq. (6.13) shows that the size of the biopolymer does not explicitly enter the calculation of retention energy. The important quantity is the change in molecular surface area ΔA, which is derived from the change in total surface area during hydrophobic binding ($A_M - A_s$), and is therefore proportional to the "hydrophobic contact surface area." Generally, the total hydrophobic surface area is expected to increase as the molecular weight of the protein increases. Of course, the hydrophobic surface area of each protein will be determined by its primary, secondary, and tertiary structure under the main experimental conditions.

As a first approximation, it can be assumed that the microthermodynamic surface tensions in the mobile phase and stationary phase domains have the same value and are given by γ, so that $\gamma \equiv \gamma'_M \equiv \gamma'_S$. On the other hand, for a salt solution, the surface tension can be expressed as a function of the salt concentration, that is, $\gamma = \gamma_0 + \sigma_s m_s$, where γ_0 is the surface tension of pure water, σ_s is the molar surface tension increment of neutral salt, and m_S is the molar salt concentration.

In the absence of a specific salt effect, the simple method outlined above allows us to express the dependence of the free energy of hydrophobic interactions on the molar surface tension increment of neutral salts, σ_S:

$$\Delta G_h^o = \Delta G_{aq}^o - \Delta A' \sigma_s m_s \tag{6.14}$$

where ΔG_{aq}^o includes all contributions to the of free energy of retention in the chromatographic system according to Eq. (6.13), except for the hydrophobic interaction mediated by the salt. ΔG_{aq}^o is given by

$$\Delta G_{aq}^o = \Delta G_{es}^o + \Delta G_{assoc}^o + \Delta G_{red}^o - N_A \Delta A' \gamma_o + RT \ln\left(\frac{RT}{P^o V}\right) \tag{6.15}$$

Therefore, the product $N_A \Delta A' \gamma_o$ represents the change in free energy required to form cavities in a pure solvent without any electrostatic effect.

6.3.3 Coupling of Electrostatic and Hydrophobic Interactions

When solute retention involves simultaneous electrostatic and hydrophobic interactions, the relationship between the retention factor and the salt concentration in the eluent can be generated by combining Eqs. (6.12) and (6.14):

$$\ln k = \ln\left(\frac{N_{\rm A}b^2\delta_{\rm p}}{1000}\right) + \left(\frac{Z_{\rm p}}{Z_{\rm s}}\right)\ln\left[\frac{1000}{(N_{\rm A}b^2\delta_{\rm s}m_{\rm s})(1 - Z_{\rm s}\xi)}\right] - \left(\frac{\Delta G^{\rm o}_{\rm aq}}{RT}\right) + \left(\frac{\Delta A'\sigma_{\rm s}m_{\rm s}}{RT}\right) + \ln\phi \tag{6.16}$$

where ϕ is the phase ratio.

The relationship between the solute retention factor and the salt concentration in the mobile phase under the combined action of hydrophobic and electrostatic interactions can be simplified as

$$\ln k = A - B\ln m_{\rm s} + Cm_{\rm s} \tag{6.17}$$

where A is a constant that contains all the characteristic parameters of the chromatographic system, including the nature, density, and distribution of the binding sites on the stationary phase, and B and C are the electrostatic and hydrophobic interaction parameters, respectively, which are closely related with the number, size, and distribution of charges and hydrophobic sites on the protein.

The physical meaning of each parameter can be seen from the following expressions:

$$A = \ln\left(\frac{N_{\rm A}b^2\delta_{\rm p}}{1000}\right) + \left(\frac{Z_{\rm p}}{Z_{\rm s}}\right)\ln\left[\frac{1000}{(N_{\rm A}b^2\delta_{\rm s}m_{\rm s})(1 - Z_{\rm s}\xi)}\right] - \left(\frac{\Delta G^{\rm o}_{\rm aq}}{RT}\right) + \ln\phi \tag{6.18}$$

$$B = Z_{\rm p}/Z_{\rm s} \tag{6.19}$$

$$C = \frac{\Delta A'\sigma_{\rm s}}{RT} \tag{6.20}$$

6.3.4 The Effect of pH on the Charge Density of the Stationary Phase

On the one hand, the pH value of the eluent will not affect the charge density ξ of the strong ion exchanger. On the other hand, weak anion and weak cation exchangers have ion-generating functions and may dissociate to varying degrees within the working pH range of the mobile phase. Therefore, their charge density value may depend on the pH value of the eluent used in the experiment. The pH dependence of the charge density of the anion exchanger is approximately expressed as

$$\xi = \frac{\xi_{\max}}{1 + 10^{\mathrm{pH} - \mathrm{p}K_{\mathrm{a}}}} \tag{6.21}$$

For cation exchanger

$$\xi = \frac{\xi_{\max} 10^{\mathrm{pH} - \mathrm{p}K_{\mathrm{a}}}}{1 + 10^{\mathrm{pH} - \mathrm{p}K_{\mathrm{a}}}} \tag{6.22}$$

where $\xi_{\max}$ is the parameter value when all fixed ionic groups on the surface are dissociated and $\mathrm{p}K_{\mathrm{a}}$ is the negative logarithm of its acid dissociation constant.

Another function of pH is to modify the value of the proton charge on biological macromolecules. Therefore, the overall effect of pH on retention is the result of changing the electrostatic potential of both the stationary phase and the biopolymer. As a result, both electrostatic and hydrophobic interactions are affected by pH changes.

6.3.5 Experimental Verification

The effect of salt concentration on protein retention has been studied under various experimental conditions. A representative graph of the measured log retention factor versus the salt concentration in the eluate is shown in Figs. 6.5, 6.6, and 6.7, where the solid line is drawn by fitting experimental data points to Eq. (6.17) to obtain appropriate coefficient.

Figure 6.5 shows the change of protein retention with salt concentration obtained in systems consisting of a weak and a strong cation exchangers and ammonium sulfate in the eluent at pH 6.0. It can be seen that on the two chromatographic columns, the isocratic elution retention factor of α-chymotrypsinogen A and lysozyme first decreases with the increase of the salt concentration in the mobile phase until it reaches the minimum value. A further increase in the concentration leads to an increase in the retention factor, which is in full agreement with the prediction of Eq. (6.17). Figure 6.6 shows the results obtained by using weak and strong anion exchange columns and salt in the eluent at pH 7.8 with another set of proteins. The B value of α-chymotrypsinogen is quite small on both the weak and strong anion exchange column because the protein is positively charged at the pH of the eluent. On the other hand, even if the concentration of ammonium sulfate in the eluate is in the range of 0.25–2.0 m, ovalbumin and bovine serum albumin are not retained on the weak anion exchanger (Fig. 6.6a). In contrast, at low and high salt concentrations, both bovine serum albumin and α-chymotrypsinogen A are retained on the

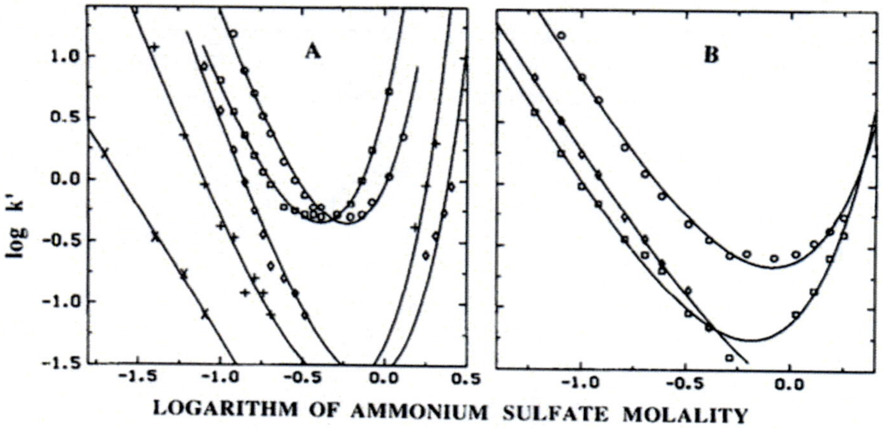

Fig. 6.5 A plot of the log retention factor of the protein obtained on the weak cation exchange (**a**) and strong cation exchange (**b**) columns versus the log molar concentration of ammonium sulfate in the mobile phase. The solid line is obtained by using Eq. (6.17). Column, 80 × 6.2 mm inner diameter; mobile phase, 20 mM phosphate buffer containing ammonium sulfate pH 6.0; flow rate 1.5 mL/min; temperature 25 °C; UV detection at 280 nm. Sample composition, (multiplication symbol) myoglobin: (plus symbol) ribonuclease; (open diamond) cytochrome c; (open square) α-chymotrypsinogen A; (open circle) lysozyme. (Reproduced from Melander et al. 1989)

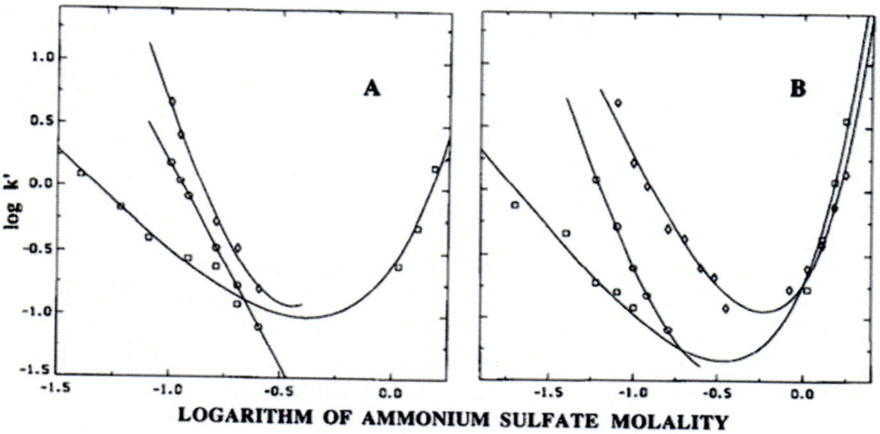

Fig. 6.6 The graph of the log retention factor of the protein obtained on the weak anion exchange (**a**) and strong anion exchange (**b**) columns versus the log molar concentration of ammonium sulfate in the mobile phase. The solid line is obtained by using Eq. (6.17). Mobile phase, 20 mM Tris-HCl, containing ammonium sulfate, pH 7.8. Sample components, (open square) chymotrypsinogen, (open circle) ovalbumin, (open diamond) bovine serum albumin. Other conditions are shown in Fig. 6.5. (Reproduced from Melander et al. 1989)

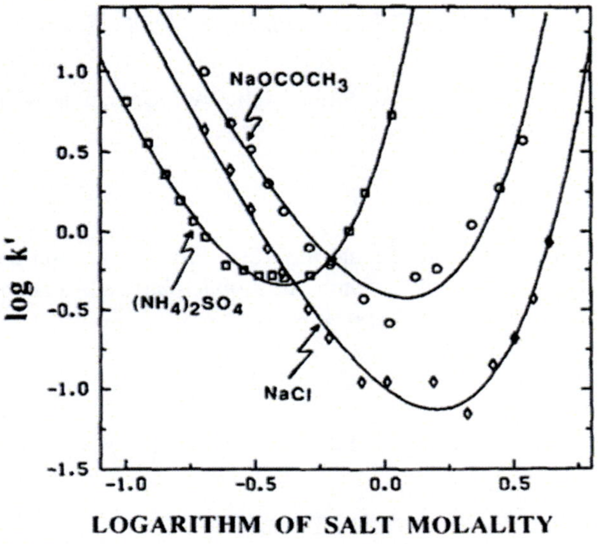

Fig. 6.7 The graph shows the log retention factor of α-chymotrypsinogen A versus various salt concentrations in the eluate. Column: WCX, 80 × 6.2 mm inner diameter; mobile phase, 20 mM phosphate buffer, pH 6.0, containing various salts; flow rate 1.5 mL/min; temperature 25 °C. (Reproduced from Melander et al. 1989)

strong anion exchanger through electrostatic and hydrophobic interactions, while ovalbumin is retained only through electrostatic interactions within the experimental salt concentration range (Fig. 6.6b).

Figure 6.7 illustrates the result of isocratic elution of α-chymotrypsinogen A on a weak cation exchange column using 25 mM phosphate buffer, pH 6.0, which contains a wide range of concentrations of sodium acetate, sodium chloride, or ammonium sulfate. As expected, since the valence Z_s of the salt counterion, sodium, is the same and equal to 1 in all cases, the B value has the same order of magnitude within the experimental error. In contrast, the value of the hydrophobic interaction parameter C is a function of the molar surface tension increment of the salt. Therefore, the descending order of C values of ammonium sulfate, sodium acetate, and sodium chloride reflects the relative magnitude of the σ_S value.

6.4 Retention Mechanism III: Carr Model

Carr and colleagues (Yang et al. 2003) studied the retention behavior of *p*-alkylbenzylammonium homologues on polybutadiene-coated zirconium oxide (PBD-ZrO_2) and octadecylsilane-bonded silica gel (ODS). Unexpectedly, they observed that on PBD-ZrO_2, solute retention decreased as the concentration of counter ions in the mobile phase increased and increased as the number of methylene groups in the molecule increased. In order to explain the observed results, they introduced a mixed-mode chromatography model based on different types of binding sites, which combined "pure" ion exchange sites, "pure" reversed-phase binding sites, and "hydrophobic-assisted ion exchange sites."

6.4.1 "Pure" Ion Exchange (IEX) Retention

The ion exchange retention of the cationic analyte A^+ is most commonly described as a stoichiometric displacement reaction:

$$(SiO^-C^+)_s + (A^+)_m = (SiO^-C^+)_s + (C^+)_m \quad (6.23)$$

where SiO^- and C^+ are the anion surface sites and counter ion, respectively. The subscripts m and s denote the mobile phase and stationary phase, respectively. The equilibrium constant of this process is:

$$K_A = [SiO^-A^+]_s[C^+]/[SiO^-C^+]_s[A^+]_m \quad (6.24)$$

After rearranging, Eq. (6.24) gives the distribution coefficient K_{IEX} of A^+:

$$K_{IEX} = [A^+]_s/[A^+]_m = [SiO^-A^+]_s/[A^+]_m = K_A[SiO^-C^+]_s/[C^+]_m \quad (6.25)$$

The retention factor k_{IEX} of solute A^+ is:

$$k_{IEX} = \Phi_{IEX}K_{IEX} = \Phi_{IEX}K_A[SiO^-C^+]_s/[C^+]_m \quad (6.26)$$

where Φ_{IEX} is the phase ratio during the ion exchange process and $[SiO^-C^+]_s$ is the total number of available silanol sites per unit area. For low buffer concentrations (less than 0.1 M), it is equal to the total ion exchange capacity of the column divided by the volume of the stationary phase. For a given column, the numerator terms in Eq. (6.26) can be combined into a single constant B_{IEX}:

$$k_{IEX} = B_{IEX}/[C^+]_m \quad (6.27)$$

It can be seen from the Eq. (6.27) that for the replacement process with a molar ratio of 1:1, when extrapolated to $1/[C^+]_m = 0$, the solute retention approaches zero ($k_{IEX} = 0$). In other words, when the counterion concentration is infinite, the ionic interaction is cancelled.

From Eq. (6.27), the well-known log-log relationship between the ion exchange retention factor of the singly charged analyte ion and the concentration of the monovalent counterion can be derived:

$$\log k_{IEX} = \log B_{IEX} - \log [C^+]_m \quad (6.28)$$

In this work, only monovalent displacement cations and monovalent cation analytes are used, so the slope of the log k_{IEX} and log $[C^+]_m$ plot should be -1.

6.4.2 "Pure" Reversed-Phase (RP) Retention

A basic feature in reversed-phase chromatography is the linear additivity of the free energy of the methylene group in the solute molecule. For the homologous series of probe solutes, there is a linear relationship between log k_{RP} and the number of methylene groups (n_{CH2}) in the probe molecule. This relationship is also called the Martin equation:

$$\log k_{RP} = \phi_{RP} K_{RP} = A + B n_{CH_2} \quad (6.29)$$

where ϕ_{RP} and K_{RP} are the phase ratio and solute distribution coefficients in the pure reversed-phase retention process, A is the intercept of the logarithm k versus the number of methylene groups of the homologue and varies with the terminal functional group of the homologue sequence, and the slope B is related to the free energy of methylene transfer from the mobile phase to the stationary phase.

6.4.3 Single-Point Model for Mixed-Mode Retention

Neue et al. proposed a single-point model of mixed-mode retention of basic solutes in reversed-phase chromatography (Neue et al. 2001), that is, reversed-phase and ion exchange processes act on the solute at the same time to cause retention. For cationic solutes, they hypothesized that reverse phase, and ion exchange interactions occur simultaneously at a single type of site on the stationary phase. Therefore, the retention k of the solute is controlled by the sum of the free energy of the two interaction types:

$$\Delta G^{o}_{RP} + \Delta G^{o}_{IEX} = -RT \ln\left(\frac{k}{\phi}\right) \quad (6.30)$$

where ΔG^{o}_{RP} and ΔG^{o}_{IEX} are the free energy contributions of the reversed-phase and ion exchange retention processes, respectively, φ is the phase ratio of a single type of site, R is the gas constant, and T is the absolute temperature.

By substituting Eq. (6.26) into Eq. (6.30), it shows the dependence of k on the counterion concentration:

$$k = k_{RP} K_{A} [SiO^{-}C^{+}]_{s} / [C^{+}]_{m} \quad (6.31)$$

The physical meaning of this equation is that the retention of basic solutes involves only one type of site, that is, the "hydrophobic-assisted ion exchange binding site." The hydrophobic and cationic moieties of the basic solute respectively interact with the alkane chain and the ionized residual silanol group on the silica-based reversed-phase packing simultaneously to form a hydrophobic ion pair.

Obviously, when $1/[C^+]_m$ approaches zero, the retention factor of basic solutes must approach zero.

Taking the logarithm of Eq. (6.31) gives

$$\log k = A' + \log k_{RP} - \log [C^+]_m \quad (6.32)$$

where A' is a constant and its value is equal to log ($K_A \cdot [SiO^- C^+]_s$). Since the two logarithmic terms on the right side are added independently, the combination of Eqs. (6.29) and (6.32) will result in an equation that clearly shows that log k versus n_{CH2} and log $[C^+]_m$'s dependence:

$$\log k = A'' + Bn_{CH_2} - \log [C^+]_m \quad (6.33)$$

where A'' is a constant related to A and A'. According to this equation, the slope of the plot of log k versus log $[C^+]_m$ for monovalent cations must be -1. The slope of the plot of log k versus n_{CH2} for the homolog series of basic compounds must be the same as that for the homolog series of neutral compounds.

6.4.4 Two-Point Model for Mixed-Mode Retention

It is assumed that the basic solute is bound to the stationary phase through interaction with discrete reversed-phase and electrostatic binding sites. In other words, the basic solute interacts with two types of binding sites sequentially but not simultaneously. In this case, the alkane chain and the silanol group on silica-based reversed-phase packing act separately and independently. Therefore, the observed solute retention factor k is equal to the sum of reversed-phase retention k_{RP} and ion exchange retention k_{IEX}:

$$k = \phi_{RP} K_{RP} + \phi_{IEX} K_{IEX} = k_{RP} + k_{IEX} = k_{RP} + B_{IEX}/[C^+]_m \quad (6.34)$$

The overall retention factor is the sum of two contribution terms, and this is equivalent to the retention observed in two columns connected in series, one of which is packed with pure reversed-phase packing and the other is packed with pure ion exchanger.

Taking the logarithm of Eq. (6.34) to obtain Eq. (6.35):

$$\log k = \log \left(k_{RP} + k_{IEX}\right) = \log \left(k_{RP} + B_{IEX}/[C^+]_m\right) \quad (6.35)$$

For convenience, we refer to this model as the "two-point model of mixed-mode retention."

Since k_{RP} has nothing to do with salt concentration, we see that the first term in the right parenthesis of Eq. (6.35) is almost a constant, while the second term is inversely proportional to the counterion concentration in the eluent. The relationship between log k and log $[C^+]_m$ will be quasi-linear, and its apparent slope depends on whether the hydrophobic or ion exchange process is dominant. When k_{RP} is dominant, the slope predicted by Eq. (6.35) is close to zero. On the other hand, when k_{IEX} is much larger than k_{RP}, the two-point model results in a slope of -1 for the plot of log k versus log $[C^+]_m$.

For the single-point model, when the salt concentration in the eluent is very high, all retention must vanish. Based on the two-point model, there is a finite value of k at very high salt concentrations, which represents the component of salt-independent reversed-phase retention.

Now we discuss the influence of n_{CH2} on k in the two-point model. For this, we must consider two different situations. It is assumed that adding methylene groups to the solute of the homologous series will not affect the size of the charge center on the solute (see below):

Case A: It is assumed that the net ion exchange equilibrium constant has nothing to do with the hydrophobicity of the solute. We don't think this situation is suitable for any ODS type stationary phase, but considering the uneven distribution of polymer in the pores, this situation may be suitable for PBD-ZrO_2 stationary phase.

Case B: It is assumed that the net ion exchange equilibrium constant varies with the hydrophobicity of the solute (hydrophobic-assisted ion exchange).

In case A, we assume that the retention only contributes to the reverse interaction, which varies with the hydrophobicity of the solute. For homologous solutes, it is consistent with Eq. (6.29). If we combine Eqs. (6.29) and (6.35) and plot the relationship between log k and n_{CH2}, when the contribution of reversed-phase retention is much greater than the contribution of ion exchange ($k_{RP} \gg k_{IEX}$), we will get a straight line with the same slope as the homologue of the nonpolar solutes.

As mentioned above, in case B, we assume that the retention in the ion exchange process is related to the hydrophobicity of the solute. In this case, in the context of the current two-point model, one can imagine an ion exchange site, the strength of which is enhanced by the hydrophobicity of the solute. Neue's single-point model is such a site. In this case, the dependence of log k on n_{CH2} cannot be used to distinguish between a single-point model and a two-point model.

6.4.5 *Analysis of Plots of log* k *vs.* $1/[C^+]_m$

The intercept of the plot of k versus the reciprocal of the counterion concentration used by Cox and Stout (1987) was used to determine the contribution of nonionic interaction to solute retention in mixed-mode chromatography. This approach has been adapted by McCalley and colleagues (McCalley 2000; Buckenmaier et al. 2002) for study of the contribution of ion exchange to solute retention in hydrophilic interaction chromatography. Equation (6.26) (for pure ion exchange model),

Eq. (6.31) (for single-point mixed-mode model), and Eq. (6.36) (for two-point mixed-mode model, case A and case B) all show that the plot of k_{IEX} versus the reciprocal of the counterion concentration of the eluent should be a straight line, and its slope B_{IEX} is proportional to the ion exchange equilibrium constant and the number of accessible silanol sites. If there is only one type of site, especially ion exchange sites, the intercept must be zero. If there are two types of sites, one of which is not caused by ion exchange, the intercept must be a positive value. The large slope value relative to the intercept indicates that ion exchange interactions are dominant.

6.4.6 Experimental Verification

With three homologous series (alkylbenzene, alkylphenone, and *p*-alkylbenzylamine) as solutes, the retention factor changes with the salt concentration in the mobile phase were measured in three different reversed-phase chromatography systems. Polybutadiene-coated zirconium oxide (PBD-ZrO_2) and octadecylsilane-bonded silica (ODS) Ace and Alltima were used as the stationary phases. Ace is a low silanol activity and Alltima is a high silanol activity reversed-phase stationary phase. Under the experimental conditions (pH 6), the organic amine is fully protonated and carries a single positive charge.

6.4.6.1 Contribution of Ion Exchange Retention

The solute retention factor was measured under the conditions of different buffer concentrations. The plot of the logarithm of the retention factor of *p*-butylbenzylamine and the logarithm of the counterion concentration is shown in Fig. 6.8.

The slope values of PBD-ZrO_2, Alltima, and Ace column are −0.954, −0.159, and −0.017, respectively, which are consistent with the prediction of the two-point model. According to this model, the apparent slope of log k and log [NH_4^+] will depend on the contribution of k_{RP} and k_{IEX} to k (see Eq. (6.35)). When ion exchange is dominant, the slope will be close to −1. On the other hand, if the reversed-phase process is dominant, the slope will be close to zero. Obviously, on the PBD-ZrO_2 phase, ionic interaction dominates because the slope is very close to −1. On the two ODS stationary phases, reversed-phase process dominates. The retention on the Ace column is actually independent of the ammonium ion concentration in the eluent. Obviously, the contribution of ion exchange to retention on a stationary phase with low silanol activity is almost negligible.

It can also be clearly seen from Fig. 6.8 that the retention of amine on PBD-ZrO_2 is much higher than that on the two ODS stationary phases, reflecting the very strong ionic interaction on this material.

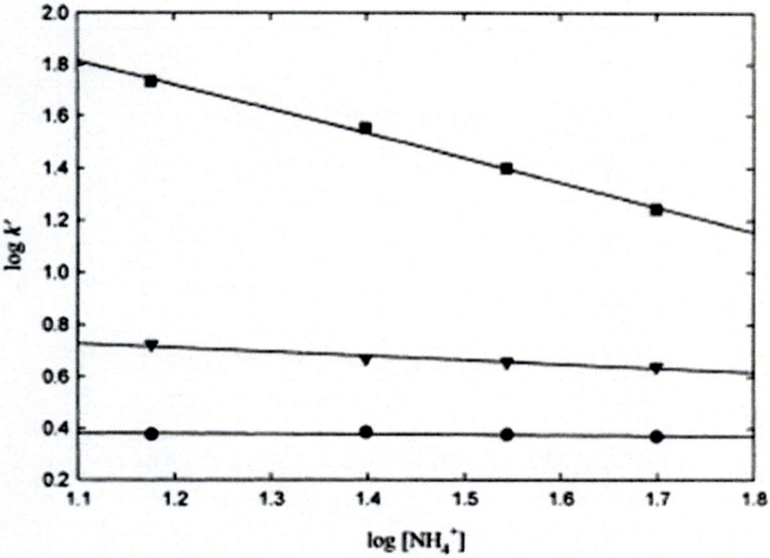

Fig. 6.8 Dependence of log k for p-butylbenzylamine on log $[NH_4^+]$. (Filled square) PBD-ZrO_2, log $k = 2.854 - 0.954 \times \log [NH_4^+]$; (inverted filled triangle) Alltima, log $k = 0.902 - 0.159 \times \log [NH_4^+]$; (filled circle) Ace, log $k = 0.401 - 0.017 \times \log [NH_4^+]$. Mobile phase: 55% MeOH + ammonium phosphate buffer adjusted to pH 6.0. Other experimental conditions: 1 mL/min; 35 °C; 254 nm. (Reproduced from Yang et al. 2003)

As mentioned above, for a pure ion exchange chromatography with a monovalent analyte and a monovalent displacer, the plot of k versus the reciprocal of the counterion concentration should be a straight line. If there is no other retention mechanism, the straight line should pass through the origin with a slope proportional to the ion exchange equilibrium constant and the number of negatively charged sites on the stationary phase (see Eq. (6.26)). The existence of other retention mechanisms will result in a positive intercept on the k-axis when $1/[NH_4^+] = 0$, which corresponds to retention without any contributions from ion exchange process (Cox and Stout 1987).

The relationship between the retention factor of p-alkylbenzylamine and the reciprocal of the ammonium ion concentration in the eluent is shown in Fig. 6.9. Only the data on the Alltima and PBD-ZrO_2 chromatographic columns are shown, because there is virtually no ion exchange interaction on the Ace chromatographic column, so the retention factor hardly changes with the salt concentration in the eluent. Obviously, the plots on the two stationary phases show positive intercepts, which increase with the length of the alkyl chain in the solute; this is the expected result for the reversed-phase retention (see Eq. (6.29)).

The slope of the relationship between the retention factor and the reciprocal of the counterion concentration is a measure of the strength of the ion exchange interaction (see Eqs. (6.28) and (6.35), and it also increases with the increase of the alkyl chain

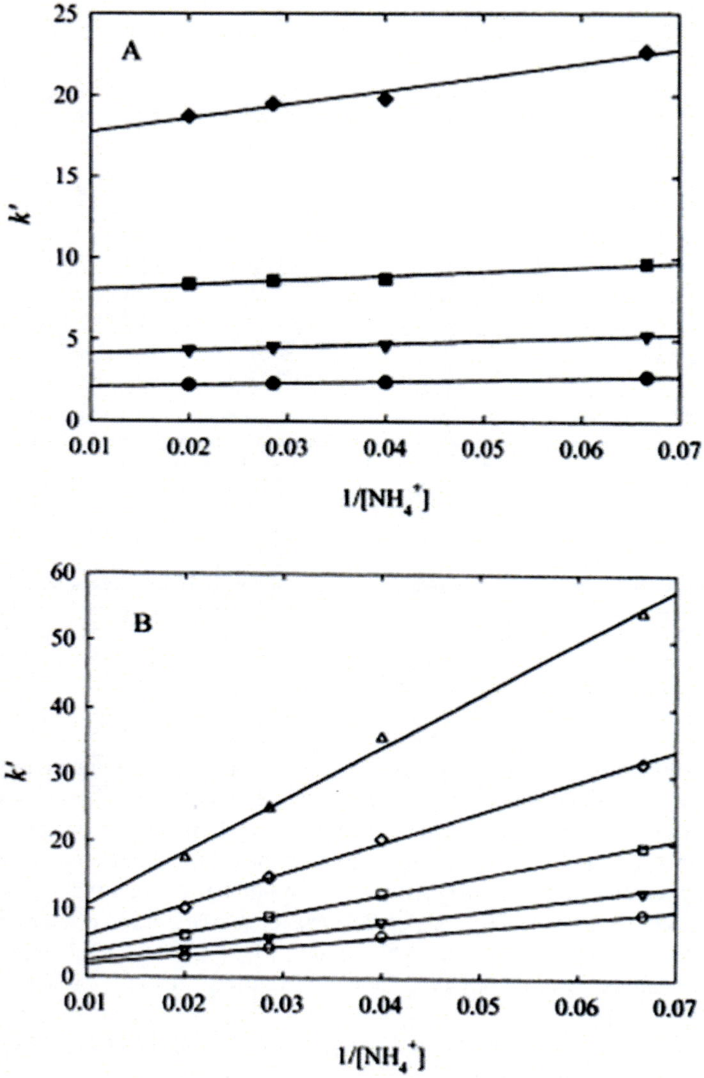

Fig. 6.9 Dependence of k' on $1/[NH_4^+]$ for *p*-alkylbenzylamine. The conditions are the same as in Fig. 6.8. (**a**) Alltima. (Filled circle) *p*-Propylbenzylamine; (inverted filled triangle) *p*-butylbenzylamine; (filled square) *p*-pentylbenzylamine; (filled diamond) *p*-hexylbenzylamine. (**b**) PBD-ZrO_2. (Open circle) Benzylamine; (inverted open triangle) *p*-methylbenzylamine; (open square) *p*-ethylbenzylamine; (open diamond) *p*-propylbenzylamine; (open triangle) *p*-butylbenzylamine. (Reproduced from Yang et al. 2003)

length in the solute, consistent with the prediction of hydrophobic-assisted ion exchange model (see Eq. (6.31)). A similar phenomenon was observed on the polymer ion exchange stationary phase. The intercepts and slopes of the k and $1/[C^+]_m$ curves for a series of phenylalkylamines increase with the increase in the number of methylene groups in the solute, indicating that reverse phase and ion exchange interactions during mixed-mode retention can be coupled to each other to form stronger retention than any single-mode retention (Rahman and Hoffman 1990).

6.4.6.2 Contribution of the Reversed-Phase Retention

In Fig. 6.10, the plots of log k value versus n_{CH2} for three homologous series are shown. On the three columns, the solute retention increased with the increase in the number of methylene groups in the molecule, reflecting the contribution of reversed-phase interaction to solute retention. On the two ODS columns, the slopes of the three solute series are not statistically different.

On the contrary, on the PBD-ZrO_2 stationary phase, the slope of the alkylbenzylamine series is significantly smaller than those of alkylbenzene and alkylphenone. In addition, the slopes of the nonionic homologues on the PBD-ZrO_2 column are almost the same and only slightly smaller than the slopes of the two ODS stationary phases.

Based on the two-point model, we know that the intercept k_{RP} in Fig. 6.9 must be the contribution of non-ion exchange retention, because it corresponds to k that is extrapolated to an infinite counterion concentration, and the slope of B_{IEX} is proportional to the equilibrium constant of the ion exchange process. In the case of the two-point model, it is assumed that the parameters k_{RP} and B_{IEX} come from independent sites. Nevertheless, we found that for Alltima and PBD-ZrO_2 stationary phases, as the number of methylene units in the alkylbenzylamines increases, the linear relationship between these two parameters is very strong. Therefore, we found that as basic solutes become more hydrophobic, the strength of hydrophobic-assisted ion exchange and hydrophobic interactions will increase.

As the alkane content of the solute increases, in reversed-phase and (hydrophobic-assisted) ion exchange modes, the solute tends to bind more firmly to the ODS stationary phase. Therefore, it is foreseeable that more hydrophobic solutes should also be more firmly bound to the stationary phase through the ion exchange process.

The intercepts of the linear plots between B_{IEX} and k_{RP} are very close to zero. If it is assumed that the intercept is negligible, for any stationary phase, k can be written as

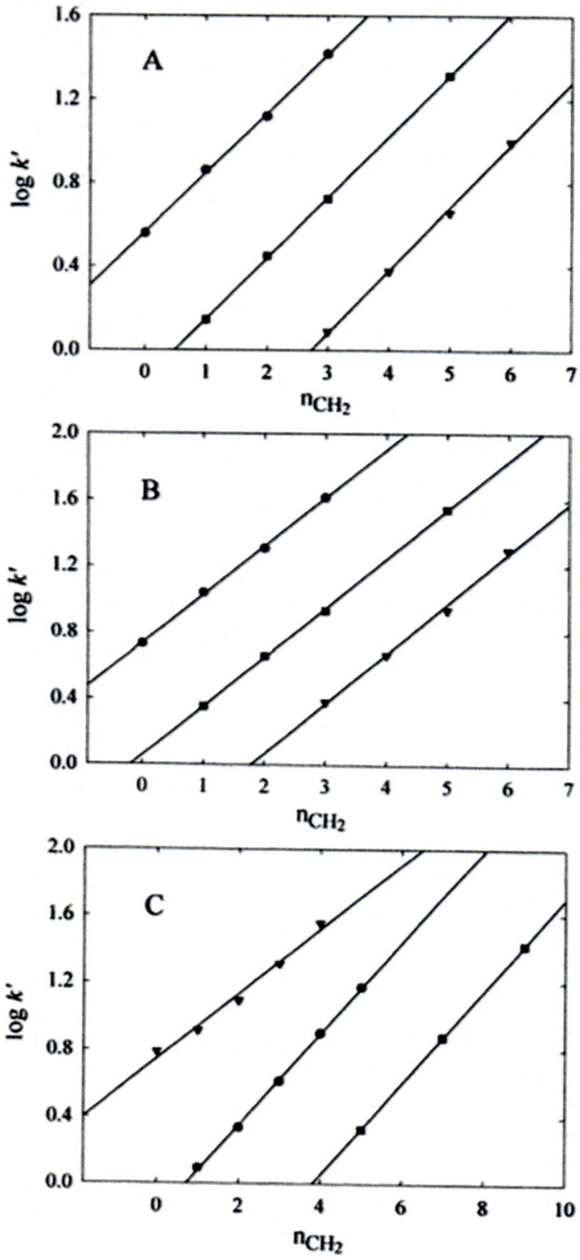

Fig. 6.10 Plots of log k versus n_{CH2} for three homologue series. Mobile phase: 55% MeOH + 25 mM ammonium phosphate buffer adjusted to pH 6.0. Other conditions are the same as in Fig. 6.8. (**a**) Ace; (**b**) Alltima; (**c**) PBD-ZrO_2. (Filled circle) Alkylbenzene homologues; (filled square) Alkylphenone homologues; (inverted filled triangle) *p*-alkylbenzylamine homologues. (Reproduced from Yang et al. 2003)

$$k = k_{RP}\left(1 + \frac{\text{const.}}{[NH_4^+]}\right) \tag{6.36}$$

For alkylbenzylamine homologues, k_{RP} varies with the different chain length of the solute on each stationary phase, but the ion exchange constant (const.) is the same. The first term (k_{RP}) on the right side of Eq. (6.36) indicates the existence of pure hydrophobic sites, while the second term (const. $k_{RP}/[NH_4^+]$) represents the hydrophobic-assisted ion exchange term.

The above conclusion can be used to explain that the slope of the plot of log k versus n_{CH2} observed on PBD-ZrO_2 is not correlated with $[NH_4^+]$. Since the hydrophobic-assisted ion exchange site is dominant on PBD-ZrO_2, as $[NH_4^+]$ increases, both k_{IEX} and k_{RP} will decrease. Therefore, according to Eq. (6.35), the slope of this plot may be the same within the specific range of $[NH_4^+]$.

On PBD-ZrO_2, the slope of the plot of log k versus n_{CH2} for alkylbenzylamines is not equal to the slope of the plot for the nonpolar homologues, in contrast to what is observed on the ODS phase. On PBD-ZrO_2, the retention of organic amines is dominated by the hydrophobic-assisted ion exchange mechanism. Although the cationic portion of the amine interacts with the anionic surface site on PBD-ZrO_2, the hydrocarbon chain of *p*-alkylbenzylamine cannot closely interact with the hydrocarbon provided by the PBD. In contrast, the ODS chain on silica gel is much denser than the PBD on zirconia and on average closer to the ion exchange site. Therefore, the hydrocarbon portion of the cationic analyte experiences an environment that is as hydrophobic as the nonionic analyte. On PBD-ZrO_2, the hydrophobic-assisted ion exchange site that plays a major role is weaker than ODS. Therefore, the slope of the plot of the log k versus n_{CH2} is lower than that on the ODS stationary phase.

6.5 Retention Mechanism IV: Wan Model

For a long time, multivalency and cooperativity have been considered as two basic principles that regulate the behavior of complex molecular systems in chemistry and biology (Badjic et al. 2005; Huskens 2006; Hunter and Anderson 2009; Ercolani and Schiaffino 2011; Fasting et al. 2012). Multivalency refers to the formation of complex between multivalent receptors and complementary ligands through multiple non-covalent interactions, such as van der Waals force, hydrophobic effect, ion pair, π-π stacking, hydrogen bond, and coordination of metal ions. Therefore, compared with the interaction between monovalent molecules, the combination of these forces greatly enhances the binding affinity between the host and the guest. Multivalent interactions play important roles in many biochemical and cellular processes, including protein folding, DNA replication, enzyme catalysis, and cell adhesion. Cooperativity is a concept closely related to multivalency, which refers to additional binding energy changes caused by the interconnection of binding sites.

The most common example of cooperativity is the complex formed between metal ions and ethylenediaminetetraacetic acid (EDTA). Compared with a single functional ligand, the chelating ring between the multifunctional ligand and the central ion increases the stability of the complex, which is usually attributed to the cooperativity effect of the multifunctional ligand. The formation of multivalent complexes can be seen as a stepwise process. The first step of complexation is the formation of monovalent complexes between host and guest molecules through intermolecular interactions. Since the host and guest molecules have redundant functional groups, multivalent complexes can be formed by intramolecular interactions, which improve the stability of the complexes. If the binding energy of the intramolecular interaction is greater than the corresponding intermolecular interaction, this is considered to be a positive cooperativity. Otherwise, it is a negative cooperativity. If there is no difference in binding energy between two events, the binding system is considered to be noncooperative, and the intermolecular and intramolecular binding occurs independently.

6.5.1 Multivalency and Cooperativity

Wan et al. (2019) introduced the concept of multivalency and cooperativity into mixed-mode chromatography to explain the retention behavior of multifunctional compounds on complementary stationary phases. They considered the two model systems shown in Fig. 6.11a, which consist of (1) benzenesulfonate as the host bonded to the porous silica and the alkylaniline homologues as the guest, labeled as

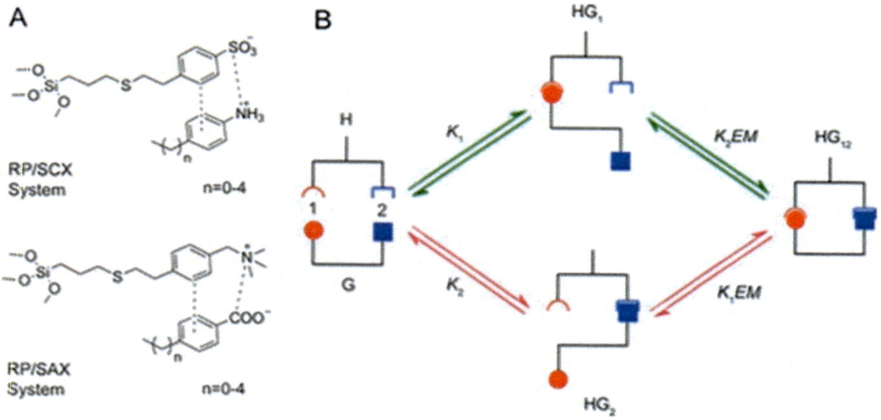

Fig. 6.11 (**a**) The host-guest complexation system studied in this work. (**b**) The transient and stable dhost-guest complexes formed during the chromatographic process. (Reprinted with permission from Wan Q-H, Wang X, Chen L (2019) Inverse mixed-mode chromatography for the evaluation of multivalency and cooperativity of host−guest complexation in porous materials. Langmuir 35: 10405–10411. Copyright {2019} American Chemical Society)

the RP/SCX system, and (2) benzyltrimethylammonium as the host bonded to porous silica and alkylbenzoic acid homologues as the guest, labeled as the RP/SAX system.

When guest molecules are introduced into the chromatographic system, they are retained by the stationary phase due to the formation of transient complexes. As shown in Fig. 6.11b, the complexation between the host and the guest can be seen as a stepwise process. A functional group of the guest molecule is first bound to the complementary functional group of the host to form a monovalent linear complex with a molar ratio of 1:1, and then another functional group of the guest molecule is bound to the host to form a cyclic bivalent complex with a molar ratio of 1:1. Guest molecules exist in four forms: unbound guest molecule G, two monovalent 1:1 complexes HG_1 and HG_2, and divalent 1:1 complex HG_{12}. The concentration of guest G bound in this supramolecular system can be derived from the mass action and related with the equilibrium constants for the binding equations as shown in Fig. 6.11b:

$$[G]_{bound} = (K_1 + K_2 + K_1K_2EM)[G][H] \tag{6.37}$$

where K_1 and K_2 are the equilibrium constants of the monovalent binding interaction at binding sites 1 and 2, respectively. [G] and [H] are the concentrations of unbound guest and host molecules. As described above, the bound guest molecule $[G]_{bound}$ has a surface concentration unit expressed in moles per unit surface area. EM is the effective molar concentration and is directly proportional to the probability of intramolecular interactions (Kirby 1980; Cacciapaglia et al. 2004; Huskens 2018). The term EM was first introduced in polymer chemistry and enzymatic catalysis to evaluate the kinetically controlled cyclization reaction rate. In this case, EM is called the kinetic effective molar concentration, which is defined as the ratio of the rate constants of the intramolecular to that of the corresponding intermolecular reactions. The concept was then transferred to supramolecular chemistry to describe the formation of thermodynamically controlled complexes. It is defined as the ratio of the equilibrium constant of the intramolecular reaction to the equilibrium constant of the corresponding intermolecular reaction. EM has a molar concentration unit, and a high value indicates that the intramolecular reaction process has priority over the intermolecular reaction process. In the literature, it is reported that the kinetic EM value is as high as 10^{13} M, but the thermodynamic EM values for most supramolecular complexes are much smaller, in the range of 1 mM to 10 M (Motloch and Hunter 2016). The reason why the upper limit of the kinetic EM value is significantly greater than the thermodynamic EM value is unclear, but it may be related to the nature of the two processes: permanent bonds are formed during the covalent cyclization process whereas the bond formation in the supramolecular complexes is reversible.

In analytical chromatography, the concentration of guest molecules in the sample injected into the chromatographic column is usually only a small fraction as compared to the host molecules bound on the silica matrix, and most of the binding sites on the stationary phase remain unbound. Therefore, under the condition of low binding site occupation, it is safe to assume that the host molecule concentration [H] can be approximated by the total concentration of host molecules immobilized on the solid substrate $[H]_0$. Equation (6.37) can be rewritten as

$$[G]_{bound} = (K_1 + K_2 + K_1K_2EM)[G][H]_0 \tag{6.38}$$

Then, the distribution coefficient D of the guest molecule between the stationary phase and the mobile phase is expressed as

$$D = \frac{[G]_{bound}}{[G]} = (K_1 + K_2 + K_1K_2EM)[H]_0 \tag{6.39}$$

It should be noted that if the surface-bound material is expressed in mole units per unit surface area, then D will not be a dimensionless parameter commonly encountered in partition chromatography but has a unit of length.

The binding strength of the solute in the chromatogram is measured by the dimensionless retention factor k, which is defined as the molar ratio of bound and unbound solutes:

$$k = \varphi D = \varphi(K_1 + K_2 + K_1K_2EM)[H]_0 = k_1 + k_2 + \frac{k_1k_2EM}{\varphi[H]_0} \tag{6.40}$$

where ϕ is the phase ratio defined by the surface area of the porous material packed in the chromatographic column divided by the void volume of the porous material, that is, the mobile phase volume. Similarly, the phase ratio ϕ is not a dimensionless parameter, but has the reciprocal unit of length. The product of ϕ and $[H]_0$ represents the molar concentration of host molecules in the packed column. Both ϕ and $[H]_0$ are characteristic parameters of porous materials, which can be determined by independent measurement.

It can be seen from Eq. (6.40) that the retention factor k_{12} of the bivalent cyclic complex HG_{12} is given by the following equation:

$$k_{12} = \frac{k_1k_2EM}{\varphi[H]_0} \tag{6.41}$$

According to Eqs. (6.40) and (6.41), the total retention factor can be decomposed into hydrophobic, ionic, and cooperative retention components k_1, k_2, and k_{12}, corresponding to the three binding species, respectively. If the values of k_1 and k_2 are known, they can be subtracted from k to obtain the value of k_{12}. Therefore, there is a need for a method to evaluate the values of k_1 and k_2 in order to quantify the cooperativity of a given binding system. Similar to the double mutation cycle analysis (Cockroft and Hunter 2007), the interaction between pairs of ions and hydrophobic molecules can be switched off step wisely to determine k_1 and k_2.

For the specific example shown in Fig. 6.11, we assume that the bivalent host molecule has hydrophobic binding site 1 and ionic binding site 2. A buffer with a high salt concentration can be used as the mobile phase to inhibit the ionic interaction at site 2. However, due to the salting-out effect (Endo et al. 2012), the hydrophobic interaction at site 1 will also be affected, which complicates the interpretation of the data. One way to overcome this difficulty is to use linear extrapolation to infinite salt concentration to evaluate the hydrophobic retention component k_1. The principle of this method can be illustrated by a typical ion exchange equilibrium:

$$\mathrm{HC} + \mathrm{G} \rightleftharpoons \mathrm{HG} + \mathrm{C} \tag{6.42}$$

where H and G represent host and guest molecules; C represents a single-charge competitive ion. For clarity, the charge signs for all ion species are omitted. The ion exchange equilibrium constant K_{ix} is given by

$$K_{\mathrm{ix}} = \frac{[\mathrm{HG}][\mathrm{C}]}{[\mathrm{HC}][\mathrm{G}]} \tag{6.43}$$

The distribution coefficient D_{ix} of the guest molecule is related to K_{ix}:

$$D_{\mathrm{ix}} = K_{\mathrm{ix}} \frac{[\mathrm{HC}]}{[\mathrm{C}]} \tag{6.44}$$

Then, the retention factor k_{ix} of the guest molecule is given by

$$k_{\mathrm{ix}} = \varphi K_{\mathrm{ix}} \frac{[\mathrm{HC}]}{[\mathrm{C}]} = \varphi K_{\mathrm{ix}} \frac{[\mathrm{H}]_0}{[\mathrm{C}]} \tag{6.45}$$

As mentioned above, in the case of low site occupancy, [HC] can be approximately regarded as equivalent to $[\mathrm{H}]_0$. Therefore, the relationship between the retention factor k_{ix} and the reciprocal of the counterion concentration 1/[C] will produce a straight line through the origin. The slope of the straight line is proportional to the ion exchange equilibrium constant K_{ix} and the molar concentration of the host molecule in the ion exchange system $\phi[\mathrm{H}]_0$. However, if there are nonionic interactions between the host and guest molecules, such as hydrophobic interactions, the straight line will deviate from the origin and a positive intercept will occur on the k-axis. This positive intercept is actually the hydrophobic retention component k_1, because under the condition of infinite competing ion concentration (1/[C] = 0), both the pure ionic interaction and the hydrophobic auxiliary ionic interaction (i.e., mixed-mode interaction) vanished (Cox and Stout 1987; Yang et al. 2003).

Similarly, the ionic retention component k_2 can be determined by extrapolating the total retention factor k to zero hydrophobic interaction. When the salt concentration does not exceed 100 mM (to reduce the salting-out effect), as described in

Eq. (6.40), the retention of guest molecules is controlled by a variety of interactions. For a series of homologs that share the same ionic head group, the retention is expected to increase as the length of the alkyl chain in the hydrophobic tail increases. Therefore, the relationship between the total retention factor k and the hydrophobic retention component k_1 should be a straight line, and its intercept and slope are both proportional to the ionic component k_2. By linearly extrapolating to $k_1 = 0$, the resulting intercept is the ionic interaction retention component k_2. In this case, both the hydrophobic interaction and the hydrophobic-assisted ionic interaction during the chromatographic retention process are cancelled.

Subtract the values of k_1 and k_2 from k to obtain k_{12}. Once the values of k_1, k_2, and k_{12} are known, the effective molar concentration EM and the cooperativity coefficient β can be estimated using the following equations:

$$\mathrm{EM} = \frac{k_{12}}{k_1 k_1}\varphi[\mathrm{H}]_0 \tag{6.46}$$

$$\beta = \frac{\mathrm{EM}}{\varphi[\mathrm{H}]_0} \tag{6.47}$$

Equation (6.47) is basically the same as that defined by Ercolani and Schiaffino (2011), except that in their work it is assumed that the guest molecule rather than the host molecule is present in excess in the binding system.

6.5.2 *Experimental Verification*

Figure 6.12 shows the separation of alkylanilines on benzenesulfonate-bonded silica (RP/SCX) and alkyl benzoic acids on benzyltrimethylammonium-bonded silica (RP/SAX). All the probe molecules were baseline separated with symmetrical peak shape and eluted in the order with increasing alkyl chain length in the molecule. These probe molecules have ionic heads and hydrophobic tails and can bind with complementary host molecules to form molecular complexes through ionic, hydrophobic, and mixed interactions.

6.5.2.1 Pure Hydrophobic Retention Component

In order to evaluate the contribution of hydrophobic interactions to retention, two series of experiments were carried out: (1) The retention factor of alkylanilines in the range of ionic strength from 30 to 70 mM in the mobile phase was measured on the RP/SCX column, and (2) the retention factor of alkyl benzoic acids in the range of ionic strength of 20–60 mM in the mobile phase was measured on the RP/SAX column. The obtained solute retention factor k is plotted against the reciprocal of the competing ion concentration, 1/[C]. As shown in Fig. 6.13, the retention factor

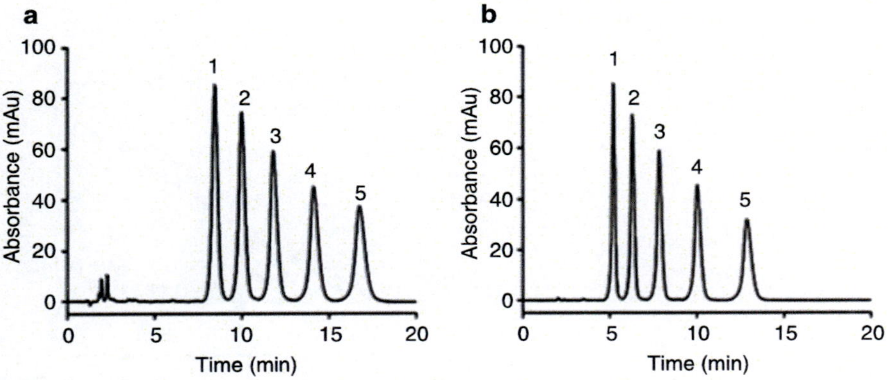

Fig. 6.12 (**a**) Representative separation of alkylanilines on the RP/SCX column. Chromatographic conditions: column, 150 × 4.6 mm; temperature, 30 °C; mobile phase, acetonitrile (ACN)/60 mM potassium phosphate, pH 2.5 (45:55, v/v); flow rate, 1.5 mL/min; UV detection, 254 nm. Solute: (1) 4-toluidine; (2) 4-ethylaniline; (3) 4-propylaniline; (4) 4-butylaniline; (5) 4-pentylaniline. (**b**) Representative separation of alkyl benzoic acids on the RP/SAX column. Chromatographic conditions: same as for **a** except that the mobile phase is ACN/50 mM ammonium formate, pH 6.3 (35:65, v/v), and the flow rate is 1.0 mL/min. Solute: (1) 4-methyl benzoic acid; (2) 4-ethyl benzoic acid; (3) 4-propyl benzoic acid; (4) 4-butyl benzoic acid; (5) 4-pentyl benzoic acid. (Reprinted with permission from Wan Q-H, Wang X, Chen L (2019) Inverse mixed-mode chromatography for the evaluation of multivalency and cooperativity of host–guest complexation in porous materials. Langmuir 35:10405–10411. Copyright {2019} American Chemical Society)

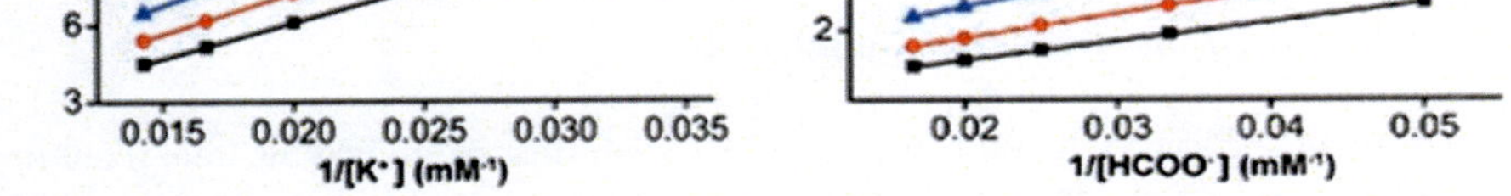

Fig. 6.13 Plots of k versus 1/[C] for two homologue series. (**a**) RP/SCX system, alkyl aniline homologues: (filled square) 4-toluidine; (filled circle) 4-ethyl aniline; (filled triangle) 4-propyl aniline; (inverted filled triangle) 4-butyl aniline; (boxed filled triangle) 4-Pentyl aniline; (**b**) RP/SAX system, alkylbenzoic acid homologues: (filled square) 4-methyl benzoic acid; (filled circle) 4-ethyl benzoic acid; (filled triangle) 4-propyl Benzoic acid; (inverted filled triangle) 4-butyl benzoic acid; (boxed filled triangle) 4-pentyl benzoic acid. (Reprinted with permission from Wan Q-H, Wang X, Chen L (2019) Inverse mixed-mode chromatography for the evaluation of multivalency and cooperativity of host–guest complexation in porous materials. Langmuir 35: 10405–10411. Copyright {2019} American Chemical Society)

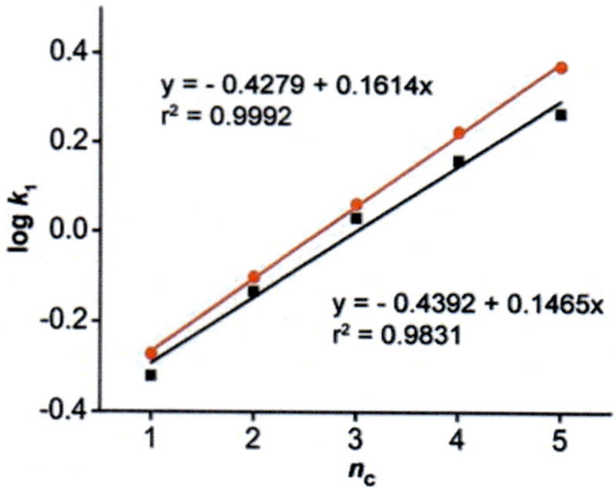

Fig. 6.14 Plots of the log k_1 versus n_C for the homologous series of (filled circle) alkylaniline and (filled square) alkylbenzoic acid

increases as the value of 1/[C] increases and the length of the alkyl chain in the hydrophobic tail increases. For all probe molecules, there is a linear relationship between k and 1/[C], and the intercept and slope of each plot are positive. As mentioned above, the positive intercept reflects the contribution of the hydrophobic interaction between the hydrophobic tail of the amphiphilic probe molecule and the spacer connected to the ionic head group on the stationary phase, because the charge on the host and guest molecules is completely shielded by high-concentration salt solution. Therefore, these intercepts can be regarded as equivalent to the hydrophobic retention component k_1, which increases as the hydrophobicity of the probe molecule increases. The positive slope indicates the strength of the ionic interaction between the probe molecule and the host molecule, including the contribution of pure ionic interaction and hydrophobic-assisted ionic interaction.

As mentioned earlier, in reversed-phase chromatography, there is a linear relationship between the logarithm of the solute retention factor and the number of methylene groups in the molecule. As shown in Fig. 6.14, in both systems, a linear graph between log k_1 and n_{CH2} with a positive slope is observed.

6.5.2.2 Pure Ionic Retention Component

To evaluate the ionic retention component k_2, we further examined the total retention factor k of the probe as a linear function of the hydrophobic component k_1. Figure 6.15 shows that good linear relationship between the values of k and k_1 was observed for both homologues, consistent with the prediction of Eq. (6.40). In both cases, positive intercept and slope values corresponding to the contribution of ionic and ionic assisted hydrophobic interactions were observed. As the salt concentration increases, the intercept and slope values will decrease because the contribution of ionic interaction will decrease at higher salt concentrations. The intercept and slope values observed on the RP/SCX column are larger than those observed on the RP/SAX column, which may be because the former has a larger ion exchange

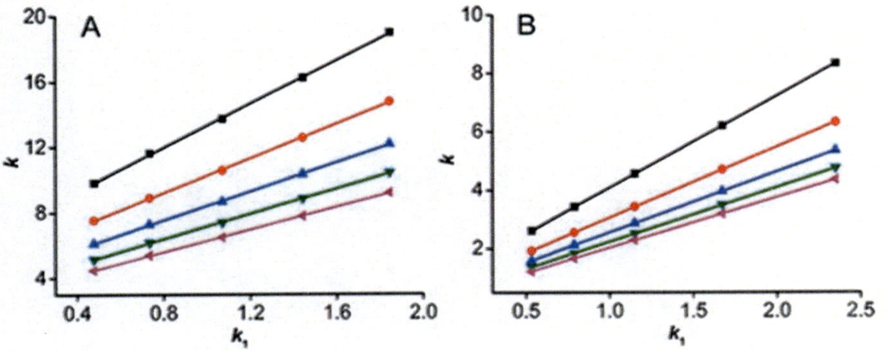

Fig. 6.15 Plots of the k versus k_1 for two homologues at different salt concentrations (mM). (**a**) RP/SCX system, alkylaniline homologues, salt concentration (mM): (filled square) 30; (filled circle) 40; (filled triangle) 50; (inverted filled triangle) 60; (boxed filled triangle) 70; (**b**) RP/SAX system, alkylbenzoic acid homologues, salt concentration (mM): (filled square) 20; (filled circle) 30; (filled triangle) 40; (inverted filled triangle) 50; (boxed filled triangle) 60. (Reprinted with permission from Wan Q-H, Wang X, Chen L (2019) Inverse mixed-mode chromatography for the evaluation of multivalency and cooperativity of host−guest complexation in porous materials. Langmuir 35: 10405–10411. Copyright {2019} American Chemical Society)

Fig. 6.16 Plots of the k_2 versus 1/[C] for (filled square) alkylaniline and (filled circle) alkylbenzoic acid homologs

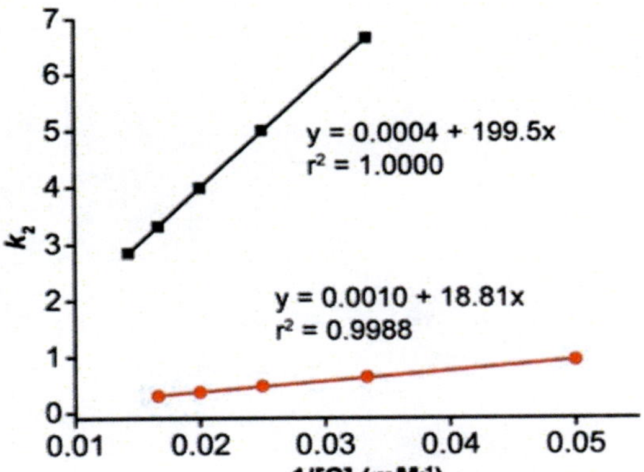

capacity, namely, $[H]_0$. The k_2 value obtained by extrapolating to $k_1 = 0$ represents the assumed retention factor for the ionic head group, amine or acid, and is the contribution of pure ionic interaction to solute retention. The linear plot of k_2 versus 1/[C] passes through the origin (Fig. 6.16), confirming the ionic characteristics of k_2.

6.5.2.3 Mixed Retention Component

The mixed retention component k_{12} is evaluated by subtracting the hydrophobic and ionic components (k_1 and k_2) from the total retention k. In the two mixed-mode

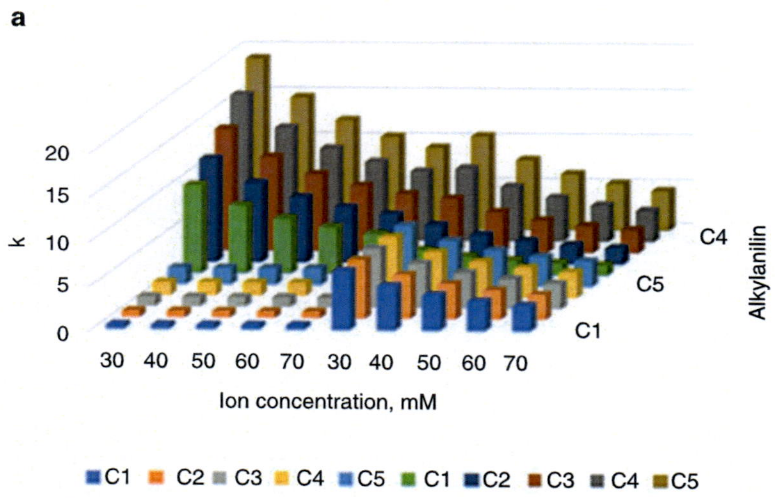

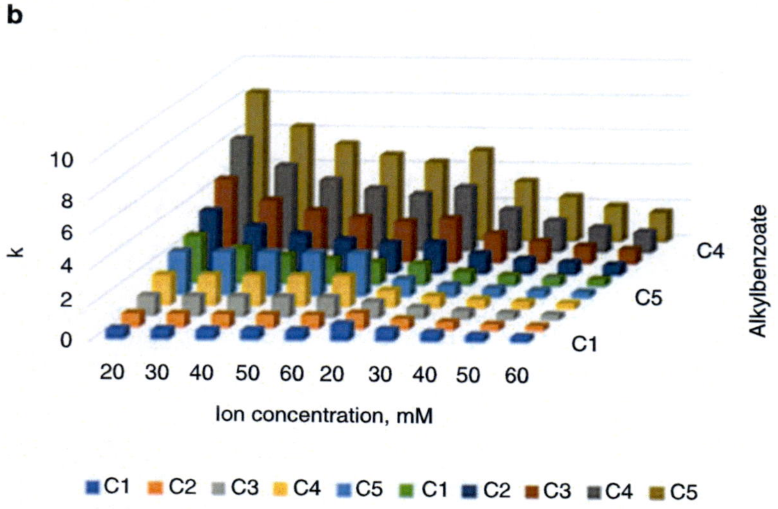

Fig. 6.17 The distribution of different retained components in a mixed-mode chromatogram. (**a**) RP/SCX system, the solute is alkyl benzylammonium homolog; (**b**) RP/SAX system, the solute is alkyl benzoic acid homolog

chromatographic systems studied, the distribution of the three chromatographic retention components k_1, k_2, and k_{12} is shown in Fig. 6.17. The k_1 values are shown in the left front area, which represents the contribution of hydrophobic interaction; the k_2 values are shown in the right front area, which represents the contribution of ionic interaction; the k_{12} values are shown in the right rear area, which represents the contribution of mixed-mode retention; and the k values are shown in the left rear area. It can be seen from the figure that the contribution of mixed-mode component to the overall retention is quite significant. As expected, due to increased hydrophobicity and decreased ionic contribution, k_{12} increased as the length of the alkyl chain increased and decreased as the concentration of competing ions increased.

6.5.2.4 Effective Molar Concentration and Cooperativity Factor

From Eq. (6.46), it can be seen that by plotting the mixed-mode retention component, k_{12}, against the product of the reversed-phase and ionic retention components, k_1k_2, a straight line passing through the origin can be obtained, with the slope corresponding to the cooperativity factor, β. As shown in Fig. 6.18, the cooperativity factors of RP/SCX and RP/SAX systems are 0.85 and 2.3, respectively. According to Eq. (6.47), the corresponding effective molar concentrations are calculated to be 0.14 M and 0.18 M, respectively.

As mentioned above, the EM value reflects the probability that the unbound part of the partially bound monovalent HG complex will find a complementary site on the host molecule. This can also be explained as the concentration of host molecules that the guest molecules can experience. EM values higher or lower than the actual host molecule concentration indicate positive or negative cooperativity and therefore provide a method for estimating whether the sequential binding steps in the formation of multivalent complexes act independently or in a synergistic manner.

The value of β is greater than or less than 1, which seems to confirm that the multivalent interactions in the two host-guest complex systems studied are all regulated by cooperativity effects. As the length of the alkyl group in the probe increases and the salt concentration increases, the EM and β values of the two systems remain unchanged. When studying the relationship between the chemical structure and the effective molar concentration for intramolecular hydrogen bonds, Hunter and colleagues (Sun et al. 2013; Adams et al. 2013) pointed out that the EM

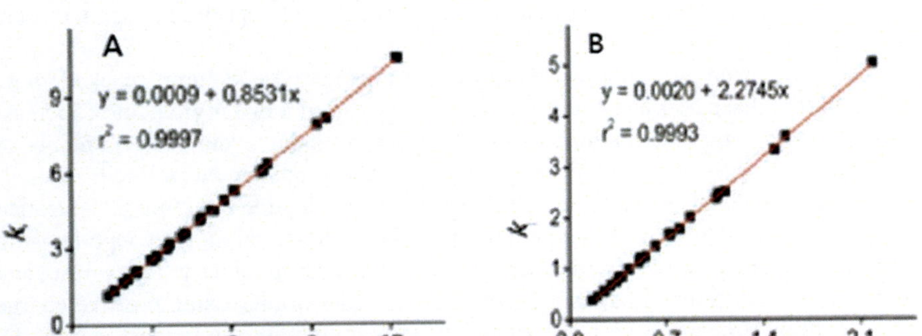

Fig. 6.18 Plots of the k_{12} versus k_1k_2 for two homolog series in mixed-mode chromatography. (**a**) Alkylanilines in RP/SCX system; (**b**) alkylbenzoic acids in RP/SAX system

value does not change with the intrinsic bonding strength but depends on the supramolecular structure and geometric complementarity. The constant EM and β values of the two systems observed in this work support their view that the cooperativity of multivalent complexes is a structural property mainly determined by geometry and charge complementarity. The positive cooperativity shown by the RP/SAX system indicates that the geometric structure between the host and guest molecules has a good match and a cyclic complex can be formed. In contrast, the negative cooperativity observed in the RP/SCX system indicates that there is some strain in the formation of cyclic complex. This may be due to the short distance between the benzene ring and the acidic part of the host, or the π-π stacking interaction will interfere with the ionic interaction between the host and the guest. In addition, the preferential solvation of the host and guest molecules in the binary solvent mixture can also lead to negative cooperativity, which can lead to conformational changes and the formation of strained cyclic complexes (Henkel et al. 2017).

6.6 Retention Mechanism V: Hydrophilic Ion Exchange Model

Zhu et al. (1992) reported the retention behavior of synthetic peptides in hydrophilic cation exchange chromatography. They used 12 synthetic peptides in three groups as model compounds (Table 6.1). Two of them are hydrophobic variants, each consisting of four decapeptides with increasing hydrophobicity. The third group is the charged hydrophobic variant, which consists of four undecapeptides with successively increasing charges and hydrophobicity. The model peptide pairs have the same amino acid sequence for a and e, b and f, c and g, and d and h. The only difference is that the model peptides a, b, c, and d contain free N-terminal amino groups, while the N-terminal amino groups of the model peptides e, f, g, and h are protected by acetylation.

On a hydrophilic cation exchange chromatography column packed with polysulfoethyl aspartamide-bonded silica gel, they used a linear gradient of increasing salt concentrations in mobile phases with different acetonitrile contents to determine the retention of model peptides. The results are shown in Fig. 6.19.

In the absence of acetonitrile in the mobile phase, peptide retention is controlled by a combination of ion exchange and hydrophobic interaction. Polypeptides with the same charge are eluted in the order of increasing hydrophobicity (Fig. 6.19a), and peptides with different charges are eluted in the order of increasing positive charge (Fig. 6.19c). No elution was observed for the peptide with the highest charge, 14, within the retention time of 60 min. When 15% acetonitrile is added to the mobile phase, the retention time of all peptides is greatly reduced, and the elution order of the peptides with different charges remains unchanged (compare Figs. 6.19c, d), but the elution order of the peptides with the same charge is reversed (compare

Table 6.1 Properties of synthetic peptides used as probes

Peptide	Peptide sequence	Positive charge	Relative hydrophobicity	
			pH 2	pH 7
a	*NH2-*Arg-Gly-**Gly-Gly**-Gly-Leu-Gly-Leu-Gly-*Lys-amide	3	16.4	20.6
b	*NH2-*Arg-Gly-**Ala-Gly**-Gly-Leu-Gly-Leu-Gly-*Lys-amide	3	17.1	21.4
c	*NH2-*Arg-Gly-**Val-Gly**-Gly-Leu-Gly-Leu-Gly-*Lys-amide	3	19.6	24.2
d	*NH2-*Arg-Gly-**Ala-Tyr**-Gly-Leu-Gly-Leu-Gly-*Lys-amide	3	22.3	27.2
e	Acetyl-*Arg-Gly-**Gly-Gly**-Gly-Leu-Gly-Leu-Gly-*Lys-amide	2	17.5	20.9
f	Acetyl-*Arg-Gly-**Ala-Gly**-Gly-Leu-Gly-Leu-Gly-*Lys-amide	2	18.3	21.8
g	Acetyl-*Arg-Gly-**Val-Gly**-Gly-Leu-Gly-Leu-Gly-*Lys-amide	2	20.8	24.4
h	Acetyl-*Arg-Gly-**Ala-Tyr**-Gly-Leu-Gly-Leu-Gly-*Lys-amide	2	23.7	27.5
i	Acetyl-Gly-Gly-Gly-Leu-Gly-Gly-Ala-Gly-Gly-Leu-*Lys-amide	1	17.7	19.2
j	Acetyl-*Lys-Tyr-Gly-Leu-Gly-Gly-Ala-Gly-Gly-Leu-*Lys-amide	2	20.4	23.2
k	Acetyl-Gly-Gly-Ala-Leu-*Lys-Ala-Leu-*Lys-Gly-Leu-*Lys-amide	3	26.0	31.7
l	Acetyl-*Lys-Tyr-Ala-Leu-*Lys-Ala-Leu-*Lys-Gly-Leu-*Lys-amide	4	27.1	35.0

Reproduced from Zhu et al. (1992)

Figs. 6.19a, b). This is because acetonitrile inhibits the hydrophobic interaction between the solute and the stationary phase, making electrostatic interaction the main retention mechanism.

The effect of different acetonitrile content in the mobile phase on the retention of peptides was systematically investigated. When the concentration of acetonitrile increases from 0% to 90%, a trend of the retention of peptides first decreasing, and then increasing can be observed. Figure 6.20 shows the retention time of the four peptides as a function of the acetonitrile content in the mobile phase. The solute retention decreases with the increase of the acetonitrile content in the mobile phase, reaching the lowest point at about 60–70%. Then, the retention of the polypeptide increases sharply with the increase of the acetonitrile concentration in the mobile phase. It reflects that under the condition of high organic solvent content, the retention is dominated by hydrophilic interaction between the solute and the stationary phase. From these experimental observations, the authors concluded that in strong cation exchange chromatography, the concentration of acetonitrile in the mobile phase has a significant impact on peptide retention. At low acetonitrile

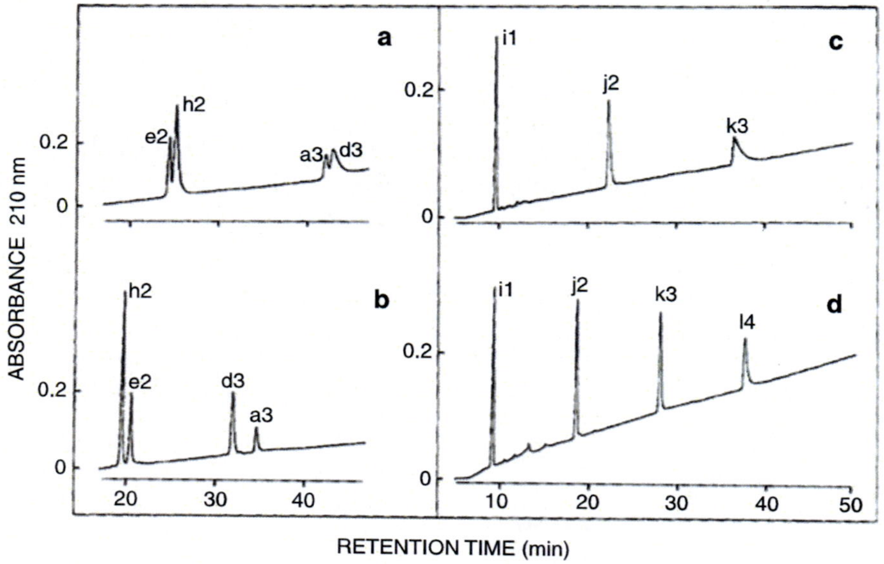

Fig. 6.19 Demonstration of hydrophobic and hydrophilic interactions between positively charged peptides and the stationary phase during strong cation exchange chromatography. Column, Polysulfoethyl A strong cation exchange column (200 × 4.6 mm I.D.). Mobile phase, panels **a** and **c**, linear buffer A–B increasing salt gradient (2%B/min, equivalent to 5 mM $NaClO_4$/min, starting with 100%A) at a flow rate of 1 mL/min, where buffer A is 5 mM aqueous TEAP, pH 7, and buffer B is buffer A plus 0.25 M $NaClO_4$, pH 7; panels **b** and **d**, linear buffer A–B increasing salt gradient (2%B/min, equivalent to 5 mM $NaClO_4$/min, starting with 100%A) at a flow rate of 1 mL/min, where buffer A is 5 mM aqueous TEAP, pH 7, and buffer B is buffer A plus 0.25 M $NaClO_4$, pH 7, both buffer A and B containing 15% (v/v) acetonitrile. Temperature, 26 °C. Peptides are denoted by both a letter and a number with the number denoting the number of potentially positively charged groups a particular peptide contains, e.g., d3 denotes peptide d has three potentially positively charged groups. The sequences of the peptides are shown in Table 6.1. (Reproduced from Zhu et al. 1992)

concentration, the retention is dominated by ion exchange and hydrophobic interaction; in the medium level of acetonitrile concentration, the retention is dominated by ionic interaction; in the high acetonitrile concentration, the retention is dominated jointly by ionic and hydrophilic interaction. Hydrophilicity can replace ion exchange as an overwhelming retention mechanism at even higher acetonitrile concentrations. Since the elution order of polypeptides is determined by the charge number and hydrophilicity of the solute molecules, it is completely different from the hydrophobicity-dependent retention mechanism in reversed-phase chromatography. Therefore, a separation method complementary to reversed-phase chromatography is provided.

Lämmerhofer et al. (2011) reported the use of hydrophilic interaction/anion exchange mixed-mode chromatography for peptide separation. They used a quinoline ring covalently connect to silica gel through a 12-carbon spacer arm to obtain a hydrophobic anion exchange stationary phase (Fig. 6.21). Since the stationary phase

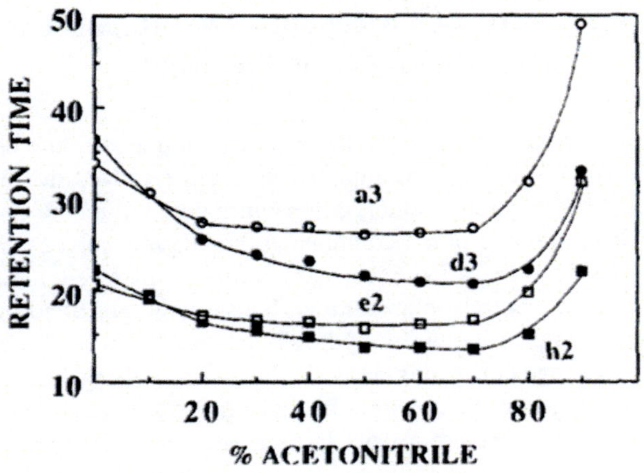

Fig. 6.20 Plots of peptide retention time versus acetonitrile concentration (v/v) in the mobile phase during strong cation exchange chromatography. Chromatographic conditions are the same as in Fig. 6.19. The sequences of peptides a3, d3, e2, and h2 are shown in Table 6.1. (Reproduced from Zhu et al. 1992)

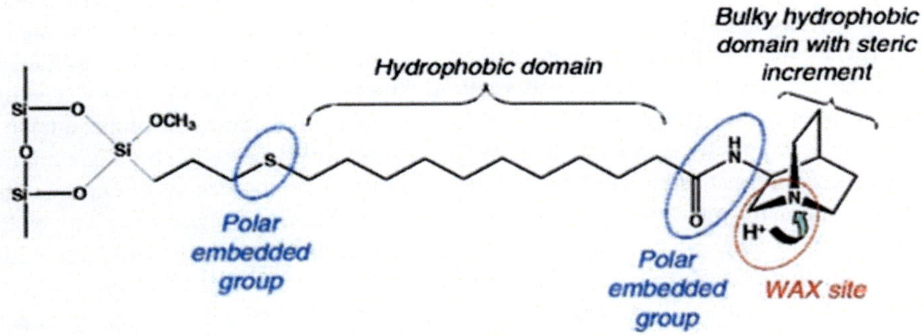

Fig. 6.21 The structure of the mixed-mode hydrophilic/weak anion exchange (RP/WAX) stationary phase (the p*K*a of the anion exchange site is 9.89, calculated by ACD/Labs 7.00). (Reproduced from Lämmerhofer et al. 2011)

contains tertiary amine cationic and alkyl chain hydrophobic groups, under acidic conditions, it can act simultaneously on the anionic and hydrophobic groups of the solute through a hydrophobic/ionic interaction hybrid mechanism. On the other hand, the stationary phase ligands also contain polar groups such as amide bonds and tertiary amines. When the mobile phase contains high concentrations of acetonitrile, these polar groups can have hydrophilic interactions with the solute. Therefore, this multifunctional stationary phase can be used to separate neutral, ionic, and polar compounds in different chromatographic modes such as reversed phase, ion exchange, ion repulsion, hydrophilic interaction, and hydrophobic interaction. According to the characteristics of the compounds to be separated, by adjusting the composition of the mobile phase, different modes of separation can be achieved on this highly versatile multimodal phase.

6.7 Retention Mechanism VI: Repulsion Hydrophilic Interaction Chromatography

Alpert (2008) observed that acidic amino acids elute before the dead volume of the cation exchange column because the electrostatic repulsion prevents the acidic amino acids from entering the entire pore volume of the stationary phase. However, if the mobile phase contains >60% organic solvent, the retention of acidic amino acids on the cation exchange column is enhanced, almost the same as the retention on the neutral polar column. This seemingly abnormal phenomenon reflects the antagonism between hydrophilic interaction and electrostatic repulsion in ion exchange chromatography. In the absence of organic solvents, the hydrophilic effect is insignificant, and the solute retention is dominated by the ionic interaction with the ion exchange groups on the stationary phase; when the organic solvent content is increased in the mobile phase, the hydrophilic interaction with the polar groups of the stationary phase gradually increased until it surpassed the electrostatic effect and became a dominant force. At this point, no matter whether the solute has the same charge as the ligand or not, they can be strongly retained on the ion exchange stationary phase. Based on this observation, Alpert (2008) proposed a new type of mixed-mode chromatography, namely, electrostatic repulsion-hydrophilic interaction chromatography, abbreviated as ERLIC. A mobile phase with a high content of organic solvent is used to elute the solute adsorbed on the ion exchange chromatography stationary phase; even if the solute and the stationary phase have the same charge, the solute can be retained by hydrophilic action. For solute mixtures with very different charges, electrostatic repulsion can be used to selectively inhibit solute retention, so that isocratic separation of mixtures that otherwise require gradient elution, including peptides, amino acids, and nucleotides, can be achieved within a reasonable time window.

Peptide standards containing acidic, neutral, and basic peptides were used as model compounds to compare their retention behavior in HILIC and ERLIC. As shown in Fig. 6.22, in the HILIC mode at pH 2.0, the acidic and neutral peptides eluted first, while the basic peptides eluted later. In a time window of less than 100 min, two widely spaced areas were formed in the elution profile. In the first area, the compounds are crowded, and the chromatographic peaks overlap each other, and then in the second area, the compounds are far apart, indicating that isocratic elution is not a good choice. In ERLIC mode, the elution time is halved, and the separation of acidic and neutral peptide standards is greatly improved. All compounds are evenly distributed throughout the elution window, showing the potential for isocratic separation of complex compounds. The chromatographic column used for ERLIC separation is a weakly basic anion exchange column with amine groups on the surface. Under separation conditions, the electrostatic interaction with acidic and basic peptides is partially inhibited, and the solute is mainly retained by hydrophilic interaction (Fig. 6.22).

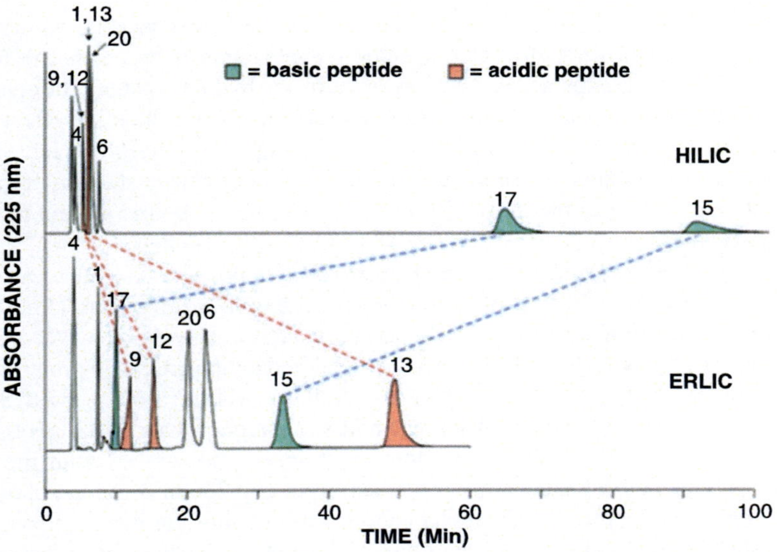

Fig. 6.22 HILIC and ERLIC separation of peptide standards. Peptide standard label: see original text. HILIC mode (top). Column: polyhydroxyethyl A, 200 × 4.6 mm; 5 μm, 300 Å. Mobile phase: 20 mM Na-MePO$_4$, pH 2.0, containing 63% ACN. Flow rate: 1.0 mL/min. ERLIC mode (bottom). Column: PolyWAX LP, 200 × 4.6 mm; 5 μm, 300 Å. Mobile phase: 20 mM Na-MePO$_4$, pH 2.0, containing 70% ACN. Flow rate: 1.0 mL/min. (Reprinted with permission from Alpert AJ (2008) Electrostatic repulsion hydrophilic interaction chromatography for isocratic separation of charged solutes and selective isolation of phosphopeptides. Anal Chem 80:62–76. Copyright {2008} American Chemical Society)

6.8 Retention Mechanism VII: Coordination Ion Exchange Chromatography

Our current understanding of the retention of biomacromolecules on (hydroxy)-apatite is largely due to the systematic study of Gorbunoff (1984a, b). The surface of hydroxyapatite is rich in positively charged calcium ions and negatively charged phosphate ions. Calcium ion sites, also known as C-sites, can mediate anion exchange retention mechanism and metal ion coordination. Phosphate sites, also known as P-sites, can mediate cation exchange retention mechanisms and hydrogen bond interactions. Gorbunoff (1984b) demonstrated through experiments that although the exchange of positively charged amino groups in basic proteins does not change its retention behavior, the retention of the acidic proteins is greatly reduced when the negatively charged carboxylic acid groups in acidic proteins are replaced by sulfonic acid groups, even though the number of charges remains unchanged. Based on these observations, Gorbunoff and Timasheff (1984) proposed a mechanism for protein binding and elution on hydroxyapatite. Basic proteins are bound through the nonspecific electrostatic interactions between the positively charged amino groups and the phosphate sites on the hydroxyapatite while acidic proteins are bound through highly specific complexation between the negatively charged carboxyl groups with calcium ion sites on the hydroxyapatite. The elution of the adsorbed protein can be achieved by nonspecific charge shielding effect or by displacement effect of a specific complexing group. The phosphate ions in the mobile phase can either act on the protein in the form of charge shielding or act on the stationary phase in the form of a displacer to elute the bound protein from the stationary phase.

The cation exchange retention mechanism of basic proteins on hydroxyapatite is supported by a large amount of experimental evidence. Aizawa et al. (1998) used human lysozyme mutants, in which the surface charge was modified by site-directed mutagenesis, to clarify the role of positively charged amine groups in proteins. Five mutant human lysozymes (K1A, K13A, K33A, R10A, R14A) are expressed in yeast (Fig. 6.23). The lysozyme dissolved in 1 mM sodium phosphate buffer pH 6.8 was loaded onto the hydroxyapatite column, equilibrated and washed with the same buffer, and then eluted using a linear gradient from 1 to 200 mM sodium phosphate buffer pH 6.8 at 1 mL/min. The molar elution concentration for these lysozymes on the hydroxyapatite column was determined. The elution molarity of native protein was 101 mM, while the elution molarity of K1A, K13A, K33A, R10A, and R14A mutants was reduced to 81, 87, 97, 86, and 76 mM, respectively. Among them, K1A and R14A mutants have the most reduction. Combined with the X-ray crystal structure of human lysozyme, it can be concluded that the adsorption site of human lysozyme is located on the back of the active site, and Arg-14, Lys-1, Arg-10, and Lys-13 play an important role in the binding.

At present, the C-site interaction of most proteins is considered to be a classic metal chelation mechanism, in which the carboxyl group of the protein is similar to the carboxyl configuration of the chelating agent (such as EDTA). Once the carboxyl group on the molecule is adsorbed by the C-site formed by two crystalline calcium ions, the calcium ions in the mobile phase can form a triangular array of calcium ions around the carboxyl group to complete the C-site chelation (Kawasaki 1991). This structure is similar to the corresponding structure inside the hydroxyapatite crystal that occurs when one (or two) hydroxyl ions are replaced by carboxyl ions, which can explain why the adsorption of carboxyl group at the C-site is more stable when calcium ions are present in the mobile phase. Similarly, it can explain why certain proteins (such as myoglobin, hemoglobin, and trypsinogen) unexpectedly behave as acidic molecules when calcium ions are present in the mobile phase (Gorbunoff 1985).

The retention of acidic protein on hydroxyapatite is dominated by calcium ion coordination. Chelation and coordination interactions are 15–60 times stronger than simple electrostatic interactions. Therefore, they are hardly affected by NaCl in the mobile phase, especially because chloride ions do not form complexes with calcium. However, in the case of purification of DNA by hydroxyapatite chromatography, the adsorption of DNA increases as the concentration of NaCl in the eluent increases. This is because the presence of salt reduces the electrostatic repulsion between the negatively charged DNA and hydroxyapatite. The coordination interaction between

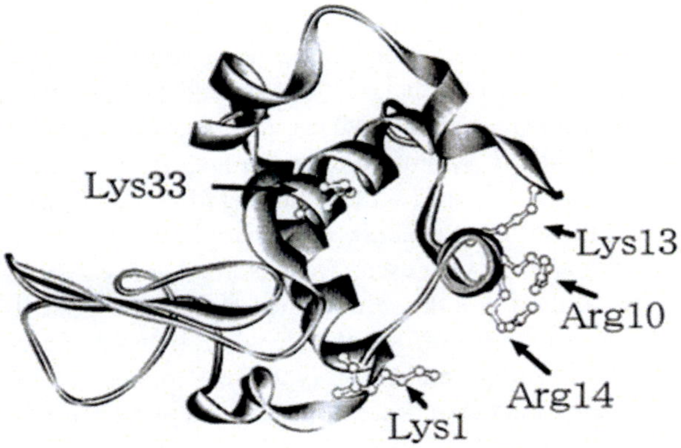

Fig. 6.23 The mutation positions of human lysozyme mutants. Plotting was performed using atomic coordinates obtained from the Brookhaven protein database (entry set 1LZ1, Artymiuk and Blake 1981) and the computer program WebLab Viewer (Molecular Simulations Inc.). (Reproduced from Aizawa et al. 1998. (Open access))

the phosphate group and the C-site is even stronger than that of the carboxyl group, which explains the strong retention of DNA (Kawasaki et al. 1986) and phosphorylated peptides (Mamone et al. 2010) on hydroxyapatite. The histidine residues in proteins are known to form complexes with immobilized transition metal ions, providing a basis for metal coordination chromatography to purify proteins. But they have no significant effect on the binding of proteins in hydroxyapatite. Ng et al. (2007) compared the retention behavior of two glutathione *S*-transferase (GST) fusion proteins (the protein has a molecular weight of 51,000 Da) on a ceramic hydroxyapatite stationary phase. One has a C-terminal poly(6)histidine tag 51 Thr, and the other does not have a tag. After 51 Thr was adsorbed on the CHT chromatographic column, it was eluted with a phosphate gradient with pH values of 5.9, 6.5, 7.0, and 8.1. The elution curve of 51 Thr at each pH value is actually the same, and it has nothing to do with whether the fusion protein has a poly(6) histidine tag. These results indicate that the binding and elution characteristics of the protein are not affected by the histidine tag, and the possibility that histidine residues compete with the acidic groups on the protein to coordinate with the metal of calcium ions is thus excluded.

Basic proteins can also interact with hydroxyapatite through a mixed mechanism of calcium ion metal affinity and phosphate cation exchange. Most recombinant monoclonal antibodies exist in the form of basic proteins. Schubert and Freitag (2007) reported that on apatite stationary phase, a double gradient (first 0–1 M NaCl and then 0.01–0.4 M phosphate) elution protocol was used to purify the recombinant human antibody rh-Anti-RhD (IgG1 subtype, p*I* 8.3) from the supernatant solution of the cultured Chinese hamster ovary (CHO) cells. In order to eliminate the coordination effect of C-sites, the elution gradient with increasing sodium chloride concentration always contains 10 mM sodium phosphate as the background electrolytes. Under this condition, the antibody elutes in the sodium chloride salt gradient and most of the contaminating protein elutes in the phosphate gradient. Therefore, through the double gradient method, the cell culture supernatant can be separated into three parts: the flow-through fraction that does not contain active antibodies, the NaCl elution fraction that contains only antibodies, and the phosphate gradient elution fractions that contain other protein contaminants.

Schubert and Freitag (2009) studied the chromatographic behavior of recombinant human antibody MDJ8 (IgG1 subtype, κ-light chain, molecular weight: 149.5 kDa, p*I*: 9.3) on fluorapatite and hydroxyapatite stationary phases, under the conditions of various pH values and buffer types (HEPES, phosphate, borate). HEPES buffer at pH 7.0, phosphate buffer at pH 8.2, and borate buffer between pH 8.5 and 11 were used. The adsorbed antibody was eluted by the double gradient method: first, use the salt gradient elution from 0 to 1 M NaCl in the corresponding buffer, and then perform the stepwise gradient elution to 0.4 M sodium phosphate. Regardless of pH and buffer type used, antibodies are eluted with a gradient of NaCl. The retention factor decreases with increasing pH. At pH 11, the antibody is eluted in the flow-through fraction. Therefore, in the whole pH range studied, solute retention is controlled by the electrostatic interaction with the P-site. Studies on antibody fragments obtained by papain digestion (Fc and Fab fragments) and deglycosylated Fc fragments show that the sugar structure has no effect on chromatographic behavior. In contrast, the chromatographic behavior of antibodies is mainly determined by the behavior of Fab fragments. The zeta potential measurement confirmed that in the pH range studied, the surface of apatite has a negative surface charge, while the surface charge of the antibody changes from positive to negative with the increase of the pH. The simulation of the molecular electrostatic potential of the antibody and the two fragments shows that there is a positively charged patch in the Fab fragment, which only disappears when the pH value is higher than 10. This patch is most likely to cause the cation exchange retention behavior of the antibody and its fragments on the hydroxyapatite stationary phase.

Gagnon et al. (2009) reported the cooperative retention behavior of monoclonal antibodies on a hydroxyapatite (HA) column. Calcium ions on the surface of hydroxyapatite crystals have a positive charge, which may mediate electrostatic interactions with protein residues, including the attraction of carboxyl groups and the repulsion of amino groups. In addition to the known ability to bind amino groups, HA can also repel protein carboxyl groups. The calcium atom on the crystal surface is surrounded by a triangular inner circle of three negatively charged oxygen atoms. The distance between the oxygen atom and calcium is about 0.12 nm and is surrounded by a triangular outer circle of oxygen atoms that is about twice the distance (Fig. 6.24). The arrangement of HA calcium and phosphate groups on the crystal surface emphasizes the possibility of synergy between different HA interaction sites. If multimodal or multivalent interactions result in solute retention higher than the sum of any individual site retention, the system is said to show positive cooperativity, otherwise the system shows negative or no cooperativity.

In order to clarify the retention mechanism of antibodies on hydroxyapatite, Gagnon et al. (2009) used a combination of sodium phosphate and sodium chloride gradient elution to elute the adsorbed antibodies. Phosphate leads to elution based on two main retention mechanisms: competition with proteins for the calcium affinity of HA and for ion exchange interactions. Sodium chloride mainly acts on ion exchange interactions and has almost no effect on calcium affinity. By systematically changing

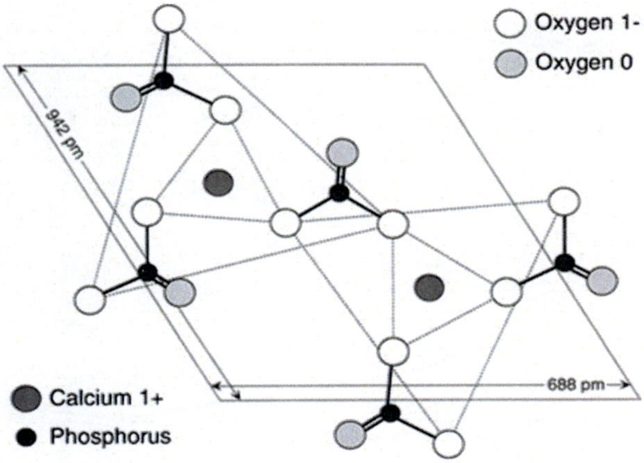

Fig. 6.24 The charge distribution on the surface of a hydroxyapatite crystal. A parallelogram represents a part of the crystal surface perpendicular to its longitudinal axis. The small triangle marks the inner circumference of the negatively charged oxygen molecules surrounding each surface calcium atom. The larger triangle marks the periphery. Note that each unit shares oxygen atoms from adjacent units, so that each calcium is surrounded by six negatively charged oxygen atoms. The net ratio in the frame in the figure is actually 5:1. (Reproduced from Gagnon et al. 2009)

the phosphate and chloride concentrations during gradient elution, the retention of the chimeric monoclonal antibody Rituxan on hydroxyapatite (HA) was mapped to characterize the interaction of solute with the calcium and phosphate residues on the solid phase. Basic protein lysozyme was used as a control to simulate the cation exchange interaction. Acidic protein bovine serum albumin was used as a control for calcium affinity interaction. Figure 6.25 shows the retention of the chimeric antibody Rituxan in the sodium chloride gradient at different phosphate concentrations (pH 7.0). The retention of lysozyme and BSA are plotted in the same graph to represent the retention behavior of proteins dominated by cation exchange and calcium affinity. The retention of BSA (p*I* ~ 4.7) is dominated by calcium affinity. Even when the phosphate concentration is as high as 40 mM, it cannot be eluted in a sodium chloride gradient. But it is eluted in the flow-through fraction when the phosphate concentration increases to 60 mM. Its steep response curve is similar to the early part of the IgG curve. The 40 mM level shift in phosphate concentration between the curves indicates that IgG has a much lower calcium affinity than BSA. This is because Rituxan has a high pI (p*I* 9.5–10), and the number of polycarboxyl groups coordinated with HA calcium is limited, and the retention of Rituxan is weaker than that of BSA. The retention of lysozyme (p*I* 11) is dominated by cation exchange. Even in the absence of phosphate, it elutes in a sodium chloride gradient. When phosphate is added to the elution buffer, its retention shows a shallow linear decrease. It is similar to the latter part of the IgG retention curve, but the retention of IgG is stronger. The p*I* of Rituxan is slightly lower than that of lysozyme. This indicates that its strong cation exchange binding reflects a larger number of

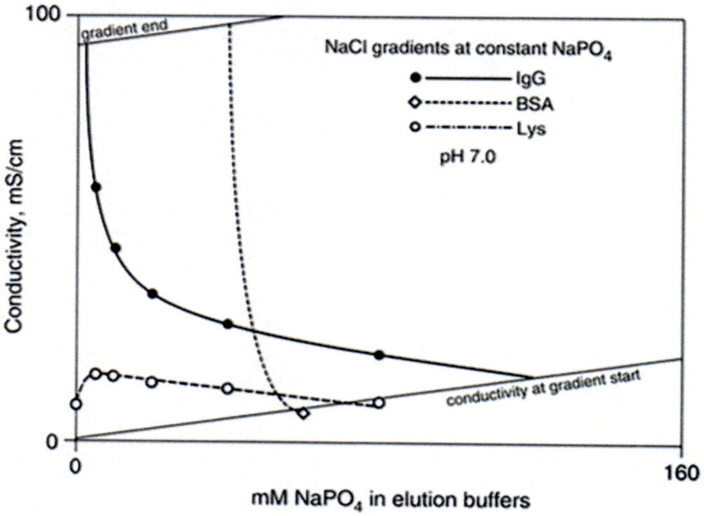

Fig. 6.25 At various phosphate concentrations (pH 7.0), the elution peak center of IgG, BSA, and lysozyme (Lys) in a sodium chloride gradient is a function of the conductivity and the phosphate concentration. A data point below the conductivity line at the beginning of the gradient indicates that the solute is not retained under that particular condition. (Reproduced from Gagnon et al. 2009)

interaction sites, consistent with the fact that the mass of IgG is more than 10 times that of lysozyme, but it may also indicate a cooperativity between the calcium affinity and ionic interaction.

Figure 6.26 shows the relationship between conductivity and sodium chloride concentration corresponding to the elution of Rituxan, BSA, and lysozyme in a phosphate gradient under different sodium chloride concentrations (pH 7.0). The elution conductivity curve of BSA is almost parallel to the conductivity curve of different sodium chloride concentrations at the beginning of the gradient, indicating that the phosphate concentration required for elution of BSA is almost constant and is not affected by the concentration of sodium chloride in the solution. These results can be explained by the calcium affinity nature of BSA's binding, to which the contribution of electrostatic interaction is not significant. The elution conductivity of lysozyme shows different trends. As the concentration of sodium chloride increases, the gap between the elution conductivity and the initial conductivity of the gradient decreases. When the NaCl concentration is 160 mM, the two overlap. It shows that the binding of lysozyme is essentially an electrostatic interaction, which decreases as the salt concentration in the solution increases. The elution conductivity curve of Rituxan is parallel to that of lysozyme but higher than that of lysozyme. It reflects that electrostatic interaction plays a very important role in the binding of antibodies. Combining with the results shown in Fig. 6.25, we can conclude that the monoclonal antibody binds to hydroxyapatite by a mechanism based on cooperativity between calcium affinity and phosphate cation exchange.

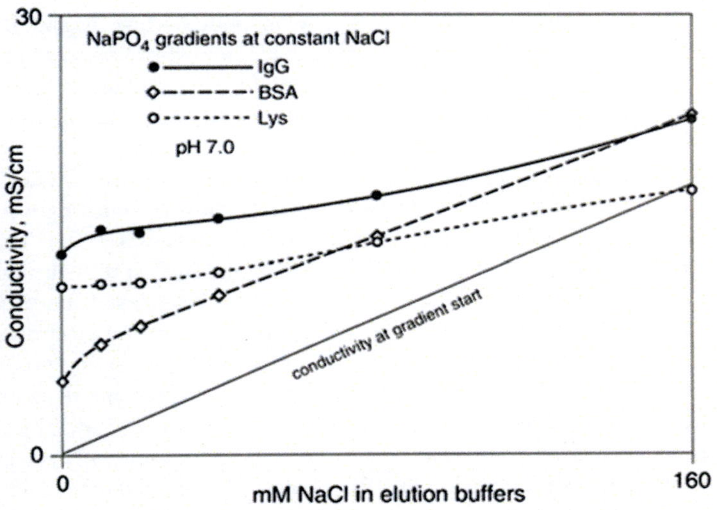

Fig. 6.26 The functional relationship between the elution peaks of IgG, lysozyme, and BSA in the sodium phosphate gradient at various sodium chloride concentrations (pH 7.0) and the concentration of sodium chloride. Marking the data points below the conductivity at the beginning of the gradient indicates that the solute is not retained under the specified conditions. (Reproduced from Gagnon et al. 2009)

Further experiments showed that the retention of Rituxan fragment Fc is dominated by calcium affinity, while the retention of Fab is mainly determined by cation exchange. $F(ab')_2$ shows a curve shape similar to Fab but retains stronger. The retention curve of intact IgG incorporates the unique elements of its fragments, but the retention is stronger than the sum of the fragment retention values. The retention curves of aggregates and non-aggregated antibodies are parallel, but the former has stronger retention, indicating that the retention mechanism is similar, but the number of binding sites for the former increases. The experimental data also provide evidence for charge repulsion between the IgG carboxyl group and HA phosphate at low conductivity values. The electrostatic repulsion of amino acid residues and the attraction of HA calcium to carboxyl groups appear to be blocked by the strong complexation of calcium with phosphate in the mobile phase.

The standard elution step for adsorbed solutes in hydroxyapatite chromatography is performed by increasing the phosphate concentration (usually 10–400 mM) gradient. This will interrupt two types of interactions, the coordination of the C-site, and the electrostatic interaction of the P-site. Almost from the beginning of hydroxyapatite chromatography, the relative insensitivity of C-site binding proteins to higher concentrations of NaCl has been used for "dual gradient" elution (Kawasaki 1991). In this scheme, the basic protein is first eluted from the hydroxyapatite column with a gradually increasing NaCl concentration gradient, and then the acidic protein is eluted with a phosphate gradient. In the above example of purification of antibodies by hydroxyapatite chromatography, gentle elution can be performed in a NaCl gradient, but a certain amount of phosphate in both buffers A and B (usually 5–10 mM) is required in order to overcome the residual C-site interaction (Gagnon et al. 2009). In fact, unless a certain concentration of phosphate is present in the mobile phase, even 1 M NaCl will not be able to elute many antibodies (Gagnon 2009).

The periodic arrangement of C- and P-positions on the surface of hydroxyapatite can be used to explain why hydroxyapatite seems to be able to distinguish between the natural and denatured forms of a given protein (Kawasaki 1991). The surface residues of the protein can bind to multiple sites on the surface of hydroxyapatite at the same time to form a multivalent complex with the strongest affinity. Since this multivalent interaction requires complementary in geometry and electrostatic interaction and the correct spatial arrangement of at least two interacting groups on the protein, it is likely that the natural form binds stronger than the denatured form. Natural lysozyme can be well retained by hydroxyapatite (Aizawa et al. 1998). Lysozyme has an isoelectric point higher than 10 and is generally regarded as a basic protein that only interacts with the P-sites. The three-dimensional form of natural lysozyme contains several lysines (K) and arginines (R) in close proximity, but there is only one pair of such lysines in the primary sequence. This may explain why the molecule is retained in its natural form (correct three-dimensional structure, with adjacent *K*/*R*) but not after denaturation. Similarly, the natural form of soybean trypsin inhibitor, chymotrypsinogen, and α-chymotrypsin are retained on hydroxyapatite while the denatured forms are not (Gorbunoff 1984a).

Kawasaki et al. (1986) provided an example of hydroxyapatite distinguishing the subtle structural differences of proteins. β-Lactoglobulin A and B variants are composed of two equal subunits, each subunit is composed of a polypeptide chain with 162 amino acid residues; the difference between the two variants is that only Asp-64 and Val-118 in the middle polypeptide chain in the A variant are replaced by Gly-64 and Ala-118 in the B variant. Figure 6.27 shows the chromatograms of a mixture of β-lactoglobulin A and B (Fig. 6.27a), β-lactoglobulin A (Fig. 6.27b), and β-lactoglobulin B (Fig. 6.27c) obtained under similar experimental conditions. Since Asp-64 is replaced by Gly-64, the B variant has one less negative charge than the A variant, and it should be eluted at a lower phosphate concentration. However, experiments show that the opposite is true. The B variant is eluted after the A variant, which means that the difference in molar elution concentration between the two proteins is not only related to the difference in the number of charges between molecules but also related to the way these charged groups are arranged in the molecules. A good match of the charged groups with the adsorption sites of the matrix will produce a stronger retention.

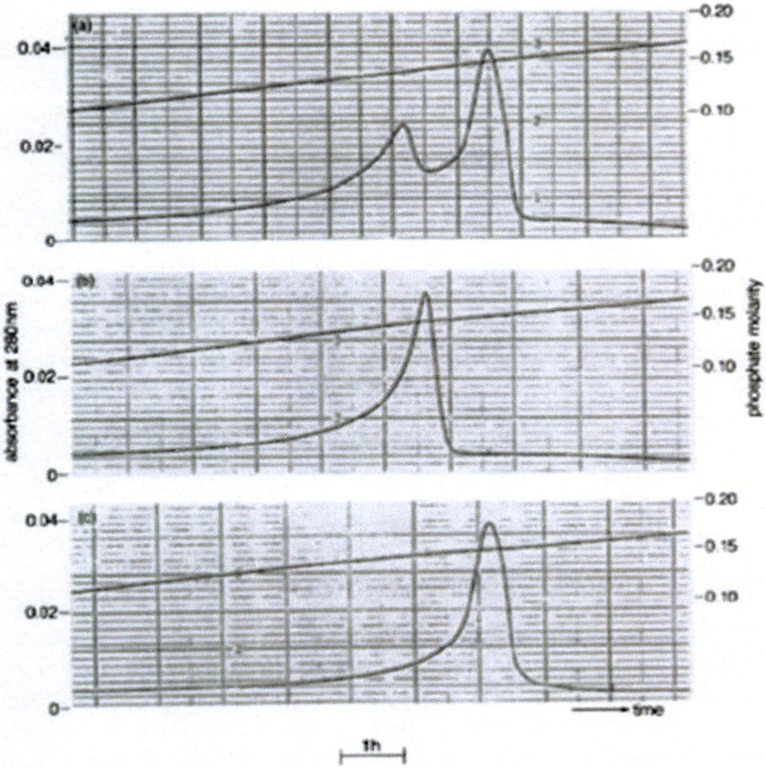

Fig. 6.27 Chromatograms of (**a**) a mixture of β-lactoglobulin A and β-lactoglobulin B, (**b**) β-lactoglobulin A, and (**c**) β-lactoglobulin B. Conditions: potassium phosphate buffer pH 6.8; column length = 2 × 3 + 7 × 30 = 216 cm, initial phosphate concentration = 1 mM, and phosphate gradient = 0.104 mM/mL. The other experimental conditions are (**a**) sample load = 0.40 + 0.40 = 0.80 mg, flow rate = 1.08 mL/min, P = 2.3–2.4 MPa, T = 18.2 °C, recovery = 115%; (**b**) sample load = 3.9 mg, flow rate = 1.08 mL/min, P = 3.7–3.8 MPa, T = 17.5 °C, recovery = 113%; (**c**) sample load = 3.9 mg, flow rate = 0.88 mL/min, P = 1.9–2.9-MPa, T = 19.1 °C, recovery = 109%. (Reproduced from Kawasaki et al. 1986)

6.9 Summary

Although Moore and Stein revealed the mixed-mode retention behavior of amino acids in ion exchange chromatography as early as 1951, it was not until about 30 years later that a quantitative description of the mixed-mode retention mechanism appeared for the first time. In 1987, Cox and Stout reported the coupling retention mechanism of ion exchange and hydrophobic interaction of basic compounds in silica-based reversed-phase chromatography. They proposed to systematically change the ionic strength in the mobile phase or the surface coverage of the hydrophobic alkyl chain in the stationary phase through the system and observe their influence on the retention of basic compounds to quantify the contribution of electrostatic and hydrophobic interactions in the coupling retention mechanism. Two years later, Horvath and colleagues explored the mixed-mode chromatographic behavior of proteins in ion exchange chromatography, describing that protein retention showed a U-shaped curve that first decreased and then increased as the salt concentration in the mobile phase increased. In the low salt concentration range, the protein retention is controlled by electrostatic and hydrophobic interaction, while in the high salt concentration range, the protein retention is solely controlled by hydrophobic interaction. Carr and colleagues further studied the mixed-mode retention behavior of basic compounds in reversed-phase chromatography and proposed a two-site interaction model consisting of independent reversed-phase interaction sites and hydrophobically assisted ion exchange sites. By analyzing the influence of ionic strength in the mobile phase on solute retention, the contribution of reverse phase and ion exchange in a given reversed-phase system can be quantitatively evaluated. Wan and colleagues further considered the complementarity between different functional groups in the stationary phase and the corresponding interaction sites of the solute in mixed-mode chromatography. They introduced the well-established concepts of multivalency and cooperativity in biological systems into the study of chromatographic mechanism and developed the solute retention model for reversed-phase/ion exchange mixed-mode chromatography. The cooperativity factors of the two model systems were determined by the proposed method. Hodges and colleagues observed that hydrophilicity has a major impact on the retention of peptides in ion exchange chromatography. With the increase in the percentage of acetonitrile in the mobile phase, the retention of peptides showed a U-shaped curve that first decreased and then increased. In the low organic solvent content range, electrostatic and hydrophobic interactions dominate solute retention, and in the high organic solvent content range, electrostatic and hydrophilic interactions dominate solute retention. The hydrophilic effect increases rapidly with the increase of the percentage of organic solvents, which can exceed the electrostatic effect and become the dominant driving force for retention. Alpert confirmed that under the condition of high organic solvent content, the charge repulsion between the chromatographic ligand and the solute can be inhibited by the hydrophilic effect, so that the compounds with the same charge are retained on the chromatographic column. Therefore, hydrophilic chromatography on ion exchange stationary phases can be used for isocratic separation of complex mixtures with different charges.

References

Adams H, Chekmeneva E, Hunter CA et al (2013) Quantification of the effect of conformational restriction on supramolecular effective molarities. J Am Chem Soc 135:1853–1863

Aizawa T, Koganesawa N, Kamakura A et al (1998) Adsorption of human lysozyme onto hydroxyapatite: identification of its adsorbing site using site-directed mutagenesis. FEBS Lett 422:175–178

Alpert AJ (2008) Electrostatic repulsion hydrophilic interaction chromatography for isocratic separation of charged solutes and selective isolation of phosphopeptides. Anal Chem 80:62–76

Artymiuk PJ, Blake CCF (1981) Refinement of human lysozyme at 1.5 Å resolution analysis of non-bonded and hydrogen-bond interactions. J Mol Biol 152:737–762

Badjic JD, Nelson A, Cantrill SJ et al (2005) Multivalency and cooperativity in supramolecular chemistry. Acc Chem Res 38:723–732

Buckenmaier SMC, McCalley DV, Euerby MR (2002) Overloading study of bases using polymeric RP-HPLC columns as an aid to rationalization of overloading on silica-ODS phases. Anal Chem 74:4672–4681

Cacciapaglia R, Di Stefano S, Mandolini L (2004) Effective molarities in supramolecular catalysis of two-substrate reactions. Acc Chem Res 37:113–122

Cockroft SL, Hunter CA (2007) Chemical double-mutant cycles: dissecting non-covalent interactions. Chem Soc Rev 36:172–188

Cox GB, Stout RW (1987) Study of the retention mechanisms for basic compounds on silica under "pseudo-reversed-phase" conditions. J Chromatogr 384:315–336

Endo S, Pfennigsdorff A, Goss K-U (2012) Salting-out effect in aqueous NaCl solutions: trends with size and polarity of solute molecules. Environ Sci Technol 46:1496–1503

Ercolani G, Schiaffino L (2011) Allosteric, chelate, and interannular cooperativity: a mise au point. Angew Chem Int Ed 50:1762–1768

Fasting C, Schalley CA, Weber M et al (2012) Multivalency as a chemical organization and action principle. Angew Chem Int Ed 51:10472–10498

Gagnon P (2009) Monoclonal antibody purification with hydroxyapatite. New Biotechnol 25:287–293

Gagnon P, Cheung CW, Yazaki PJ (2009) Cooperative multimodal retention of IgG, fragments, and aggregates on hydroxyapatite. J Sep Sci 32:3857–3865

Gorbunoff MJ (1984a) The interaction of proteins with hydroxyapatite. I. Role of protein charge and structure. Anal Biochem 136:425–432

Gorbunoff MJ (1984b) The interaction of proteins with hydroxyapatite. II. Role of acidic and basic groups. Anal Biochem 136:433–439

Gorbunoff MJ (1985) Protein chromatography on hydroxyapatite columns. Methods Enzymol 117: 370–380

Gorbunoff MJ, Timasheff SN (1984) The interaction of proteins with hydroxyapatite. III. Mechanism. Anal Biochem 136:440–445

Henkel S, Misuraca MC, Ding Y et al (2017) Enhanced chelate cooperativity in polar solvents. J Am Chem Soc 139:6675–6681

Hunter CA, Anderson HL (2009) What is cooperativity? Angew Chem Int Ed 48:7488–7499

Huskens J (2006) Multivalent interactions at interfaces. Curr Opin Chem Biol 10:537–543

Huskens J (2018) Chapter 2 Models and methods in multivalent systems. In: Multivalency: concepts, research & applications. Wiley, Hoboken, NJ

Jonsson T, Christensen CB, Jordening H et al (1988) The bioavailability of rectally administered morphine. Pharmacol Toxicol 62:203–205

Kawasaki T (1991) Hydroxyapatite as a liquid chromatographic packing. J Chromatogr 544:147–184

Kawasaki T, Ikeda K, Takahashi S et al (1986) Further study of hydroxyapatite high-performance liquid-chromatography using both proteins and nucleic acids, and a new technique to increase chromatographic efficiency. Eur J Biochem 155:249–257

Kirby A (1980) Effective molarities for intramolecular reactions. Adv Phys Org Chem 17:183–278

Lämmerhofer M, Nogueira R, Lindner W (2011) Multi-modal applicability of a reversed-phase/weak-anion exchange material in reversed-phase, anion-exchange, ion-exclusion, hydrophilic interaction and hydrophobic interaction chromatography mode. Anal Bioanal Chem 400:2517–2253

Mamone G, Picariello G, Ferranti P et al (2010) Hydroxyapatite affinity chromatography for the highly selective enrichment of mono- and multi-phosphorylated peptides in phosphoproteome analysis. Proteomics 3:380–393

Manning GS (1969) Limiting laws and counterion condensation in polyelectrolyte solutions. I. Colligative properties. J Chem Phys 51:924–933

McCalley DV (2000) Effect of temperature and flow-rate on analysis of basic compounds in high-performance liquid chromatography using a reversed-phase column. J Chromatogr A 902:311–321

McCandless D (2010) Thiamine deficiency and associate clinical disorders. Humana, New York, pp 157–159

Melander WR, El Rassi Z, Horvath C (1989) Interplay of hydrophobic and electrostatic interactions in biopolymer chromatography: effect of salts on the retention of proteins. J Chromatogr 469:3–27

Motloch P, Hunter CA (2016) Thermodynamic effective molarities for supramolecular complexes. Adv Phys Org Chem 60:77–118

Neue UD, Phoebe CH, Tran K et al (2001) Dependence of reversed-phase retention of ionizable analytes on pH, concentration of organic solvent and silanol activity. J Chromatogr A 925:49–67

Ng PK, He J, Gagnon P (2007) Mechanistic model for adsorption of immunoglobulin on hydroxy-apatite. J Chromatogr A 1142:13–18

Rahman A, Hoffman NE (1990) Retention of organic cations in ion exchange chromatography. J Chromatogr Sci 28:157–161

Schubert S, Freitag R (2007) Comparison of ceramic hydroxy- and fluoroapatite versus Protein A/G-based resins in the isolation of a recombinant human antibody from cell culture supernatant. J Chromatogr A 1142:106–113

Schubert S, Freitag R (2009) Investigation of the interaction mechanism of the recombinant human antibody MDJ8 and its fragments with chromatographic apatite phases. J Chromatogr A 1216: 3831–3840

Sinanoğlu O, Fernández A (1985) The denaturation maxima of proteins and of drug-biomolecule complex formation in a wide range of methanol/water mixtures: solvophobic theory predictions as compared to experiments. Biophys Chem 21:157–162

Sun H, Hunter CA, Navarro C et al (2013) Relationship between chemical structure and supramolecular effective molarity for formation of intramolecular H-bonds. J Am Chem Soc 135:13129–13141

Wan Q-H, Wang X, Chen L (2019) Inverse mixed-mode chromatography for the evaluation of multivalency and cooperativity of host–guest complexation in porous materials. Langmuir 35: 10405–10411

Wright GA, Baker DD, Palmer MJ et al (2013) Caffeine in floral nectar enhances a pollinator's memory of reward. Science 339:1202–1204

Yang X, Dai J, Carr PW (2003) Analysis and critical comparison of the reversed-phase and ion-exchange contributions to retention on polybutadiene coated zirconia and octadecyl silane bonded silica phases. J Chromatogr A 996:13–31

Zhu B-Y, Mant CT, Hodges RS (1992) Mixed-mode hydrophilic and ionic interaction chromatography rivals reversed-phase liquid chromatography for the separation of peptides. J Chromatogr 594:75–86

Chapter 7
Stationary Phases

Abstract The mixed-mode stationary phase contains at least two functional groups with different properties, allowing simultaneous two interaction modes with the solute, and both can be controlled by the selection of the mobile phase composition. Compared with single-mode chromatographic methods, the multiple interaction mechanisms in mixed-mode chromatography allow the stationary phase to have a wider application range and higher selectivity, such as simultaneous analysis of polar and nonpolar, ionizable and neutral, and organic and inorganic compounds and separation of target compounds from mixtures with closely related structures. In this chapter, we will review the design, classification, and synthesis of mixed-mode ligands.

7.1 Introduction

Mixed-mode chromatography has become a powerful tool for separating various molecules, ranging from small inorganic ions of dozens of Daltons to large organisms like virus of millions of Daltons, in which the stationary phase plays a fundamentally important role. The mixed-mode stationary phase contains at least two functional groups with different properties, allowing two interaction modes with the solute at the same time, and both can be controlled by the selection of the mobile phase composition. Compared with single-mode reversed-phase or ion-exchange methods, the multiple interaction mechanisms in mixed-mode chromatography allow the stationary phase to have a wider application range and higher selectivity, such as simultaneous analysis of polar and nonpolar, ionizable and neutral, and organic and inorganic compounds and separation of target compounds from mixtures with closely related structures. In addition, mixed-mode chromatography can directly extract the target substance from the high-conductivity cell culture medium without dilution, simplifying the operation and saving costs. Therefore, in recent years, mixed-mode chromatography has become more and more widely accepted and applied in the fields of environment, food, pharmaceuticals, and biomedicine.

Q.-H. Wan, *Mixed-Mode Chromatography*,
https://doi.org/10.1007/978-981-16-5485-5_7

7.2 Design Principles

Today, the vast majority of chromatographic separations are achieved in single-mode chromatography. Traditional chromatographic methods can provide suitable separation schemes for compounds with simple structures and meet the requirements of compound separation analysis and structure identification. However, the separation of many mixtures with variable structures and complex components requires the combined use of a variety of different chromatographic methods. For example, the purification of proteins from cell lysates usually requires capture, intermediate purification, and polishing steps. These steps usually involve the use of different modes of chromatographic methods. One of the goals of mixed-mode chromatography is to achieve separation that usually requires multiple conventional separation modes in a single chromatographic system through the combination of different retention mechanisms, in order to achieve better solute retention characteristics, separation capabilities, and selectivity. The diversity of compound structure and properties determines the diversity of mixed-mode stationary phases. Polar compounds containing multiple functional groups, such as biologically active molecules, natural products, and drug metabolites, are either weakly retained or poorly resolved in conventional chromatographic modes, posing a significant challenge to biological processing. Therefore, it is necessary to design and synthesize corresponding mixed-mode ligands according to the structural characteristics of different compounds to maximize the separation efficiency. Starting from the retention mechanism of mixed-mode chromatography, we can summarize some empirical rules that can be used as a guideline for ligand design.

7.2.1 *Multivalent Binding*

Let us discuss the role of multivalent binding in ligand design starting from the functional properties of the solute. It is assumed that the analyte is an amphiphilic molecule, that is, one end of the molecule is a hydrophobic part, and the other end is an ionic group. We can design mixed-mode ligands with hydrophobic parts and oppositely charged ionic groups. It can bind amphiphilic molecule through simultaneous hydrophobic and ion pair interactions to obtain higher retention and selectivity than in the corresponding single-mode chromatography. The increase in retention of amphiphilic compounds is due to the multivalent effect, which makes the compound and the stationary phase ligand bind together in an ion pair stacking manner, just as the oppositely charged surfactants form mixed micelles in water. The increase in selectivity is because both functional groups of the amphiphilic compound participate in the binding. Therefore, it can be distinguished not only from compounds with only a single property but also from compounds with different functional groups. By analogy, if the target molecule contains more than two different functional groups, a mixed-mode chromatography ligand with more than two functional groups should be designed accordingly to take advantage of the multivalent effect.

7.2.2 *Functional Complementarity*

As mentioned above, multivalent effect is an important principle governing the retention in mixed-mode chromatography. In order to maximize the multivalent effect, it is necessary to consider that the functional group configuration of the ligand matches that of the corresponding solute. When hydrophobic and ion pair interactions work alone, they are all point-to-point interactions, and the selectivity is not strong. However, when the two functional groups are connected together, an amphiphilic molecule with a dipole moment is formed, which can interact with another amphiphilic molecule with an opposite charge, similar to stacking of dipole moments, resulting in a complex formed by amphiphilic molecule stacking. If the charge of the amphiphilic molecule is opposite to that of the ligand, the amphiphilic molecule will bind to the amphiphilic ligand immobilized on the solid substrate. However, if the charge of the ligand is the same as the charge of the amphiphilic solute, charge repulsion will occur between them, destroying the multivalent effect and greatly reducing retention and selectivity. The principle of functional complementarity requires that the ligand is complementary to the functional group of the solute, such as hydrophobic group-hydrophobic group, cationic group-anionic group, π-electron donor-π-electron acceptor, metal ion-ligand, crown ether-primary amine groups, and so on.

7.2.3 *Geometric Fit*

The principle of geometric fit is derived from the concept of lock and key and the principle of induced fitting. When studying the catalytic mechanism of enzymes, Fischer proposed that the binding of an enzyme with a substrate is like the relationship between a lock and a key, and the substrate is like a key embedded in the enzyme as a lock. Later, Koshland extended the concept of lock and key. During the enzymatic reaction, the conformation of the enzyme is variable. The binding of the substrate at the active site of the enzyme may cause the conformation of the enzyme to change and make the binding tighter. This is "induced fitting." Ligands and solutes meet each other in space, and the distance between different interaction sites in the ligand can be increased or reduced to improve the geometrical fit and maximize the multivalent effect.

The importance of spatial matching can be illustrated by inclusion chromatography. Only solute molecules whose molecular volume is equivalent to the cavity of inclusion ligands such as cyclodextrin can be retained. Solute molecules that are too large to enter the cavity and interact with the hydrophobic group inside the cavity will not be retained; too small solute molecules will not be retained strongly because it can't effectively interact with the hydrophobic group inside the cavity. If the retention of the protein in affinity chromatography is due to the interaction between the affinity ligand and the amino acid residue in the binding pocket of the protein, then the fitting of the affinity ligand into the protein binding pocket is the key factor. In addition, the free stretch length of the spacer arm of the ligand also plays an important role. If it is too short, the functional group cannot effectively contact the amino acid residue and cannot play an effective role.

7.2.4 Ligand-Solute Reciprocity

If the amphiphilic solute molecule A is effectively retained and separated by the ligand B during mixed-mode chromatography, then the amphiphilic solute molecule A can also be used as a mixed-mode ligand to separate compounds with the corresponding structure similar to ligand B. This is because no matter whether the ligand A or ligand B is used, the structure of the instantaneous supramolecular complex formed with the solute during the chromatographic process is the same, so it has similar retention and selectivity.

7.3 Classification of Ligands Used in Mixed-Mode Chromatography

According to the principle of multivalent interaction, functional complementarity, and geometric fitting, different functional groups are covalently bonded to form a series of compounds with specific retention characteristics, which are used as ligands for mixed-mode chromatography. The combination of different functional groups can produce a large number of ligands. Therefore, it is not difficult to imagine that there are many kinds of mixed-mode ligands and their number is only limited by our imagination. Below, we divide the ligands into binary and multiple mixed-mode ligands according to their structural characteristics. The binary mixed-mode ligand is mainly composed of two functional groups. During the chromatographic separation process, it binds to the solute molecule with two different types of noncovalent interactions. Therefore, its retention mechanism is controlled by the corresponding divalent effect. The multiple mixed-mode ligands contain three or more different functional groups, and their structural features are more complex than binary ligands, and it is more difficult to decompose the role of different functional groups in the ligands. At present, most of the mixed-mode chromatography research focuses on the binary mixed-mode system, and we will give a detailed introduction about the classification of binary phases and only a brief introduction to the ternary mixed-mode system.

7.3.1 Binary Mixed-Mode Chromatography Ligands

Reversed-phase (hydrophobic)-ionic interaction mixed-mode ligands Amphiphilic ligands with hydrophobic and ionic moieties are the most important type of ligands in mixed-mode chromatography. Their structural characteristics are that they have a hydrophobic part and an ionic part. The hydrophobic part can be an alkyl chain, an aromatic ring, or a heterocyclic ring, and the ionic part can be a sulfonic acid group, a carboxyl group, a primary amine, a secondary amine, a tertiary amine, and a quaternary amine. In recent years, nitrogen-containing heterocycles have also been widely used as basic groups for the construction of amphiphilic ligands.

Metal chelation-ionic interaction mixed-mode ligand The metal chelation-ionic mixed-mode ligand is composed of a dissociable compound with chelating function. Representatives of this type of ligands are hydroxyapatite and iminodiacetic acid (IDA) immobilized on a solid matrix. Under neutral conditions, the calcium ions in the hydroxyapatite coordinate with the phosphoric acid or carboxylic acid in the solute molecule, and the phosphate ion can form an ion pair with the positively charged amino group in the solute. Iminodiacetic acid can form binary coordination complexes with metal ions such as Cu(II) and nickel(II) and can further combine with histidine, cysteine, or tryptophan in proteins to form a ternary Complex. At the same time, the free iminodiacetic acid not chelated with metal ions can act as a cation exchanger to interact with the basic groups in the protein. Therefore, the ligand can have multivalent interactions with proteins mediated by metal coordination and ion pairing.

Metal chelation-hydrophobic interaction mixed-mode ligand The ligand contains a chelating agent and a hydrophobic group and binds solute through metal coordination and hydrophobic interaction. Representatives of this type of ligands are *N*-hexadecyl-L- or -D-hydroxyproline immobilized on a solid matrix. Under neutral conditions, the optically active N-alkylated hydroxyproline can bind copper ions to form a blue 1:1 binary complex, which can further bind α-amino acids to form a 1:1:1 ternary complex. Since the ternary complexes with two optical isomers of α-amino acid have different stability constants, they elute with different retention times during the chromatographic process. The alkyl chain in the ligand will have a hydrophobic interaction with the side chain of the α-amino acid, resulting in the retention and separation of hydrophobic amino acids. Therefore, this chromatographic process is actually dominated by multivalent interactions mediated by metal coordination and hydrophobic interactions.

Inclusion-hydrophobic interaction mixed-mode ligands The ligand is composed of cavity, cage, or cryptate and hydrophobic moieties. It can form an inclusion complex with one part of the solute molecule and exert hydrophobic interaction on the other part of the solute molecule. Therefore, this chromatographic process is dominated by multivalent interactions mediated by inclusion complexation and hydrophobic interaction. A representative of this type of ligands is the crown ether immobilized on a solid matrix. The crown ether ligand can be used to separate metal ions as well as primary amines. The six oxygen atoms on the crown ether and the three hydrogen atoms of the protonated primary amine form a tripartite hydrogen bond. The methylene group on the crown ether has a hydrophobic interaction with the alkyl chain of the primary amine, which enhances the binding of the primary amines on the crown ether.

Inclusion-ionic interaction mixed-mode ligands The ligands are composed of a cavity, cage, or cryptate compound and ionic moieties. They can form an inclusion complex with one part of the solute molecule and exert an ion pairing effect on the other part of the solute molecule. Therefore, this chromatographic process is a multivalent process mediated by inclusion complexation and ion pairing. A representative of this type of ligands is cyclodextrin containing dissociable side chains. The complexation of solute molecules with cyclodextrin is enhanced by the additional ion pairing. The chromatographic process is a multivalent process mediated by inclusion complexation and ion pairing.

Inclusion-inclusion mixed-mode ligands The chimeric ligands are composed of two or more cage-like or cryptate compounds with different properties, such as crown ethers and cyclodextrins. One part of the solute molecule is complexed with the crown ether, and the other part is complexed with the cyclodextrin. The two complexing events reinforce each other, so this multivalent effect enhance the solute retention.

Inclusion-metal chelation mixed-mode ligands The ligands are composed of cage or cryptate compound and metal chelating groups. One part of the solute molecule is complexed with the crown ether or cyclodextrin of the ligand, and the other part is coordinated with the metal chelate to form a multivalent complex with greatly enhanced stability. A representative of this type of ligands is that the hydroxyl group at the large-mouth end of the cyclodextrin is replaced by an amino acid. After the amino acid is loaded with metal ions, a bifunctional multivalent ligand is formed. The solute retention is enhanced by dual action of inclusion and chelation.

Hydrophilic interaction-ion exchange mixed-mode ligands The ligands are composed of ionic groups or a polar group covalently bonded to an ionic group. The ionic group of the ligand can simultaneously exert polar and electrostatic forces on the solute, thereby enhancing solute retention. Almost all ion exchangers behave like this type of ligand in eluents containing high concentrations of organic solvents. For solutes with polar and ionic groups separated, ligands containing polar and ionic parts are useful because the polar part of the solute molecule interacts with the polar group of the ligand and the ionic portion of the solute interacts with oppositely charged ionic groups on the ligand. It is possible to improve solute retention and separation selectivity through multivalent effects involving the superposition of polarity and ionic interactions. This has been confirmed by separation of peptides.

Hydrophilic interaction-metal chelation mixed-mode ligands The ligands are composed of a polar group and a chelating group. After the chelating group is loaded with metal ions, a ligand with coordination function is formed. The combination of polar groups and coordinating groups not only enhance the interaction with the solute, thereby increasing the solute retention, but also improve the selectivity, because multivalent interactions involving hydrogen bonding, dipole stacking, and coordination have a high degree of discrimination. A matrix that binds metal ions, such as hydroxyapatite, when used in hydrophilic interaction chromatography, will show a different selectivity from silica gel, because in addition to polar and ionic interactions, it can also bind solutes through metal coordination.

Hydrophilic interaction-inclusion mixed-mode ligands The binding of a crown ether with primary amines can also be regarded as an example of hydrophilic-inclusion mixed-mode chromatography. In the polar organic solvent mobile phase, the hydrogen bond between the primary amine and the oxygen of the crown ether is greatly enhanced, resulting in an enhanced inclusion effect.

π-π Interaction-hydrophilic interaction mixed-mode ligands The ligands are composed of a π-electron donor or π-electron acceptor and a polar group. In chemistry, the π-effect or π-interaction is a noncovalent interaction involving the π system. Just like in the electrostatic interaction, the negatively charged region interacts with the positive charge; the electron-rich π system can interact with the electron-deficient metal (cation or neutral), another molecule, or even another π system. The π-interacting part of the ligand binds to the π-interacting part of the solute molecule, and the polar part of the ligand binds to the polar part of the solute. Both π-π stacking and dipole stacking have high directivity. Therefore, this mode separation has high selectivity. In chiral chromatography, there are a large number of examples of this type of ligands, especially Pirkle-type chiral chromatography ligands (Pirkle and Pochapsky 1989), almost all can be considered as π-interaction-hydrophilic interaction mixed-mode ligands.

π-Interaction-ion exchange mixed-mode ligands π-interaction-ion exchange mixed-mode ligand is characterized by including π-electron donor or π-electron acceptor, ion exchange site, or zwitterion in the ligand, which can interact with solute molecules through π-stacking, dipole stacking, van der Waals force, and single or double ion pair interaction. The anion exchange-type, cation exchange-type, and amphoteric ion exchange-type chiral chromatographic stationary phases based on cinchona alkaloid derivatives (mainly carbamate compounds of quinine and quinidine) was developed by Lindner and colleagues (Lämmerhofer and Lindner 1996). The phase shows a wide range of chiral recognition capabilities for ionic chiral compounds, including chiral acids, chiral amines, and amphoteric compounds such as amino acids and small peptides. Interestingly, the amphoteric ion exchange ligand can realize three ion exchange modes under the conditions of polar organic mobile phase: anion exchange mode is used to separate chiral acids, and cation exchange mode is used to separate chiral amines, and the zwitterion exchange mode is used to separate amphoteric chiral compounds. Therefore, according to the characteristics of the solute molecule, the ligand not only act on the solute molecule according to the mixed-mode retention mechanism but also realize automatic switching of multimode separation at the same time, which is a true multifunctional chromatographic stationary phase.

π-Interaction-metal chelation mixed-mode ligands The ligands are composed of π-electron donor or π-electron acceptor coupled with chelating group. The π-electron donor or π-electron acceptor of the ligand is stacked with the π-electron acceptor or π-electron donor of the solute molecule, and the immobilized metal on the ligand is coordinated with the corresponding group of the solute molecule. The ligand exchange chiral chromatography developed by Davankov and his colleagues (1988) uses a large number of such ligands. They prepared a ligand exchange chiral stationary phase by chloromethylation of phenyl group, bonding L-proline on polystyrene-divinylbenzene particles and loading metal ions. The ligand contains a phenyl group rich in π-electrons and a metal chelating group. Therefore, it can multivalently bind phenyl ring-containing amino acids, such as tryptophan, tyrosine, and phenylalanine, so that the ligands have greatly improved the retention and chiral recognition of such amino acids.

π-Interaction-inclusion mixed-mode ligands π-Interaction-inclusion mixed-mode ligand is composed of π-electron donor or acceptor connected with cavity, cage, or cryptand. The π-donor or acceptor part of the ligand is π-stacked with the corresponding part of the solute, and the inclusion part of the ligand is complexed with the corresponding group of the solute. The retention of the solute is determined by π-stacking and inclusion. The optically active bis(1,1′-binaphthyl) 22-crown-6 synthesized by Cram and colleagues (Dotsevi et al. 1975) and the optically active (3,3′-diphenyl-1,1′-binaphthyl)-20-crown-6 are representatives of this type of ligands. The aromatic ring on the ligand can interact with the electron-containing moieties of the solute molecule, and the crown ether can interact with the primary amine group of the solute molecule, thus greatly enhancing the ligand's ability to retain and recognize the solute molecule containing the π-electrons.

7.3.2 Multiple Mixed-Mode Ligands

The multiple mixed-mode ligands contain three or more different structural modules or structural features, allowing three or more different types of interactions between the ligand and the solute molecule to occur at the same time. Therefore, the chromatographic separation process is actually controlled by multivalent interactions. Compared with divalent interactions involving bifunctional groups on chimeric ligands, multivalent interactions require a higher degree of complementarity between the ligand and the solute for them to work. The geometry (size), charge (positive and negative), and distribution (distance) of the functional group, as well as the surface topology of the ligand (concave and convex), all have a great influence on the stability of the molecular complex formed between the ligand and solute. This means that the binding specificity is better on multiple ligands than on binary

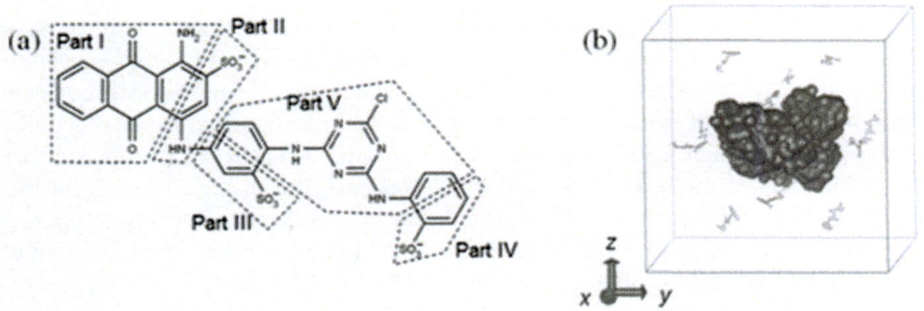

Fig. 7.1 (**a**) The structure of Cibacron Blue is divided into five main parts. (**b**) A snapshot of the initial configuration of the HSA simulation with 12 CB molecules. For better viewing, water and ion particles are not shown here. (Reprinted with permission from Liang-Schenkelberg J, Fieg G, Waluga T (2017) Molecular insight into affinity interaction between Cibacron Blue and proteins. Ind Eng Chem Res 56:9691–9697. Copyright {2017} American Chemical Society)

ligands. Therefore, it is more suitable for applications in chiral chromatography and affinity chromatography with high selectivity requirements.

Cibacron Blue F3GA is the most commonly used ligand in dye affinity chromatography and can be used as an example to illustrate how this type of ligand works (Liang-Schenkelberg et al. 2017). As shown in Fig. 7.1, the structure of Cibacron Blue F3GA can be divided into five parts, sulfonated anthraquinone ring (part I and II), phenylenediamine sulfonic acid (part III), aminobenzenesulfonic acid (Part IV), and chlorinated triazine ring (part V). Parts I and II contain π-electrons and ionic groups, which can interact with hydrophobic patches and ionic groups of proteins. The sulfonic acid groups of parts III and IV can interact with amino-containing residues of the protein, while part V is connected to the carrier through a spacer and may also participate in the hydrophobic interaction with the protein. According to reports in the literature, this dye ligand can separate more than 60 proteins and enzymes such as dehydrogenases, kinases, hydrolases, serum proteins, and interferons. The mechanisms of action of the biomimetic ligands developed in recent years all involve heterogeneous multivalent interactions, and most of these ligands can be classified as multiple mixed-mode ligands.

7.4 Silica-Based Stationary Phases

Preparation of mixed-mode chromatographic stationary phases is more complicated than that of single-mode counterparts. There are many synthetic methods available in the literature. An appropriate method should be selected according to the material of the carrier, the nature of the ligand, the conditions of use, and the purpose of application.

As mentioned earlier, the surface of the silica gel matrix contains a large number of highly reactive silanol groups. They can be used to couple the ligand to the silica matrix directly or indirectly. The direct approach includes the one-step covalent immobilization of previously synthesized ligands, containing both the functional groups and an additional reactive anchor group. Typical anchor groups are trimethoxysilane, triethoxysilane, and trichlorosilane. The major advantage of this direct immobilization approach is its simplicity and resultant uniformity of binding sites due to the presence of only one type of ligand at the surface. The limitation of direct immobilization is often complicated synthesis and purification of the ligand, which in many instances is quite complicated, if not impractical.

For the preparation of many multifunctional mixed-mode ligands, an indirect approach involving a sequence of reactions at the surface is employed, which is also known as the "surface assembly method". The advantage of this method is the possibility to synthesize surface anchored ligands of a relatively complex structure without involving extensive purification of the intermediates and end products. The surface assembly method, similar to solid-phase synthesis of peptides and oligonucleotides, makes it possible to construct very complex structures at the silica surface, but the active concentration of bonded groups is generally lower than that achievable through a more direct immobilization method, as the yields of each step reaction are often not quantitative. As a result of such incomplete reactions, the presence of unwanted residual functional groups can negatively impact the properties of the resultant mixed-mode ligands.

There are a large number of commercially available silane coupling agents that can be used for surface assembly of mixed-mode ligands. The most common silane coupling agents include γ-glycidoxypropyltrimeth(or eth)oxysilane, 3-aminopropyltrimeth(or eth)oxysilane, 3-chloropropyltrimeth(or eth)oxysilane, and 3-mercaptopropyltrimethyl(eth)oxysilane. After the silane coupling agent is bonded to the surface of the silica gel, the silica is "activated." The immobilized end groups, such as epoxy, amino, chlorine, sulfhydryl groups, can be used to react with ligands or with the structural units of more complex ligands.

Reversed-phase ion exchange stationary phases Reversed-phase ion exchange mixed-mode stationary phase is mainly used for the separation of hydrophobic compounds with dissociable groups, such as basic drugs and their metabolites, peptides, nucleotides, etc. The hydrophobic part of the stationary phase ligand is composed of alkyl chains or aromatic hydrocarbons, while the ion exchange part is composed of dissociable or ionic groups embedded in the end, middle, or vicinity of the hydrophobic part. In this mixed-mode chromatography, four ion exchange (IEX) modes can be distinguished: strong cation exchange (SCX), weak cation exchange (WCX), strong anion exchange (SAX), and weak anion exchange (WAX). Similar to traditional ion exchange stationary phases, carboxyl groups are usually used as weakly acidic groups and sulfonic acids as strongly acidic groups; primary, secondary, or tertiary amines are used as weakly basic groups, and quaternary amines are used as strongly basic groups. The mechanism of action of this type of stationary phase is to form a divalent complex through the interaction with the hydrophobic and

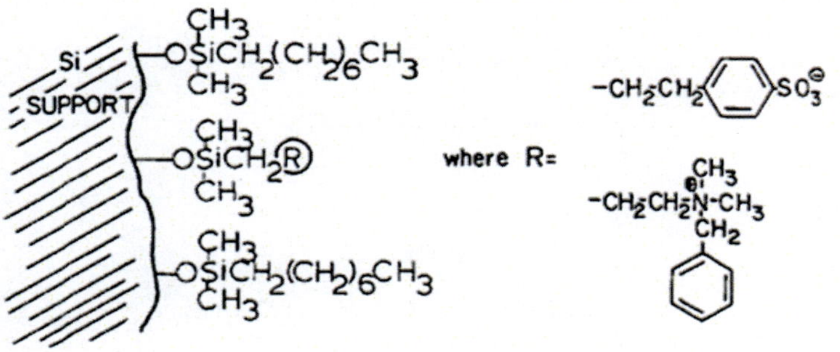

Fig. 7.2 A schematic diagram of a "mixed-mode" stationary phase synthesized by combining functionalized ionic (R) groups and reversed-phase groups on a silica support, resulting in a mixed-mode stationary phase with well-defined hydrophobic and ionic property. (Reproduced from Crowther and Hartwick 1982)

charged parts of the solute molecule. Therefore, the retention of amphiphilic solutes on the mixed-mode stationary phase is determined by both hydrophobic and ionic properties of the solute.

Hartwick and colleagues (Crowther and Hartwick 1982) first reported the synthesis and application of reversed-phase ion exchangers based on silica gel. They mimicked the dynamic adsorption of surfactants on the reversed-phase stationary phase in ion pair chromatography and simultaneously bonded octyl groups and ion exchange groups, propyl sulfonic acid, or propyl dimethyl benzyl amine on silica (Fig. 7.2). The first-generation high-efficiency mixed-mode chromatography stationary phase with surfactant immobilized on a silica matrix was prepared and used for the separation of nucleotides and catecholamines.

McLaughlin and colleagues (McLaughlin 1989) reported that on aminopropyl bonded silica, some amino groups reacted with alkyl acyl chlorides of different chain lengths to prepare a mixed-mode stationary phase containing both alkyl amide chains and unreacted amino groups for the separation of polynucleotides (Fig. 7.3). The first-generation of high-efficiency mixed-mode stationary phases is characterized by mixed ligands, that is, functional groups with different properties are adjacent to each other but exist independently.

Wongyai (1994) synthesized a new integrated high-efficiency mixed-mode stationary phase by coupling phenylpropanolamine to the surface of epoxy-silane-modified silica gel (Fig. 7.4). The stationary phase has reverse phase, hydrogen bonding, and anion exchange sites, where the anion exchange site is embedded in the middle of the hydrophobic group and can separate neutral, acidic, and basic compounds and inorganic anions at the same time. The separations can be optimized by adjusting the pH of the mobile phase and the concentration of organic modifiers. The chromatographic behavior on this phase is similar to that on a mixed bed column containing a mixture of reversed-phase and strong anion exchange particles. Possible applications include direct serum injection analysis of drugs.

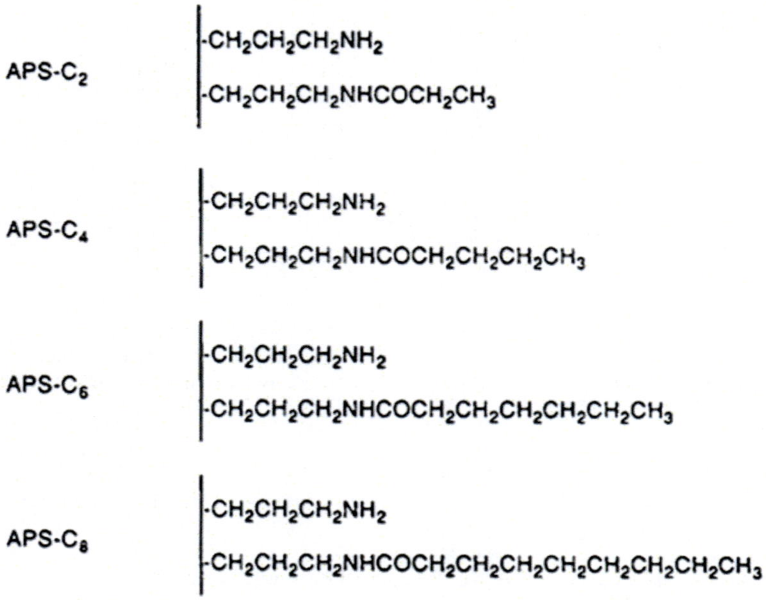

Fig. 7.3 A mixed-mode chromatographic stationary phase obtained by partially modifying the amino group of propylamino silica gel with an alkyl carboxylic acid. (Reprinted with permission from McLaughlin LW (1989) Mixed-mode chromatography of nucleic acids. Chem Rev 89:309–319. Copyright {1989} American Chemical Society)

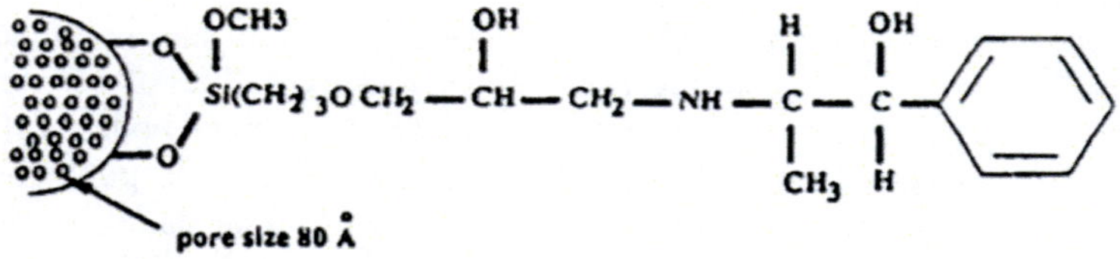

Fig. 7.4 Structure of phenylpropanol amino-bonded silica. (Reproduced from Wongyai 1994)

Muenter et al. (1999) reported the simultaneous separation of ions and neutral organic compounds on several single-ligand high-efficiency mixed-mode chromatographic stationary phases (Fig. 7.5). Among them, 8-hydroxyquinoline is bonded to the surface of silica gel through a spacer arm, and the formed ligand contains multiple interaction sites, such as hydrogen bonding, polarity, hydrophobicity, and metal ion coordination. Under the condition of mobile phase pH 2.8, when the nitrogen atom on the heterocyclic ring is protonated ($pK_a \approx 3.3$), there is an anion exchange site on the ligand, and both inorganic anions and polycyclic aromatic hydrocarbons are separated. In addition, similar chromatographic behavior was observed on commercial mixed-mode chromatographic stationary phases

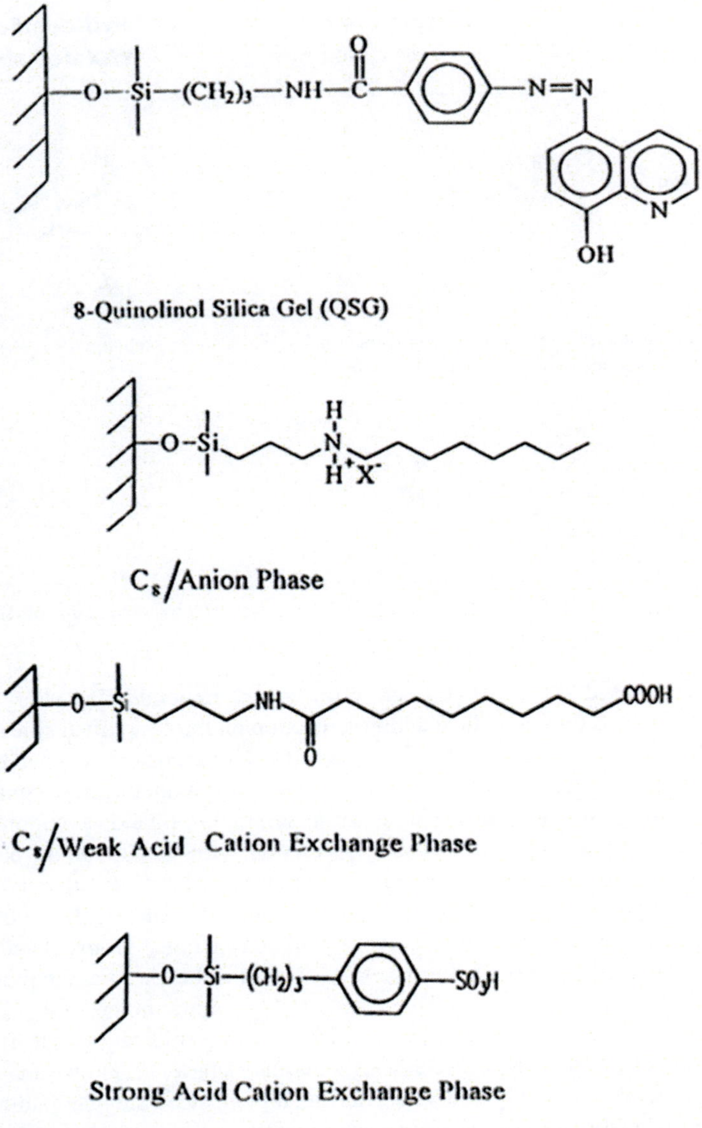

Fig. 7.5 The ligand structure of the mixed-mode chromatographic stationary phase. (Reproduced from Muenter et al. 1999)

(C8/anion exchanger and C8/cation exchanger) and even on propanesulfonic acid-bonded silica gel as a strong cation exchanger. Both ionic and neutral compounds can be separated in a single run.

Nogueira et al. (2005) used a new mixed-mode reversed-phase/weak anion exchange (RP/WAX) stationary phase for the separation and purification of peptides.

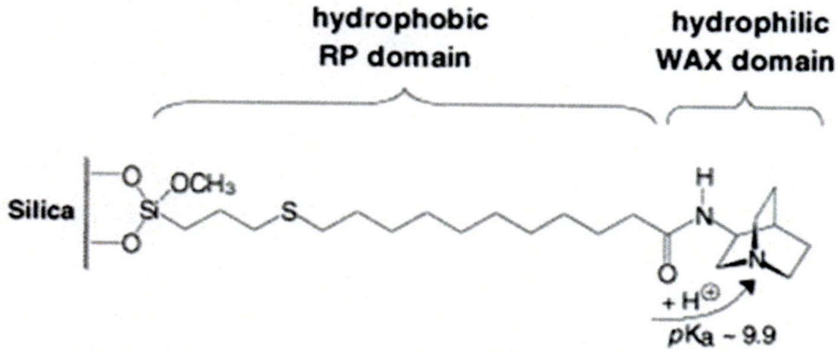

Fig. 7.6 The structure of a mixed-mode reversed-phase/weak anion exchange (RP/WAX) stationary phase. (Reproduced from Nogueira et al. 2005)

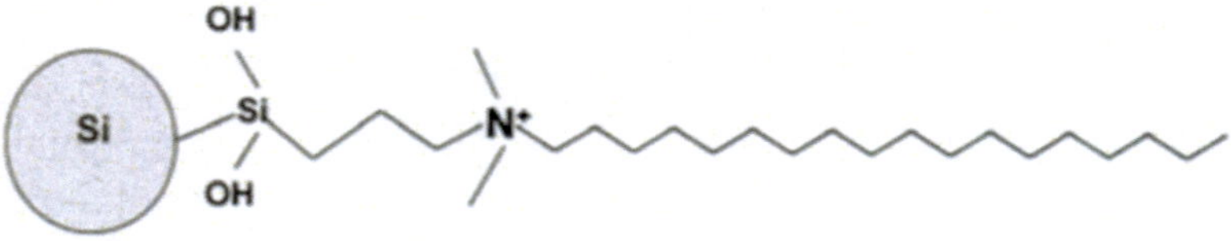

Fig. 7.7 Structure of BS C23 for mixed-mode chromatography stationary phase stability. (Reproduced from Abbood et al. 2009)

RP/WAX is based on *N*-(10-undecenoyl)-3-aminoquinuclidine selector, which is covalently bonded by free radical addition reaction on thiol-modified silica particles (5 μm, pore size 100 Å) (Fig. 7.6). The remaining thiol groups are capped by radical addition with 1-hexene. This newly developed separation material contains two different binding domains in a single chromatographic ligand: a lipophilic alkyl chain, which is used for hydrophobic interaction with the lipophilic part of the solute, such as in reversed-phase chromatography; and cation sites are used for anion exchange chromatography of oppositely charged solutes. This also allows repulsive ionic interactions with positively charged functional groups, leading to ion repulsion. The bivalent separation material is used for the chromatographic separation of the tetrapeptide *N*-acetyl-Ile-Glu-Gly-Arg-*p*-nitroaniline and its reaction by-products with N- and C-terminal protection. The results show that by using a single chromatographic step of RP/WAX stationary phase, all known major impurities are removed, showing better selectivity and enhanced sample loading ability than conventional reversed-phase chromatography.

Abbood et al. (2009) studied the retention behavior of various peptides on stationary phases with C23 alkyl quaternary ammonium groups (BS C23) (Fig. 7.7) and compared it with the separations on conventional reversed-phase C18 (RP C18) chromatography under isocratic conditions. BS C23 provides a different selectivity from RP C18. Between the peptide and the permanently positively charged group embedded in BS C23, electrostatic attraction or repulsion is clearly observed, depending on the pH of the solution. If a mobile phase with a neutral pH and low ionic strength is used, the retention of the peptides on the

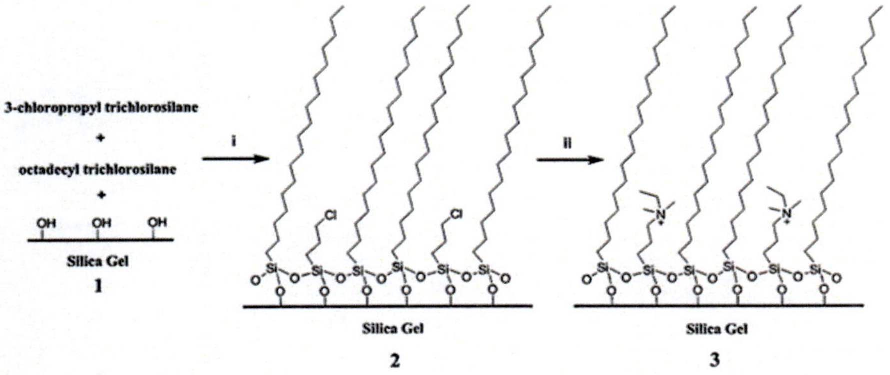

Fig. 7.8 Synthesis procedure of C18SAX stationary phase. (i) Anhydrous toluene, room temperature, 24 h; (ii) $CH_3CH_2N(CH_3)_2$, DMF, 80 °C, 24 h. (Reproduced from Wei et al. 2012)

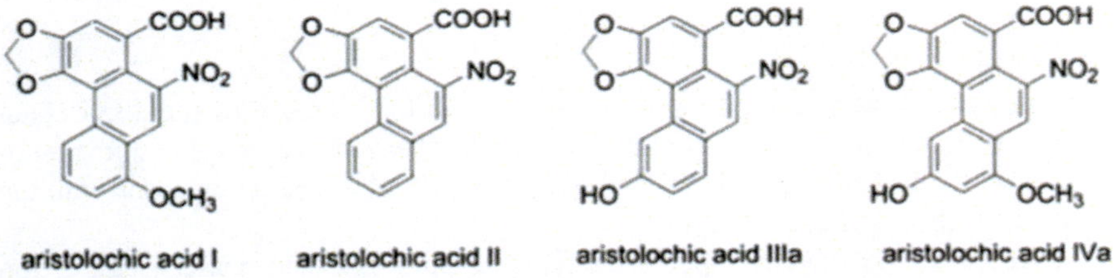

Fig. 7.9 The structure of four aristolochic acids. (Reproduced from Wei et al. 2012)

stationary phase embedded with quaternary ammonium is strongly influenced by anion exchange mechanism.

Wei et al. (2012) used a two-step method involving copolymerization and modification of chlorosilanes (Fig. 7.8) to synthesize a new reverse phase/strong anion exchange mixed-mode stationary phase C18SAX and use it for separation and enrichment of aristolochic acid in chestnut grass (Fig. 7.9). The results show that the C18SAX material can effectively enrich aristolochic acid at neutral pH and elution can be achieved under acidic conditions, showing the great potential of enriching acidic compounds from complex matrices.

Wan et al. (2019) reported the simple synthesis of two mixed-mode stationary phases of reversed-phase strong cation exchange (RP/SCX) and reversed-phase strong anion exchange (RP/SAX). First, the mercaptopropyl silane was chemically bonded to the silica gel particles by vapor precipitation, and then benzenesulfonic acid and benzyltrimethylamine were, respectively, connected to the mercaptopropyl group of the silica gel through the mercapto-ene click chemistry reaction (Fig. 7.10). The resulting stationary phase ligand contains both a hydrophobic part (benzene ring and its linking arm) and an ionic group (sulfonic acid group or quaternary amine group). RP/SCX shows the retention of alkyl aniline homologues under acidic

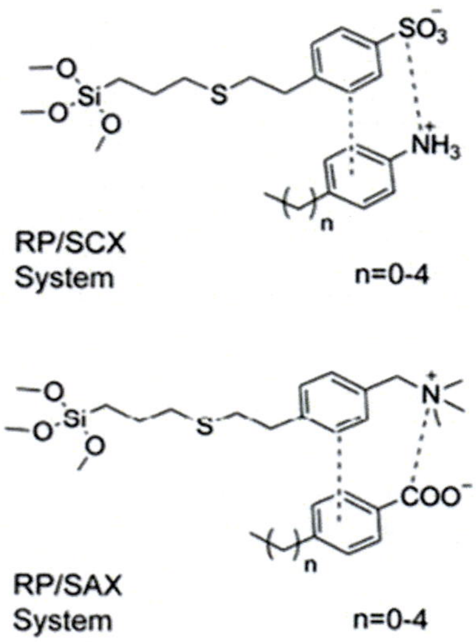

Fig. 7.10 The structure of RP/SCX and RP/SAX stationary phases synthesized by two-step method and their combination with alkyl aniline and alkyl benzene carboxylic acid. (Reprinted with permission from Wan Q-H, Wang X, Chen L (2019) Inverse mixed-mode chromatography for the evaluation of multivalency and cooperativity of host−guest complexation in porous materials. Langmuir 35:10405–10411. Copyright {2019} American Chemical Society)

conditions, and the retention value increases with the increase of the alkyl chain length. RP/SAX shows the retention of polyalkylbenzene carboxylic acid homologues under near-neutral pH conditions, and the retention value increases with the increase of the alkyl chain length.

Zhang et al. (2019) developed a new mixed-mode reversed phase/strong cation exchange stationary phase preparation method for the separation of fixed-dose combination drugs. The epoxy-silane-bonded silica gel prepared by vapor deposition was used as the starting material to prepare a mixed octadecyl/sulfonate group-bonded silica gel stationary phase (Fig. 7.11). The synthesized octadecyl/sulfonate bonded silica has a remarkable mixed-mode retention behavior and has excellent retention and selectivity for alkyl aniline homologs. The mixed-mode retention is affected by the ionic strength or solvent strength in the mobile phase, so that the separation can be optimized by fine-tuning these parameters. The mixed-mode stationary phase is used to separate two fixed-dose combination drugs: compound reserpine tablets and compound anisidine capsules. The results show that multiple active ingredients in compound preparations can be separated at the same time on the mixed-mode stationary phase, while traditional methods require multiple cycles of analysis of multiple components, which is time-consuming and labor-intensive.

Hydrophilic interaction-ion exchange stationary phases In this mixed-mode chromatography, hydrophilic interaction is superimposed on the ion exchange retention mechanism, making the retention of charged compounds stronger than any single-mode retention. Zhu et al. (1992) first reported this mixed-mode retention behavior of peptides on a strong cation exchange column, polysulfoethyl aspartamide-bonded silica gel (Fig. 7.12). The retention mechanism of peptides changes with the increase in the percentage of acetonitrile in the mobile phase,

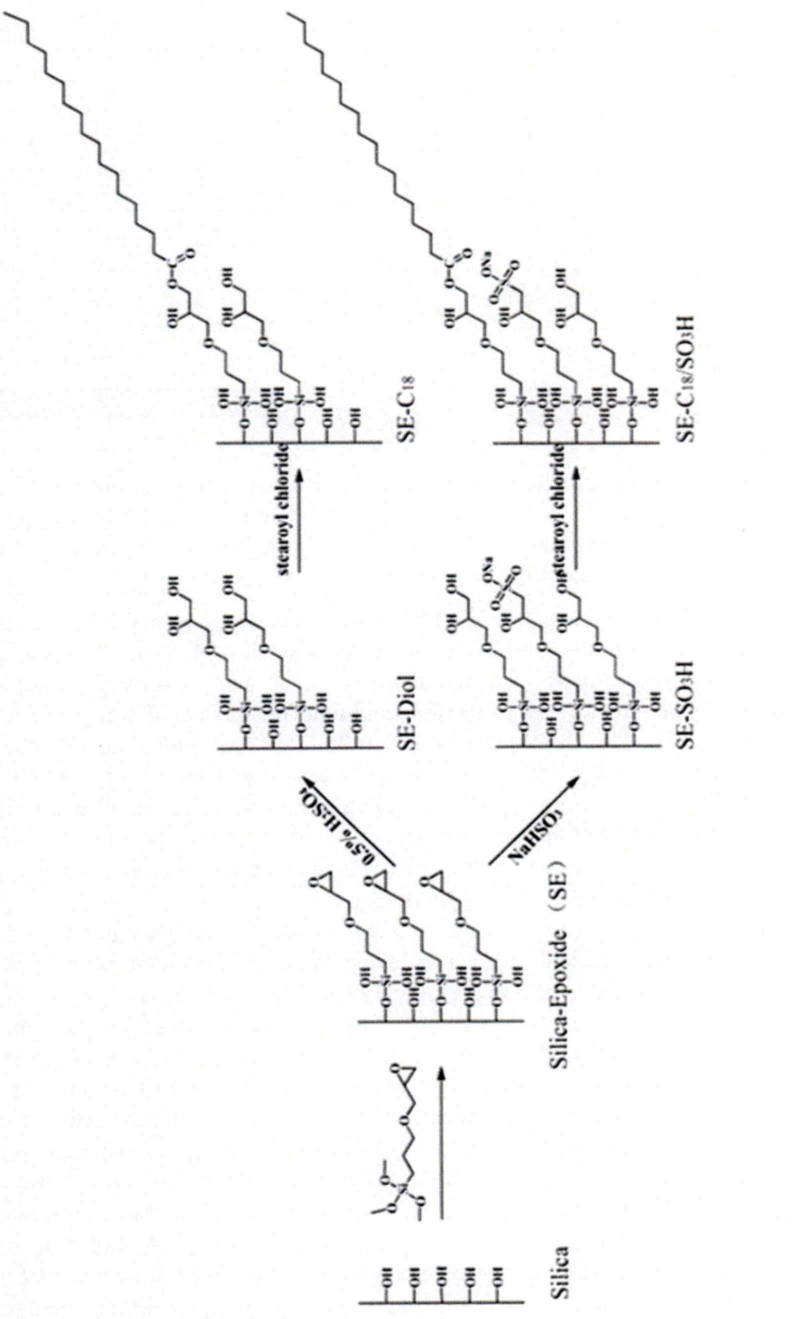

Fig. 7.11 Schematic diagram of mixed-mode chromatographic stationary phase synthesis. (Reproduced from Zhang et al. 2019)

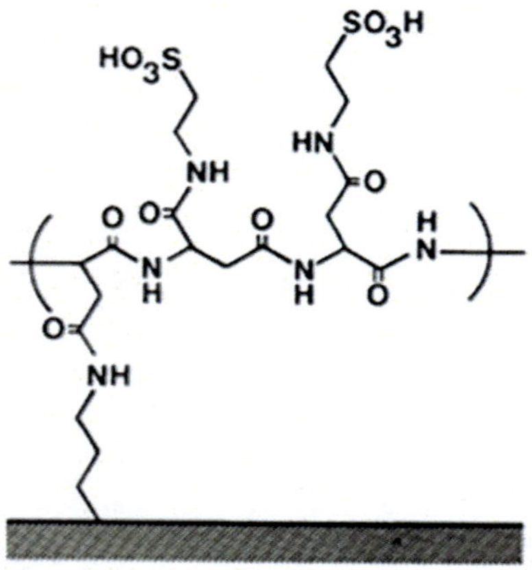

Fig. 7.12 Structure of polysulfoethyl aspartamide-bonded silica gel stationary phase

from the hydrophobic ion exchange at low percentage of acetonitrile to the hydrophilic ion exchange at high percentage of acetonitrile. At >90% acetonitrile, the retention of the peptides is dominated by hydrophilic interactions with negligible contribution from electrostatic interactions.

Lämmerhofer and Lindner (1996) used two different types of spacers and/or grafting modes to immobilize quinine carbamate and quinidine carbamate on porous silica gel, respectively, and obtained four weak anion exchange chiral stationary phases (CSP) (Fig. 7.13). Under the conditions of aqueous mobile phases containing buffers, these chiral stationary phases successfully separated chiral aryl-, aryloxy-, and arylthiocarboxylic acids, as well as N-derivatized amino acids (DNB-, B-, Ac-, F-, DNP-, DBD-, Fmoc-, Boc-, Z-, DNS-amino acid, etc.) and many other chiral acids, including sulfonic acid, phosphonic acid, and phosphoric acid. Chiral molecular recognition mainly involves multiple interactions such as ion pairs, hydrogen bonding, π-stacking, and steric hindrance (Fig. 7.14).

Hoffmann et al. (2007) prepared a chiral cation exchange stationary phase based on β-aminocyclohexane sulfonic acid and evaluated the enantioselectivity of chiral amines. Using different configurations of β-aminocyclohexanesulfonic acid, chiral stationary phases CSP1 and CSP2 with opposite elution order were synthesized. As shown in Fig. 7.15, CSPs 1 and 2 are covalently linked to the silica support by chiral β-aminocyclohexane sulfonic acid. The β-amino acid motif has two chiral centers, where the sulfonate anion is the site of ionic interaction with the cationic analyte. In addition, β-amino acids are part of the conformationally restricted cyclohexane. Cyclohexane scaffolds with structural rigidity and good conformation may be beneficial for enantioselective molecular recognition in liquid chromatography. The chiral amine compounds, which are mainly amino alcohol drug molecules, were selected, and the synthesized chiral stationary phase was evaluated under the conditions of polar organic mobile phase. The operating conditions including proton activity, the types of common ions and counter ions, the concentration of counter ions, and the types of polar organic solvents have been studied in detail. The results

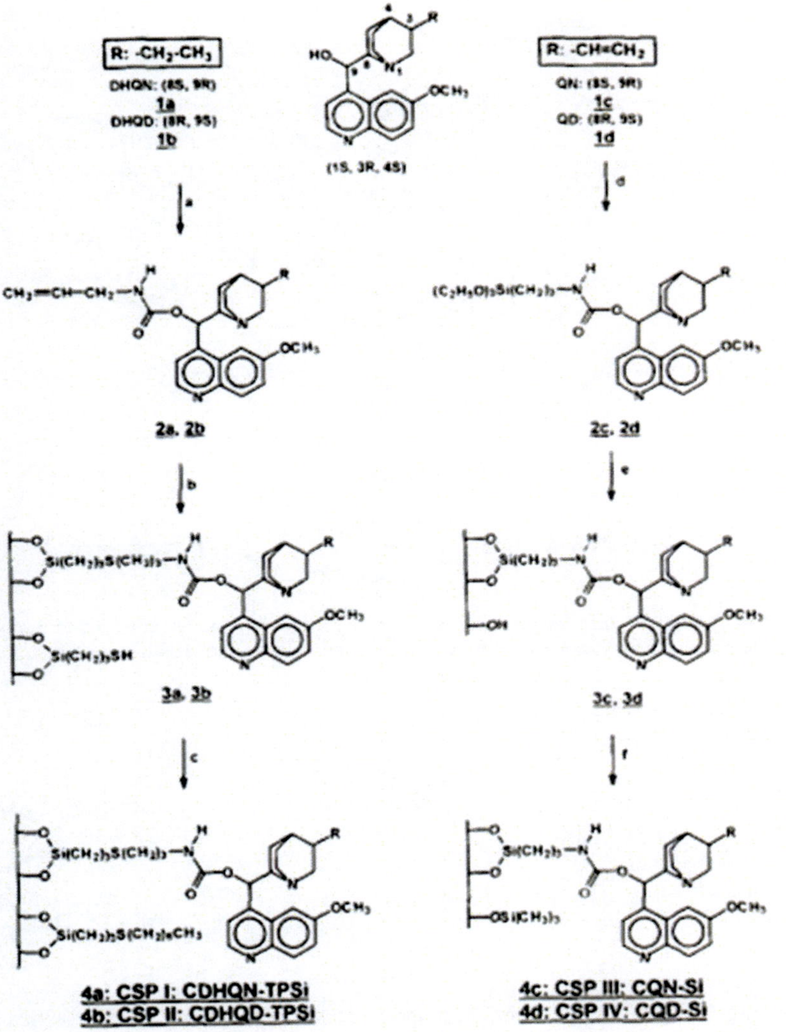

Fig. 7.13 Reaction scheme for the synthesis of four chiral stationary phases (CSP): (a) Allyl isocyanate, dibutyltin dilaurate, toluene, reflux; (b) 3-mercaptopropyl silanized silica, AIBN, chloroform, reflux, N_2; (c) 1-hexene, AIBN, chloroform, reflux, N_2; (d) 3-isocyanatopropyltriethoxysilane, dibutyltin dilaurate, toluene, reflux; (e) silica, toluene, reflux; (f) *N,O*-bis-(trimethylsilyl)acetamide, toluene, reflux. (Reproduced from Lämmerhofer and Lindner 1996)

showed that hydrophilic interaction, ion pairing, and ion exchange were identified as the main processes controlling solute retention.

Shen et al. (2013) immobilized glutathione on the surface of silica through a "thiol-ene" click chemistry reaction and prepared a hydrophilic interaction/cation

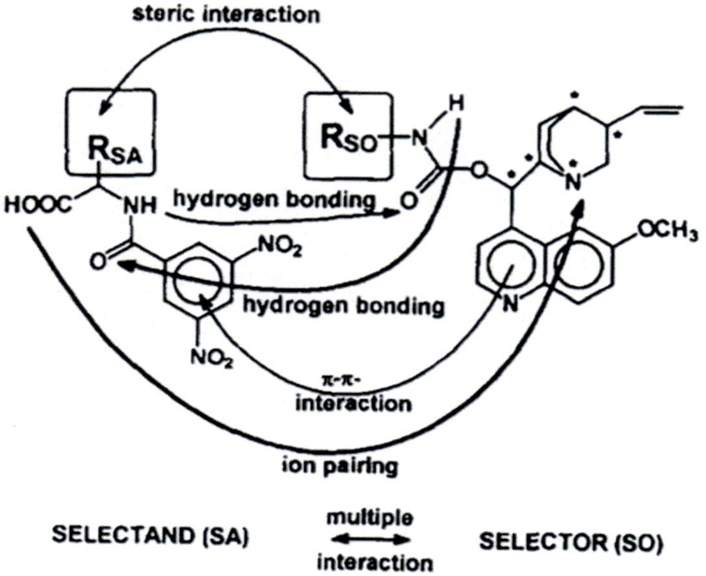

Fig. 7.14 Tentative chiral selection-selection and recognition model of *N*-3,5-dinitrobenzoyl amino acid derivatives. (Reproduced from Lämmerhofer and Lindner 1996)

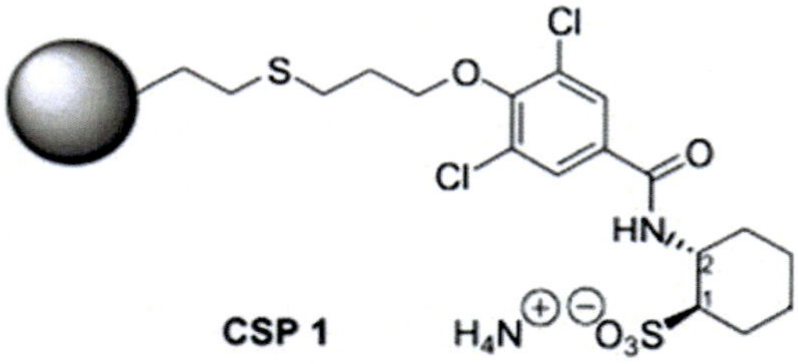

Fig. 7.15 The chemical structure of a new type of strong chiral cation exchanger material CSP 1 with (1*R*, 2*R*) configuration. CSP 2 has the opposite configuration. (Reproduced from Hoffmann et al. 2007)

exchange mixed-mode chromatography stationary phase (GSH silica gel) (Fig. 7.16). By taking advantage of the good hydrophilicity and inherent cation exchange properties of GSH bonded silica, separations of neutral fructose of high degree of polymerization, basic chitosan oligosaccharide, and strong acid carrageenan oligosaccharide were successfully achieved in hydrophilic interaction chromatography (HILIC), hydrophilic interaction/weak cation exchange mixed-mode chromatography (HILIC/WCX), cation exchange chromatography (WCX), and electrostatic repulsion/hydrophilic interaction chromatography (ERLIC). On the other hand, the high-efficiency separation of standard peptides with different hydrophobicity/hydrophilicity and charge were achieved in both WCX and HILIC/WCX modes. The separation of human serum albumin (HSA) tryptic digest was achieved in HILIC/WCX mode, and as many as 86 HSA peptides were identified, with a

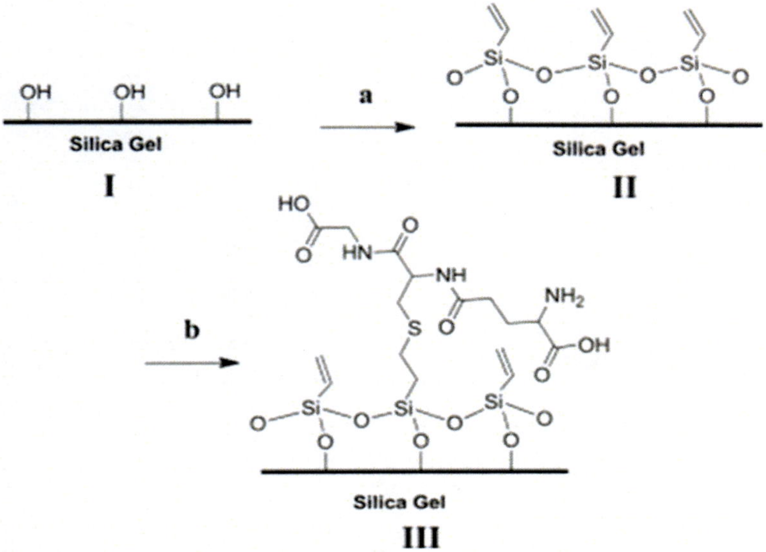

Fig. 7.16 Preparation of glutathione-bonded silica gel stationary phase by thiol-ene click chemistry. (a) Toluene, trichlorovinylsilane, room temperature; (b) MeOH/H_2O (5:7, v/v), glutathione, AIBN, 65 °C. (Reproduced from Shen et al. 2013)

sequence coverage of 85%. These results demonstrate the potential of the GSH bonded material in glycomics and proteomics.

Qiao et al. (2014) synthesized an ionic liquid *N*,*N*-diallyl-*N*-methyl-*d*-glucosamine bromide based on ammonium gluconate and then used the "thiol-ene" click chemistry to bond it to the surface of the 3-mercaptopropyl-modified silica (Fig. 7.17). The prepared glucosamine-bonded silica gel is used to separate nucleotide mixtures, including cyclic nucleotides, nucleoside monophosphate and nucleoside diphosphate, and adenosine mono/di/triphosphate mixtures. Under the mixed-mode retention mechanism of HILIC and anion exchange, most nucleotides are well separated in the order of cyclic nucleotides, nucleoside monophosphate, and nucleoside diphosphate. In these three types, the elution order is mainly based on the hydrophilicity of the nucleotide nucleoside portion. For adenosine nucleotide mixtures, AMP, ADP, and ATP elute in the order of increasing phosphate groups. It suggests the potential of nucleotide separation on the HILIC/weak anion exchange mixed-mode stationary phase.

Hydrophilic zwitterionic ion exchange stationary phases In order to expand the application range of chiral weak anion exchangers (WAX) based on cinchona alkaloids and chiral strong cation exchangers (SCX) based on sulfamic acid, Hoffmann et al. (2008) proposed a strategy to fuse their respective chiral selective subunits into a single zwitterionic chiral selective ligand by combinatorial synthesis method (Fig. 7.18). Using the polar organic mobile phase elution mode, the

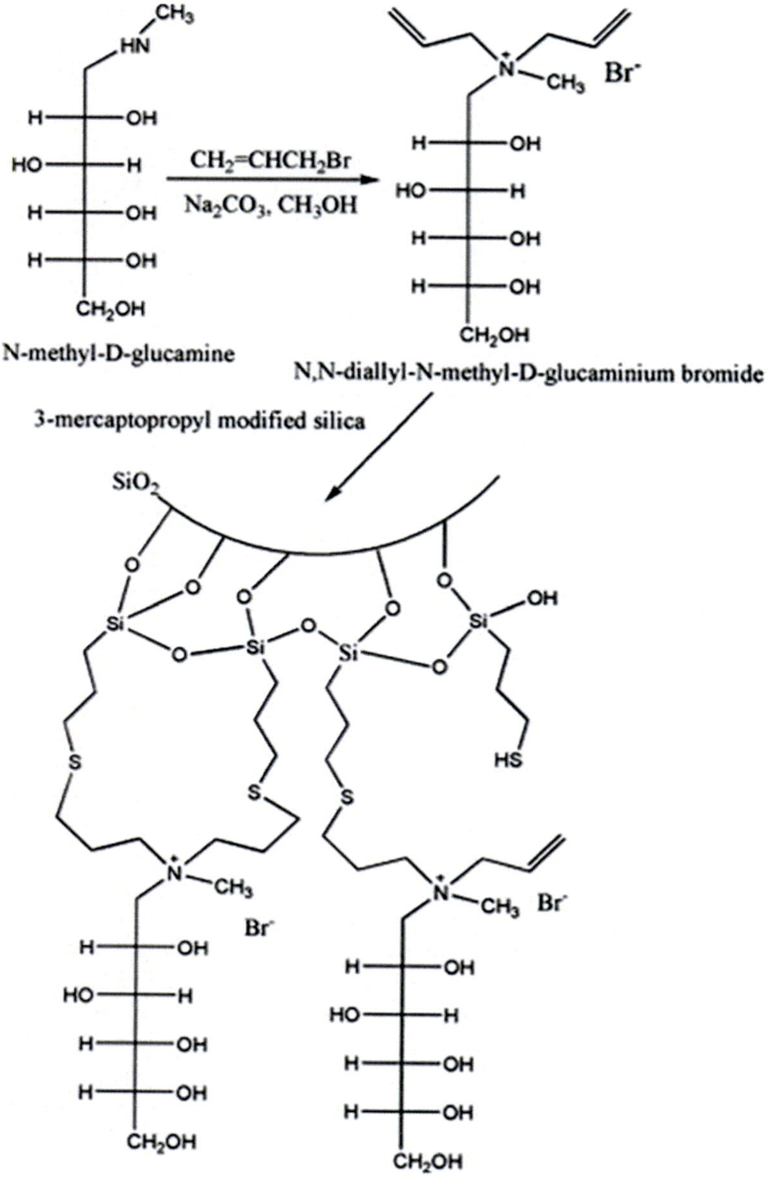

Fig. 7.17 Preparation of glucosamine-bonded silica gel. (Reproduced from Qiao et al. 2014)

synthesized zwitterionic ion exchange chiral stationary phase (CSP) not only retains the performance of the parent ligand, it can chiral separation of acidic and amine solutes but also displays new functions. Various zwitterionic compounds, such as α- and β-amino acids and polypeptides, can be chirally separated. The separation of

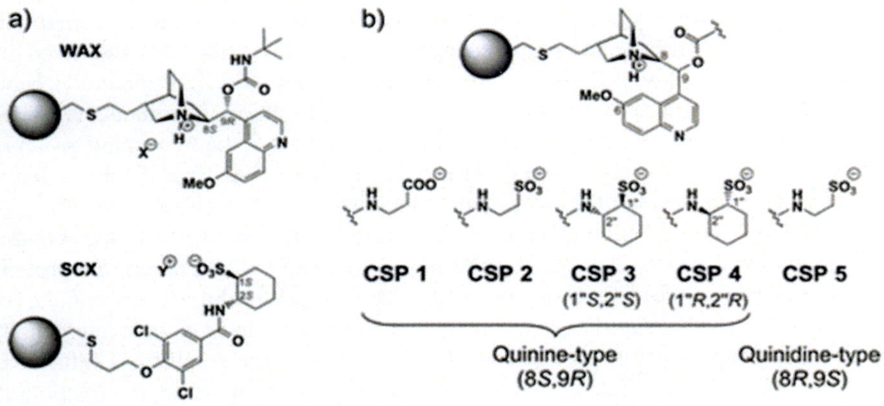

Fig. 7.18 (**a**) The chemical structure of the weak anion exchanger QN-AX (WAX) and strong cation exchanger (SCX) as the parent chiral stationary phase and (**b**) the newly prepared zwitterionic chiral stationary phase 1–5's lead structure. (Reproduced from Hoffmann et al. 2008)

zwitterions is obviously due to the cooperative effect of two oppositely charged ion exchange subunits, involving molecular recognition between zwitterionic ligands and zwitterionic analytes driven by diionic complexation.

7.5 Polysaccharide-Based Stationary Phases

The polysaccharide matrix mainly includes agarose, dextran, and cellulose microspheres. The surface of these substrates is rich in hydroxyl groups, which can be bonded with various ligands after activation. The use of cyanogen bromide (CNBr) to activate the polysaccharide matrix and then react with the ligand to prepare mixed-mode chromatography packing is a classic method that is often used. The advantage is that the reaction conditions are mild and the conversion rate is high; the disadvantage is that there are many reaction by-products and the introduction of additional positive charges. Cyanogen bromide reacts with the hydroxyl groups on the surface of polysaccharides to generate two reactive intermediates, namely, cyanate ester and imino carbonate, which react with the amino group of the ligand to generate three products: isourea derivatives, N-substituted carbamate esters, and N-substituted imine carbonates. Among them, the isourea derivative is a basic group, which is protonated and positively charged at a neutral pH. Now, the epoxy activation method is a common method for preparing mixed-mode chromatography packing. The activated matrix can be coupled with amino, hydroxyl, or sulfhydryl groups. Unlike cyanogen bromide activation, the mixed-mode chromatography packing prepared by epoxy activation does not introduce anion exchange groups.

Hydrophobic ion exchange stationary phases Yon (1972) first reported the preparation of hydrophobic ion exchange packings. Sepharose 6B dispersed in water was activated with CNBr, and then coupled with excess 1,10-diaminodecane in 0.1 M sodium tetraborate buffer (pH 9.5) at 5 °C, resulting in aminodecyl (AD)-agarose. When this material is used at a pH value below 9.5, the free amino group is expected to be positively charged. The AD-Sepharose was treated with a large excess of succinic anhydride (0.2 g/mL precipitated gel) at pH 6, and 5 °C. *N*-(3-carboxypropionyl)aminodecyl (CPAD)-agarose was obtained. Using bovine serum albumin with an isoelectric point of 4.7–4.9 as a model protein, the protein adsorption behavior of CPAD-agarose under different pH conditions was studied. At pH 5 and 6.5, the protein is almost completely absorbed by hydrophobic interaction. However, when the pH is increased to 8.0 and 9.5, the charge repulsion is enhanced, and the fraction of the eluted protein increases, and at the same time, the fraction of the hydrophobically bound protein decreases. The charge repulsion can be used for the selective elution of hydrophobic proteins in complex protein mixtures. This technique was used to purify aspartate carbamoyltransferase from wheat germ (Yon and Simmonds 1975) and to purify human erythrocyte glycoprotein (Simmonds and Yon 1976).

Shaltiel and colleagues (Er-El et al. 1972; Shaltiel and Er-El 1973) used cyanogen bromide activation and α-alkylamine or α,ω-alkyldiamine coupling steps to synthesize a series of hydrophobic anion exchangers with varying alkyl chain lengths on agarose gel. Their ability to bind glycogen phosphorylase increases with the increase of alkyl chain length, showing the important role of hydrophobic interaction in protein retention. At the same time, the binding of enzyme on the ligand is also affected by ionic interaction. For example, in the α-aminoalkyl agarose series, the butyl side chain is sufficient to retain glycogen phosphorylase, but in the α,-ω-alkyldiamine-agarose series, the side chain above the hexyl group is required to retain the protein on the packing.

At about the same time, Hofstee (1973a, b) used the cyanogen bromide activation method to bond hydrophobic alkylamines or arylamines on agarose gels to prepare hydrophobic anion exchangers. The enzyme is adsorbed to the chromatographic packing at neutral pH and eluted at increasing salt concentration. In addition, adding ethylene glycol to the eluent helps the elution of adsorbed proteins. Serum albumin, β-lactoglobulin, ovalbumin, and ovalbumin are strongly bound to phentermine-bonded agarose at pH 8 and ionic strength of about 0.05 but are eluted by 1 M NaCl to varying degrees. In each case, the elution by the salt is enhanced by the presence of a polarity reducing agent such as ethylene glycol. Under the same conditions, negatively charged proteins (chymotrypsinogen X, serum albumin, ovalbumin, and β-lactoglobulin) and positively charged proteins (chymotrypsinogen A, α-chymotrypsin and lysozyme), on the contrary, have a strong affinity for more hydrophobic adsorbents, namely, butyl and octyl ligands. These findings indicate that the binding of these proteins to hydrophobic anion exchangers depends on the combined effect of hydrophobic and electrostatic forces (perhaps a mutually reinforcing synergy) (Hofstee 1974).

Burton and Harding (1997) reported the use of various chemical methods to prepare hydrophobic ion exchange mixed-mode adsorbents based on cellulose, agarose, and methacrylate matrix and apply them to the one-step purification of chymosin. These matrices contain hydrophobic (aliphatic and/or aromatic) and ionic (carboxylate or alkylamine) groups (Burton et al. 1997). The hydrophobic amine ligand is attached to epichlorohydrin-activated agarose gel (single-ligand hydrophobic anion exchanger). The other mixed-mode matrix is through the incomplete reaction (0–80%) of the same amine ligand with the carboxylate matrix, resulting in a two-ligand hydrophobic weak cation exchanger composed of hydrophobic chains and residual carboxyl groups. Unmodified or partially modified aminocaproic acid-bonded agarose (Sepharose) and cellulose (Perloza) were used for one-step purification of chymosin for the best results.

Hydrophobic charge induction chromatography stationary phases Hydrophobic interaction chromatography is a common method for protein purification. In a high-salt concentration solution, the protein is adsorbed to the alkyl chain or aromatic ring bonded on the polysaccharide matrix by hydrophobic interaction, and after washing to remove impurities, the targeted protein is eluted and recovered by reducing the salt concentration. In order to avoid denaturation of the adsorbed protein and facilitate recovery, a stationary phase with intermediate hydrophobicity is usually used, and the adsorption and elution behavior of the protein can be optimized by adjusting the hydrophobicity of the ligand or reducing the density of the ligand. In hydrophobic interaction chromatography, the use of electrostatic repulsion to promote the elution of adsorbed proteins is also a common strategy in protein chromatographic purification.

As early as 1972, Yon (1972) proposed a protein purification strategy using hydrophobic ion exchanger adsorption and electrostatic repulsion desorption. Later in 1979, Sasaki et al. (1979, 1982) proposed the concept of hydrophobic ion chromatography and applied it to the purification of porcine pancreatin. All tested porcine pancreatic enzymes, regardless of their p*I*, were adsorbed on the polymer based weakly acidic ion exchanger Amberlite CG-50 at pH 4. The adsorption of proteins is hardly affected by ionic strength in the eluent. The adsorbed enzyme can be effectively eluted by increasing the pH from 4 to 6. The adsorption-elution mechanism of protein on Amberlite CG-50 may be as follows. When the pH is less than or equal to 4.5, the protein is adsorbed on the active site of the resin by hydrophobic interaction, and each active site of the resin contains at least one carboxyl group. As the pH value increases, the carboxyl group dissociates, resulting in a decrease in hydrophobicity and an increase in the negative charge of the active site. Therefore, the protein is now retained by the remaining hydrophobic affinity minus the increased electrostatic repulsion. Kasche et al. (1990) demonstrated that at neutral pH, penicillin amidase is adsorbed on phenylbutylamine (PBA)-bonded Eupergit C, PBA-bonded agarose, and phenyl-bonded agarose. By lowering the pH value of the eluent and increasing the electrostatic repulsion between the adsorbed protein and the adsorbent, the protein can be selectively desorbed.

The strategy of hydrophobic adsorption and desorption by electrostatic repulsion has attracted widespread attention since Burton and Harding (1998) introduced the concept of hydrophobic charge induction chromatography in 1998. The weak acid or weak base ligand is covalently bonded to the chromatographic matrix with high density to prepare a mixed-mode chromatographic stationary phase. The pH of the solution is adjusted so that the protein is adsorbed to the surface of the uncharged stationary phase. After the column is washed to remove impurities, the pH of the solution is readjusted so that the ligand is dissociated and carries the same charge as the protein. The electrostatic repulsion between the adsorbed protein and the ligand promotes the desorption of the protein. The hydrophobicity and titration range of the matrix can be matched to the requirements of the target protein by selecting the appropriate ligand type and ligand density. Because the adsorption is carried out on an uncharged surface, and the elution is induced by titrating the ionizable group into a charged form by the change of pH, this process is called hydrophobic charge induction chromatography.

They determined the pK_a values for weakly basic ligands (14 compounds in total) and weakly acidic ligands (9 compounds in total) bonded on the polysaccharide matrix (Tables 7.1 and 7.2) and compared them with those for free ligands. They found that the pK_a values of solid-phase bonded basic ligands decreased slightly, and the pK_a values of acidic ligands increased slightly. The pK_a values are also affected by the type of chemical bonds that link the ligands to the solid support. The pK_a values of 2-mercaptopyridine and 2-mercaptobenzimidazole derivatives bonded to

Table 7.1 Titration data for weak base matrices

Ligand	Start pH	pK_a	pH 90% titrated
Diethylaminopropylamine	10.5	9.3	8.4
Aminopropylmorpholine	8.9	7.1	6.2
Aminopropylmorpholine (10 mM)	8.05	5.8	4.8
Aminopropylimidazole	8.0	6.25	5.2
Histamine	8.4	6.85	5.6
Aminomethylbenzimidazole	6.75	4.8	4.0
4-Aminomethylpyridine	7.2	4.7	3.7
3-Aminomethylpyridine	6.7	4.2	3.3
2-Aminomethylpyridine	6.1	4	3.0
Mercaptomethylimidazole	6.9	5.3	3.8
4-Mercaptopyridine	7.2	4.8	3.7
2-Mercaptopyridine	5.4	2.8[a]	1.9[a]
4-Mercaptoethylpyridine	7.0	4.8	3.7
2-Mercaptobenzimidazole	6.2	4.1	3.0

Reproduced from Burton and Harding (1998)

All matrix samples (1 g) were suspended in 5 mL of 0.5 M NaCl except that marked 10 mM for which 10 mM NaCl was used. Titration below pH 4 was corrected by subtraction of values obtained for an unmodified Perloza control

[a]Accuracy of values is likely to be lower due to the relatively large correction for dilution required at very acidic pH

Table 7.2 Titration data for weak acid matrices

Ligand	Start pH	pK_a	pH 90% titrated
Aminocaproic acid	3	5.2	6.5
Aminocaproic acid (10 mM)	3.3	6.1	7.5
Nitrohydroxybenzoic acid	4	6.4	7.5
Nitrotyrosine/ethanolamine	5	7.2	9.0
Dichlorosalicylic acid	5	7.2	9.3
Dibromotyramine	5	7.7	9.3
Chlorohydroxyphenylacetic acid	6.5	9.8	10.8
Hydroxyphenylacetic acid	7.5	10.7	11.2
Tyramine	7.5	10.7	11.3

Reproduced from Burton and Harding (1998)
Matrix samples were suspended in NaCl solutions as described for Table 7.1. The start pH was determined after washing with 0.1 M HCl and water. Titration from pH 12 down to 4 was corrected for an unmodified Perloza control

the epoxy-activated polysaccharide matrix are lower than the pK_a values of their 2-aminomethyl analogs. In addition, for nitrogen-containing heterocycles, the pK_a value of the ligand is also related to the bonding position. When pyridine is bonded to the substrate at the para position of the ring nitrogen, its pK_a is the highest; when bonded at the ortho position of the ring nitrogen, it has the lowest pK_a. Among the tested ligands, the pK_a values are arranged at regular intervals in the preferred pH range of 4–10. The titration range of pyridyl and imidazole-based matrices is considered to be particularly useful, as the uncharged form is obtained between pH 7 and 8.5, and most are ionized at pH $\leq$ 5. Therefore, a suitable ligand can be selected according to the isoelectric point of the target protein to meet the requirements of hydrophobic adsorption and electrostatic repulsion-induced elution under neutral conditions.

Schwartz et al. (2001) first demonstrated the use of hydrophobic charge induction chromatography (HCIC) to achieve effective harvesting and recovery of high-purity monoclonal antibodies. The adsorbent MEP HyperCel used has a ligand derived from 4-mercaptoethylpyridine (4-MEP) (Fig. 7.19). The presence of a thioether functional group at the proximal end of the heterocyclic ring can provide additional selectivity. The pK_a of 4-MEP is 4.8. Therefore, under near-neutral or alkaline conditions, the ligand is uncharged and behaves much like a phenyl group, binding to the antibody through hydrophobic interactions. The desorption of the adsorbed protein is achieved by lowering the pH of the mobile phase so that both the ligand and the antibody have a net positive charge. In the pH range of 4.0–4.5, the ligand and most antibodies have a predominantly positive charge. Therefore, high-activity recovery of protein can be achieved under fairly mild conditions.

MEP HyperCel was used for the purification of monoclonal antibodies from protein-free cell culture supernatant (CCS) and clarified/concentrated milk of transgenic goats. Figure 7.20 shows a typical chromatogram of a monoclonal antibody (mouse IgG1) harvested from protein-free CCS. Before loading, the CCS containing

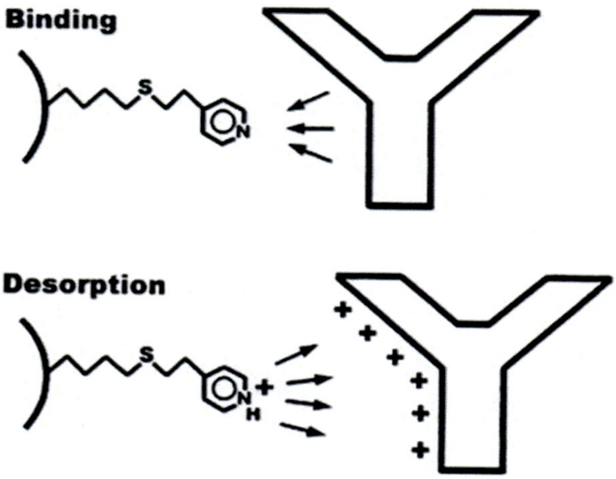

Fig. 7.19 Schematic diagram of the mechanism of hydrophobic charge induction chromatography and the key structural features of the ligand. The antibody selective ligand is derived from 4-mercaptoethylpyridine (4-MEP; pK_a 4.8). Hydrophobic binding under near-physiological conditions is illustrated. The desorption usually performed at pH 4 is based on electrostatic repulsion. (Reproduced from Schwartz et al. 2001)

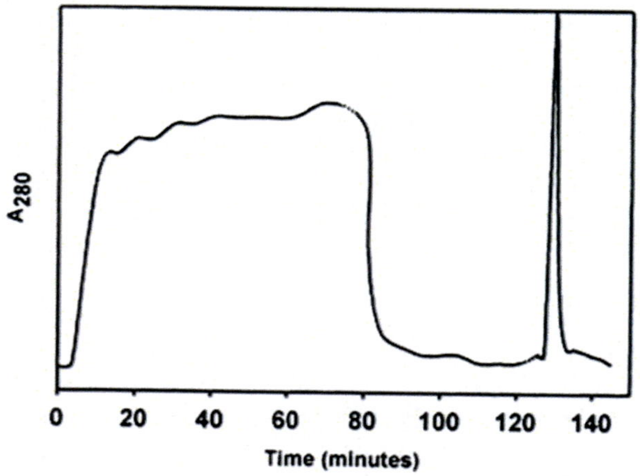

Fig. 7.20 Harvesting and purification of murine IgG1 from protein-free cell culture supernatant. A 9.0 × 2.5 cm MEP HyperCel column was used for chromatography, and the whole process was run at a speed of 150 cm/h. The detector sensitivity is 0.5 AUFS at 280 nm. The column was equilibrated with 50 mM Tris-HCl, pH 8.0. A 1000 mL of clarified cell culture supernatant (containing 200 μg IgG/mL) was applied to the column, and then the column was washed with equilibration buffer until the absorbance returned to the baseline. The IgG fraction was then eluted under the influence of 50 mM sodium acetate, pH 4.0. (Reproduced from Schwartz et al. 2001)

200 μg IgG/mL was clarified by microfiltration, but without adjustment of pH or ionic strength. After loading the sample, the column was washed with equilibration buffer (50 mM Tris-HCl buffer, pH 8.0) until the absorbance returned to the baseline. The IgG fraction was then eluted under the influence of 50 mM sodium acetate buffer, pH 4.0. The purity of the product is ≥95%, and the recovery is about 92%.

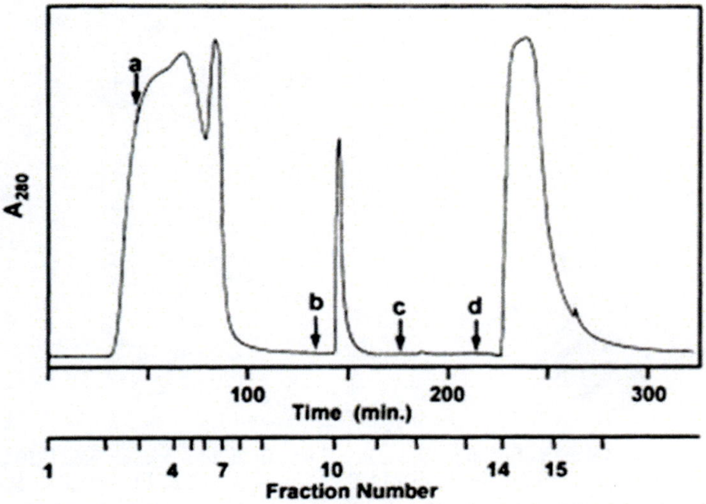

Fig. 7.21 Capture and purification of IgG1 from clarified concentrated transgenic milk. Chromatography was performed using a 8.0 × 1.1 cm column of MEP HyperCel. During the first 30 min, the column was run at 47 cm/h and thereafter at 147 cm/h. The detector sensitivity is 1 AUFS at 280 nm. The column was equilibrated with 50 mM Tris-HCl, pH 7.4, and then feed (15.2 mL, containing 154 mg IgG/mL) was applied to the column. The column was washed with the following solutions successively: (a) equilibration buffer; (b) purified water; (c) 25 mM sodium caprylate in equilibration buffer at pH 7.4. The IgG fraction was eluted under the influence of 50 mM sodium citrate (pH 4.0) (d). (Reproduced from Schwartz et al. 2001)

Transgenic goat milk contains humanized monoclonal IgG1. Figure 7.21 shows the chromatogram of the monoclonal antibody purified from the clarified/concentrated raw material. The chromatographic sequence includes two additional washing steps. After loading and washing with the equilibration buffer, the column was washed with pure water first and then washed with 25 mM sodium caprylate in the equilibration buffer. These washing steps can be used to promote the desorption of impurity components that are less hydrophobic than the target antibody. IgG binding does not depend on salt under a variety of conditions. However, reducing the ionic strength of the mobile phase from 50 mM buffer to pure water can be a useful technique to promote the desorption of less hydrophobic proteins. The mild detergent-like properties of sodium caprylate can also promote the desorption of this type of protein. Product purity was >95%.

Girot et al. (2004) reported the use of 2-mercapto-5-benzimidazole sulfonic acid (MBISA) (Fig. 7.22) as a mixed-mode chromatography ligand for the separation of antibodies. Compared with MEP, the sulfonic acid substituent on the nitrogen-containing heterocycle obviously has an important influence on the adsorption behavior of proteins. At pH 5.0–5.5, antibodies are adsorbed to the resin, while most other proteins will not interact with the resin, especially when the ionic strength is similar to physiological conditions. The elution of the adsorbed antibody is

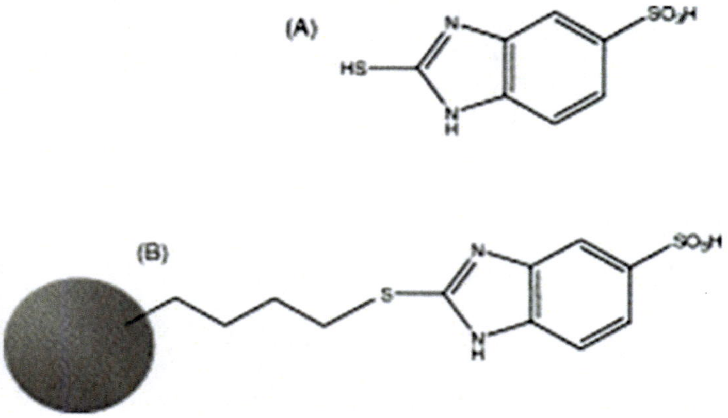

Fig. 7.22 (**a**) The structure of natural 2-mercapto-5-benzimidazole sulfonic acid (MBISA; **a**) and (**b**) the schematic diagram of the resin of MBISA covalently attached to cellulose microspheres (MBI HyperCel). (Reproduced from Girot et al. 2004)

performed by raising the pH of the solution to 9. Experiments showed that using the synthesized mixed-mode resin, immunoglobulins protein G (IgG) can be separated from crude biological fluids (such as ascites, cell culture supernatant, and Cohn fraction II + III from human plasma fractionation, where the amount of protein impurities is particularly high). The purity of one-step separation of antibodies is in the range of 85–98%, depending on the nature of the raw materials.

Figure 7.23 shows the chromatogram of the separation of monoclonal IgG1 from crude rat ascites fluid. The sample was first adjusted with 1 M acetic acid to lower the pH to 5.2 and then directly loaded onto the MBI HyperCel column that had been equilibrated with 50 mM acetate buffer containing 0.14 M sodium chloride pH 5.2. A large amount of impurity protein appeared in the flow-through fraction. After washing with equilibration buffer to further remove impurities, the adsorbed antibody was eluted with a carbonate solution of pH 9.0, and the purity of the obtained antibody is >85%.

Zhao et al. (2008) reported the preparation and evaluation of 5-aminoindole-bonded agarose gel (AI-Sepharose) (Fig. 7.24). In contrast to the abovementioned 2-mercapto-5-benzimidazole sulfonic acid, the synthesized new ligand introduces amino substituents on the nitrogen-containing heterocycle. The adsorption equilibrium and kinetics of lysozyme and bovine serum albumin (BSA) on AI-Sepharose were determined by batch adsorption experiments under different conditions. The effects of ligand density, salt type, and salt concentration on the adsorption of proteins on hydrophobic charge-inducing adsorbents are consistent with the trends observed in hydrophobic interaction chromatography. When the pH decreased from 7 to 3, the adsorption capacity of alkaline protein lysozyme decreased rapidly due to the increase of electrostatic repulsion, while the acidic protein BSA showed the largest adsorption capacity near its isoelectric point. By adsorbing at pH 7.10 and eluting at pH 3.0, protein recovery can be achieved under different NaCl

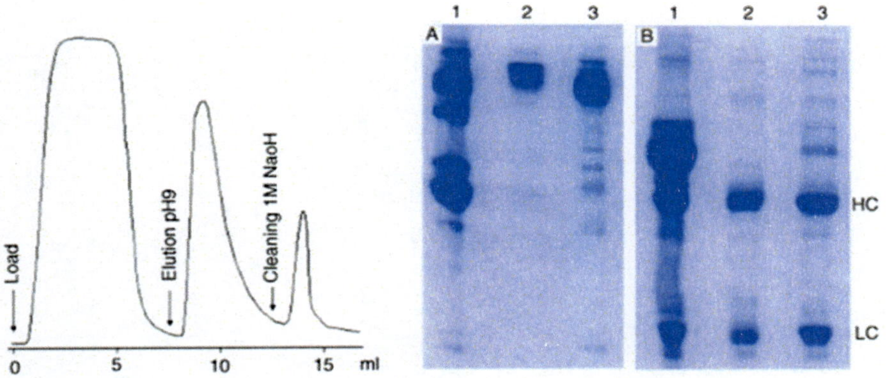

Fig. 7.23 The chromatogram of using MBI HyperCel column to separate monoclonal IgG1 from mouse ascites fluid. Column: 6 mm inner diameter × 50 mm (1.4 mL resin); equilibration buffer: 50 mM acetate, NaCl 0.14 M, pH 5.2. Sample: The concentration of IgG1 in mouse ascites is 3.5 mg/mL; injection volume: 6.6 mL; linear flow rate: 26 cm/h; antibody elution: 50 mM carbonate buffer, 0.14 M NaCl, pH 9.0. Insertion: SDS polyacrylamide gel electrophoresis was performed under non-reducing (A) and reducing (B) conditions. From left: crude sample (1); pure IgG standard (2); elution fraction (3). *LC* light chain, *HC* heavy chain. (Reproduced from Girot et al. 2004)

Fig. 7.24 The structure of 5-aminoindole-bonded agarose gel. (Reproduced from Zhao et al. 2008)

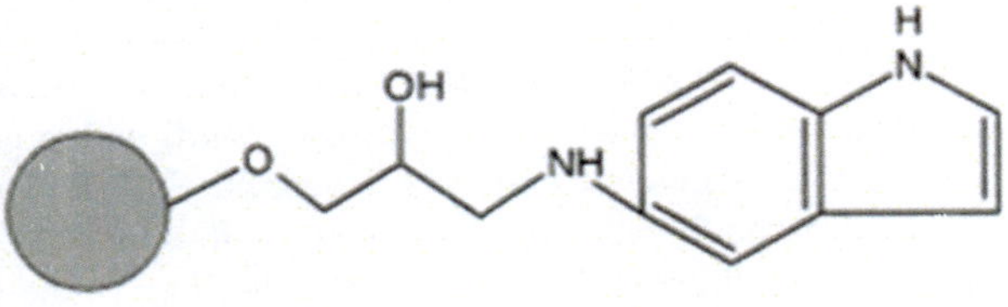

concentrations. The results show that AI-Sepharose has the typical characteristics of HCIC adsorbent and has potentials in protein purification.

Multimodal ion exchange stationary phases As the name implies, the process of multimodal ion exchange is to increase the type of interaction on the basis of hydrophobic ion exchange, typically hydrogen bonds and π-π stacking. Burton et al. (1997) used phenylalanine, phenylpropanolamine, and aminophenyl propanediol (Fig. 7.25) as ligands and epichlorohydrin-activated Sepharose to synthesize multimodal ion exchange chromatography stationary phases for one-step mixed-mode chromatographic purification of chymosin. These ligands all contain phenyl, amino, and alcohol groups and have the potential to form hydrophobic, ionic, hydrogen bonding, and π-π stacking with the target protein.

Johansson et al. (2003a, b) introduced the concept of multimodal ion exchange chromatography in the process of screening mixed-mode ligands for protein capture under high-salt conditions. They found that nonaromatic multimodal anion exchange ligands based on primary or secondary amines (or both) or aromatic multimodal cation exchange ligands based on carboxylic acids are most suitable for capturing

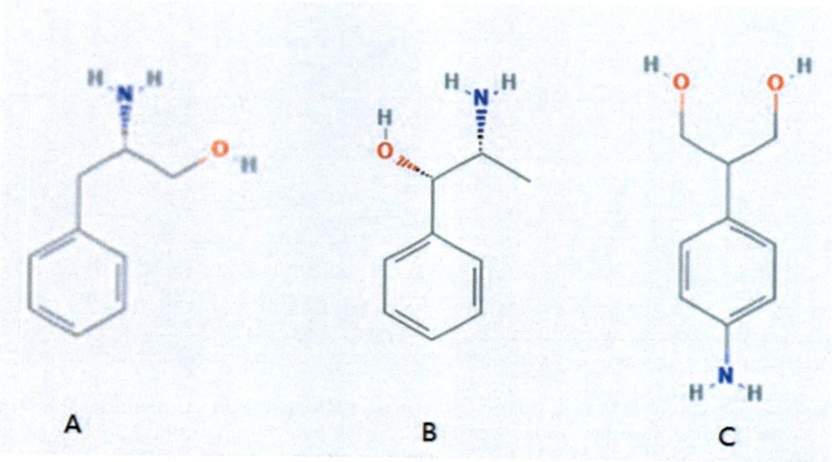

Fig. 7.25 Structural formulas of (a) phenylalaninol, (b) phenylpropanolamine, and (c) aminophenyl propanediol

proteins under high-salt conditions. In addition, the introduction of hydroxyl groups or amides near ionic groups can further enhance protein adsorption, because these new multimodal anion exchange ligands not only use hydrophobic and electrostatic interactions, but can also form hydrogen bonds with proteins. Compared with single-mode ion exchangers (such as Q Sepharose Fast Flow or SP Sepharose Fast Flow), the breakthrough capacity of the new multimodal ion exchanger when the mobile phase is 28 mS/cm and pH is 6.0 is 20–30 times of the former. The new multimodal ion exchanger can also be used under normal ion exchange conditions, and protein elution can be accomplished through the salt step or the pH step gradient.

7.6 Commercial Mixed-Mode Chromatography Stationary Phases

Bio-Rad (www.bio-rad.com) **CHT™** (n.d.) ceramic hydroxyapatite is a spherical, macroporous form of hydroxyapatite. Unlike natural or chemically synthesized crystalline hydroxyapatite, the material has been sintered at high temperatures to convert it from a crystalline form to a ceramic form. The ceramic material overcomes the limitations of traditional crystalline hydroxyapatite of low mechanical strength and fragility during use, and promotes its use in industrial-scale applications. The ceramic material retains the unique separation characteristics of crystalline hydroxyapatite but can be reused for many cycles at high-flow rates and large columns. The two types of CHT ceramic hydroxyapatite are type I and type II, with particle sizes of 20, 40, and 80 μm, respectively. The elution characteristics of both types are similar to crystalline hydroxyapatite, but there are some important differences. Type I CHT has higher protein binding capacity and better acidic protein absorption capacity. CHT type II has lower protein binding capacity but has better resolution for nucleic acids and certain proteins. Type II materials also have low affinity for albumin and are particularly suitable for purifying many types of immunoglobulins.

Thermo Fisher Scientific (www.thermofisher.com) **Acclaim** mixed-mode columns provide two functions (reversed phase + ion exchange) and three functions (reversed phase or HILIC + ion exchange). The biggest advantage of using a mixed-mode column is the optimization of selectivity, simply by adjusting the ionic strength, pH, and/or organic solvent concentration of the mobile phase. With adjustable selectivity, it is also possible to separate analytes with significantly different hydrophobicity and charge states, for example, simultaneous separation of active pharmaceutical ingredients (API) and counterions in a single analysis.

The Acclaim column series includes three bimodal mixed-mode columns and two trimodal columns. A dual-modal column has both hydrophobic arms (providing reversed-phase retention) and ion exchange or glycol end-groups (providing ion exchange or HILIC retention).

Thermo Fisher Scientific Acclaim Trinity P1 and P2 columns are based on Thermo Fisher Scientific Nano Polymer Silica Hybrid (NSH) technology. These are high-purity porous spherical silica particles coated with charged nano-polymer particles. The inner pore is modified with covalently bonded bimodal functional groups, while the outer surface is modified with nanopolymers with different ion exchange functional groups. The spatial separation of the two ion exchange zones allows the two retention mechanisms to operate simultaneously and independently controlled.

The inner bore of the Acclaim Trinity P1 column has dual-mode reverse phase and anion exchange functions. The outer surface is modified with cation exchange nano polymer beads.

The inner pore area of Acclaim Trinity P2 is composed of a bimodal hydrophilic surface and cation exchange retention, while the outer surface is modified with anion exchange nanopolymer beads.

SIELC Technologies (**www.sielc.com**) **Primesep** HPLC mixed-mode columns provide several improved forms of stationary phases, including reversed-phase cation exchange mode (Primesep A, Primesepep 100, Primesep 200) and reversed-phase anion exchange mode (Primesep B). In addition, a chromatographic column with unique properties is also provided, which can switch the retention time of neutral and ionizable compounds (Primesep SWITCH) by changing the pH value of the mobile phase. The Primesep C column forms a stable complex with amines. The stability of the complexes increases in the order from the tertiary amine to the secondary amine and the primary amine, and the pK_a value of the amine generally decreases in the same order. It was observed that the elution order of substituted amines on the Primesep C column was opposite to the elution order on the ion exchange chromatography column.

A chromatographic column based on SWITCH Phase technology will change its performance according to the pH of the mobile phase. The embedded carboxylic acid is completely ionized at a pH higher than its pK_a and loses charge when the pH of the mobile phase is lower than its pK_a. Therefore, by controlling the pH of the mobile phase, the polarity of the stationary phase can be changed to adjust the elution order of the analytes.

SIELC Technologies (https://www.sielc.com/) offers the Obelisc Series of mixed-mode stationary phases composed of zwitterions. Obelisc R with reverse phase characteristics and Obelisc N with normal phase characteristics are different in the type and proximity of their charged groups and the hydrophobicity of their long chains. In Obelisc R, the cationic and anionic groups of the ligand are separated by a hydrophobic chain, the cationic group is embedded in the hydrophobic chain close to the silica gel, and the anionic group is located at the far end of the hydrophobic chain. In contrast, in Obelisc N, the anionic and cationic groups of the ligand are separated by a hydrophilic chain. The anionic group is embedded near the hydrophilic chain of the silica gel while the cationic group is located at the far end of the hydrophilic chain.

Obelisc R has reverse phase characteristics and can be used in traditional reverse phase applications. Due to the presence of ionic groups and long hydrophobic chains, Obelisc R can provide retention and adjustment functions that are different from traditional reversed-phase chromatography stationary phases.

Obelisc N is extremely polar and can be used to separate polar and charged compounds. In ion exchange mode, charged analytes interact with oppositely charged groups on the stationary phase. Obelisc N provides positively and negatively charged groups to interact with positively or negatively charged analytes. In the traditional HILIC mode, charged or neutral polar analytes interact with hydrophilic groups on the surface of the polar stationary phase. Compared with traditional HILIC silica gel columns, on Obelisc N, the amphoteric charges in the ligands are greatly separated and can act independently, resulting in different selectivities.

Imtakt USA (www.imtaktusa.com) provides three multimode reversed-phase stationary phases, Scherzo SS-C18, Scherzo SM-C18, and Scherzo SW-C18. Scherzo SS-C18 consists of ionic ligands bonded to octadecylsilane-bonded silica in a high proportion to enhance the retention of zwitterions or weak ionic compounds. Scherzo SM-C18 binds weak ionic ligands to octadecylsilane-bonded silica in a moderate proportion, aiming at the separation of basic or acidic compounds under neutral pH conditions. Scherzo SW-C18 is composed of strong ionic ligands bonded to octadecylsilane-bonded silica at a low ratio, so that formic acid can be used as a mobile phase for elution of relatively strong ionic or basic compounds. Since the Scherzo family of multimodal reversed-phase chromatography columns contain octadecyl chains and ion exchange groups, compounds can be separated by various mechanisms such as reversed phase, ion exchange, and hydrophilic interaction chromatography. It has unique advantages in the analysis of small molecule drugs and their metabolites.

Daicel Chiral Technologies (chiraltech.com) provides chiral reversed-phase anion exchange stationary phases CHIRALPAK® QN-AX and CHIRALPAK® QD-AX and chiral zwitterionic ion exchange stationary phases CHIRALPAK® ZWIX(+) and CHIRALPAK® ZWIX(−). CHIRALPAK®QN-AX and CHIRALPAK®QD-AX are enantioselective weak anion exchange (AX) HPLC columns based on quinine (QN) and quinidine (QD) derivatives, respectively, which are most suitable for the separation of acidic chiral compounds. It has excellent enantiomeric separation ability for acidic chiral compounds with carboxyl, phosphonic, phosphinic, phosphoric, or sulfonic acid groups. In some cases, it is also possible to separate weakly acidic compounds, such as phenol and coumestrol. The elution order of the enantiomers is opposite to each other on the two phases. They are compatible with all common HPLC solvents (such as methanol, acetonitrile, tetrahydrofuran, dioxane or chloroform) and are compatible with a wide pH range of 2–8. Typical buffers used in the water-organic solvent mode are acetate, formate, citrate, and phosphate.

CHIRALPAK®ZWIX (+) and CHIRALPAK®ZWIX (−) are zwitterionic chiral stationary phases, which are mainly used for the chiral separation of free amino acids. They show remarkable stereoselectivity for zwitterionic molecules (especially amino acids and peptides) without the need for derivatization. They are compatible with all common HPLC solvents (e.g., methanol, acetonitrile, tetrahydrofuran, water).

Bio-Rad (www.bio-rad.com) Nuvia cPrime is a hydrophobic cation exchange stationary phase specially designed for protein purification. It is prepared by bonding a multifunctional ligand to highly cross-linked, macroporous polymer microspheres with an average particle size of 70 μm (Fig. 7.26). The ligand contains phenylamino, carboxyl, and amide groups, which can bind to proteins through hydrophobic, hydrogen, and ionic bonds. The most significant advantage of the multimodal ion exchange stationary phase is that it can effectively capture and recover the target protein in a wide range of pH and salt concentration. Many cell culture fluids can be directly loaded onto the Nuvia cPrime column for separation and purification without dilution.

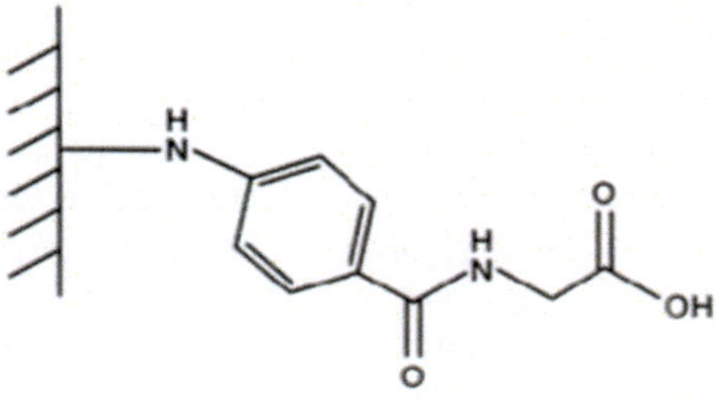

Fig. 7.26 Nuvia cPrime ligand structure

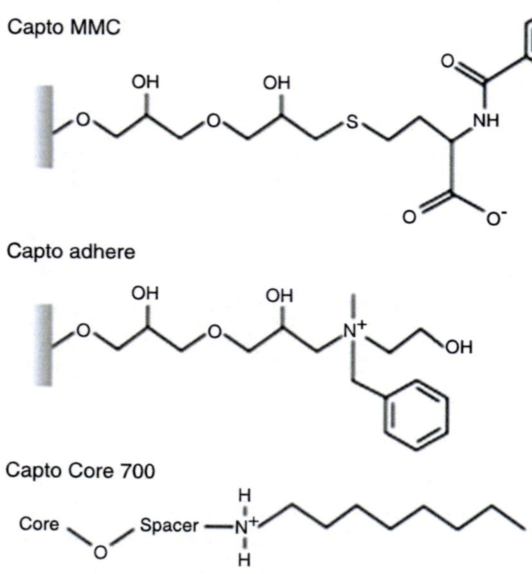

Fig. 7.27 Capto MMC, Capto adhere, and Capto Core are shown as examples. Note that the Capto Core 700 ligand is found only in the core of the bead

Cytiva (www.cytiva.com) currently offers multimodal chromatography stationary phases including Capto MMC, Capto MMC ImpRes, Capto adhere, Capto adhere ImpRes, and Capto Core 700 (Fig. 7.27).

The first multimodal matrix introduced by Cytiva was Capto MMC and Capto adhere soon after. Like all Capto matrices, the selected ligands are coupled to a high-flow agarose base matrix with excellent pressure-flow properties. The highly cross-linked agarose base matrix gives the media high chemical and physical stability. The high-flow rate increases the productivity of large-scale biological processing operations and allows large amounts of organisms to be processed in one work shift. Multimodal ligands in Capto adhere and Capto MMC media have been developed to provide new selectivity different from traditional anion exchangers and cation exchangers. Now, coupling multimodal ligands to Capto ImpRes high-resolution base matrix provides a new generation of multimodal stationary phases Capto MMC ImpRes and Capto adhere ImpRes.

Cytiva also introduced Capto Core 700, which combines size exclusion and multimodal anion exchange characteristics. The Capto core concept is based on microspheres with a nonfunctional outer layer (no ligand) and a functional core with coupled ligands. This design combines the characteristics of exclusion and adsorption chromatography. The exclusion limit of the outer layer pores is about 700 kDa, which can prevent macromolecular entities such as viruses, DNA, large proteins, or protein complexes from entering the internal space, but impurities with relatively small molecular weights can enter the pores and interact with ligands. In this way, an efficient flow-through purification step can be achieved, which is usually used for virus purification in vaccine production.

Table 7.3 Characteristics, mechanisms, and applications of multimodal chromatography media from Cytiva

Medium	Structure	Mechanism	Application
Capto adhere	Ligand: *N*-benzyl methyl ethanolamine Base matrix: Capto	Electrostatic interaction, hydrogen bonding, and hydrophobic interaction	Intermediate purification and polishing of MAbs after capture on protein A. Traditionally used in flowthrough mode Purification of other target proteins from capture to polishing steps
Capto adhere ImpRes	Ligand: *N*-benzyl methyl ethanolamine Base matrix: Capto ImpRes	Same as Capto adhere	Efficient MAb polishing, removal of aggregates and HCP, and separations of charge variants. Polishing resulting in smaller elution volumes The properties of the small Capto ImpRes particle are best utilized in bind/elute mode
Capto MMC	Ligand: *N*-benzoyl-homocysteine Base matrix: Capto	Thiophilic interaction, hydrophobic interaction, hydrogen bonding, and electrostatic interaction	Capture and intermediate purification of proteins from large feed volumes by packed bed chromatography Purification can be performed at the conductivity of the feed material
Capto MMC ImpRes	Ligand: *N*-benzoyl-homocysteine Base matrix: Capto ImpRes	Same as Capto MMC	Efficient MAb polishing, removal of aggregates and HCP, and separations of charge variants. Polishing resulting in smaller elution volumes The properties of the small Capto ImpRes particle are best utilized in bind/elute mode
Capto Core 700	Ligand: octylamine Base matrix: high-flow agarose	GF, AIEX, and HIC Core bead technology with ligand-activated core and nonfunctionalized shell allows efficient capture of contaminants while target molecules are collected in flowthrough	Purification of viruses and other large target molecules

Table 7.3 provides an overview of the multimodal chromatography products provided by Cytiva, adapted from Cytiva MMC handbook.

Sartorius (www.sartorius.com) provides mixed-mode stationary phases and columns such as MEP Hypercel, CMM HyperCel, PPA Hypercel, and HEA HyperCel (Table 7.4). They were composed of 4-mercaptoethylpyridine (4-MEP), aminobenzoic acid (CMM), phenylpropylamine (PPA), and hexamine (HEA) coupled to HyperCel cellulose microspheres with an average particle size of 90 μm. They can achieve excellent flow performance under low column back

Table 7.4 Characteristics, mechanisms, and uses of MEP, CMM, PPA, HEA 和 HyperCel

Medium	Structure	Mechanism	Application
MEP HyperCel	Ligand: 4-mercaptoethylpyridine Base matrix: Cellulose	Thiophilic interaction, hydrophobic interaction, hydrogen bonding, charge repulsion	Capture and intermediate purification of proteins from large feed volumes by packed bed chromatography. Purification can be performed at the conductivity of the feed material
CMM HyperCel	Ligand: Aminobenzoic acid Base matrix: Cellulose	Hydrophobic interaction, electrostatic interaction, hydrogen bonding	Capture, intermediate purification and polishing
PPA HyperCel	Ligand: Phenylpropylamine Base matrix: Cellulose	Hydrophobic interaction, electrostatic interaction, hydrogen bonding, π-π interaction	Capture, intermediate purification and polishing
HEA HyperCel	Ligand: Hexamine Base matrix: Cellulose	Hydrophobic interaction, electrostatic interaction, hydrogen bonding,	Capture, intermediate purification and polishing

Table 7.5 Characteristics, mechanisms, and uses of Eshmuno HCX and CMX

Medium	Structure	Mechanisms	Applications
Eshmuno HCX	Ligands: Sulfo, carboxy, and phenyl groups Base matrix: 85 μm Polyvinylether polymer microspheres	Hydrophobic interaction, electrostatic interaction, hydrogen bonding	Monoclonal antibody and fusion protein purification; ADC polishing; removal of low molecular weight impurities and HCPs
Eshmuno CMX	Ligands: Carboxy and alkyl-functional groups Base matrix: 50 μm Polyvinylether polymer microspheres	Hydrophobic interaction, electrostatic interaction, hydrogen bonding	Monoclonal antibody and fusion protein purification; ADC polishing; removal of low molecular weight impurities and HCPs

pressure and are compatible with large-scale production. MEP, CMM, HEA, and PPA HyperCel resins are chromatographic resins used in industrial processes. MEP Hypercel operates in hydrophobic charge induction chromatography mode, while CMM, HEA, and PPA HyperCel operate in hydrophobic ion exchange mode. MEP, CMM, HEA, and PPA HyperCel resins have unique selectivity and are combined with conventional ion exchange or hydrophobic interaction chromatography for protein capture and impurity removal in biopharmaceutical environments.

MilliporeSigma (www.milliporesigma.com) provides Eshmuno HCX, CMX, and other multimodal ion exchange stationary phases (Table 7.5). The multimodal cation exchanger Eshmuno®HCX medium is specifically designed for direct capture of recombinant proteins at higher salt concentrations. Coupling ligand tentacles containing sulfonic, carboxylic acid groups, and phenyl groups in hydrophilic polyvinyl ether-based microspheres to form a three-dimensional hydrophobic ion exchange network, making the ligands easy to contact with proteins. It can achieve rapid flow in biopharmaceutical operations and achieve higher binding capacity through multivalent interactions with biopharmaceuticals. The resin can be used in binding/elution (capture) or flow-through (polishing) mode. The unique functional properties of this resin allow it to capture monoclonal antibodies from high-salt, neutral pH solutions (such as bioreactor harvest materials).

Eshmuno® CMX chromatography resin is a member of the Eshmuno® high-performance chromatography resin family, designed to meet the needs of high-yield downstream purification processes. Eshmuno®CMX resin combines surface tentacle structures with carboxyl and alkyl chains on 50 μm polymer microspheres and forms multivalent interactions with biological macromolecules, which can bind target substances more effectively. In this way, higher binding capacity and shorter processing time can be achieved. It can be used to remove retrovirus and parvovirus from the clarified protein A capture monoclonal antibody feed stream.

Tosoh Bioscience (www.tosohbioscience.com) offers TOYOPEARL MX-Trp-650 M (Fig. 7.28), which is a multimodal cation exchange resin with unique selectivity and high recovery. It provides high protein binding capacity and can tolerate high-conductivity raw materials. In addition to ionic groups, the ligand also has a hydrophobic region. Therefore, the binding of target molecules is determined by electrostatic and hydrophobic interactions. TOYOPEARL MX-Trp-650M is especially suitable for the purification of target molecules that are difficult to purify using conventional purification platforms. TOYOPEARL MX-Trp-650M is a highly rigid methacrylic polymer microsphere with an average particle size of 75 μm coupled with tryptophan as the active ligand (Fig. 7.28). This amino acid has weak carboxyl cation exchange and indole hydrophobic functional groups. The selectivity of the resin can be adjusted by controlling the binding or elution pH, ionic strength, salt type, and additives.

The application of TOYOPEARL MX-Trp-650M includes mAb polishing, because the resin can separate closely related protein species (such as aggregates and monomers). Antibodies captured by protein A affinity chromatography may be

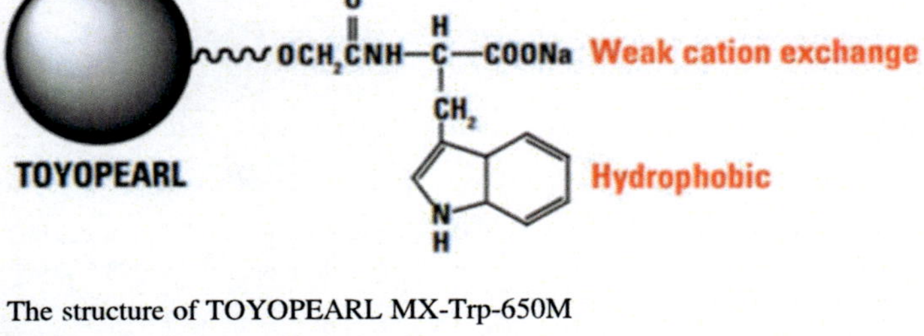

Fig. 7.28 The structure of TOYOPEARL MX-Trp-650M

loaded at pH 4.5 and eluted as the salt concentration and pH increase. Using a total salt concentration of 0.5 mol/L and pH 6 can successfully elute various IgG monomers and their well-separated aggregates and obtain monomers with a purity of up to 98.5%.

7.7 Summary

The mixed-mode chromatographic stationary phase contains at least two functional groups with different properties, which can interact with the complementary groups of the solute at the same time. The design of mixed-mode ligands can roughly follow four principles, namely, multivalent interaction, functional complementarity, geometric fit, and reciprocity between ligand and solute. Multivalent effect is to increase the strength and selectivity of binding by increasing the number of interaction sites between the ligand and the solute. The nature of the functional group of the ligand must be complementary to the nature of the functional group of the solute in order to maximize the interaction. The interaction between the ligand and the solute should match each other geometrically, and there is no steric hindrance; otherwise, the binding affinity will be affected. Finally, the ligand and the solute can replace each other without having too much influence on the strength and selectivity of the binding. According to the type of functional group in the ligand, the mixed-mode stationary phase can be divided into binary or multiple ligands. The binary ligand combines two different functional groups, which can be understood as a chimeric ligand that combines the two different single-mode ligands, for example, hydrophobic and ionic groups, hydrophobic and coordination groups, hydrophobic and inclusion groups, ions and coordination groups, ions and inclusion groups, ions and π-groups, etc. Multivariate ligands can be understood as further hybridization on the basis of binary ligands to form multifunctional multisite ligands. According to the matrix of the mixed-mode stationary phases, they can be divided into three categories: silica-based, polymer-based, and polysaccharide-based mixed-mode stationary phases. Among them, silica-based and polysaccharide-based stationary phase are the most common. The former is mainly used for the separation of small molecules, oligonucleotides, and peptides in high-performance liquid chromatography, while the latter is mainly used for the preparation, separation, and purification of nucleic acids, proteins, and viruses. Both types of stationary phases can be obtained from commercial sources, which facilitates the development of chromatographic methods.

References

Abbood A, Smadja C, Herrenknecht C et al (2009) Retention mechanism of peptides on a stationary phase embedded with a quaternary ammonium group: a liquid chromatography study. J Chromatogr A 1216:3244–3251

Bio-Rad CHT™ ceramic hydroxyapatite instruction manual (n.d.) Bio-Rad Laboratories, Inc. https://www.bio-rad.com/webroot/web/pdf/lsr/literature/LIT611E.PDF

Burton SC, Harding DRK (1997) High-density ligand attachment to brominated allyl matrices and application to mixed mode chromatography of chymosin. J Chromatogr A 775:39–50

Burton SC, Harding DRK (1998) Hydrophobic charge induction chromatography: salt independent protein adsorption and facile elution with aqueous buffers. J Chromatogr A 814:71–81

Burton SC, Haggarty NW, Harding DRK (1997) One step purification of chymosin by mixed mode chromatography. Biotechnol Bioeng 56:45–55

Crowther JB, Hartwick RA (1982) Chemically bonded multifunctional stationary phases for high-performance liquid chromatography. Chromatographia 16:349–353

Davankov VA, Navratil JD, Walton HF (1988) Ligand exchange chromatography. CRC press, Boca Raton, FL

Dotsevi G, Sogah Y, Cram DJ (1975) Total chromatographic optical resolutions of α-amino acid and ester salts through chiral recognition by a host covalently bound to polystyrene resin. J Am Chem Soc 97:1259–1261

Er-el Z, Zaidenzaig Y, Shaltiel S (1972) Hydrocarbon-coated Sepharoses. Use in the purification of glycogen phosphorylase. Biochem Biophys Res Commun 49:383–390

Girot P, Averty E, Flayeux I et al (2004) 2-Mercapto-5-benzimidazolesulfonic acid: an effective multimodal ligand for the separation of antibodies. J Chromatogr B 808:25–33

Hoffmann CV, Laemmerhofer M, Lindner W (2007) Novel strong cation-exchange type chiral stationary phase for the enantiomer separation of chiral amines by high-performance liquid chromatography. J Chromatogr A 1161:242–251

Hoffmann CV, Pell R, Lämmerhofer M et al (2008) Synergistic effects on enantioselectivity of zwitterionic chiral stationary phases for separations of chiral acids, bases, and amino acids by HPLC. Anal Chem 80:8780–8789

Hofstee BHJ (1973a) Hydrophobic affinity chromatography of proteins. Anal Biochem 52:430–448

Hofstee BHJ (1973b) Protein binding by agarose carrying hydrophobic groups in conjunction with charges. Biochem Biophys Res Commun 50:751–757

Hofstee BHJ (1974) Non-specific binding of proteins by substituted agaroses. In: Dunlap RB (ed) Immobilised biochemicals and affinity chromatography. Plenum Press, New York, pp 43–59

Johansson BL, Belew M, Eriksson S et al (2003a) Preparation and characterization of prototypes for multi-modal separation media aimed for capture of negatively charged biomolecules at high salt conditions. J Chromatogr A 1016:21–33

Johansson BL, Belew M, Eriksson S et al (2003b) Preparation and characterization of prototypes for multi-modal separation aimed for capture of positively charged biomolecules at high-salt conditions. J Chromatogr A 1016:35–49

Kasche V, Löffler F, Scholzen T et al (1990) Rapid protein purification using phenylbutylamine—Eupergit: a novel method for large-scale procedures. J Chromatogr A 510:149–154

Lämmerhofer M, Lindner W (1996) Quinine and quinidine derivatives as chiral selectors I. Brush type chiral stationary phases for high-performance liquid chromatography based on cinchonan carbamates and their application as chiral anion exchangers. J Chromatogr A 741:33–48

Liang-Schenkelberg J, Fieg G, Waluga T (2017) Molecular insight into affinity interaction between Cibacron Blue and proteins. Ind Eng Chem Res 56:9691–9697

McLaughlin LW (1989) Mixed-mode chromatography of nucleic acids. Chem Rev 89:309–319

Muenter MM, Stokes KC, Obie RT et al (1999) Simultaneous separation of inorganic ions and neutral organics on ion-exchange stationary phases. J Chromatogr A 844:39–51

Nogueira R, Lämmerhofer M, Lindner W (2005) Alternative high-performance liquid chromatographic peptide separation and purification concept using a new mixed-mode reversed-phase/weak anion-exchange type stationary phase. J Chromatogr A 1089:158–169

Pirkle WH, Pochapsky TC (1989) Considerations of chiral recognition relevant to the liquid chromatography separation of enantiomers. Chem Rev 89:347–362

Qiao L, Wang S, Li H et al (2014) A novel surface-confined glucaminium-based ionic liquid stationary phase for hydrophilic interaction/anion-exchange mixed-mode chromatography. J Chromatogr A 1360:240–247

Sasaki I, Gotoh H, Yamamoto R et al (1979) Hydrophobic-ionic chromatography. Its application to purification of porcine pancreas enzymes. J Biochem 86:1537–1548
Sasaki I, Gotoh H, Yamamoto R et al (1982) Hydrophobic-ionic chromatography: its application to microbial glucose oxidase, hyaluronidase, cholesterol oxidase, and cholesterol esterase. J Biochem 91:1555–1561
Schwartz W, Judd D, Wysocki M et al (2001) Comparison of hydrophobic charge induction chromatography with affinity chromatography on protein A for harvest and purification of antibodies. J Chromatogr A 908:251–263
Shaltiel S, Er-El Z (1973) Hydrophobic chromatography: use for purification of glycogen synthetase. Proc Natl Acad Sci U S A 70:778–781
Shen A, Li X, Dong X et al (2013) Glutathione-based zwitterionic stationary phase for hydrophilic interaction/cation-exchange mixed-mode chromatography. J Chromatogr A 1314:63–69
Simmonds RJ, Yon RJ (1976) Protein chromatography on adsorbents with hydrophobic and ionic groups. Chromatography of dodecyl sulphate-solubilized proteins of the human erythrocyte membrane on N-(3-carboxypropionyl)aminodecyl-sepharose. Biochem J 157:153–159
Wan Q-H, Wang X, Chen L (2019) Inverse mixed-mode chromatography for the evaluation of multivalency and cooperativity of host−guest complexation in porous materials. Langmuir 35: 10405–10411
Wei J, Guo Z, Zhang P et al (2012) A new reversed-phase/strong anion-exchange mixed-mode stationary phase based on polar-copolymerized approach and its application in the enrichment of aristolochic acids. J Chromatogr A 1246:129–136
Wongyai S (1994) Synthesis and characterization of phenylpropanolamine bonded silica for multimode liquid chromatography of small molecules. Chromatographia 38:485–490
Yon RJ (1972) Chromatography of lipophilic proteins on adsorbents containing mixed hydrophobic and ionic groups. Biochem J 126:765–767
Yon RJ, Simmonds RJ (1975) Protein chromatography on adsorbents with hydrophobic and ionic groups. Some properties of N-(3-carboxypropionyl)aminodecyl-sepharose and its interaction with wheat-germ aspartate transcarbamoylase. Biochem J 151:281–290
Zhang S, Wan Q-H, Li Y (2019) Epoxide-derived mixed-mode chromatographic stationary phases for separation of active substances in fixed-dose combination drugs. J Sep Sci 42:2796–2804
Zhao G, Peng G, Li F et al (2008) 5-Aminoindole, a new ligand for hydrophobic charge induction chromatography. J Chromatogr A 1211:90–98
Zhu B-H, Mant CT, Hodges RS (1992) Mixed-mode hydrophilic and ionic interaction chromatography rivals reversed-phase liquid chromatography for the separation of peptides. J Chromatogr A 594:75–86

Chapter 8
Mobile Phases and Elution

Abstract The mobile phase plays an important role in controlling solute retention and elution in mixed-mode chromatography because it participates in the interaction between solute molecules and ligands on the stationary phase. In order to achieve a good separation, we must carefully select and optimize the mobile phase parameters, including pH, ionic strength, organic solvent percentage, additives, and elution mode. In this chapter, we will review the characteristics of mobile phase composition and different elution methods and focus on displacement elution.

8.1 Introduction

In mixed-mode chromatography, the mobile phase composition competes with solute molecules for active sites on the stationary phase and participates in the interaction between solute molecules and stationary phase ligands, so it plays a very important role in solute retention and elution. In fact, considerable efforts have been spent on the selection of mobile phase parameters in the process of method development for mixed-mode chromatography. Mobile phase parameters include pH, ionic strength, organic solvent content, additives, and elution mode. The pH value of the mobile phase not only affects the dissociation state of the solute molecules, thereby affecting the charge characteristics and affinity/hydrophobicity of the solute molecules, but also affects the dissociation state of the stationary phase ligands and related binding and elution behaviors. The ionic strength of the mobile phase can inhibit the ionic interaction between the solute molecules and the stationary phase and enhance the hydrophobic interaction between them, but it has little effect on the coordination of metal ions. Similarly, the organic solvent in the mobile phase has opposite roles in the reverse phase and hydrophilic interaction chromatography, which can inhibit the reverse phase binding capacity of the stationary phase and increase its hydrophilic binding capacity. The additives in the mobile phase can selectively affect the properties of the solute or the ligand and can play a role similar to the specific elution in affinity chromatography. Therefore, according to the characteristics of a specific separation system, suitable mobile phase parameters can be selected to adjust and optimize the separation selectivity in mixed-mode

Q.-H. Wan, *Mixed-Mode Chromatography*,
https://doi.org/10.1007/978-981-16-5485-5_8

chromatography. In this chapter, we will first review the general properties of solvents commonly used in mixed-mode chromatography and then buffers and salts. We will give a detail review on the role of mobile phase additives in elution of proteins.

8.2 The Composition of the Mobile Phase

8.2.1 Polar Organic Solvents

In mixed-mode chromatography involving reversed-phase and hydrophilic interaction mechanisms, the mobile phase used is usually composed of a polar organic solvent plus water or buffer. In the reverse phase retention mechanism, water is the strongest binding solvent and the weakest elution solvent. Organic solvents are the weakest binding solvents and the strongest elution solvents. Therefore, a mixed solution of water and organic solvent is usually used as the mobile phase to control the retention and elution of solutes (Moldoveanu and David 2013, 2017). In the hydrophilic retention mechanism, contrary to the reversed-phase mode, organic solvents, especially aprotic organic solvents such as acetonitrile, are the strongest binding solvent, the weakest elution solvent, while water is the weakest binding solvent and the strongest elution solvent. Generally, a high content of organic solvent is used as the mobile phase to promote the retention of the hydrophilic solute.

Depending on the type of mixed-mode chromatography, the specific physical and chemical properties of the solvent used as the mobile phase can have an important impact on the efficiency and separation selectivity of the chromatographic column. These properties include (1) solvent viscosity, (2) dielectric constant, (3) dipole moment, and (4) surface tension. These parameters may not be the key factors in mobile phase selection. However, a thorough understanding of the meaning of these parameters will help us better adjust the choice of mobile phase. The properties of commonly used solvents are shown in Table 8.1.

Solvent viscosity The dynamic viscosity of a solvent is a measure of the resistance of a fluid to shear flow. The ratio between dynamic viscosity and fluid density is called kinematic viscosity. The solvent dynamic viscosity is an important parameter of the mobile phase, which affects the separation in various ways. One effect is related to the back pressure of the chromatographic column. The increase in the viscosity of the mobile phase leads to an increase in back pressure. Secondly, mobile phase viscosity is related to column efficiency. A solvent with a higher viscosity will reduce the diffusion coefficient of the solutes and slow down the mass transfer process, resulting in a decrease in column efficiency.

Since solvent mixtures are often used as mobile phases, the change in viscosity with solvent composition is particularly important, especially when separation is carried out under gradient conditions, the viscosity will change during

Table 8.1 Parameters of polar organic solvents commonly used in mixed-mode chromatography

Solvent	Molecular weight	Bp (°C)	Density (g/mL)	Viscosity (mPa s)	Dielectric constant	Dipole moment (D)	Surface tension (10^{-5} N/cm)	UV cutoff (nm)	Reflex index
Water	18	100	0.998	1.00	78.5	1.84	73	170	1.333
Methanol	32	65	0.792	0.58	32.7	1.66	22	205	1.326
Ethanol	46	78	0.789	1.19	24.5	1.68	22	205	1.359
Acetonitrile	41	82	0.787	0.36	38.8	3.27	29	190	1.342
Isopropanol	60	82	0.785	2.39	19.9	1.68	21	205	1.375
Tetrahydrofuran	72	66	0.889	0.51	7.58	1.70	28	210	1.404
Dioxane	88	101	1.034	1.26	2.21	0.45	33	215	1.420

chromatographic operation. The viscosity can vary greatly with the composition of the mobile phase.

Surface tension In reversed-phase chromatography, the importance of solvent surface tension for solute retention has been established. According to a retention model based on the solvophobic theory, the retention of a solute, k, can be related to surface tension, γ:

$$\log k = a + b \tag{8.1}$$

where a and b are parameters, which depend on multiple properties of the separation system. Equation (8.1) shows that the higher surface tension of the mobile phase leads to stronger solute retention.

Dielectric constant and dipole moment The dielectric constant, ε, and dipole moment, μ, of the solvent are important physical constants that characterize the polar nature of the solvent. The dielectric constant is the ratio of the dielectric constant of a substance to the dielectric constant of a vacuum (the dielectric constant is a measure of the resistance encountered when an electric field is formed in a medium). The dielectric constant is dimensionless (the vacuum dielectric constant is $\varepsilon_0 = 8.854 \times 10^{-12}$ C/V m). The dielectric constant of the solvent affects the interaction between ions and polar molecules in the solution. When the dielectric constant increases, the energy of the electrostatic interaction between molecules will decrease. Polar compounds are generally more soluble in solvents with higher dielectric constants. Moreover, it will affect the solvophobic interaction. In reversed-phase chromatography, solvents with higher ε are usually weaker eluents.

The dipole moment of a molecule is a parameter related to the polarity of the molecule, which is defined as the product of the distance between the positive and negative charge centers and the amount of charge in the charge center. It is a vector whose direction is specified from the positive center to the negative center. The dipole moment of solvent molecules can be used to characterize solvent polarity, which is expected to have significant effects on the molecular interactions involving polarity of the solute such as in hydrophilic interaction chromatography.

Hydrogen bonding Hydrogen bonding is based on the electrostatic attraction between a hydrogen atom bonded to a highly electronegative atom and another nearby electronegative atom. Polar protic solvents are both hydrogen bond donors and acceptors, because they contain at least one hydrogen atom directly connected to the electronegative atom (such as OH or NH bond) and electronegative atom containing a lone pair of electrons. They can effectively dissolve cations and anions. Typical polar protic solvents are water, ethanol, methanol, ammonia, and acetic acid. Polar aprotic solvents do not contain hydrogen atoms directly connected to electronegative atoms, so they have no hydrogen bond donor capacity and cannot form hydrogen bonds with negatively charged ions. Representatives of this type of solvents include acetonitrile, acetone, dimethyl sulfoxide, DMF (*N*,*N*-dimethylformamide), HMF (hydroxymethyl furfural), and crown ether.

Boiling point Chromatographic separation performed at room temperature or below usually does not need to consider the solvent boiling point. However, it is not uncommon to perform the mixed-mode separation at higher than the ambient temperature. This is especially the case for the separation of peptides and oligonucleotides as the high temperature offers several advantages including increased separation speed, reduced mobile phase viscosity, increased efficiency, and selectivity changes with temperature. High boiling point solvents are not conducive to some detection methods that require evaporation of solvents, such as mass spectrometry and evaporative light scattering detection.

Ultraviolet cutoff wavelength The ultraviolet spectroscopic detector based on the measurement of the ultraviolet absorption of the solute is perhaps the most commonly used detection technique in mixed-mode chromatography. This technique has been successfully used to measure analytes containing unsaturated bonds and aromatic groups and functional groups containing heteroatoms. To reduce interference with the measurement, the solvent should be "transparent" (absorb very little light) in the region where the analyte is measured. In other words, the measurement wavelength should be set above the UV cutoff value of the solvent. If the light absorption of the solvent is too strong, it will cause high background absorption, making it impossible to use the solvent as a mobile phase.

Refractive index The refractive index of a medium is defined as the ratio of the speed of light in vacuum to the speed of light in the medium. A refractive index detector (RI or RID) is used to measure the refractive index of an analyte relative to a solvent. They are often used as an alternative to UV detectors to detect solutes that do not contain chromophores, such as sugars, polysaccharides, and synthetic polymers. They are considered universal detectors because they can detect anything with a refractive index different from that of the solvent, but their sensitivity is relatively low. When the differential refractive index detector is used in mixed-mode chromatography, it should be considered to select mobile phase components with a large difference in refractive index from the solute in order to maximize the detection sensitivity.

8.2.2 Buffers

A buffer solution (more precisely, a pH buffer or a hydrogen ion buffer) is an aqueous solution composed of a mixture of a weak acid and its conjugate base and vice versa. When a small amount of strong acid or alkali is added, the pH value changes little. In many chemical applications, buffer solutions are used as a means to keep the pH at an almost constant value. In nature, there are many systems that use buffers to adjust pH. For example, bicarbonate buffer is used to adjust the pH of the

Table 8.2 Commonly used buffers for mixed-mode anion exchange chromatography

Substance	pH range	pK_a (25 °C)	Common concentration (mM)	Countering ion
N-Methylpiperazine	4.5–4.5	4.75	20	Cl^-
Piperazine	4.0–6.5	4.68	20	Cl^-, $HCOO^-$
L-Histidine	4.5–6.0	4.96	20	Cl^-
Bis(2-hydroxyethyl)amino(trihydroxy)-methane(BIS-TRIS)	4.8–6.4	6.46	20	Cl^-
1,3-bis[tris(hydroxymethyl)-methylamino]propane(BIS-TRIS Propane)	6.4–7.3	6.8	20	Cl^-
Triethanolamine	7.3–7.7	7.76	20	Cl^-, $HCOO^-$
TRIS	7.6–8.0	8.06	20	Cl^-
N-Methyldiethanolamine	8.0–8.5	8.52	50	Cl^-, $HCOO^-$
Diethanolamine	8.4–8.8	8.88	20–50	Cl^-
1,3-Diaminopropane	8.5–9.0	8.64	20	Cl^-
Ethanolamine	9.0–9.5	9.50	20	Cl^-
Piperazine	9.5–9.8	9.73	20	Cl^-
1,3-Diaminopropane	9.8–10.3	10.47	20	Cl^-
Piperazine	10.6–11.6	11.12		

blood, and bicarbonate can also be used as a buffer in the ocean. There are many types of buffers and substances that can play a buffering role can be divided into two categories: The first category is a system composed of weak acids (acetic acid, phosphoric acid, etc.) or weak bases (NH_3H_2O) and corresponding salts (such as HOAc-NaOAc; NH_4Cl). The second category is organic amines or amphoteric organic compounds.

For the first type of buffer substances, if the charge of the buffer ions acting as a buffer is opposite to the ion functional groups on the mixed-mode ion exchanger, they will participate in the ion exchange process and cause local pH changes. Therefore, buffer ions with the same charge group as that of the ion exchanger should be used as much as possible. That is, when using anion exchangers (positively charged functional groups), positively charged buffer ions are selected; when using cation exchangers (negatively charged functional groups), negatively charged buffer ions are selected. However, this choice is not absolute. For example, phosphate buffer is often used in the anion exchange process, but in this case, special care should be taken to fully equilibrate the system before sample loading to ensure that the pH and ionic strength of the chromatographic system are compatible with the starting buffer.

The second type of buffer material can be used in both anion and cation exchange. Tables 8.2 and 8.3 list the buffers commonly used in anion exchange chromatography and cation exchange chromatography, respectively. The buffering agent is

Table 8.3 Commonly used buffers in mixed-mode cation exchange chromatography

Substance	pH range	pK_a (20 °C)	Common concentration (mM)	Countering ion
Maleic acid	1.5–2.5	2.00	20	Na^+
Malonate	2.38–3.38	2.88	20	Na^+, Li^+
Citric acid	2.63–3.63	3.13	20	Na^+
Lactic acid	3.6–4.3	3.81	50	Na^+
Formic acid	3.8–4.3	3.75	50	Na^+, Li^+
Succinic acid	4.3–4.8	4.21	50	Na^+
Acetic acid	4.8–4.2	4.76	50	Na^+, Li^+
Malonate	4.0–6.0	4.68	50	Na^+, Li^+
Phosphoric acid	6.7–7.6	7.2	50	Na^+
N-2-Hydroxyethylpiperazine-*N′*-2-Ethanesulfonic acid (HEPES)	7.6–8.2	7.55	50	Na^+, Li^+
N,N-Dihydroxyethylglycine (BICINE)	8.2–8.7	8.35	50	Na^+

selected according to the required pH buffering range. Table 8.4 lists the structures of commonly used buffer compounds.

In some cases, volatile buffer substances are preferred because they can be easily removed from the pooled fractions by freeze-drying. This is especially true when the amount of protein collected after chromatographic purification is very limited, and any dialysis or desalting method usually further reduces the recovery. By directly freeze-drying the pooled fractions to remove volatile buffer substances, the post-processing of collected protein can be simplified. Common volatile buffer substances are listed in Table 8.5.

8.2.3 Salts

In mixed-mode chromatography, it is usually necessary to add a buffer to the mobile phase to control the pH of the solution, and sometimes it is also necessary to add a certain concentration of salts to adjust the ionic strength of the mobile phase. In the ion exchange mode, the solute is adsorbed to the stationary phase through electrostatic interaction with the ligand, and then the adsorbed solute is eluted by increasing the salt concentration in the mobile phase. In this case, sodium chloride is usually used. In hydrophilic ion exchange chromatography, the percentage of organic solvents in the mobile phase is relatively high. Salts with higher solubility in organic solvents can be used, such as sodium perchlorate, ammonium perchlorate, and salts composed of organic acids or bases. In the hydrophobic interaction mode, the protein is adsorbed in the mobile phase containing high concentration of salts and eluted in the mobile phase with low concentration of salts. In this case, salting-out salts such as ammonium sulfate and sodium sulfate are usually used to stabilize the protein conformation, promote the hydrophobic interaction between the protein itself and

Table 8.4 Structures of commonly used buffer compounds

Common name (chemical name)	Structure	pK_a, 25 °C	Temp. effect, dpH/dT (K^{-1})	Mol. weight
TAPS, ([tris(hydroxymethyl)-methylamino]propanesulfonic acid)	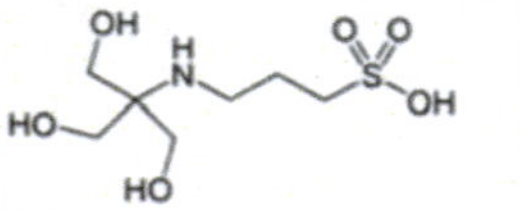	8.43	−0.018	243.3
Bicine, (2-(bis(2-hydroxyethyl) amino)acetic acid)	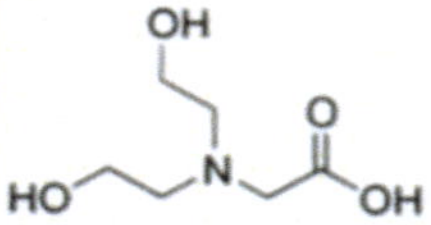	8.35	−0.018	163.2
Tris, (tris(hydroxymethyl)-aminomethane, or 2-amino-2-(hydroxymethyl)propane-1,3-diol)	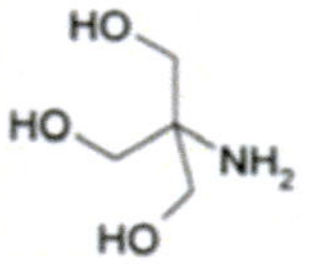	8.07[a]	−0.028	121.14
Tricine, (*N*-[tris(hydroxymethyl)-methyl]glycine)	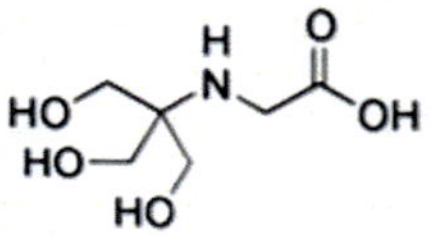	8.05	−0.021	179.2
TAPSO, (3-[*N*-tris (hydroxymethyl)methylamino]-2-hydroxypropanesulfonic acid)	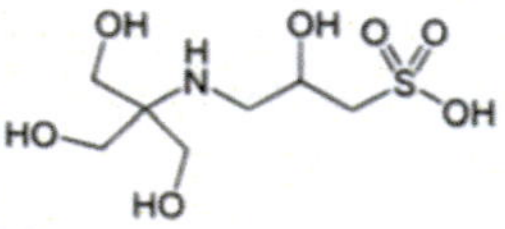	7.635		259.3
HEPES, (4-(2-hydroxyethyl)-1-piperazineethanesulfonic acid)	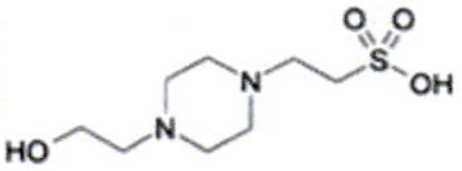	7.48	−0.014	238.3
TES, (2-[[1,3-dihydroxy-2-(hydroxymethyl)propan-2-yl] amino]ethanesulfonic acid)	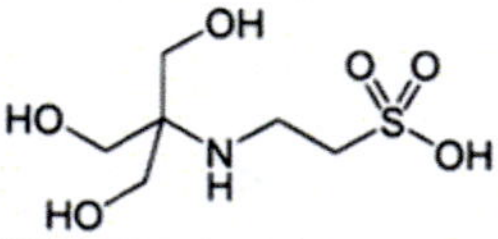	7.40	−0.020	229.20
MOPS, (3-(*N*-morpholino) propanesulfonic acid)	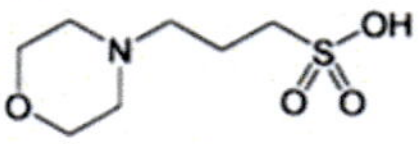	7.20	−0.015	209.3
PIPES, (piperazine-*N*,*N*′-bis (2-ethanesulfonic acid))	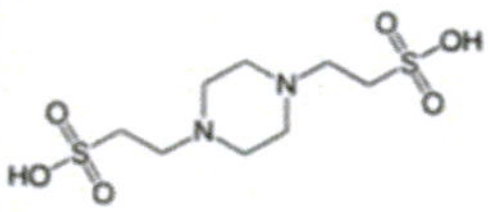	6.76	−0.008	302.4

(continued)

Table 8.4 (continued)

Common name (chemical name)	Structure	pK_a, 25 °C	Temp. effect, dpH/dT (K^{-1})	Mol. weight
Cacodylate, (dimethylarsenic acid)	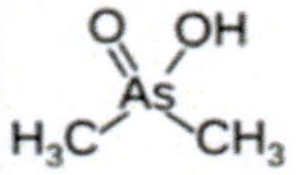	6.27		138.0
MES, (2-(*N*-morpholino) ethanesulfonic acid)	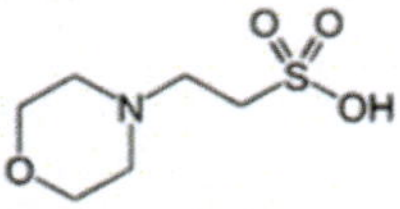	6.15	−0.011	195.2

Reproduced from Wikipedia at https://en.wikipedia.org/wiki/Buffer_solution
[a]Tris is a base, the pK_a = 8.07 refers to its conjugate acid

Table 8.5 Commonly used volatile buffers in mixed-mode chromatography

Substance	pH	Countering ion	Substance	pH	Countering ion
Formic acid	2.5	H^+	Ammonia/acetic acid	8.5–10	CH_3COO^-
Pyridine/formic acid	2.5–3.5	$HCOO^-$	Trimethylamine/CO_2	7–12	CO_3^-
Trimethylamine/formic acid	3.0–4.0	$HCOO^-$	Ammonium bicarbonate	7.9	HCO_3^-
Pyridine/acetic acid	3.0–6.0	CH_3COO^-	Ammonium carbonate/ammonia	8.0–9.5	CO_3^-
Trimethylamine/HCl	6.8–8.8	Cl^-	Ethanolamine/HCl	8.5–10.5	Cl^-
Ammonia/formic acid	7.0–8.5	$HCOO^-$	Ammonium carbonate	8.9	CO_3^-

precipitate, or promote the hydrophobic interaction between the protein and the ligand on the stationary phase. Salting-out salt is characterized by a large molar surface tension increment, which can increase the cohesion of water after being dissolved in water, showing a high hydrophobic effect. At the same time, the use of salt-in salts, such as thiocyanate, guanidine hydrochloride, is avoided, because they will increase the water solubility of the protein and denature the protein at the same time. There is also a type of salts such as calcium chloride, magnesium chloride, that can interact specifically with proteins and have a different behavior from general salting-out salts.

The ability of salt to promote hydrophobic retention is consistent with the ability to precipitate proteins and macromolecules from aqueous solutions, usually following the Hofmeister ion sequence. According to the salt's ability to stabilize the protein in solution, the Hofmeister sequence of the anion is as follows: $CO_3^{2-} > SO_4^{2-} > S_2O_3^{2-} > H_2PO_4^- > F^- > Cl^- > Br^- > NO_3^- > I^- > ClO_4^- > SCN^-$

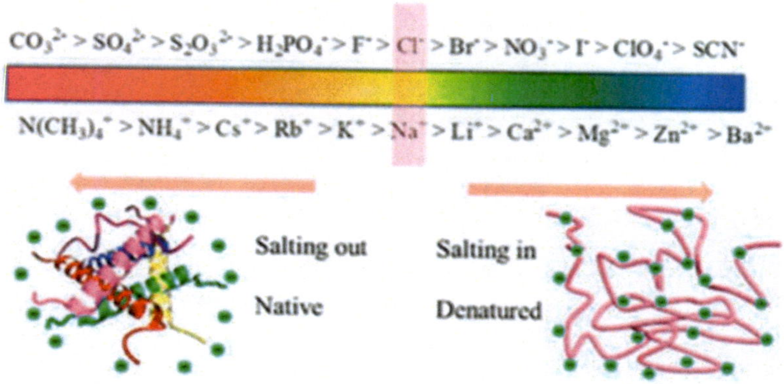

Fig. 8.1 Hofmeister sequence reflecting ion salting-out. (Adapted with permission from Kang B, Tang H, Zhao Z et al. (2020) Hofmeister series: insights of ion specificity from amphiphilic assembly and interface property. ACS Omega 5:6229–6239. Copyright {2020} American Chemical Society)

(Fig. 8.1). The series includes two different groups: strong hydrated ions and weakly hydrated ions. The anion on the left side of Cl^- has good hydration and is classified as kosmotrope (water structure builder) and tends to stabilize the natural folding structure of the protein, leading to salting-out behavior. At the same time, the anion on the right side of Cl^- has low hydration capacity, which is called chaotrope (water structure destroyer), and it is easy to promote protein denaturation and unfolding, showing salt-in behavior. Cl^- is the boundary between two different ions and has no obvious effect on the stability of the protein. The Hofmeister sequence of the cation is $N(CH_3)_4^+ > NH_4^+ > Cs^+ > Rb^+ > Na^+ > Li^+ > Ca^{2+} > Mg^{2+} > Zn^{2+} > Ba^{2+}$ (Fig. 8.1). The weakly hydrated cations on the left side of Na^+ tend to stabilize the protein, while the divalent cations on the right side can increase the solubility of the protein and promote its unfolding and denaturation.

8.2.4 *Additives*

Mixed-mode chromatography has been widely used in the separation and purification of proteins. In some cases, the adsorbed protein can be eluted by changing the pH or ionic strength of the eluent. However, in other cases, the protein is firmly bound to the ligand on the stationary phase, causing problems such as protein conformation changes, denaturation and inactivation, and low recovery. In order to overcome this difficulty and further improve the purity and activity of the recovered protein, an appropriate concentration of additives is usually added to the mobile phase during the washing and elution steps to selectively remove impurities or selectively elute the target protein. Low-concentration additives (such as water-miscible alcohols, detergents, chaotropic salts, arginine) can effectively compete with proteins for binding sites on mixed-mode ligands, weaken protein-ligand interactions, and lead to displacement or elution of bound protein.

Chaotropic salts As mentioned above, the anions after Cl^- and the cations after Na^+ in the Hofmeister sequence reflecting ion salting-out ability are chaotropic ions, which have the ability to destroy water structure and promote protein dissolution and denaturation. Potassium thiocyanate and magnesium chloride are the most commonly used chaotropic salts. At a suitable concentration, the adsorbed protein can be eluted from the chromatographic support and maintain activity. Durkee et al. (1993) found that 3.5 M $MgCl_2$ in 0.05 M Tris·HCl is an effective non-denaturing eluent in the affinity system, in which the monoclonal antibody immobilized on agarose column is used to separate enzymes. More than 90% of the adsorbed protein is recovered. Thompson et al. (1992) reported that a buffer containing 3.5 M $MgCl_2$, 0.05 M Tris·HCl, and 0.1 M EDTA was ineffective when used alone with 0.75 M NaCl. However, in the presence of 30–40% ethylene glycol (EG), a good recovery of the protein with high specific enzyme activity was achieved by dissociation from the immobilized monoclonal antibody.

Polyols Polyols, such as ethylene glycol, glycerol, and polyvinyl alcohol, are commonly used as cosolvents to enhance the stability of proteins in aqueous solutions. It is known that glycerol can transform the whole natural protein into a more compact state. Glycerol also inhibits protein aggregation during refolding of many proteins. Vagenende et al. (2009) studied the mechanism of protein stabilization induced by glycerol through molecular simulation. Contrary to the general concept that the preferential hydration of proteins in polyol/water mixtures is determined by the molecular size of polyols and protein surface area, they proposed that the preferential hydration of proteins in glycerol/water mixtures is mainly due to electrostatic interactions that induce the orientation of glycerol molecules on the protein surface, thereby further excluding glycerol. These interactions shift the natural protein to a more compact conformation. In addition, glycerol preferentially interacts with large continuous hydrophobic patches on the protein surface, where glycerol acts as an amphiphilic interface between the hydrophobic surface and the polar solvent. Therefore, glycerol stabilizes proteins by inhibiting protein unfolding and by preferential interaction with hydrophobic surface regions that tend to be oriented toward the amphiphilic interface of glycerol, preventing their aggregation. These mechanisms are also applicable to other cosolvents to a large extent, including polyols, arginine, and urea.

Denaturants Chemical denaturants, such as urea and guanidine hydrochloride, are the most commonly used protein unfolding agents and are believed to work by better dissolving the unfolded state of the protein. With the increase of the denaturant concentration, the solvation of the hydrophobic residues on the protein is improved, which helps to stretch the unfolded chain, thereby providing a sensitive probe for the action of the denaturant. Zheng et al. (2016) used all-atom molecular simulations to study the effects of urea and guanidine hydrochloride on the structure of the intrinsic disordered protein ACTR and compared with the experimental results in a wide range of denaturant concentrations. Using unbiased molecular simulations of a carefully calibrated denaturant model, they found that protein chains swell as the denaturant concentration increased. This is because urea or guanidine hydrochloride can bind to the main chain as well as the side chains of almost all residues, and the free energy of water transfer inferred from the binding is reasonably close to the experimental estimate. The binding of the denaturant to the main chain is mainly through hydrogen bonding, and the binding to the side chain also includes other effects such as electrostatic interaction.

Substance P(SP) is an undecapeptide neuromediator, which stimulates the function of human T lymphocytes by binding to stereospecific membrane receptors. About 20,000 SP receptors are expressed on the surface of B lymphoblasts cultured in human IM-9, with a K_d of 0.3 nM. [^{125}I] SP is specifically cross-linked with 78, 58, and 33 kDa IM-9 cell membrane proteins via suberic acid disuccinimidyl group. McGillis et al. (1997) reported an indirect immunoaffinity purification procedure based on the immunization of detergent-soluble [^{125}I] SP-labeled IM-9 cell membrane proteins with anti-SP antibodies coupled to chromatographic carriers via epoxides. Using acidic 8 M urea for elution, the 58-kDa and 33-kDa SP receptor complexes were purified to an apparent uniformity, and the yields obtained by three consecutive elution cycles were 63%, 11%, and 6%, respectively.

Kim et al. (1999) examined the reusability of avidin-biotinylated IgY columns in immunoaffinity chromatography through repeated use and regeneration. The Actigel avidin-biotinylated IgY column retains about 90% of the initial IgG binding capacity after 50 cycles, using 0.1 M glycine-HCl buffer (pH 2.8) as the eluent without regeneration. However, using the commercial eluent Actisep, the IgG binding capacity of the UltraLink avidin-biotinylated IgY column gradually decreased to 75% and 65% after 10 and 20 cycles, respectively. The results of the CovaLink NH system are consistent with the results of the UltraLink avidin-biotinylated IgY column. The UltraLink avidin-biotinylated IgY column was regenerated twice by using 8 M guanidine hydrochloride with a pH of 1.6 to dissociate the biotinylated IgY antibody from the column. After the first and second regeneration, about 40% and 25% of the IgG binding capacity are still retained. By applying the new biotinylated IgY to the treated column, approximately 95% and 90% of the initial IgG binding capacity can be recovered before any treatment. These results indicate that depending on the substrate to which the avidin is immobilized, the avidin-biotinylated IgY column can be reused with or without regeneration.

Nord et al. (2000) used two ligands that selectively bind bacterial Taq DNA polymerase and human apolipoprotein A-1 (M) to recover the two target proteins with high selectivity and efficiency from cell lysates. After repeated cycles of sample loading, washing, and low pH elution, no loss of column capacity or selectivity was observed. These studies showed that the reusability of a chromatographic column may depend on the physical properties of the solid phase and the biological properties of the receptor and ligand, and these factors may vary greatly between different systems (Downham et al. 1992; Uemura and Hirai 1999).

Detergents Detergent is a surfactant or a mixture of surfactants that has cleaning properties in a dilute solution (Linke 2009). A typical representative among them is alkylbenzene sulfonate, which is a class of compounds similar to soap, but more soluble in hard water, because the polar sulfonate of detergent is less likely to complex with the calcium and other ions in hard water than the polar carboxylate of soap. In biochemistry, detergents (surfactants) are key reagents for the extraction and purification of intact membrane proteins. Solubilization of cell membranes and membrane proteins with detergents is usually the first step in the purification of intact membrane proteins (Neugebauer 1990).

Like soap, detergents are amphiphilic compounds: part of the molecule is hydrophilic (polar), and the other part is hydrophobic (nonpolar). Their dual nature promotes the association of hydrophobic compounds (such as oils and greases) with water, thereby helping them dissolve in water. Detergents are also lipid-like substances. As the main component of the membrane, the phospholipid molecule also contains a hydrophilic head and a hydrophobic tail. Amphiphilic detergents can compete with bilayer lipids and are more hydrophilic than lipids. As a result, the detergent-protein complex is soluble in aqueous solutions, and the detergent molecules mimic lipid molecules, helping to maintain the natural configuration of the membrane protein during the purification process. Due to the amphiphilic nature of detergents, when they are dissolved in an aqueous solution and reach a certain concentration, they will also self-assemble like phospholipids in cells. The hydrophobic tails are close together, and the hydrophilic heads are close together to form a spherical structure of micelles, with a hydrophobic core on the inside and a hydrophilic shell on the outside. This concentration is called the critical micelle concentration (cmc). The cmc of a detergent is not only related to the nature of the detergent itself, but also affected by factors such as solvent, temperature, pH, and ionic strength.

The aggregation number (*N*) is the average number of monomers forming each micelle. The techniques commonly used to determine the size of biological macromolecules are also generally applicable to micelles. There may be differences in the number of aggregates of detergents determined by different technologies. In addition, some detergents exhibit concentration-dependent micellar growth, so the number of aggregates measured in the case where the detergent concentration is much greater than cmc may be different from the number of aggregates at cmc.

Critical micelle temperature (cmt) is the temperature at which the detergent/solvent system changes from a hydrated crystalline state to an isotropic micelle solution. Some detergents obviously do not have cmt, but for these compounds, cmt is just below the freezing point of the solvent. The Krafft point (TK) specified in some documents is the cmt at cmc. Some detergents exhibit a lower solid solution temperature or cloud point. At the cloud point, the detergent solution transfers from the isotropic micellar system to the two-phase system: one phase is depleted of detergent, and the other phase is rich in giant micelles. The cloud point normally recorded is the temperature at which a warm and turbid 1% solution becomes clear

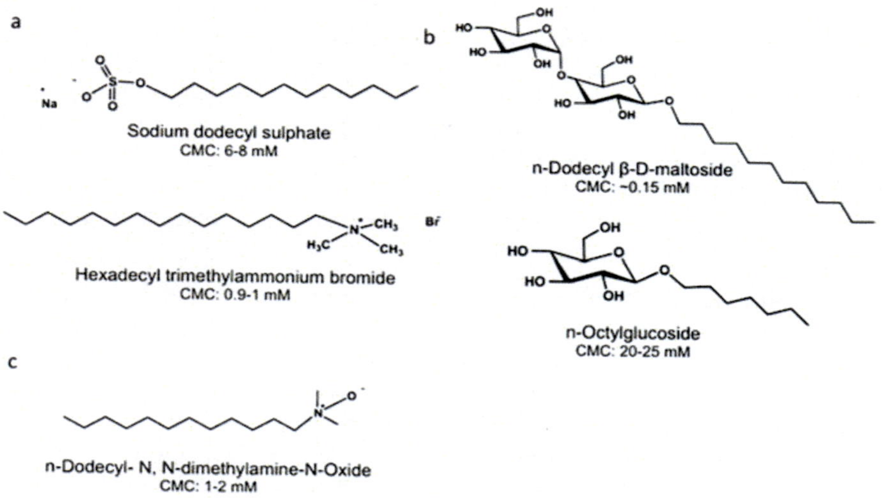

Fig. 8.2 Chemical structures of conventional detergents. (**a**) Ionic detergents, e.g., sodium dodecyl sulphate (anionic detergent) and hexadecyl trimethylammonium bromide (cationic detergent). (**b**) Nonionic detergents, e.g., *n*-dodecyl-β-D-maltoside (DDM) and *n*-octylglucoside (OG). (**c**) Zwitterionic detergent, e.g., *n*-dodecyl-*N,N*-dimethylamine-*N*-oxide (LDAO). (Reproduced from Anandan and Vrielink 2016)

after cooling. The hydrophilic-lipophilic balance (HLB) of a detergent is an indicator of its hydrophilicity, and the larger its value, the more hydrophilic the detergent.

Depending on the charge of the surfactant, detergents can be divided into four categories: anionic detergents, cationic detergents, amphoteric detergents, and nonionic detergents. A typical anionic detergent is alkylbenzene sulfonate. The alkylbenzene portion of these anions is lipophilic, and the sulfonate is hydrophilic. Bile acids, such as deoxycholic acid (DOC), are an anionic detergent produced in the liver, which helps in the digestion and absorption of fats and oils. Cationic detergents are similar to anionic detergents in that they have hydrophilic components, but quaternary ammonium salts serve as polar ends instead of anionic sulfonate groups. Zwitterionic detergents have a net zero charge due to the presence of equal numbers of +1 and −1 electrochemical groups. Examples include CHAPS. Nonionic detergents are characterized by an uncharged hydrophilic head group. Typical nonionic detergents are based on polyoxyethylene or glycosides. Common examples of the former include the Tween, Triton, and Brij series. Glycosides have sugars as their uncharged hydrophilic head groups. Examples include octyl glucosinolate and maltoside. HEGA and MEGA series detergents are similar, with sugar alcohol as the head base.

The structures of detergents commonly used in protein purification are shown in Fig. 8.2. A number of new generation detergents for membrane protein solubilization are shown in Fig. 8.3.

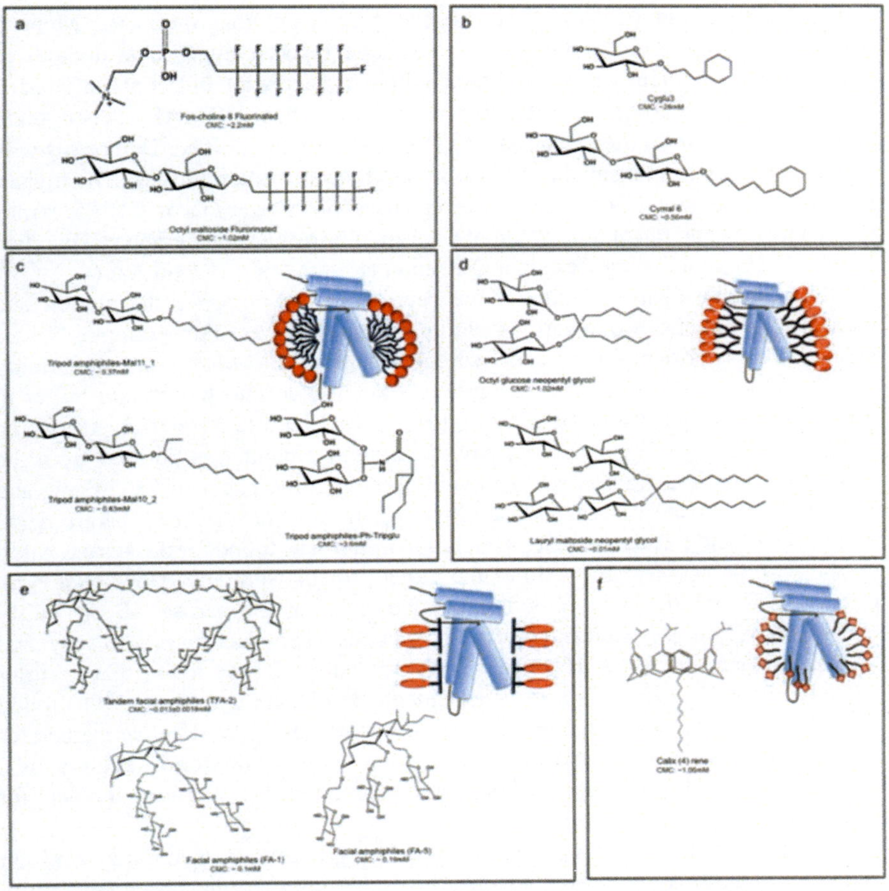

Fig. 8.3 Chemical structure of "new" detergents. Examples of hydrophobic tail modified detergents (**a**) fluorine substituted alkyl chain, (**b**) cyclic molecules incorporated in the tail (**c**) branched alkyl chain, e.g. tripods. Example of lipid-like detergents (**d**) and steroid based facial amphiphiles (**e**). Structure of calix[4]arene (**f**). Detergent cmc values in water are listed under the respective chemical structures. Cartoon representation of hypothetical micelle shape for different classes of detergents is also shown. (Reproduced from Anandan and Vrielink 2016)

Detergents (surfactants) are key reagents for the extraction and purification of intact membrane proteins. Welling-Wester et al. (1998) used zwitterionic and nonionic detergents to extract recombinant glycoprotein D (gD-1) of type 1 herpes simplex virus (HSV-1) from insect cells infected with recombinant baculovirus. The two zwitterionic detergents, (*N*-dodecyl-*N*,*N*-dimethylammonium) undecanoate [DDMAU, cmc = 0.13 mM] and (*N*-dodecyl-*N*,*N*-dimethylammonium)butyrate (DDMAB, cmc = 4.3 mM), were used as additives to elution buffers in ion exchange chromatography to purify HSV-1 gD-1 from the extract, and the results were compared with the nonionic detergent pentaethylene glycol monodecyl ether

(C10E5) (Damhof et al. 1994). Two chromatographic columns were used, Mono Q and Resource Q, with flow rates of 1 and 5 mL/min, respectively. The results showed that when the detergent was used as the additive of the elution buffer at 0.2% (w/v), the detergents DDMAU and C10E5 were better than DDMAB. When 0.2% DDMAB was used in the eluent, HSV gD-1 could not be purified. Detergents with higher cmc may not be suitable for use as additives in elution buffers. The highest recovery of gD-1 was achieved by chromatographic separation of C10E5 extract using Mono Q column at a flow rate of 1 mL/min or by chromatographic separation of DDMAU extract using Resource Q column at a flow rate of 5 mL/min.

Many mobile phase additives have been shown to improve the elution and recovery of proteins from various chromatography resins. For example, Santos et al. (1985) purified D-lactate dehydrogenase from *E. coli* by immunoaffinity chromatography. They titrated guanidine hydrochloride into the elution buffer to determine the concentration that the target protein could be effectively recovered without inhibiting the enzyme activity. The target protein was purified from the membrane vesicle extract in one step, with a purification factor of 230. Tsang and Wilkins (1991) tested 13 buffers that differ in pH, salt concentration, chaotropicity, or hydrophobicity. The most effective elution buffer was found for the system based on the specific activity and yield of the target protein. Similarly, Downham et al. (1992) used monoclonal antibodies immobilized on Sepharose 4B to purify overexpressed C_{κ}-β-galactosidase fusion proteins. The maximum recovery and β-galactosidase activity of 16 buffers were compared. It was found that a buffer containing ethylene glycol was a better eluent. Ben-David and Firer (1996) found that 4 M $MgCl_2$ dissolved in 25% ethylene glycol was the most effective method for recovering active monoclonal antibodies from the immunoaffinity column in a variety of elution buffers tested. This solution can be used as a universal eluent for monoclonal antibody purification.

The antibody is usually purified from the antigen affinity column by acid/salt elution. The antibody recovered through this step affinity purification retains activity, but the recovery of protein is usually low. Narhi et al. (1997) studied the effects of acid and other denaturing or chaotropic solvents on the conformation of monoclonal antibodies (mAb) made against the extracellular region of the Her2 receptor (sHer2) derived from Chinese hamster ovary cells. In the 0.1 M glycine (pH 2.9) commonly used for elution, the mAb remained almost completely folded and had a complete β-sheet secondary structure, and only small changes were detected in the tryptophan environment. In 7 M urea, 50 mM NaAc pH 4.0, the antibody was partially unwound, the Trp environment was further disturbed, and part of the β-sheet structure was transformed into a disordered structure. In 6 M guanidine hydrochloride, 50 mM NaAc, pH 4.0, the antibody was completely unfolded, and there was no secondary or tertiary structure. The antibody exposed to glycine or urea was refolded in phosphate buffered saline (PBS) by dialysis, while the antibody denatured by guanidine hydrochloride was dialysized into 7 M urea (pH 4.0) to refold, and then infiltrated into PBS. The refolded antibody can form an antigen-antibody complex, which can be separated by gel filtration chromatography.

Therefore, for antibodies that cannot be recovered with glycine eluent, they can be eluted with denaturant and then dialyzed into an appropriate solution for renaturation.

RVV-X is a factor X activator derived from Russell viper venom. Durkee et al. (1993) used affinity chromatography to separate the enzyme-specific monoclonal antibody on an agarose column. By testing acid, base, and high concentration of $MgCl_2$ for elution, it was found that high concentration of $MgCl_2$ was most effective to elute the active RVV-X most effectively, with 90% of homogenous enzymes recovered.

8.3 Effect of Mobile Phase Parameters on Binding and Elution

8.3.1 *Effect of pH*

In mixed-mode chromatography, the role of mobile phase pH in the retention of solute can be understood from both the influence on the solute and the stationary phase. The solutes frequently encountered in mixed-mode chromatography are mostly amphiphilic ions, that is, solutes have both polar or ionized parts and nonpolar parts. Amino acids, proteins, and nucleotides are examples of this type of molecules. Take protein as an example. In different pH environments, the charged properties of protein are different. The type and number of charges carried by protein molecules are not constant, but are directly related to the pH of the solution. When the pH of the solution is lower than the p*I,* the protein's isoelectric point, the protein is positively charged. The farther the pH value is from the p*I*, the more positive charge the protein carries, and the stronger the hydrophilicity of the protein. When the pH of the mobile phase is higher than p*I*, the situation is just the opposite. The protein is now negatively charged and becomes more hydrophilic as the pH increases further. When the pH of the mobile phase is equal to the p*I* of the protein, the net charge of the protein is zero, and the protein as a zwitterion has the greatest hydrophobicity. However, due to uneven distribution of the surface charge of the protein, local areas may be positively charged, while other areas are negatively charged.

In addition to affecting the dissociation equilibrium of the solute, the pH value of the solution also affects the nature of the ligand and the matrix in the stationary phase. For example, the silanol group on the surface of the silica gel substrate is weakly acidic, and its pK_a is in the range of pH 3–6. When the pH value is lower than the pK_a of silica gel, the silica gel does not dissociate and is neutral; however, when the pH value is higher than the pK_a of silica gel, the silanol groups dissociate, making the surface of the silica gel negatively charged and the surface hydrophilicity increases. Similarly, if the mixed-mode ligand contains weakly acidic groups, when the pH of the solution is lower than the pK_a of the ligand, the ligand is neutral and has the greatest hydrophobicity; when the pH of the solution is higher than pK_a, the acid group of the ligand is dissociated, the ligand is negatively charged, and the hydrophilicity increases. When the ligand contains a basic group, the abovementioned trend of change is opposite. At low pH, the ligand is positively charged and the hydrophobicity decreases; at high pH, the ligand is neutral and the hydrophobicity increases.

During the elution process of mixed-mode chromatography, the pH value of the mobile phase can be kept constant, and the adsorbed solute can be eluted by changing the ionic strength or the polarity of the mobile phase. In the case that either changing the ionic strength or changing the polarity of the eluent is not effective, the method of changing the pH of the mobile phase can also be used for elution. By adjusting the pH of the mobile phase, so that the solute and the ligand have the same charge, the resulting charge repulsion will cause the solute to be eluted.

From the above discussion, it can be seen that selecting the pH value of the mobile phase requires consideration of many factors, and several conditions must be met at the same time:

1. First the appropriate pH is selected according to the isoelectric point of the amphiphilic solute component and the pK_a of the ionic group of the ligand. In the adsorption stage, in order to charge the solute so as to be adsorbed by the mixed-mode ligand, the pH should be about 1–2 pH units lower than the pI on the cation exchanger while the pH should be 1–2 pH units higher than the pI on the anion exchanger. In the elution stage, it is necessary to select an appropriate pH value to minimize the interaction between the solute and the ligand, which is conducive to the elution and activity recovery of the solute.
2. The selected pH value is within the pH range of the mixed-mode stationary phase. It is necessary to ensure that the packing has a certain degree of dissociation and a certain amount of adsorption. At the selected pH value, the packing should be stable. For example, silica-based packings should not be used in basic solution and hydroxyapatite-based packings should not be use in acidic solution; otherwise, it will cause degradation of the chromatographic packing framework and shorten the service life.
3. At the selected pH value, the solute can exist stably and the biological activity is retained.

In mixed-mode chromatography, pH step changes or gradient changes are used to promote the elution of target compounds. For example, in hydrophobic charge induction chromatography, pyridine derivatives are used as chromatographic ligands, and proteins are adsorbed at a pH higher than the pK_a of the ligand. After washing and removing the adsorbed impurities on the chromatographic column, the pH value of the eluent is lowered, so that the protein is positively charged as with the ligand, and eluted by electrostatic repulsion. At high pH values, the protein is strongly adsorbed due to hydrophobic interaction. If an organic solvent is used as a polarity regulator for elution, it may cause denaturation and inactivation of the protein. It has been observed that on the mixed-mode anion exchanger, the elution sequence of monoclonal antibodies with different p*I* values is reversed around pH 6; therefore, the adjustment of the mobile phase pH value becomes an effective tool for removing impurities. Further studies have shown that the adsorption of monoclonal antibodies on mixed-mode ion exchangers is mainly through the interaction of the ligand with the charges in the vicinity of the hydrophobic patch. The change of pH value may affect the charge distribution in these areas, thereby changing the elution order.

Robinson et al. (2020) further examined the effect of mobile phase pH on mAb surface properties and retention based on previous studies on the contribution of mAb domains to multimodal (MM) chromatographic retention. In four multimodal cation exchange resins with two different ligand densities (eight in total), several mAbs with different p*I* values and surface hydrophobicities were tested with a linear salt gradient between pH 5 and 7. mAb retention is inversely proportional to pH and has a nonlinear relationship. Changing the pH value will affect the elution order, resulting in a unique selectivity window in each multimodal resin. One mAb shows a pH-dependent domain contribution profile, which indicates that pH can be used to adjust the relative importance of the $(Fab)_2$ and Fc domains in the retention of mAbs in the MM system. The positively charged, negatively charged, and hydrophobic patches were calculated between the pH 5–7 for the mAb. Visualizing these patches on the protein surface shows that each mAb has a unique distribution of surface charge and hydrophobicity that changes with pH. The total patch area in this pH range was tracked to quantify how the pH affects these important surface properties. Then the scope of quantitative analysis was narrowed down, and only the patches in the complementarity determining domain (CDR) loops were considered. These patches are considered to be important interaction sites for certain mAbs in these systems. Interestingly, the difference in the titer of CDR loop patches of each mAb was shown to be the result of histidine titration, and the patches in this area were qualitatively related to the experimental trend (including the observed elution order reversal). These results indicate that pH can be used as a lever for strategy design in multimodal step to create flow-through, binding, and elution or weak partition operations, which has important implications for the design of integrated and/or continuous downstream purification processes.

8.3.2 Effect of Ionic Strength

In mixed-mode chromatography, the ionic strength of the mobile phase has an important influence on solute retention. The ionic strength mainly affects the electrostatic interaction and hydrophobic interaction between the solute and the ligand. If the mixed-mode retention mechanism involves an ion exchange process, the increase in ionic strength in the mobile phase will cause the solute molecule to weaken the binding of the charged group of the ligand, and the retention will be reduced or even not retained. If the mixed-mode retention mechanism involves hydrophobic interactions, the increase in the ionic strength of the mobile phase will facilitate the binding of the solute to the hydrophobic part of the ligand in the stationary phase, resulting in enhanced solute retention.

The ionic strength of the solution has little effect on the coordination or complexation of the metal ions. In hydroxyapatite chromatography, the presence of neutral salts such as sodium chloride in the mobile phase promotes the retention of acidic proteins or polynucleotides. This is because the ionic strength in the mobile phase shields the solute from the stationary phase, and the elimination of the electrostatic repulsion strengthens the coordination of metal ions between the solute and the bound calcium ions. The same phenomenon has been observed in immobilized metal ion affinity chromatography, in which eluents with high ionic strength are usually used to eliminate the ionic interaction between protein and ligand, making metal ion coordination the dominant mechanism for solute retention.

In recent years, the application of chromatography to directly capture target proteins from cell culture supernatants has become more and more common. For example, protein A affinity chromatography used to separate and enrich monoclonal antibodies from Chinese hamster ovary cell cultures has become standard procedure in the production of therapeutic antibodies. The ionic strength in the cell culture medium is at a moderate level, approximately equivalent to 150 mM NaCl solution. For ion exchange chromatography, this ionic strength is too high for the target protein to be retained, and the feed solution must be diluted 3–5 times, which greatly increases the volume of the feed solution. For hydrophobic interaction chromatography, this ionic strength is too low to bind the target protein to the hydrophobic ligand, and 1–3 M salting-out salt must be added. Therefore, salt-tolerant chromatographic system is highly desirable for biopharmaceutical processing, especially in the capture step.

In ion exchange chromatography, one way to improve salt tolerance is to add hydrophobic components to ion exchange ligands, that is, multimodal ion exchange chromatography. The hydrophobic components enhance the retention of the protein and at the same time increase the salt tolerance of the ligand. The so-called salt tolerance means that the solute retention in the chromatographic system is not sensitive to the change of the salt concentration in the mobile phase, and the change of the salt concentration within a certain range will not cause the desorption of the solute. In mixed-mode chromatography with hydrophobic ion exchanger, the hydrophobic and ionic parts of the ligand act on the solute simultaneously, and the two retention mechanisms are coupled to achieve a synergistic enhancement effect. In this cooperative retention mechanism, the ion exchange process is strongly affected by the hydrophobic effect. The greater the contribution of the hydrophobic effect, the smaller the influence of the mobile phase ionic strength on the ion exchange process. Therefore, the system exhibits a high digress of salt tolerance.

In hydrophobic interaction chromatography, proteins are adsorbed under high-salt concentration conditions and desorbed under low-salt conditions. The ligand density on the stationary phase is low, so that the adsorbed protein in the low-concentration salt solution is eluted. The weak hydrophobic effect caused by the low ligand density means that a high concentration of salting-out salt is needed to promote the adsorption of proteins on the hydrophobic ligand. The use of a large amount of salting-out salt on an industrial scale results in high disposal costs and environmental burden. Therefore, there is a need to reduce the use of high-concentration salts in hydrophobic interaction chromatography. One way to solve this problem is to increase the density of hydrophobic ligands so that proteins could be adsorbed at low-salt solutions and desorbed by electrostatic repulsion. Burton and Harding (1998) proposed the concept of hydrophobic charge induction chromatography, using nitrogen-containing heterocyclic compounds as hydrophobic ligands, coupled to the polysaccharide matrix with a high ligand density. Because nitrogen-containing heterocyclic rings are weak bases, they act as hydrophobic ligands at neutral pH to adsorb proteins from cell lysates without the need for conditioning the feed. The adsorbed proteins are then eluted by lowering the pH of the eluent to induce electrostatic repulsion.

8.3.3 Effect of Organic Solvent

There are several reasons for the use of polar organic solvents (methanol, acetonitrile, ethanol, tetrahydrofuran, etc.) as mobile phase additives in mixed-mode chromatography. The most common reason is to increase the solubility of the solute and control the retention of the solute in the chromatographic system. In reversed-phase ion exchange mixed-mode chromatography, organic solvents are usually used as strong eluotropic components in the mobile phase, and solute retention decreases with the increase of the concentration of the organic solvent in the mobile phase. The order of the eluotropic strength of different solvents is as follows: water < methanol < acetonitrile < ethanol < propanol < isopropanol < tetrahydrofuran. When choosing an organic solvent, not only the eluotropic strength of the solvent must be considered but also the type of stationary phase and the characteristics of the solute molecule. For example, in peptide and protein chromatography, acetonitrile is preferred over alcohols due to its low viscosity in the mixture with water and high solubility of peptides and proteins. The use of methanol in protein purification is quite limited because of its low solubility for some proteins. In contrast, *n*-propanol and isopropanol are more frequently used as higher protein recovery can be obtained.

In hydrophilic ion exchange mixed-mode chromatography, organic solvents are used as components of weak eluotropic strength in the mobile phase, and solute retention increases with the increase of the organic solvent content in the mobile phase. Acetonitrile is the preferred organic solvent, because acetonitrile is an aprotic solvent and does not act as a hydrogen bond donor to compete with solute molecules for the ligand. Therefore, in mixed-mode chromatography involving hydrophilic interaction, acetonitrile has a very good retention promotion effect. Obviously, in the presence of high levels of acetonitrile (up to 90% v/v), the hydrophilic interaction between the solute and the ligand may exceed the electrostatic interaction and become the dominant factor controlling solute retention.

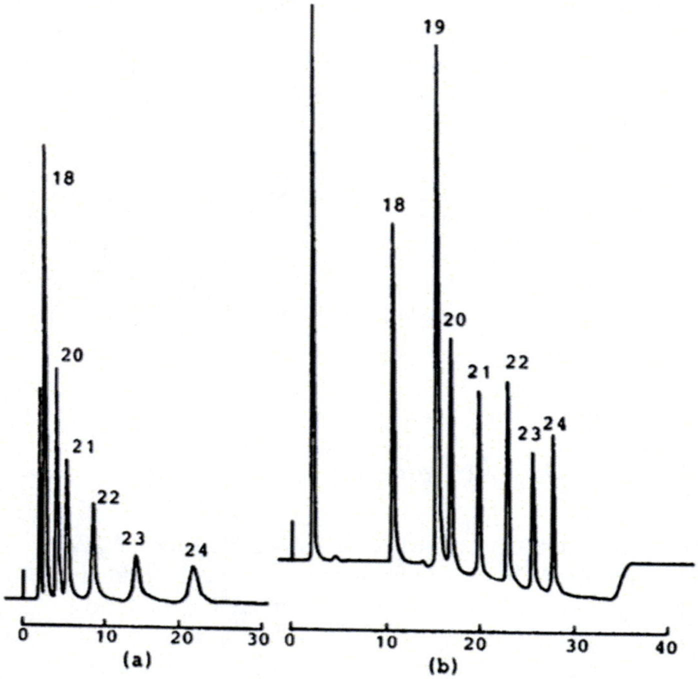

Fig. 8.4 Chromatographic separation of Panax ginseng saponin obtained on an HPLC column packed with spherical HA (Asahi Optical). The column size is 7.5 mm × 10 cm, and chromatography was performed in the presence of (**a**) 80% acetonitrile aqueous solution and (**b**) a linear gradient of 90–70% acetonitrile aqueous solution within 30 min. The time scale in minutes. (Reproduced from Kawasaki 1991)

The effect of high concentration of organic solvent in the mobile phase on the retention of hydrophilic solutes was also observed in hydroxyapatite chromatography. A variety of triterpene and diterpene glycosides, including ginsenosides and stevia glycosides, can be separated on a hydroxyapatite chromatography column using an aqueous acetonitrile eluent (Kasai et al. 1987). Figure 8.4 shows a typical chromatographic separation of ginsenosides obtained on a hydroxyapatite column in the presence of (a) 80% acetonitrile aqueous solution and (b) a linear gradient from 90% to 70% acetonitrile aqueous solution. Obviously, in the presence of a high percentage of acetonitrile in the mobile phase, solutes are adsorbed to the surface of hydroxyapatite through polar interactions. The adsorbed solute desorbs when the acetonitrile content is reduced, which is consistent with the behavior observed with other polar adsorbents such as silica gel.

8.3.4 *Effect of Additive*

A small amount of organic compounds, such as urea, ethylene glycol, octyl carboxylic acid, and sucrose, can be added to the mobile phase to reduce the polarity of the mobile phase, weaken the hydrophobic retention of the solute, and make it easier to be eluted. Surfactants are also commonly used as mobile phase modulators.

Hou and Cramer (2011) reported the effects of ethylene glycol, urea, and arginine on protein retention on the multimodal anion exchanger Capto adhere. It is well known that ethylene glycol mainly reduces hydrophobic interactions and causes a slight increase in electrostatic interactions. Urea can affect hydrophobic and hydrogen bonding interactions by directly interacting with the polar side chain and backbone of the protein and/or by changing the solvation of the protein. Urea may also cause a slight reduction in electrostatic interactions. Arginine can interact with protein side chains through hydrogen bonds and aromatic moieties through π-electron-cation interactions, which can reduce hydrophobic and electrostatic interactions and hydrogen bonding. The retention time of proteins in the presence of these mobile phase modifiers was determined in a series of linear gradient experiments. Figures 8.5, 8.6, and 8.7 show the experimental results of ethylene glycol, urea, and

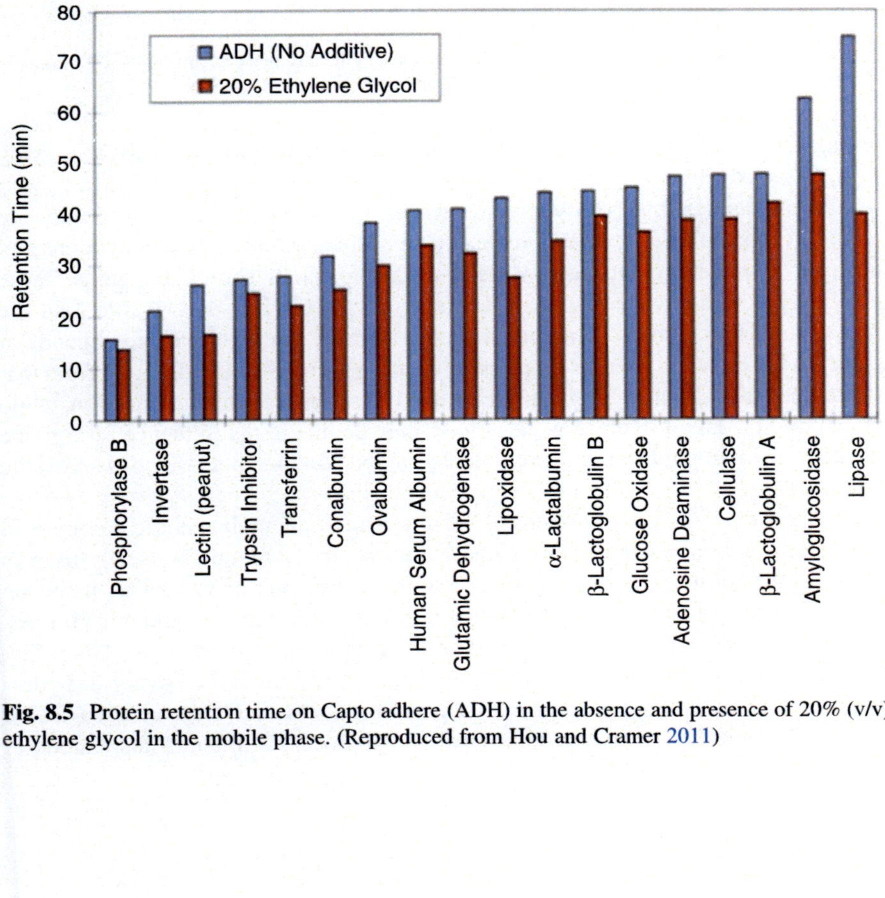

Fig. 8.5 Protein retention time on Capto adhere (ADH) in the absence and presence of 20% (v/v) ethylene glycol in the mobile phase. (Reproduced from Hou and Cramer 2011)

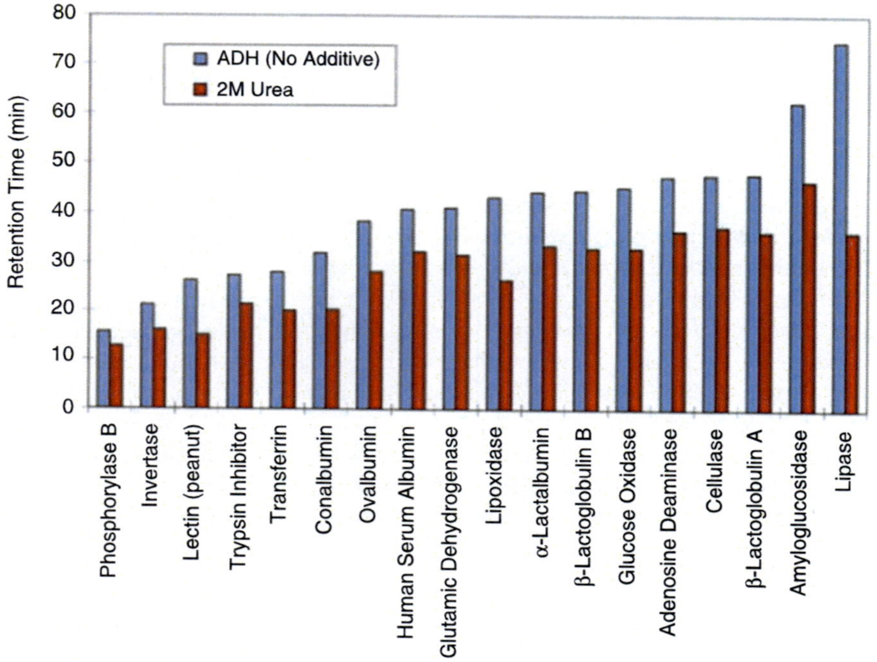

Fig. 8.6 Protein retention time on Capto adhere (ADH) in the absence and presence of 2 M urea in the mobile phase. (Reproduced from Hou and Cramer 2011)

arginine, respectively. It can be seen that the mobile phase modifier leads to varying degrees of reduction in protein retention.

The effects of ethylene glycol and urea on the retention of protein library are qualitatively similar. Both modifiers resulted in similar retention time reductions and changes in selectivity between various protein pairs. It shows that under these conditions, the hydrophobic interaction may play a more important role than the hydrogen bonding in the binding of the protein on the Capto adhere ligand. In contrast, the effect of arginine on protein retention is significantly different from that of ethylene glycol or urea. Some proteins have a greater reduction in retention, while others do not. The difference in selectivity may be due to the influence of arginine not only on hydrophobic interactions and hydrogen bonding, but also on electrostatic interactions.

Holstein et al. (2012a) investigated the role of multiple mobile phase modifiers in multimodal cation exchange chromatography systems. Arginine is usually used to improve the refolding efficiency of recombinant proteins expressed as inclusion bodies in bacteria and is also used for protein purification in conventional chromatography to improve the activity recovery of hydrophobic proteins. Guanidine hydrochloride is also often used to enhance the separation and recovery of hydrophobic proteins. Sodium caprylate is an anionic surfactant, usually used in the washing step of the chromatographic purification process of monoclonal antibodies

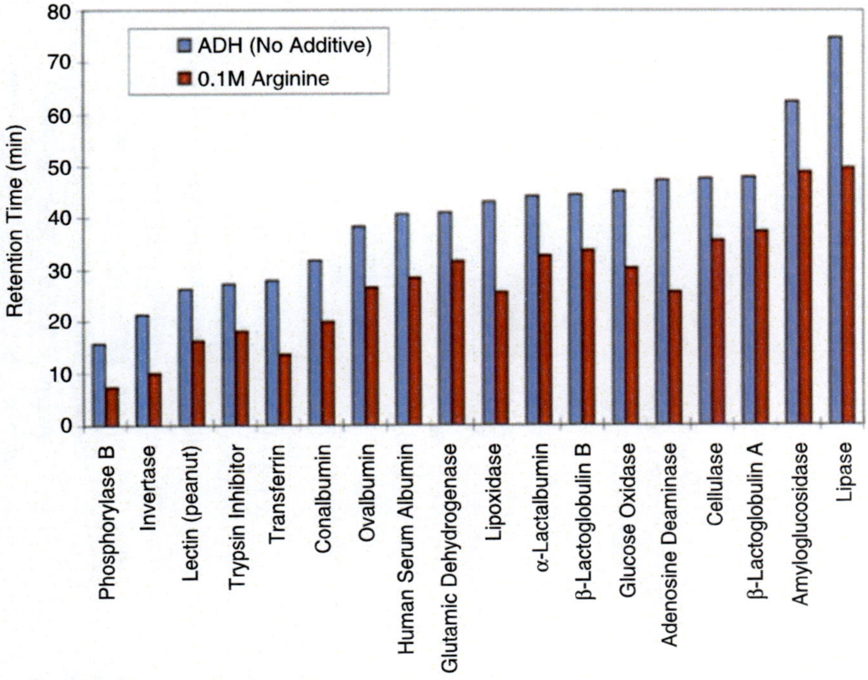

Fig. 8.7 Protein retention time on Capto adhere (ADH) in the absence and presence of 0.1 M arginine in the mobile phase. (Reproduced from Hou and Cramer 2011)

to remove contaminated albumin from the cell culture fluid. Polyols (such as ethylene glycol and glycerol) are commonly used as protein stabilizers and can also significantly change the properties of the solution.

First, a library containing a variety of proteins with different properties was used to determine their linear salt gradient elution concentration on a multimodal cation exchanger (Capto MMC); then, in the presence of a mobile phase regulator, the above experiments were repeated. The results show that mobile phase modifiers have an important effect on protein retention and selectivity.

Figure 8.8 shows that arginine can cause a decrease in the retention of all tested proteins, but the degree of decrease varies with different types of proteins. For example, on the one hand, for some proteins, such as lysozyme and avidin, there is a small decrease in retention in the presence of arginine, but still a high concentrations of salt is required to elute. On the other hand, the retention of some proteins, such as ubiquitin and papain, is greatly reduced in the presence of arginine. Their retention is completely lost in 0.1 M arginine. Due to the different degrees of reduction in the retention of different proteins, arginine can be used to improve the selectivity of protein separation. For example, equine cytochrome c and human serum albumin have similar retention and co-elute in the absence of modulators. However, after adding 0.1 M arginine, the retention of horse cytochrome c was only

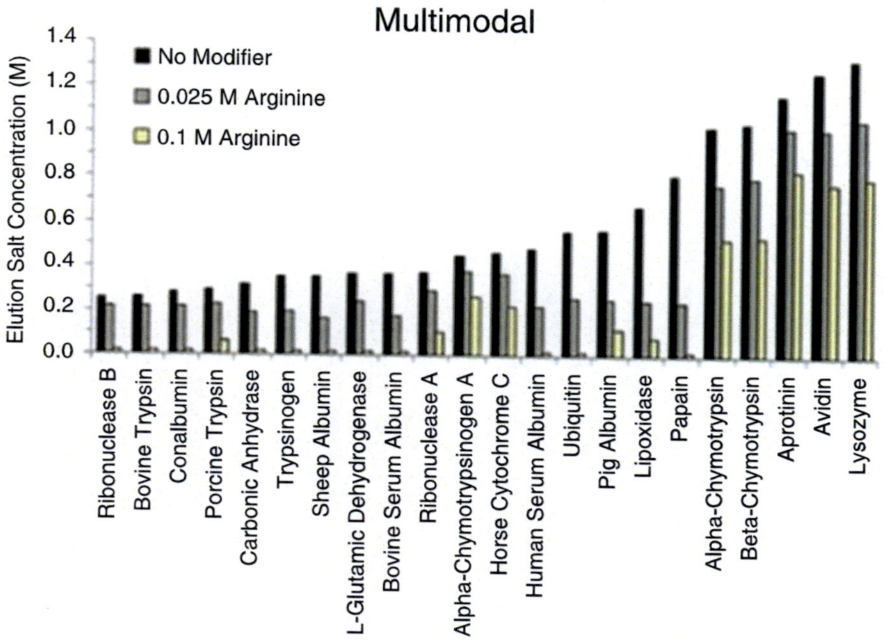

Fig. 8.8 Chromatographic retention data of a commercial protein library on a multimodal cation exchange resin (Capto MMC) in the absence and presence of various arginine concentrations in the mobile phase. Linear salt gradient conditions: buffer A (20 mM sodium phosphate, pH 6) to buffer B (20 mM sodium phosphate, 1.5 M chlorine) in the presence of 0, 0.025, and 0.1 M arginine over 45 column volumes of sodium chloride, pH 6, for elution. (Reproduced from Holstein et al. 2012a)

reduced by half, while the retention of human serum albumin was almost completely lost. Therefore, the addition of arginine produced a dramatic effect, from the original co-elution to the baseline separation, which greatly increased the selectivity of the ligand for the two proteins. The unique retention change pattern of these proteins may be caused by the preferential interaction of arginine with specific areas on the protein surface.

As shown in Fig. 8.9, guanidine hydrochloride has a similar effect to arginine. In the presence of guanidine hydrochloride, many proteins showed a decrease in retention, but the magnitude of the decrease was much smaller than that observed with the same concentration of arginine. In addition, increased retention of several proteins in the presence of guanidine hydrochloride was also observed. They are mainly trypsin family, including bovine trypsin, porcine trypsin, chymotrypsin, and β-chymotrypsin. Similarly, because guanidine hydrochloride has different degrees of influence on the retention of different types of proteins, the selectivity of ligands for proteins is enhanced. For example, in the absence of guanidine hydrochloride, α-chymotrypsinogen A and horse cytochrome c co-elute, but in the presence of 0.1 M guanidine hydrochloride, the retention of α-chymotrypsinogen A is almost

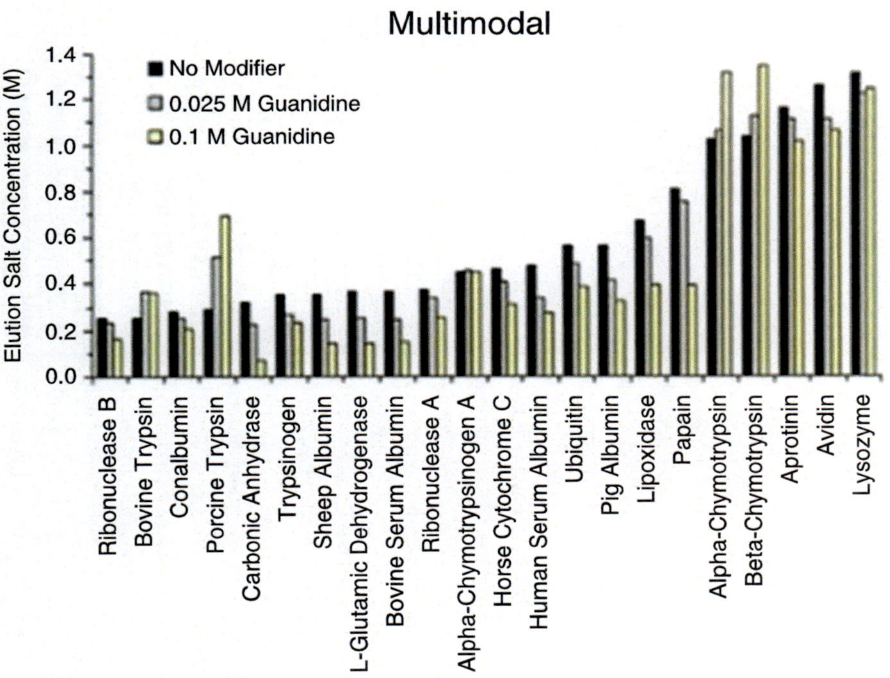

Fig. 8.9 Chromatographic retention data of a commercial protein library on a multimodal cation exchange resin (Capto MMC) in the absence and presence of various guanidine concentrations in the mobile phase. Linear salt gradient conditions: within 45 column volumes, in the presence of 0, 0.025, and 0.1 M guanidine, buffer A (20 mM sodium phosphate, pH 6) was changed to buffer B (20 mM sodium phosphate, 1.5 M chlorinated sodium, pH 6) for elution. (Reproduced from Holstein et al. 2012a)

unchanged. The retention of horse cytochrome c decreased by 30%. A reversal of the protein elution order was also observed. For example, in the absence of guanidine hydrochloride, aprotinin eluted after β-chymotrypsin, but after the addition of guanidine, the elution order of the two was reversed. Like arginine, guanidine hydrochloride causes unique changes in protein retention, which may be caused by changes in the solvent accessible area or water structure of proteins with different characteristics in the presence of guanidine.

As shown in Fig. 8.10, in the presence of sodium caprylate (10 and 25 mM, pH 6), all tested proteins showed a decrease in retention, but for the albumin family, including bovine, human, pig, and sheep protein, retention declined sharply. In addition, ubiquitin and L-glutamate dehydrogenase also showed a significant loss of retention. Caprylic acid may bind to the region composed of the positively charged residues of the protein and the adjacent hydrophobic patch, and this region is also the binding site of the multimodal cation exchange ligand. Therefore, caprylic acid causes a large decline in the retention of certain proteins by a competitive binding mechanism.

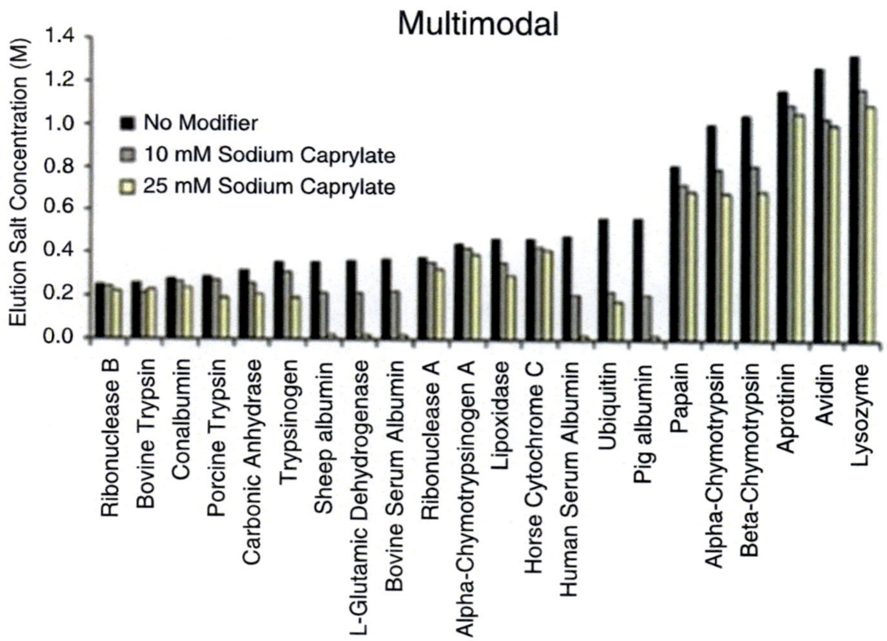

Fig. 8.10 Chromatographic retention data of a commercial protein library on a multimodal cation exchange resin (Capto MMC) when the mobile phase contains different concentrations of sodium caprylate. Linear salt gradient conditions: buffer A (20 mM sodium phosphate, pH 6) to buffer B (20 mM sodium phosphate, 1.5 M sodium chloride, pH 6) in the presence of 0, 10, and 25 mM sodium octanoate over 45 column volumes. (Reproduced from Holstein et al. 2012a)

In order to verify the competitive binding mechanism, Holstein et al. (2012a) used ubiquitin as a model protein to determine the NMR saturation transfer difference spectrum of the protein in the coexistence of ligand and caprylic acid. Saturation transfer difference (STD) spectrum is an effective method to study the binding of ligands to proteins. The basic principle is to irradiate the sample with an appropriate frequency to saturate some of the protons in the sample. During the resonance saturation period, the spin diffusion leads to the complete saturation of the protons in the entire protein, and the protein protons also transfer the saturation signal to the ligand protons immediately adjacent to the protein surface, causing its resonance signal to decrease. Therefore, if there is a competitive binding mechanism with ubiquitin between the multimodal ligand and caprylic acid, then caprylic acid will result in a further decrease in the resonance signal of the ligand. The results indicate that both ligand and caprylic acid can bind to ubiquitin, confirming that the competitive binding between caprylic acid and ligand is the main mechanism leading to decreased protein retention.

The influence of ethylene glycol and propylene glycol (20%, pH 6) on protein retention in multimodal chromatography shows a clear trend. Proteins with weaker retention show enhanced retention, while proteins with stronger retention show decreased retention. Both of these modifiers can increase the electrostatic interaction by reducing the dielectric constant of the mobile phase and weaken the hydrophobic interaction by changing the water structure and reducing the polarity of the mobile phase. The retention of weakly retained proteins is likely to be dominated by electrostatic interactions. Therefore, increasing the dielectric constant of the solution is conducive to the increase in retention. On the contrary, in the protein with strong retention, the hydrophobic effect plays a major role. Reducing the polarity of the solution helps the elution of the protein and reduces the retention. Propylene glycol is more hydrophobic than ethylene glycol, so its effect of reducing protein retention is more obvious. Like other mobile phase modifiers, alcohol modifiers can change the elution order of proteins and therefore affect the selectivity of mixed-mode chromatography.

Holstein et al. (2013) conducted an in-depth study on the role of urea in multimodal cation exchange chromatography Capto MMC. First, traditional cation exchangers and multimodal cation exchangers were used to examine the difference in retention behavior of various protein libraries in the presence of urea. Ubiquitin was used as a model protein to perform nuclear magnetic resonance spectroscopy titration experiments, and molecular dynamics simulations were used to determine the effect of urea on protein binding to ligands. The experimental results show that in the presence of 2 M urea, the retention of almost all test proteins on the traditional ion exchanger will slightly decrease, which can be attributed to the increase in the dielectric constant of the solution caused by urea, which reduces the electrostatic effect. Under 2 M urea, the dielectric constant of the solution increases by 7%. Proteins show significantly different responses to urea on multimodal ion exchange ligands. The retention of almost all test proteins decreased in the presence of urea, but different types of proteins had different decline rates, resulting in a significant change in the elution order, which resulted in a change in selectivity. The effect of urea on selectivity is mainly through weakening the hydrophobic interaction between ligand and protein. Both NMR titration experiments and molecular dynamics simulations using ubiquitin as a model protein show that urea regulates the interaction between multimodal ligands and proteins in a "regioselective" manner. In the presence of urea, the binding constants of different amino acid residues change differently, as observed in NMR experiments. Molecular dynamics simulation results show that in the presence of urea, the total probability of the interaction between ligand and ubiquitin is reduced, but the decline between different residues is different, resulting in the "regioselectivity" of urea.

8.4 Washing and Elution

The elution of the adsorbed solute from the mixed-mode chromatographic stationary phase essentially changes the distribution equilibrium of the solute between the two phases. Changing the composition of the mobile phase makes the solute adsorption equilibrium move to the desorption equilibrium. The mode of action of small molecule compounds and stationary phase ligands is relatively simple, so the elution of solutes can be easily achieved by changing or not changing the composition of the mobile phase. For example, in the reversed-phase ion exchange mixed-mode and the hydrophilic ion exchange mixed-mode, the elution of the solute can usually be achieved in an isocratic or gradient elution manner. In the isocratic elution mode, the composition of the mobile phase remains unchanged, and the solute molecules are eluted sequentially according to their hydrophobicity, hydrophilicity, or small differences in charge characteristics. In the gradient elution mode, a component in the mobile phase composition, such as organic solvent content or salt concentration, increases or decreases in chronological order, causing solutes with large differences in properties to elute within a reasonable time window. Another way to achieve gradient elution is segmented elution. The difference with linear gradient elution is that the change of the elution components in the mobile phase is stepwise or segment-wise. This method is simple and easy to implement and is widely used in large-scale preparation, separation, and purification.

In the mixed-mode chromatography of biological macromolecules, gradient elution or segmented elution is the more common way of solute desorption. First of all, the heterogeneity of biological samples is significant, and the difference between molecules is relatively large. This is often the case where one molecule appears in the flow-through fluid while the other molecule remains firmly in the column when an isocratic elution is used to elute the molecules. This situation is very common in single mode chromatography such as ion exchange and hydrophobic interaction chromatography, let alone mixed-mode chromatography. Secondly, the multivalent interaction between biomacromolecules and multifunctional ligands leads to complex and variable solute retention behavior. The interaction between proteins and mixed-mode ligands is usually based on a combination of electrostatic, hydrophobic, and hydrogen bonding. In order to disrupt or weaken this interaction, a large change in the elution composition in the mobile phase is often required. Careful consideration of the relative importance of these three interactions and the stability of the binding protein will help to select the appropriate eluent. Effectively eluting the adsorbed protein and maintaining protein activity may be the most challenging problem in mixed-mode chromatography. In addition, when the binding constant of the protein is very high, peak tailing and broadening of the desorption peak are usually more pronounced. This may be due to slow diffusion and mass transfer or slow adsorption kinetics or extensive multivalent binding forces in the system. The use of gradient elution can overcome the problems of peak tailing and peak broadening caused by multivalent effects.

8.4.1 Elution by pH Change

The change in pH of the eluent has two major effects on the adsorbed protein: electrostatic repulsion and hydrophilicity. In ion exchange chromatography, the pH change of the eluent can induce electrostatic repulsion between the protein and the ion exchanger. In hydrophobic interaction chromatography, the pH change of the eluent can increase the net charges on the protein and thus hydrophilicity of the protein. Both effects will contribute to the elution of the protein in mixed-mode chromatography.

Eriksson et al. (2009) compared the elution pH values of five different monoclonal antibodies on the multimodal anion exchanger Capto adhere and the traditional strong anion exchanger Capto Q. A descending pH gradient in bind-elute mode was used for elution. The elution order of monoclonal antibodies on the two stationary phases was the same, but the pH values at which the different antibodies from Capto adhere were much lower than those from Capto Q. This indicates that the retention of proteins on Capto adhere involves additional interactions and that multimodal functions can lead to different selectivity.

Holstein et al. (2012b) studied the effect of pH gradient on the binding and elution behavior of proteins on the multimodal cation exchanger Capto MMC. α-Lactalbumin, β-lactoglobulin A, and trypsin inhibitor with pIs of 4.5, 5.1, and 4.5, respectively, were used as model proteins to evaluate the separation efficiency of linear salt and pH gradients on the mixture. First, a linear gradient with the sodium chloride concentration increased from 0 to 1.5 M in 45 min was used. The salt gradient caused the three proteins to elute as a single large peak at 20 min (0.58 M sodium chloride), indicating that there was no separation of three proteins under these conditions. Adjusting the parameters of the salt gradient, including gradient slope, range, and pH value, produced no improvement in protein separation. Then an increasing pH gradient elution was used. The different gradient slopes and the influence of the background salt concentration were examined. Figure 8.11 shows the separation of three protein mixture using a linear pH gradient from pH 3.8 to 8.0 (0.05 pH units/min) in the absence (Fig. 8.11a) and the presence (Fig. 8.11b) of 50 mM sodium chloride. As can be seen from Fig. 8.11a, two different peaks are observed. The first peak contains α-lactalbumin and trypsin inhibitor, and the second peak consists of β-lactoglobulin A. Figure 8.11b shows the results of the same gradient (pH 3.8–8.0) in the presence of 50 mM sodium chloride. In this experiment, a relatively low salt concentration was used to weaken some electrostatic interactions while minimizing the impact on hydrophobic interactions. As shown in the figure, although the separation of α-lactalbumin and trypsin inhibitor is slightly improved in the presence of sodium chloride, the peaks are still not completely separated. In addition, the presence of sodium chloride reduces the retention of β-lactoglobulin A.

A further improvement was to change the slope of the pH gradient from 0.05 pH unit/min to 0.01 pH unit/min and reduce the high limit of the pH gradient from pH 8 to 5.5. Figure 8.12 shows the results of this gradient for a protein mixture. Three different protein elution peaks can be clearly seen from the figure. Although a shallower gradient reduces the sharpness of the peaks, it achieves better protein resolution. These results show that through the proper use of pH gradients, separations that cannot be achieved with salt gradients can be achieved.

Bacillus licheniformis 9945a α-amylase is an effective raw starch hydrolyzing enzyme. One feature of raw starch digestive enzymes is that they are adsorbed on raw starch granules through carbohydrate binding domains or surface binding sites.

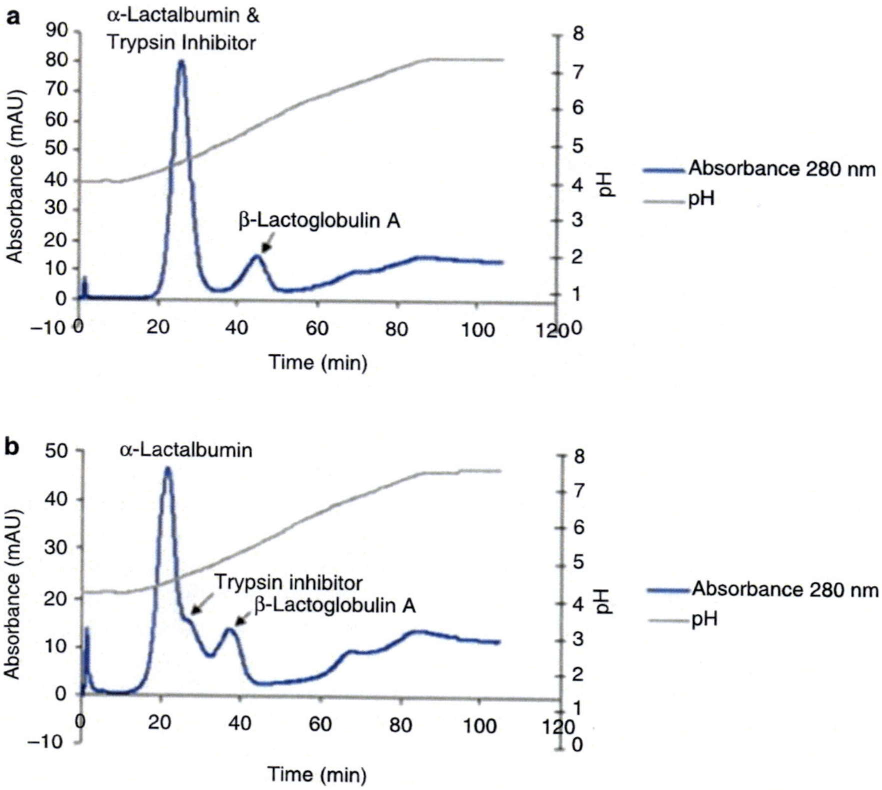

Fig. 8.11 Chromatographic separations of a mixture of α-lactalbumin, β-lactoglobulin A, and trypsin inhibitor on a multimodal resin using a pH gradient (pH 3.8–8.0, slope of 0.05 pH units/min) for elution in the absence (**a**) and presence (**b**) of 50 mM sodium chloride. (Reproduced from Holstein et al. 2012b)

In most cases, the surface binding site consists of tyrosine and tryptophan residues exposed on the surface of the enzyme. Hydrophobic interaction chromatography is usually destructive to the target protein and leads to a decrease in yield. However, in the case of raw starch digestive amylase (RSDA), hydrophobic interaction is a property of substrate binding, so it is expected that the recovery rate from mixed-mode resins is high. Lončar et al. (2015) used mixed-mode resin Nuvia cPrime™ combined with an elution strategy of simultaneously changing pH and salt concentration to purify recombinant starch hydrolyzing enzymes from *E. coli* cell culture fluid in one step. The crude enzyme preparation with pH 5.3, and conductivity of 18.3 mS/cm was loaded on the hydrophobic cation exchange column, washed to baseline with 50 mM, pH 5.3 sodium acetate equilibration buffer containing 150 mM NaCl, and then with 30 mM TrisHCl buffer, pH 8.0 containing 0.5 M NaCl to elute the adsorbed amylase. A 96% yield was obtained directly from the

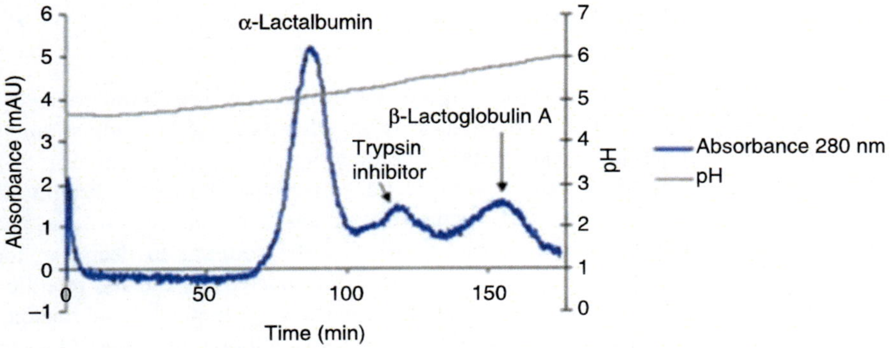

Fig. 8.12 Chromatographic separation of a mixture of α-lactalbumin, β-lactoglobulin A, and trypsin inhibitor on a multimodal resin using a pH gradient (pH 3.8–5.5, slope of 0.01 pH units/min) for elution. (Reproduced from Holstein et al. 2012b)

fermentation broth. This single-step process can simultaneously concentrate protein, remove pigments, and purify amylase.

8.4.2 *Elution by Ionic Strength Change*

Under the condition that the pH value of the mobile phase remains unchanged, increasing the ionic strength of the solution will elute the proteins that are mainly bound by electrostatic interactions. As mentioned earlier, in mixed-mode chromatography, protein adsorption usually occurs through a combination of charge, hydrophobic, and hydrogen bonding interactions. Charge interactions can be suppressed by increasing the ionic strength of the mobile phase and can be cancelled from the coupling with other nonionic interactions, thereby promoting the desorption of adsorbed proteins. 1 M NaCl is usually sufficient to elute the adsorbed protein.

In hydrophobic interaction chromatography, the protein is adsorbed to the stationary phase at high salt concentration and eluted at low salt concentration, so the elution mode with reduced salt concentration is mainly used. This elution mode is not commonly used in mixed-mode chromatography. The main reason is that in mixed-mode chromatography, the ligand density of the stationary phase is relatively high and the hydrophobic effect is relatively strong. Even at low salt concentrations, it cannot induce protein desorption. Reducing the salt concentration in the mobile phase is not an effective strategy for elution in mixed-mode chromatography.

8.4.3 Elution by Polarity Change

In reversed-phase ion exchange or hydrophilic ion exchange mixed chromatography, changing the polarity of the mobile phase is an effective means to adjust solute retention. In reversed-phase ion exchange chromatography, a binary mixture of a polar organic solvent and a buffer with a certain pH value and ionic strength are commonly used. Among them, the organic solvent acts as a polarity regulator, and the increase in its content causes the polarity of the solution to decrease, the hydrophobic interaction of the solute and the ligand decreases, and the retention decreases. The most commonly used polarity modifiers are methanol and acetonitrile. Methanol is cheap, but the viscosity of the mixed solution with water is higher; acetonitrile is slightly more expensive, but the viscosity of the solvent is low, showing excellent mass transfer characteristics.

In hydrophilic ion exchange mixed mode chromatography, the mobile phase is also composed of a binary mixture of organic solvent and buffer. Acetonitrile is the most commonly used polarity modifier. Increasing the content of organic solvents reduces the polarity of the solution, strengthens the hydrophilic interaction of solutes and ligands, and enhances retention. Decreasing the content of organic solvents will result in weaker hydrophilicity and solute elution.

In mixed-mode chromatography of proteins, when the solutes are strongly bound to the stationary phase and are affected by hydrophobic interactions, a more severe elution method than reducing the salt concentration must be used, such as reducing the polarity of the mobile phase. Arakawa et al. (2010) described the use of this type of elution method in hydrophobic charge induction chromatography to recover adsorbed monoclonal antibodies. Figure 8.13 shows the chromatogram of the anti-IL8 monoclonal antibody (IL8-mAb) adsorbed on MEP HyperCel resin eluted with

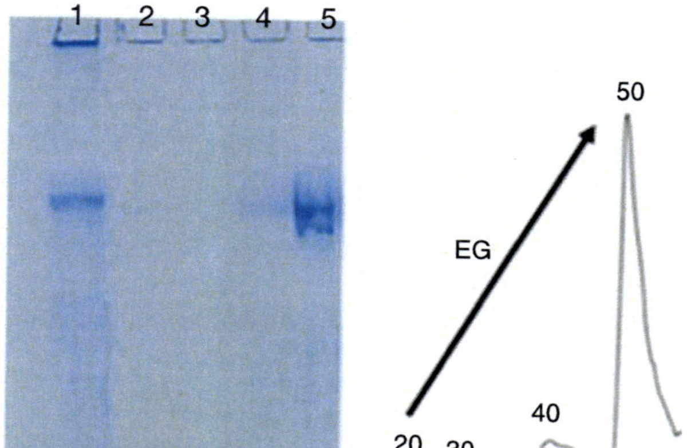

Fig. 8.13 Chromatogram of IL8-mAb eluted from MEP HyperCel. The bound protein was eluted using increasing concentrations of ethylene glycol (EG) in 10 mM phosphate pH 7.0. Lanes in SDS-PAGE: 1, load; 2, 20% EG elution fraction; 3, 30% EG elution fraction; 4, 40% EG elution fraction; 5, 50% EG elution fraction. The concentration of EG is shown in the chromatogram. (Reproduced from Arakawa et al. 2010)

10–50% ethylene glycol (EG). When the EG content in the mobile phase is 20–30%, only a small amount of protein is eluted. However, as the EG content in the mobile phase increases, the band of the eluted protein gradually increases. At 50%, a high intensity elution peak can be observed. Gel electrophoresis showed that the elution peaks of 20% and 30% EG contained no protein (lanes 2 and 3) and higher concentrations of EG could elute the antibody IL8-mAb (lanes 4 and 5). The absorption peaks without protein bands observed in 20–30% of EG indicate that low concentrations of EG cannot effectively elute proteins but may effectively remove ultraviolet absorbing substances present in cell culture. Since high concentration of EG can elute IL8-mAb, its lower concentration (less than 40%) can be used to wash contaminants.

8.4.4 Elution by Additives

In addition to polar modifiers, chaotropic salts, denaturants, or aggregation inhibitors are also commonly used elution additives in protein chromatography to inhibit nonspecific hydrophobic interactions between solutes and ligands. High concentrations of chaotropic salts and denaturants can cause protein unfolding and denaturation, but low to moderate concentrations of chaotropic salts and denaturants have the effect of increasing the solubility of proteins and stabilizing protein structure. Commonly used chaotropic salts are KSCN and KI, with a concentration range of 1–3 M. Commonly used denaturants are urea and guanidine hydrochloride at a moderate concentration (4–6 M).

Arginine is a basic amino acid, which has an inhibitory effect on protein aggregation, but does not have a denaturing effect. Arakawa et al. (2009, 2010) reported that arginine can effectively elute adsorbed antibodies from MEP HyperCel under neutral pH conditions. MEP HyperCel contains a pyridine ligand, which is uncharged at neutral pH and binds proteins through hydrophobic interactions. At weakly acidic pH, pyridyl groups are dissociated, and the adsorbed proteins are eluted through charge repulsion. The bound antibody and Fc fusion protein can also be eluted using 1 M arginine solution at pH 7.0. In addition, at this pH, ethylene glycol, glycerol, isopropanol, and urea can also elute the bound protein to varying degrees. Therefore, low concentrations of ethylene glycol, glycerol, isopropanol, urea, and arginine are effectively used to wash the loaded resin and remove the adsorbed impurities. Molecular dynamics simulations showed that the interaction between the alkyl chain of the arginine side chain and the end ring structure of the MEP resin ligand resulted in the decrease of the hydrophobicity of the MEP ligand and the elution of the adsorbed protein (Hirano et al. 2014).

Ren et al. (2014) developed a new competitive elution method, using clathrates for competitive elution in hydrophobic charge induction chromatography (HCIC), and applied it to antibody elution under neutral pH conditions. The interaction between 4-mercaptoethylpyridine (MEP), a typical ligand of HCIC, and four different types of cyclodextrins (CD) was studied by molecular docking. CD's immunoglobulin G (IgG) elution capacity is characterized on MEP-based HCIC medium. The results show that the CD-based eluent has a general effect on HCIC. This displacement eluent can effectively elute the bound antibody in a wide range of pH and ionic strength. Using 15 mM β-CD, human IgG can be eluted at physiological pH, with an average IgG recovery of 87%. When using this elution strategy to directly separate antibodies from human serum, the bound IgG can be fully eluted at pH 7.4, and the product purity is equivalent to the traditional method using acidic buffers. This strategy for elution in HCIC is particularly useful for proteins that are sensitive to acidic conditions, such as antibodies.

Wolfe et al. (2014) used a set of four monoclonal antibodies and two model proteins (RNase and lysozyme) to characterize the interaction of the protein with the multimodal cation exchanger Capto MMC. Under a linear gradient (0–2.2 M) of sodium chloride containing various mobile phase modifiers (ammonium sulfate, urea, potassium thiocyanate, ethylenediamine, arginine), the elution concentrations for these proteins were measured in mixed-mode chromatography. With the exception of sodium sulfate, almost all modulators reduce the binding affinity of proteins to ligands. In the presence of 1 M ammonium sulfate, lysozyme and all four mAbs bound to the resin very firmly and did not elute even at the highest salt concentration (2.2 M). It shows that in the presence of high concentration of salting-out salt, the mixed-mode binding of the proteins is converted to hydrophobic binding. On the contrary, under the same conditions, RNase always eluted in the flow-through fraction, indicating that the binding of the protein to the ligand was driven by electrostatic interaction. They further evaluated the use of mobile phase modifiers for intermediate washing to remove host cell protein (HCP) contaminants bound to the target antibody. Single or several mobile phase regulators are added to the washing buffer, the adsorbed target antibody is washed and eluted, and then the HCP content in the product eluate pool is analyzed. The results show that when the optimal selective washing method is incorporated into the antibody purification process, the HCP level can be reduced while maintaining the step yield.

Hirano et al. (2015) used bovine serum albumin (BSA) and anti-interleukin 8 monoclonal antibody (mAb-IL8) as model proteins to examine the effect of arginine on the elution behavior of the proteins on multimodal anion exchanger Capto adhere. The negatively charged BSA binds to the positively charged Capto adhere and can be easily eluted from the column by stepwise or gradient elution using 1 M NaCl at pH 7.0. For heat-treated BSA, NaCl gradient can also be used to elute small oligomers and remaining monomers, while larger oligomers require arginine to be effectively eluted, indicating that arginine has a stronger elution ability. Positively charged mAb-IL8 binds to Capto adhere at pH 7.0. The adsorbed antibody cannot be eluted by 1 M NaCl but can be completely eluted by 1 M arginine (Fig. 8.14). The excellent elution ability of arginine means that it interacts with the Capto adhere ligand, which reduces the hydrophobic interaction between the protein and the ligand and increases the electrostatic repulsion. The free energy of binding between arginine molecule and Capto adhere ligand was calculated by molecular dynamics simulation, and the mechanism of interaction between arginine and Capto adhere was studied. The affinity of arginine for Capto adhere can be attributed to the

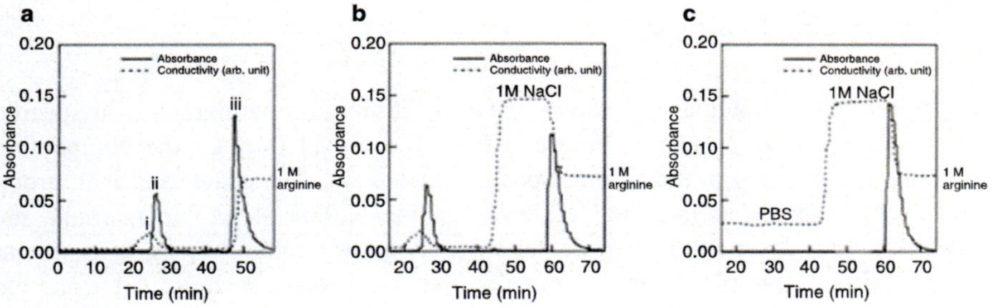

Fig. 8.14 Chromatograms of monoclonal antibody mAb-IL8 on Capto adhere chromatography. Component ii is the washing peak of bound antibody, and component iii is the elution peak of bound antibody. The solid line represents the absorbance of the solution, and the dotted line represents the conductivity of the solution. (Reproduced from Hirano et al. 2015)

hydrophobic and π-π interactions between the side chain of arginine and the aromatic part of the ligand, as well as the hydrogen bond between arginine and the hydroxyl group of the ligand.

Parimal et al. (2017) reported that the addition of arginine or guanidine has a significant effect on the retention of proteins in a multimodal cation exchange chromatography system. The experimental results of linear salt gradient obtained on the Capto MMC resin system for commercially available protein libraries show that the addition of arginine significantly reduces the elution salt concentration of all proteins, but the degree of reduction varies with the protein. The effect of guanidine presents a different trend, some proteins have a decreased elution salt concentration while others have an increased elution salt concentration. To clarify the mechanism of action of these additives, they performed molecular dynamics (MD) simulations on the interaction of guanidine and arginine with three model proteins (ubiquitin, cytochrome C, and α-chymotrypsinogen A). The results showed that both guanidine and arginine mainly interact with negatively charged regions on the protein, which can be easily predicted using electrostatic potential diagrams. Then efficient and fast coarse-grained technology was used to perform protein surface characterization on a larger set of proteins. The data shows that the protein exhibiting increased retention in the presence of guanidine has a hydrophobic patch adjacent to the negatively charged patch on its surface. Guanidine neutralizes the negative charge on the protein through electrostatic interaction, which increases the hydrophobicity of the protein and enhances retention. In contrast, the protein exhibiting reduced retention in the presence of guanidine does not have a hydrophobic patch adjacent to the negatively charged patch. Guanidine neutralizes the negative charge of the protein, but the hydrophobicity of the protein does not increase at the same time, so the retention is reduced. On the contrary, arginine leads to a significant reduction in protein retention time, which is mainly due to arginine's competitive binding to resin ligands and the spatial position effect caused by binding to the protein.

8.4.5 Specific Elution

Specific elution strategies are widely used in affinity chromatography. In specific elution, the bound protein is desorbed from the ligand through the competitive binding of the eluent to the ligand. Specific eluents are most often used with group specific adsorbents to increase selectivity in the elution step. For example, the glycoprotein is desorbed from the lectin chromatography column by using the eluent of competitive carbohydrate; the histidine-labeled protein is desorbed from the immobilized metal affinity chromatography column by using the imidazole eluent. Elution is usually carried out with a lower concentration of eluent (5–100 mM). The concentration of a single eluent or pulsed concentrations and gradients of several different eluents can be used. Specific elution is usually carried out at neutral pH, so it is a gentle desorption method that hardly causes denaturation. If the eluent binds to the protein, it can be dissociated by desalting on a gel filtration column or by dialysis when the binding is strong.

The use of phosphate to elute acidic proteins in hydroxyapatite chromatography is another example of specific elution methods used in mixed-mode chromatography. Acidic protein is bound by metal coordination between the rich carboxyl group in its molecule and the calcium ion site on hydroxyapatite. The bound protein cannot be eluted even in the presence of 2 M NaCl. However, the mobile phase contains a certain concentration of phosphate, and elution can be achieved at a neutral pH. This is because phosphorate has stronger calcium ion binding capacity than carboxylate and effectively competes for calcium ion binding sites to replace proteins. Therefore, phosphoratge displacement elution can be regarded as an example of specific elution in mixed-mode chromatography, which is similar to imidazole elution in immobilized metal affinity chromatography.

8.4.6 Elution by Selective Displacement Chromatography

Displacement chromatography is a chromatographic technique in which a sample is placed on the head of a column and then displaced with a solute that is more strongly adsorbed than the components of the original mixture (Tugcu et al. 2003; Tugcu 2008). The result is that these components are resolved into continuous "rectangular" regions of high-concentration pure substances, rather than "peaks" separated by solvents. It is mainly a preparation technology; compared with other chromatographic methods, higher product concentration, higher purity, and higher throughput can be obtained. Chromatographic purification of proteins from complex mixtures can be very challenging, especially when the mixture contains similar retained proteins or when there is a need to enrich for trace components in the feed. In addition, when using traditional chromatography modes (such as linear gradients, isocratic chromatography) that require high resolution, the column loading capacity is usually limited. In these cases, displacement chromatography is an effective technique for purifying proteins from complex mixtures with high column loads in a variety of applications (Shukla et al. 2000a, b).

An important development of displacement chromatography is the development of low-molecular-weight displacers for protein purification in ion exchange systems (Cramer and Jayaraman 1993; Jayaraman et al. 1993, 1995). This research is significant because it is very different from the traditional view, which believes that high-molecular-weight polyelectrolyte polymers are required to displace proteins in the ion exchange system. Compared with high-molecular-weight polyelectrolyte displacement agents, low-molecular-weight displacement agents have obvious operational advantages. For example, if there is any overlap between the displacer and the target protein, standard size-based purification methods (e.g., size exclusion chromatography, ultrafiltration) can be used to easily extract the purified protein during the post-displacement process. In addition, the salt-dependent adsorption behavior of these low-molecular-weight displacers greatly promotes the regeneration of the chromatographic column. These displacers have been used in ion exchange systems for various high-resolution separations (Kundu et al. 1995a, b, 1998, 1997a, b; Kundu and Cramer 1997; Shukla et al. 1998a). In addition, displacement chromatography is also used to purify recombinant growth factors (Barnthouse et al. 1998), antigen vaccine proteins (Shukla et al. 1998b), and antisense oligonucleotides (Tugcu et al. 2001). In a few examples, displacement chromatography has used ion exchange, hydrophobic interaction, and reverse phase chromatography to purify proteins (Freitag and Breier 1995).

Morrison et al. (2010) conducted a detailed study on the elution method using selective displacers on ceramic hydroxyapatite (HA) chromatographic columns. From the linear gradient elution experiments involving sodium chloride and sodium phosphate, respectively, ribonuclease A (RNaseA), α-chymotrypsinogen A, myoglobin (myo), cytochrome C, and lysozyme (lys) were selected as the model proteins, parallel batch screening experiments were carried out on a displacer library composed of 55 small molecule compounds. Figure 8.15 shows the structure of some representative displacer molecules. According to the possible interaction with active sites on ceramic hydroxyapatite, 55 displacers are divided into different categories. For example, the category shown in Fig. 8.15a is composed of molecules that primarily interact with HA through cation exchange, while the category shown in Fig. 8.15b is composed of molecules that can interact through cation exchange and/or metal chelation. The library also includes molecular analogs (represented by red or blue arrows in Fig. 8.15). For example, the displacer 37 is an analog of the displacer 17, in which the former is capable of cation exchange and metal chelation, and the latter (having a methylated carboxyl group) is capable of only cation exchange interaction. By systematically studying the selective elution effect of displacers with different chemical structures, the nature of the interaction between displacers and HA is revealed.

By introducing positively charged groups such as amino groups, metal chelating groups such as carboxyl groups and/or hydrogen bonding groups such as hydroxyl groups into the design of the displacer, specific interaction sites on HA can be targeted, thereby improving the selectivity of the separation. High-throughput

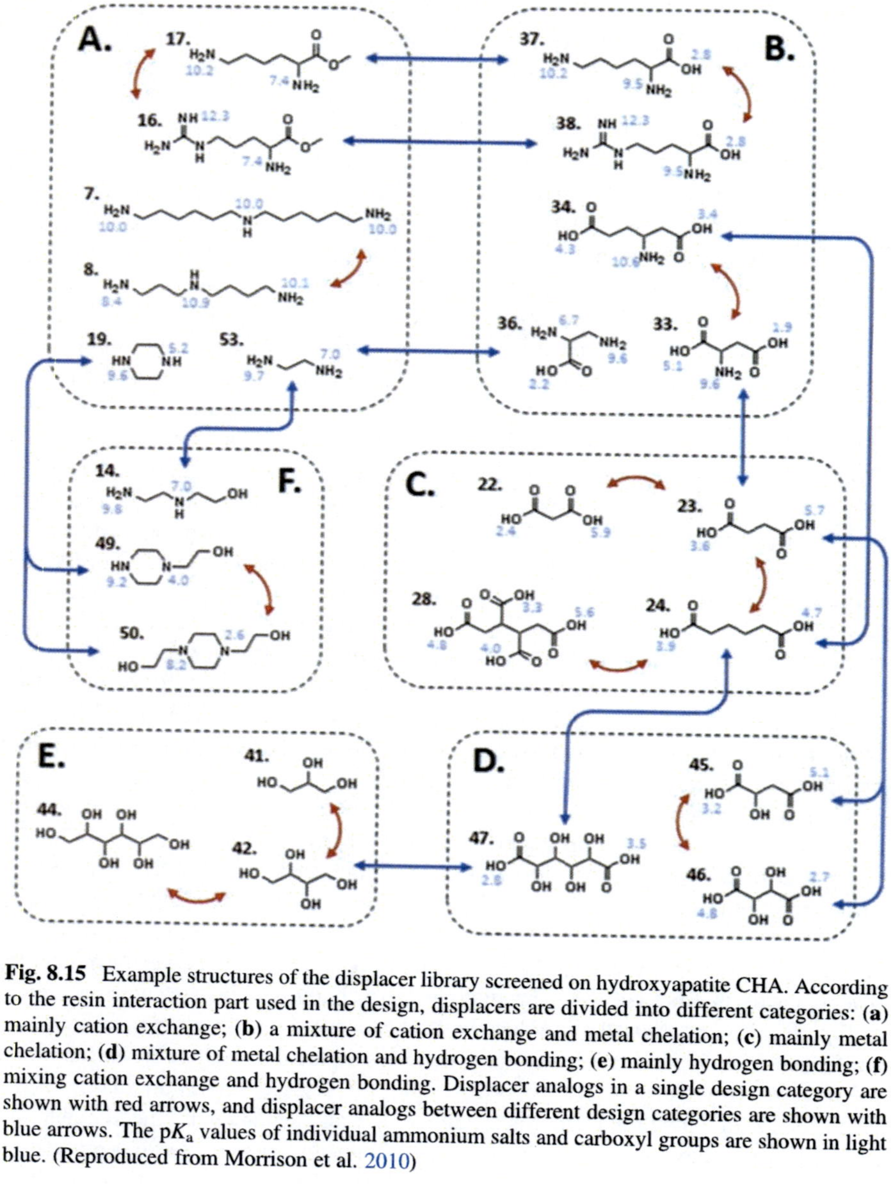

Fig. 8.15 Example structures of the displacer library screened on hydroxyapatite CHA. According to the resin interaction part used in the design, displacers are divided into different categories: **(a)** mainly cation exchange; **(b)** a mixture of cation exchange and metal chelation; **(c)** mainly metal chelation; **(d)** mixture of metal chelation and hydrogen bonding; **(e)** mainly hydrogen bonding; **(f)** mixing cation exchange and hydrogen bonding. Displacer analogs in a single design category are shown with red arrows, and displacer analogs between different design categories are shown with blue arrows. The pK_a values of individual ammonium salts and carboxyl groups are shown in light blue. (Reproduced from Morrison et al. 2010)

batch adsorption experiments found that the displacement efficiency of the displacer containing two carboxyl groups increases with the increase of the separation distance between the two carboxyl groups, due to the better geometric matching of the displacer and the bound calcium sites on the HA surface (Fig. 8.16). On the other hand, the displacement efficiency of the displacer containing three amino groups

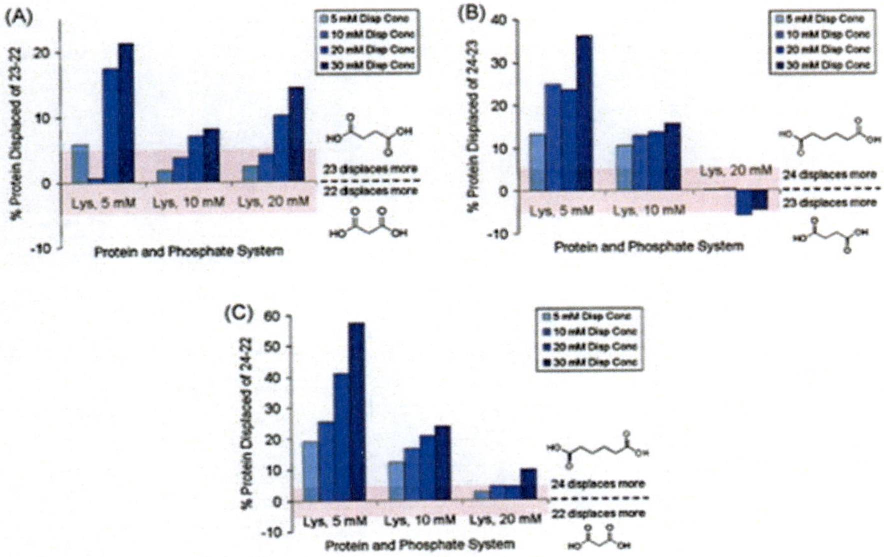

Fig. 8.16 Comparison of displacer efficacy, used to analyze the different separation distances between two carboxyl groups to replace lysozyme under different phosphate concentrations (5, 10, and 20 mM). The pink bar represents ±5% error of the displacement value. (**a**) Comparison of the separation distance between one carbon (displacer 22, malonic acid) and two carbons (displacer 23, succinic acid); (**b**) comparison of the separation distance between two carbons (displacer 23, succinic acid) and four carbons (displacer 24, adipic acid); (**c**) comparison of the separation distances between one carbon (displacer 22, malonic acid) and four carbons (displacer 24, adipic acid). (Reproduced from Morrison et al. 2010)

increases as the separation distance between the positively charged groups decreases, which may be due to the increase in the charge density on the displacer, which promotes its electrostatic interaction with the phosphate groups on the HA surface (Fig. 8.17). It is found that the displacer that uses both cation exchange and metal chelation interaction is less efficient than the displacer that uses only one interaction mode, which is probably due to the charge repulsion effect.

Using a suitable displacer/eluent, selective batch separation can be achieved between multiple protein pairs that cannot be separated by HA combined with a linear gradient. The use of displacer 8 with sodium phosphate resulted in the displacement of Lys and the elution of RNaseA. The synergistic use of displacer and phosphoric acid eluent results in a baseline separation between these proteins. However, if only the phosphoric acid linear gradient method or only the displacing agent is used, the baseline separation of these proteins cannot be obtained.

Using displacer 28 (butane tetracarboxylic acid) and borate as mobile phase modifiers will cause the above elution order to be reversed. RNaseA is displaced first, followed by Lys elution. Displacer 28 is a strong calcium ion complexing agent, and borate does not participate in the complexation with calcium ions. The binding of RNaseA and calcium ions on HA is cancelled due to the competition of the

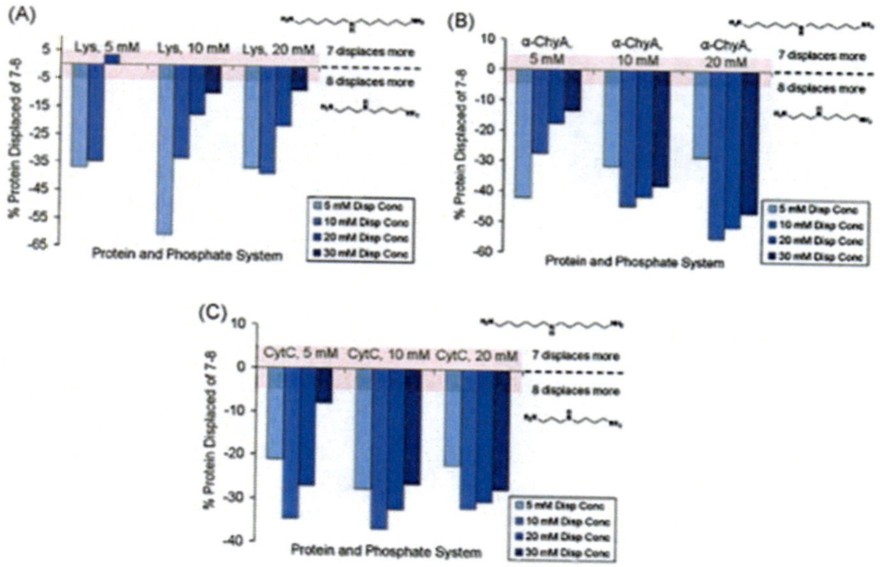

Fig. 8.17 Comparison of the efficacy of displacers with different separation distances between three amine groups, using different proteins and phosphate concentrations (5, 10, and 20 mM). Displacer 7, bis(hexamethylene)triamine and 8, spermidine, are used. The pink bar represents the ±5% error of the displacement value. (**a**) Lysozyme is used for comparison; (**b**) α-chymotrypsinogen A is used for comparison; (**c**) cytochrome C is used for comparison. (Reproduced from Morrison et al. 2010)

displacer, while Lys binds to the P-site of HA mainly through electrostatic interaction, it is eluted due to electrostatic shielding in high concentration of cations. This work proves that selective displacement/elution chromatography mode can effectively improve the selectivity of mixed-mode chromatography. Reasonable selection of displacer and buffer components can change the elution order of the protein while keeping the baseline separation unchanged.

Sheth et al. (2011) evaluated the effect of displacer libraries with different degrees of electrostatic, hydrophobic, and hydrogen bonding on the selectivity of the multimodal cation exchanger Capto MMC under high loading conditions. α-Chymotrypsinogen A (α-ChyA), cytochrome C (CytC), and ribonuclease A (RNaseA) were selected as model proteins, and an automated liquid handling system was used to screen a displacer library consisting of 38 commercially available compounds. Based on the possible interactions with mixed-mode ligands, displacers are divided into several different categories. For example, category 1 is mainly composed of molecules that bind to resin through electrostatic interactions; category 2 includes molecules that can bind to resin through electrostatic and hydrogen bonding interactions; category 3 is composed of molecules that may bind to the resin through hydrophobic, π-π, and electrostatic interactions. The selection of displacer analogues with different degrees of charge, charge density, hydrophobicity, and hydrogen bonding potential will help to determine the effect of subtle

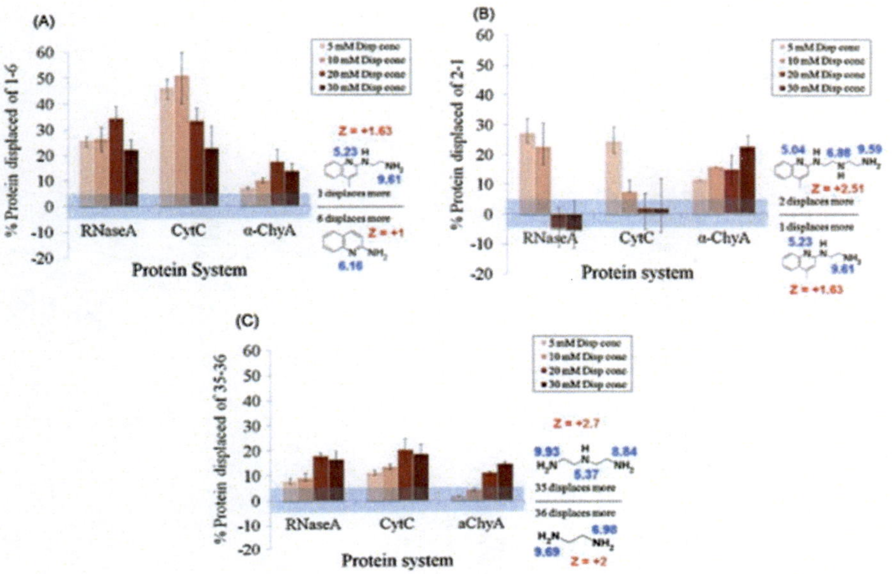

Fig. 8.18 Comparison charts of displacer efficacy, used to analyze the effect of net charge on displacer on RNaseA, CytC, and α-ChyA. The *y*-axis of each graph represents the difference in percentage of displaced protein between the two displacer analogs. The blue bar represents the ±5% error of the displacement value. The net charge (Z) of each displacer is indicated in red with the appropriate symbol. (**a**) Comparison between displacer 1 (Z = +1.63) and displacer 6 (Z = +1); (**b**) comparison between displacer 2 (Z = +2.51) and displacer 1 (Z = +1.63), (**c**) comparison between displacer 35 (Z = +2.7) and displacer 36 (Z = +2). (Reproduced from Sheth et al. 2011)

differences in these compounds on the efficiency of protein replacement and interaction selectivity on Capto MMC resin. For example, the displacers 36 and 23 are analogs, the former can only carry out electrostatic interactions, while the latter has an additional alkyl tail, which can carry out coupled electrostatic interaction-hydrophobic interactions.

First, the effect of the charge of the displacer on the displacement of protein from the multimodal cation exchanger was investigated. Figure 8.18 shows the results of the displacement of three model proteins by three analogue pairs with different charges. The displacement ability of the displacer increases with the increase of its charge. Figure 8.18a shows the displacer efficacy diagrams of displacer 1 and 6, where displacer 1 has an elongated amine tail and an additional positive charge relative to displacer 6. Under the condition of pH 5, the net charges of displacer 1 and 6 are +1.63 and +1, respectively. At all concentrations, the three proteins displaced by displacer 1 were higher than those displaced by displacer 6. For these experiments (and most screens), CytC was the most severely displaced protein, followed by RNaseA, and then α-ChyA.

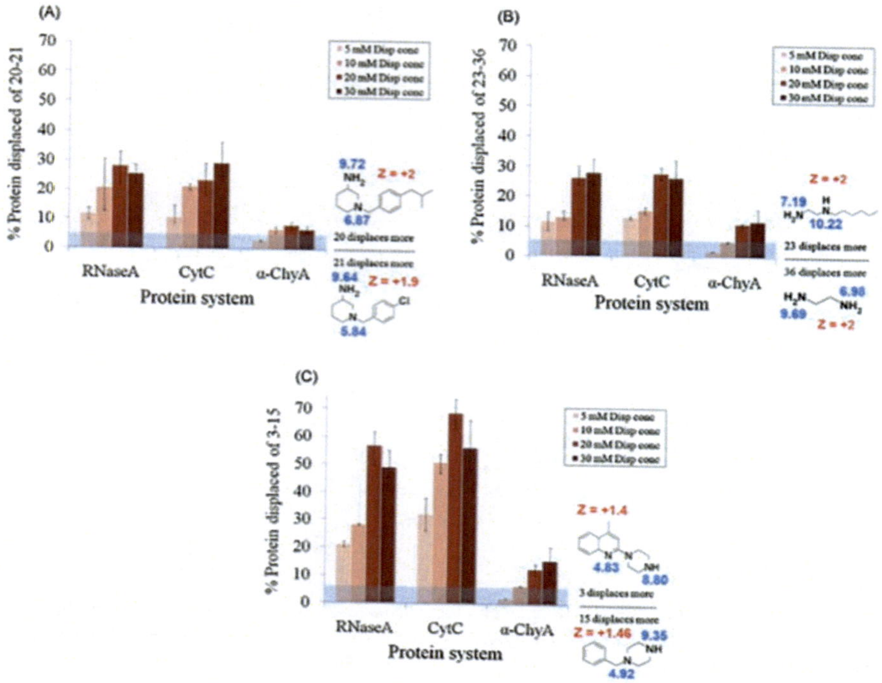

Fig. 8.19 Comparison charts of displacer's efficacy, used to analyze the influence of displacer's hydrophobicity on the displacement of RNaseA, CytC, and α-ChyA. The *y*-axis of each graph represents the difference in protein displacement percentage between the two replacer analogs. The blue bar indicates that the error of the replacement value is ±5%. The net charge (*Z*) of each displacer is indicated in red with the appropriate symbol. (**a**) Comparison between displacer 20 ($Z = +2$) and displacer 21 ($Z = +1.9$); (**b**) comparison between displacer 23 ($Z = +2$) and displacer 36 ($Z = +2$); (**c**) comparison between displacer 3 ($Z = +1.4$) and displacer 15 ($Z = +1.46$). (Reproduced from Sheth et al. 2011)

Figure 8.18b compares displacer 2 and 1, where displacer 2 again has an elongated amine tail and an additional positive charge, and the net charge of +2.51, high than +1.64 for the displacer 1. Although similar protein displacement trends were observed with this group of displacers, once the displacer concentration reached 20 mM, there was basically no difference in the relative efficacy of displacing RNaseA and CytC. Figure 8.18c compares displacer 35 and 36, where displacer 35 (+2.7) has a higher net positive charge than displacer 36 (+2). Although the chemical properties of displacers 35 and 36 are different from those of displacers 1, 2, and 6, the same trend is observed again, with all three proteins being displaced by displacers with a higher net charge.

Next, the influence of the hydrophobicity of the displacer on protein was investigated. Three displacer analogs with similar net charges, but different hydrophobicity were selected for comparison. The results are shown in Fig. 8.19. The displacement ability of the displacer increases with the increase in hydrophobicity, and the displacer with higher hydrophobicity can displace more protein than its counterpart with lower hydrophobicity.

As shown in Fig. 8.19a, the displacers 20 and 21 have similar net charges (+2 and +1.9, respectively), but the former has an additional isobutyl tail, so it has higher hydrophobicity and increased hydrophobic interactions, leading to enhanced protein displacement efficiency. As shown in Fig. 8.19b, the displacer 36 mainly binds to the ligand through electrostatic interaction, and the displacer 23 has one more *n*-hexyl group than the displacer 36, which can perform electrostatic and hydrophobic interactions at the same time, resulting in stronger binding force and improved displacement effect. Finally, Fig. 8.19c compares displacer 3 with 15, where displacer 3 has an additional aromatic ring. Displacer 3 can more effectively displace all three proteins under all conditions.

Finally, the influence of the charge density of the displacer on the efficacy of the displacer was investigated. The results of the two analogue pairs are shown in Fig. 8.20. Figure 8.20a compares displacer 12 and displacer 13, where the charge density of displacer 13 is lower than that of displacer 12, and the net charge of the two compounds is the same (+1.94). The displacer 13 has a greater displacement effect on all three proteins than the displacer 12. The authors proposed three possible explanations. Increasing the alkyl chain length between the charged parts from 3 carbons to 5 carbons can lead to an increase in the flexibility of the displacer. This allows it to have hydrophobic and electrostatic interactions with a single ligand on the surface of the resin, thereby obtaining a more favorable conformation to bind to the Capto MMC ligand. In addition, compared with the displacer 12, the longer chain length and improved flexibility of the displacer 13 can make the H bond interaction with the resin a more favorable conformation. Another possible explanation for the increased efficacy of this displacer may be that the alkyl group of the displacer is longer, resulting in increased hydrophobic interaction with the Capto MMC ligand, thereby increasing its affinity for the resin. Finally, the increased effective length of the displacer 13 can also play a role in the process of interacting with two adjacent MM ligands on the surface. The results shown in Fig. 8.20a indicate that the synergistic interaction between the displacer and the resin may play an important role in the overall performance of the displacer.

Figure 8.20b compares displacers 32 and 34, where displacer 32 has a lower charge density than displacer 34, and both compounds have the same net charge. Contrary to the results of displacers 12 and 13, displacers 32 and 34 showed mixed displacement behavior of these three model proteins. The degree of displacement of α-ChyA by the displacer 34 is slightly more, while the degree of displacement of RNaseA by the displacer 32 is greater, and there is basically no difference in the results of CytC displacement. This indicates that the reduced electrostatic interaction with the displacer 32 due to the low charge density can be compensated by the increase in the hydrophobic interaction induced by the six carbon chains between the amine groups in the displacer. Due to these synergistic effects, the displacer 32 is more firmly bound to the resin than the displacer 34. The reason why α-ChyA is displaced by displacer 34 may be more due to the "chemoselective" displacer effect, in which the displacer directly interacts with the protein itself, thereby increasing the

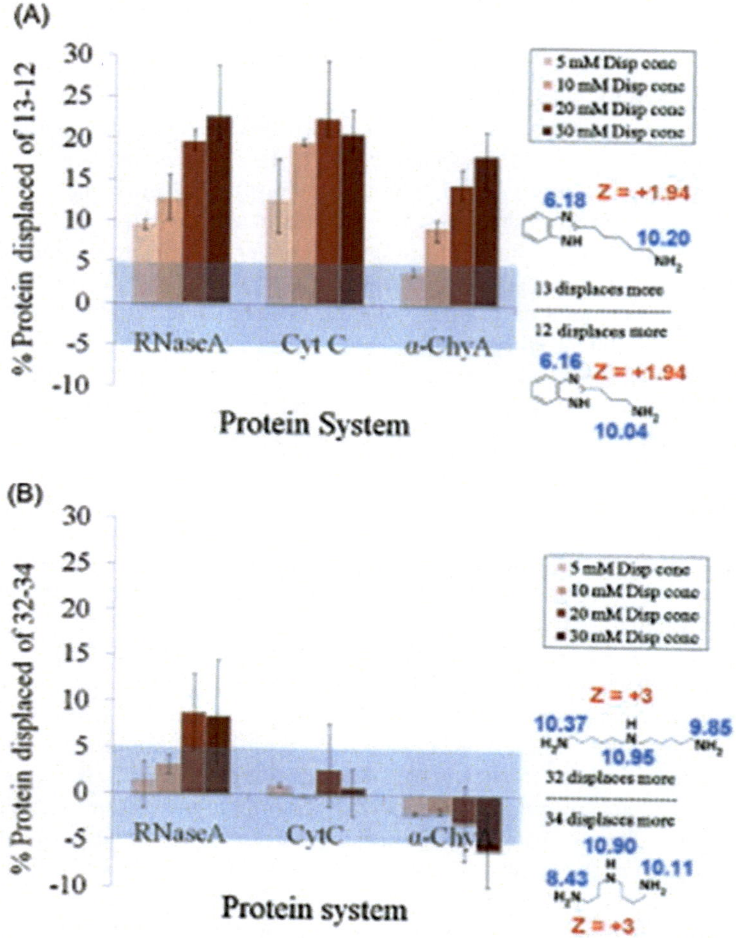

Fig. 8.20 Comparison of displacer's efficacy, used to analyze the effect of displacer's charge density on displacement of RNaseA, CytC, and α-ChyA. The *y*-axis of each graph represents the difference in protein displacement percentage between the two displacer analogs. The blue bar indicates that the error of the replacement value is ±5%. The net charge (*Z*) of each displacer is indicated in red with the appropriate symbol. (**a**) Comparison between displacer 13 ($Z = +1.94$) and displacer 12 ($Z = +1.94$); (**b**) comparison between displacer 32 ($Z = +3$) and displacer 34 ($Z = +3$). (Reproduced from Sheth et al. 2011)

binding affinity of the protein to the resin and reducing "displacement." The comparison in Fig. 8.20a, b reveals the relative importance of net charge, hydrophobicity, and charge density in the design of displacers for MM systems.

8.5 Summary

In mixed-mode chromatography, the composition of the mobile phase plays an important role in controlling solute binding, washing, and elution. The pH of the mobile phase can affect the dissociation state of the solute and the stationary phase ligand, thereby changing the adsorption and elution behavior of the solute. Both the ionic strength and the organic solvent in the mobile phase have a dual role. Ionic strength can shield electrostatic interactions and enhance the elution ability in ion exchange mode and the binding ability in hydrophobic interaction mode. The organic solvent can reduce the polarity of the solution and enhance the elution ability in reversed-phase mode and the binding ability in hydrophilic mode. In addition, detergents, chaotropic salts, denaturants, and aggregation inhibitors can be used flexibly as mobile phase additives to adjust solute retention, remove pollutants, and improve protein activity recovery. The displacement chromatography elution mode is an effective way to improve the selectivity of biomacromolecules in conventional chromatography, and it has a wide range of applications in ion exchange chromatography, hydrophobic interaction chromatography, and affinity chromatography. This elution mode can also play an important role in mixed-mode chromatography. Existing studies have shown that choosing the right displacer is the key to the success of the experiment. The structural characteristics of the displacer should be compatible with those of the mixed-mode ligand, so as to maximize the interaction between them. These structural features include hydrophobicity, molecular rigidity, and charge density and distribution.

References

Anandan A, Vrielink A (2016) Detergents in membrane protein purification and crystallisation. In: Moraes I (ed) The next generation in membrane protein structure determination, Advances in experimental medicine and biology, vol 922. Springer, Cham. https://doi.org/10.1007/978-3-319-35072-1_2

Arakawa T, Kita Y, Sato H et al (2009) MEP chromatography of antibody and Fc-fusion protein using aqueous arginine solution. Protein Expr Purif 54:110–116

Arakawa T, Futatsumori-Sugai M, Tsumoto K et al (2010) MEP HyperCel chromatography II: binding, washing and elution. Protein Expr Purif 71:168–173

Barnthouse KA, Trompeter W, Jone R et al (1998) Cation-exchange displacement chromatography for the purification of recombinant protein therapeutics from variants. J Biotechnol 66:125–136

Ben-David A, Firer MA (1996) Immunoaffinity purification of monoclonal antibodies: in search of an elution buffer of general applicability. Biotechnol Tech 10:799–802

Burton SC, Harding DRK (1998) Hydrophobic charge induction chromatography: salt independent protein adsorption and facile elution with aqueous buffers. J Chromatogr A 814:71–81

Cramer SM, Jayaraman G (1993) Preparative chromatography in biotechnology. Curr Opin Biotechnol 4:217–225

Damhof RA, Feijlbrief M, Welling-Wester S et al (1994) Purification of the integral membrane glycoproteins D of Herpes simplex virus types 1 and 2, produced in the recombinant baculovirus

expression system, by ion-exchange high-performance liquid chromatography. J Chromatogr A 676:43–49
Downham M, Busby S, Jefferis R et al (1992) Immunoaffinity chromatography in biorecovery: an application of recombinant DNA technology to generic adsorption processes. J Chromatogr B Biomed Sci App 584:59–67
Durkee KH, Roh BH, Doellgast GJ (1993) Immunoaffinity chromatography purification of Russell's viper venom factor X activator using elution in high concentrations of magnesium chloride. Protein Expr Purif 4:405–411
Eriksson K, Ljunglöf A, Rodrigo G et al (2009) MAb contaminant removal with a multimodal anion exchanger. BioProcess Int 7:52–56
Freitag R, Breier J (1995) Displacement chromatography in biotechnological downstream processing. J Chromatogr A 691:101–112
Hirano A, Maruyama T, Shiraki K et al (2014) Mechanism of protein desorption from 4-mercaptoethylpyridine resins by arginine solutions. J Chromatogr A 1373:141–148
Hirano A, Arakawa T, Kameda T (2015) Effects of arginine on multimodal anion exchange chromatography. Protein Expr Purif 116:105–112
Holstein MA, Parimal S, McCallum SA et al (2012a) Mobile phase modifier effects in multimodal cation exchange chromatography. Biotechnol Bioeng 109:176–186
Holstein MA, Nikfetrat AAM, Gage M et al (2012b) Improving selectivity in multimodal chromatography using controlled pH gradient elution. J Chromatogr A 1233:152–155
Holstein MA, Parimal S, McCallum SA et al (2013) Effects of urea on selectivity and protein-ligand interactions in multimodal cation exchange chromatography. Langmuir 29:158–167
Hou Y, Cramer SM (2011) Evaluation of selectivity in multimodal anion exchange systems: a priori prediction of protein retention and examination of mobile phase modifier effects. J Chromatogr A 1218:7813–7820
Jayaraman G, Gadam S, Cramer SM (1993) Ion-exchange displacement chromatography of proteins. Dextran-based polyelectrolytes as high affinity displacers. J Chromatogr A 630:53–68
Jayaraman G, Li Y, Moore JA et al (1995) Ion-exchange displacement chromatography of proteins dendritic polymers as novel displacers. J Chromatogr A 702:143–155
Kang B, Tang H, Zhao Z et al (2020) Hofmeister series: insights of ion specificity from amphiphilic assembly and interface property. ACS Omega 5:6229–6239
Kasai R, Yamaguchi H, Tanaka O (1987) High-performance liquid chromatography of glycosides on a new type of hydroxyapatite column. J Chromatogr A 407:205–210
Kawasaki T (1991) Hydroxyapatite as a liquid chromatographic packing. J Chromatogr 544:147–184
Kim HO, Durance TD, Li-Chan EC (1999) Reusability of avidin-biotinylated immunoglobulin Y columns in immunoaffinity chromatography. Anal Biochem 268:383–397
Kundu A, Cramer SM (1997) Low-molecular-weight displacers for high-resolution protein separations. Anal Biochem 248:111–116
Kundu A, Vunnum S, Jayaraman G et al (1995a) Protected amino acids as novel low-molecular-weight displacers in cation-exchange displacement chromatography. Biotechnol Bioeng 48:452–460
Kundu A, Vunnum S, Cramer SM (1995b) Antibiotics as low-molecular-mass displacers in ion-exchange displacement chromatography. J Chromatogr A 707:57–67
Kundu A, Barnthouse K, Cramer SM (1997a) Selective displacement chromatography of proteins. Biotechnol Bioeng 56:119–129
Kundu A, Shukla AA, Barnthouse KA et al (1997b) Displacement chromatography of proteins using sucrose octasulfate. BioPharm 10:64–70+95
Kundu A, Vunnum S, Cramer SM (1998) Displacement chromatography of proteins using low molecular weight anionic displacers. Adsorption 4:3–4

Linke D (2009) Chapter 34 Detergents: an overview. Methods Enzymol 463:603–617
Lončar N, Slavić MŠ, Vujčić Z et al (2015) Mixed-mode resins: taking shortcut in downstream processing of raw-starch digesting α-amylases. Sci Rep 5:15772
McGillis JP, Organist ML, Payan DG (1997) Immunoaffinity purification of membrane protein constituents of the IM-9 lymphoblast receptor for substance P. Anal Biochem 164:502–513
Moldoveanu SC, David V (2013) Essentials in modern HPLC separations. Elsevier, Amsterdam
Moldoveanu SC, David V (2017) Chapter 13 Solvents, buffers, and additives used in the mobile phase, selection of the HPLC method in chemical analysis. Elsevier, Amsterdam
Morrison CJ, Gagnon P, Cramer SM (2010) Unique selectivity windows using selective displacers/eluents and mobile phase modifiers on hydroxyapatite. J Chromatogr A 1217:6484–6495
Narhi LO, Caughey DJ, Horan T et al (1997) Effect of three elution buffers on the recovery and structure of monoclonal antibodies. Anal Biochem 253:236–245
Neugebauer JM (1990) Chapter 18 Detergents: an overview. Methods Enzymol 182:239–253
Nord K, Gunneriusson E, Uhlen M et al (2000) Ligands selected from combinatorial libraries of protein A for use in affinity capture of apolipoprotein A-1 M and taq DNA polymerase. J Biotechnol 80:45–54
Parimal S, Garde S, Cramer SM (2017) Effect of guanidine and arginine on protein–ligand interactions in multimodal cation-exchange chromatography. Biotechnol Prog 33:435–447
Ren J, Yao P, Cao Y et al (2014) Application of cyclodextrin-based eluents in hydrophobic charge-induction chromatography: elution of antibody at neutral pH. J Chromatogr A 1352:62–68
Robinson J, Roush D, Cramer SM (2020) The effect of pH on antibody retention in multimodal cation exchange chromatographic systems. J Chromatogr A 1617:460838
Santos E, Tahara SM, Kaback HR (1985) Monoclonal antibodies against the membrane-bound, flavin-linked d-lactate dehydrogenase of Escherichia coli: preparation, characterization and the use of immunoaffinity chromatography. Biochemistry 24:3006–3011
Sheth RD, Morrison CJ, Cramer SM (2011) Selective displacement chromatography in multimodal cation exchange systems. J Chromatogr A 1218:9250–9259
Shukla AA, Barnthouse KA, Bae SS et al (1998a) Structural characteristics of low-molecular-mass displacers for cation-exchange chromatography. J Chromatogr A 814:83–95
Shukla AA, Hopfer RL, Chakravarti DN et al (1998b) Purification of an antigenic vaccine protein by selective displacement chromatography. Biotechnol Prog 14:91–101
Shukla AA, Sunasara KM, Rupp RG et al (2000a) Hydrophobic displacement chromatography of proteins. Biotechnol Bioeng 68:672–680
Shukla AA, Deshmukh RR, Moore JA et al (2000b) Purification of oligonucleotides by high affinity, low molecular weight displacers. Biotechnol Prog 16:1064
Thompson NE, Hager DA, Burgess DR (1992) Isolation and characterization of a polyol-responsive monoclonal antibody useful for gentle purification of Escherichia coli RNA polymerase. Biochemistry 31:7003–7008
Tsang VCW, Wilkins PP (1991) Optimum dissociating condition for immunoaffinity and preferential isolation of antibodies with high specific activity. J Immunol Methods 138:291–299
Tugcu N (2008) Purification of proteins using displacement chromatography. Methods Mol Biol 421:71–79
Tugcu N, Deshmukh RR, Sangvic YS et al (2001) Purification of an oligonucleotide at high column loading by high affinity, low-molecular-mass displacers. J Chromatogr A 923:65–73
Tugcu N, Deshmukh RR, Sanghvi YS et al (2003) Displacement chromatography of anti-sense oligonucleotide and proteins using saccharin as a non-toxic displacer. React Funct Polym 54:37–47
Uemura T, Hirai K (1999) Purification of L-kynurenine-3-monooxygenase from mitochondrial outer membrane of pig liver. Adv Exp Med Biol 467:619–623

Vagenende V, Yap MGS, Trout BL (2009) Mechanisms of protein stabilization and prevention of protein aggregation by glycerol. Biochemistry 48:11084–11096

Welling-Wester S, Feijlbrief M, Koedijk GAM et al (1998) Detergent extraction of herpes simplex virus type 1 glycoprotein D by zwitterionic and non-ionic detergents and purification by ion-exchange high-performance liquid chromatography. J Chromatogr A 816:29–37

Wolfe LS, Barringer CP, Mostafa SS et al (2014) Multimodal chromatography: characterization of protein binding and selectivity enhancement through mobile phase modulators. J Chromatogr A 1340:151–156

Zheng W, Borgia A, Buholzer K et al (2016) Probing the action of chemical denaturant on an intrinsically disordered protein by simulation and experiment. J Am Chem Soc 138:11702–11713

Chapter 9
Method Development

Abstract Method development in chromatography involves selecting the appropriate operating parameters for a given separation, including stationary phase, mobile phase, and elution method. Because mixed-mode chromatography is based on multivalent interactions between solutes and ligands, its method development is more complicated than single-mode chromatography. This is especially true when developing elution methods for strongly retained solutes. In this chapter, we will review the strategies used in hydroxyapatite chromatography, hydrophobic charge induction chromatography, and multimode ion exchange chromatography.

9.1 Introduction

Method development in mixed-mode chromatography is more complicated than in single-mode chromatography. This is because the nature of solute retention is multipotent, that is, the solute can have multiple intermolecular interactions with the chromatographic ligand or stationary phase. Canceling one of the interactions by changing the composition of the mobile phase may weaken the retention of the solute, but does not necessarily lead to the complete elution of the solute. Therefore, in the process of method development, it is usually necessary to conduct a comprehensive investigation on solute properties, ligand structure and mobile phase composition, and their influence on solute retention and separation and establish key process control parameters around key quality indicators, so as to better understand the chromatographic mechanism. On the basis of this, the mixed-mode chromatographic method is developed and optimized. In this chapter, we will give a general introduction to the method development process for several types of mixed-mode chromatography, mainly through a large number of examples to help understand the principles of method development.

Q.-H. Wan, *Mixed-Mode Chromatography*,
https://doi.org/10.1007/978-981-16-5485-5_9

9.2 Method Development in Hydroxyapatite Chromatography

The surface of hydroxyapatite is rich in positively charged calcium ions (C-site) and negatively charged phosphate ions (P-site). When they are in contact with biological macromolecules, both electrostatic interaction and metal chelation or coordination can occur depending on the functional groups in the solute molecule. In addition, the calcium ions and phosphate ions on the surface of hydroxyapatite are distributed in a regular pattern, which can interact simultaneously with multiple functional groups in biological macromolecules. Through multi-site (C-site and P-site) and multi-valence (multiple C-sites and P-sites), hydroxyapatite chromatography can distinguish subtle differences in the geometric arrangement of functional groups in a local area on the molecular surface, showing high affinity and selectivity for biological macromolecules. The C-site on the surface of hydroxyapatite is a positively charged calcium ion, which can chelate with the vicinal carboxyl groups in the protein and coordinate with the phosphate groups in the nucleic acid. The P-site on the surface is a phosphate ion, which can be electrostatically attracted to the positively charged amino groups in the protein, or electrostatically repelled with the phosphate groups in the nucleic acid. Studies have shown that under conditions of neutral pH, neutral and basic proteins bind to hydroxyapatite mainly through the P-site interaction, while acidic proteins bind mainly through the C-site interaction.

9.2.1 General Strategy

In the downstream process of biopharmaceuticals, hydroxyapatite offers some unique advantages. It can adsorb some difficult impurities from the feed, such as endotoxins, nucleic acids, virus, and aggregates. Chromosomal DNAs bind strongly to hydroxyapatite and may require up to 500 mM phosphate for elution. Endotoxins have many phosphate groups on their polysaccharide and lipid A parts. They also bind well to hydroxyapatite and may require up to 1 M phosphate to be completely eluted, although some subgroups of endotoxins can be eluted in the range of 0–0.5 M phosphate (Gagnon et al. 2006). When the phosphate gradient elution is performed at an elevated NaCl concentration, the retention of endotoxins will be reduced, which indicates that the binding of endotoxins contains a cation exchange component. Therefore, the NaCl gradient elution is usually more effective in removing endotoxins than the phosphate gradient elution in antibody purification (Gagnon 2009). The lipid-encapsulated virus also has a strong affinity for hydroxyapatite. It is reported that after the phosphate or NaCl gradient was used for the hydroxyapatite purification step, more than 1000-fold clearance of xenotropic murine leukemia virus (X-MuLV) and 100-fold clearance of minute virus of mice (MVM) were realized (Ng et al. 2006). In some cases, its performance may be better than virus removal steps including gel filtration and ion exchange chromatography.

In addition to removing viruses from biopharmaceuticals, hydroxyapatite chromatography is also used to purify viruses from cell culture fluids to prepare vectors for gene therapy and vaccines to prevent infectious diseases. The capsid protein on the surface of the virus is rich in carboxyl groups and is firmly bound to the surface of the matrix by the metal coordination of calcium ions on the surface of hydroxyapatite. When eluting with an increasing concentration of phosphate buffer, the host cell protein is usually washed out in the flow-through fraction, the target virus is eluted at an elevated salt concentration in the gradient, and the chromosomal nucleic acid of the host cell remains on the column, desorbs when using the stripping solution. The content of residual impurities in the target virus can be further reduced by optimizing the composition of the phosphate gradient solution. For example, if the impurity double-stranded nucleic acid is co-eluted with the target virus, the neutral salt NaCl can be added to the phosphate buffer to enhance the retention of the nucleic acid and increase the separation from the target virus. This is because in the neutral pH phosphate solution, there is electrostatic repulsion between nucleic acid and hydroxyapatite. The presence of sodium chloride produces electrostatic shielding effect, eliminating the electrostatic repulsion. The metal coordination between nucleic acid and hydroxyapatite is almost unaffected. Therefore, after adding a high-concentration sodium chloride solution to the eluent, the retention of the target virus is almost unchanged but the retention of nucleic acid is greatly enhanced.

As mentioned above, the retention of biomacromolecules on hydroxyapatite chromatography is dominated by mixed-mode interactions. However, both metal chelation and electrostatic interaction can be cancelled by the phosphate ions in the mobile phase. Therefore, the adsorbed solute molecules can be eluted using a gradient of increasing phosphate concentration, for example, from 10 mM sodium phosphate, pH 6.8 to 500 mM sodium phosphate, pH 6.8. This phosphate gradient elution mode is a good starting point for the development of protein separation on hydroxyapatite. For antibody purification, most of the antibodies described so far have an isoelectric point in the alkaline pH range and belong to basic proteins. Standard hydroxyapatite purification schemes have been developed for these proteins. As an alternative to these solutions, hydroxyapatite can usually retain acidic antibodies at higher sodium chloride concentrations, and high concentrations of sodium chloride can dissociate ionic complexes between antibodies and certain contaminants (including DNA), thereby improving purification performance (Josic et al. 1991).

The conventional phosphate gradient elutes most proteins, first the acidic protein retained by the C-site interaction and then the basic protein retained by the P-site interaction. The slope and amplitude of the gradient can be adjusted according to the initial result. Once the elution concentration of the target molecule is known, the linear gradient can be converted into a step gradient. To optimize the separation, the pH of the buffer can be changed in subsequent experiments (Schröder et al. 2003). For basic proteins, the pH of the solution can be increased to improve selectivity. Above the range of pH 6.0–7.5, the buffer capacity of phosphate buffer will decrease. It is recommended to switch to borate buffer. If the mobile phase pH to be lower than 6 is desired in a particular separation, fluoroapatite should be used instead of hydroxyapatide, because the working pH range of the former can be reduced to 5.0.

In the case of separation of basic proteins (most antibodies are basic proteins), a dual gradient elution scheme using sodium chloride and phosphate may produce better results than a single phosphate gradient. In this case, a dual gradient usually improves the resolution while eluting the target molecule (antibody) before the impurities. The feature of the dual gradient elution scheme is to use a linear gradient with increasing sodium chloride concentration for elution under the background of low phosphate concentration such as 5 mM. After the target protein is eluted, a high concentration of phosphate such as 500 mM phosphate is used for elution. This is a good starting scheme for the separation of P-site interacting molecules through a double gradient. The slope and amplitude of the gradient can be adjusted according to the initial result. It can also be converted to a step elution format. If the target protein fails to elute in the NaCl gradient, it may be due to residual C-site interactions. In this case, the background phosphate concentration in the elution buffer may be increased to 15 mM. However, further increasing the concentration usually reduces the resolution, especially the efficiency of removing genomic DNA and endotoxin (Gagnon et al. 2006).

If necessary, a certain concentration of polyethylene glycol (PEG) can be added to the mobile phase to increase the resolution and further improve the removal of aggregates and high molecular weight impurities (Gagnon 2008). In the presence of PEG, the retention of antibody monomers and aggregates increased, but the increase in aggregates was greater than the monomers. The retention enhancement effect of PEG may be caused by two effects. First, PEG forms a PEG-free solution layer on the surface of the protein or matrix, forming a concentration gradient in the interface with the bulk PEG solution; secondly, the specific surface area of the complex between the protein and the matrix is smaller than the sum of the two before the binding; therefore, these two factors are both conducive to binding. The retention enhancement effect of PEG also exists in other chromatographic modes, but the effect is especially obvious in hydroxyapatite chromatography. It is speculated that both the interactions at the P-sites and the C-sites will be affected, because the influence of PEG in the phosphate gradient is stronger than that in the NaCl gradient (Gagnon 2008). Screening the appropriate mobile phase composition should start with 10% PEG-400 or PEG-600 and then gradually reduce it to the lowest level to avoid the increase in column pressure drop due to the increase in solution viscosity.

Polynucleotides (such as DNA) are another type of biological macromolecules that can be purified very efficiently on hydroxyapatite (Freitag and Hilbrig 2012). Unlike proteins, the binding of this type of molecule is primarily through the metal coordination of calcium ions on hydroxyapatite mediated by the phosphate groups in its backbone. This coordination geometry is not affected by the sodium chloride concentration in the solution. Each nucleic acid molecule contains a large number of nucleotide monomers, which will establish multivalent interactions, resulting in strong binding. Most polynucleotides can be eluted using a phosphate gradient, which elutes later in the standard phosphate gradient. If elution is not possible under these conditions, the phosphate concentration at the upper limit of the gradient can be further increased.

Adding salt (up to 1 M NaCl) to the mobile phase can increase the binding strength of nucleic acids, because the charge-repulsive interaction between DNA and phosphate groups on hydroxyapatite is shielded. At the same time, the hydrodynamic radius of DNA molecules is reduced, which can promote the molecules to enter the pores of the stationary phase. However, the hydrophobic interaction (the binding between biomolecules is also strengthened) may be strengthened when the salt concentration is too high. This also explains the usual observation that when the concentration of phosphate used in the mobile phase exceeds 500 mM, the binding of DNA to hydroxyapatite starts to increase again. It has also been shown that adding up to 10 mM $CaCl_2$ can increase the dynamic binding capacity and to a lesser extent the yield of DNA in hydroxyapatite chromatography (Aberin and Franklin 2003). This effect is most likely due to chelation between plasmid DNA and P-sites or bridging between plasmid molecules. If the mobile phase becomes turbid after adding $CaCl_2$ (formation of calcium phosphate), adding (some) EDTA may make the solution clear. Especially for the purification of plasmid DNA, a linear gradient elution of phosphate with 1 mM EDTA added to the mobile phase was used as the starting point for method development. The slope and amplitude of the gradient can be adjusted according to the initial result and can be converted to a step elution format. In some cases, it may be useful to switch to a mobile phase that contains borate. According to Aberin and Franklin (2002), using this protocol, nearly pure plasmid DNA can be obtained from alkaline lysates in one step.

Commercially available CHT ceramic hydroxyapatite is the spherical hydroxyapatite with macropores. It is calcination at high temperature to convert nanocrystals of hydroxyapatite into ceramic form. The ceramic material maintains the unique separation characteristics of crystalline hydroxyapatite. The methods developed on crystalline materials can often be directly transferred to ceramic materials without the need for a complicated development process. Like silica gel in normal phase chromatography, CHT ceramic hydroxyapatite is both a supporting matrix and a ligand. During the separation process, calcium ions may fall off the matrix and enter the product, which can be removed by means such as dialysis. Therefore, there is no problem of product contamination due to ligand leakage as with many other chromatographic media.

Commercial CHT ceramic hydroxyapatite has two specifications, type I and type II. The CHT I type material is calcined at 400 °C. It has a large specific surface area, high protein binding capacity, and acidic protein binding capacity. The CHT II type material is calcined at 700 °C. Compared with type I, its specific surface area and protein binding capacity are reduced, but it has better nucleic acid separation ability. CHT II has a low affinity for albumin and is especially suitable for the purification of immunoglobulins. In addition, there is a fluorapatite matrix, in which the hydroxyl group in hydroxyapatite is replaced by fluorine, resulting in a material with higher stability under acidic conditions. The minimum usable pH of hydroxyapatite is 6.5, and the fluoroapatite can be used as low as pH 5.0.

Applications of commercial hydroxyapatite in the purification of biological macromolecules are illustrated as follows to help understand the method development process.

9.2.2 *Examples of Method Development*

Purification of Moloney murine leukemia virus Moloney murine leukemia virus (MoMuLV) is a retrovirus, named for its ability to cause cancer in a murine (mouse) host. Some MLVs may infect other vertebrates. As a type C retrovirus, the replicated murine leukemia virus produces a virus particle that contains a spherical nucleocapsid (viral genome and viral protein complex) surrounded by a lipid bilayer derived from the host cell membrane. The lipid bilayer contains integrated host and viral proteins with interspersed carbohydrate molecules. The diameter of the virus particle is approximately 90 nm. The viral glycoprotein is expressed on the membrane as a trimer of the precursor Env, which is cleaved into SU and TM by the host furin or furin-like proprotein convertase. This cleavage is essential for the incorporation of Env into viral particles.

Traditionally, the purification of MoMuLV relied on methods such as density centrifugation and gel permeation chromatography. These processes are time-consuming and not easy to scale up. Kuiper et al. (2002) reported the two-step method of ultrafiltration membrane concentration and hydroxyapatite chromatography to purify viruses with transfection ability from cell culture media. The crude cell culture supernatant was clarified by filtration, and the clarified material was directly loaded into the hydroxyapatite chromatography column, washed with a low-concentration phosphate buffer, and eluted with a linear gradient of increasing phosphate concentration. Then any strongly adsorbed impurities were stripped off by 1 M NaOH. Figure 9.1a shows the separation of the viruses on hydroxyapatite. As the main contaminant, most of the bovine serum protein from the culture medium is eluted in the flow-through fraction. The viruses are eluted in a linear gradient, and nucleic acids are eluted in the stripped fraction. Gel electrophoresis experiments (Fig. 9.1b, c) showed that contaminating proteins and DNA in the virus-containing eluate fraction were significantly removed, and the virus titer increased up to 42 times. The recovery of viruses with transfection capability is between 18% and 31%.

Purification of dengue virus Dengue virus (DENV) infection has a global distribution. The infection is transmitted by mosquitoes and can cause dengue fever and more severe dengue hemorrhagic fever. It is estimated that 50 million people worldwide are infected with dengue fever each year and 2.5 billion people worldwide are currently at risk of infection. Although all four DENV serotypes can cause disease, the most serious and fatal infections are usually associated with DENV-2 and DENV-3. Dengue virus particles are composed of envelope and nucleocapsid. These virus particles are spherical and small in size (approximately 40–60 nm in

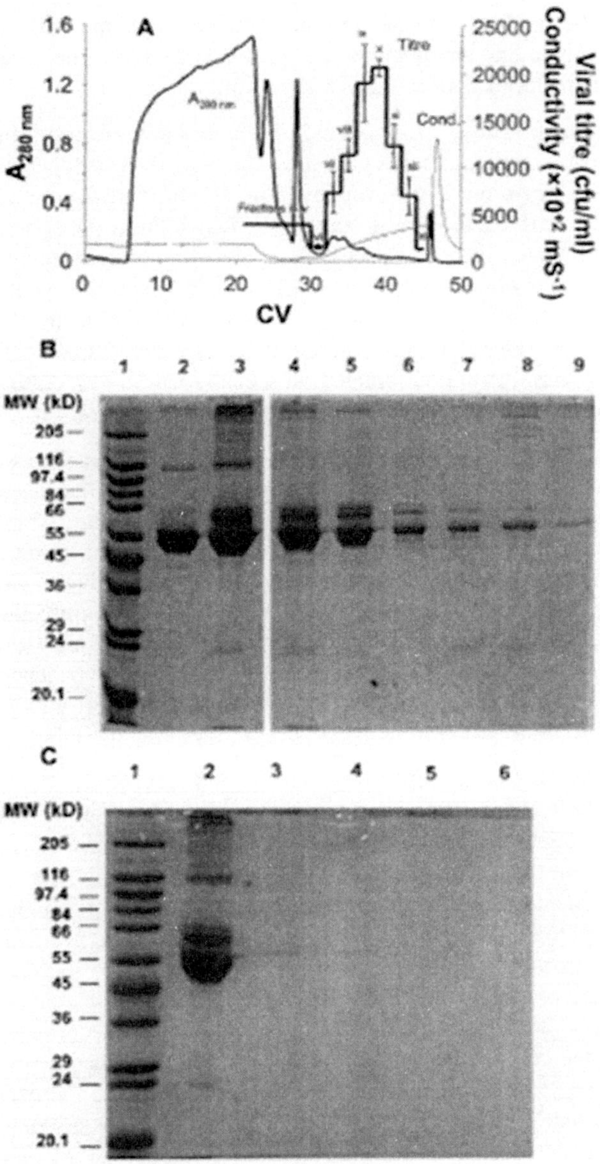

Fig. 9.1 (**a**) Chromatographic separation using a 2.5 mL hydroxyapatite (MCHT2) column containing 40 mL of clarified (but not concentrated or diafiltered) crude cell culture supernatant. The production cell line contains $6.7 \pm 1.3 \times 10^4$ MoMuLV cfu/mL. After loading the sample, the column was washed with PBS, and then with an additional washing step: 3 mM phosphate buffer pH 6.8, gradient elution range from 3 to 500 mM phosphate buffer pH 6.8. Finally, the column was stripped with 1 M NaOH and re-equilibrated with PBS. The eluate fractions were collected, and the virus titers were determined (thick line, right ordinate). The UV absorbance was continuously monitored (line indicating protein, medium concentration, left ordinate) and conductivity (thin line; right ordinate) of the column eluate at 280 nm. The data has been corrected for the amount of delay. (**b**, **c**) SDS-PAGE analysis of 10 mL feed and elution fractions of (**a**), lanes **b**1 and **c**1, sigma M4038

diameter). The surface of the virion contains envelope (E) and membrane (M) proteins, including small spikes and protruding edges. The cores are isometric and have a diameter of approximately 25–30 nm.

Kurosawa et al. (2009, 2014, 2012a, b) reported a one-step purification method for a number of mammalian virus particles. The dengue virus type 2 culture medium was loaded on a ceramic hydroxyapatite column and eluted with a linear gradient (10 mM NaPB, pH 7.2–600 mM NaPB, pH 7.2) sodium phosphate buffer within 15 min. As shown in Fig. 9.2a, the culture medium and protein contaminants derived from the host cells eluted first, followed by dsDNA, and the type 2 dengue virus eluted in about 33 min. In the hemagglutination test (HA), the recovery of type 2 dengue virus was 64%. The virus infectivity is 60%. No characteristic peaks for the virus were observed in the chromatograms of the C6/36 cell culture medium that were not infected with the virus (Fig. 9.2b) and the medium itself (Fig. 9.2c).

Purification of poliovirus There are currently two polio vaccines on the market: Salk and Sabin vaccines. The Salk vaccine is an injectable vaccine that uses a virulent poliovirus strain inactivated by formalin, while the Sabin vaccine is an oral vaccine that uses a live attenuated virus strain. The Sabine oral vaccine occasionally causes vaccine-related paralytic poliomyelitis due to circulating poliovirus derived from polio. Therefore, the development of noninfectious vaccination methods has become a top priority. Poliomyelitis virus belongs to the acid-stable *Picornaviridae* family and is still infectious at pH 3 and lower. The spherical virus particle does not contain a lipid envelope, has a diameter of about 30 nm, and is composed of four structural proteins: VP1, VP2, VP3, and VP4.

The standardized one-step virus purification procedure based on ceramic hydroxyapatite (CHAp) chromatography is not suitable for polio virus because the double-stranded DNA of the host cell cannot be separated from the virus. Kurosawa et al. (2019) combined ceramic fluoroapatite (CFAp) (applicable pH range 5–14) chromatographic capture and ceramic hydroxyapatite (CHAp) (applicable pH range 6.5–14) chromatographic polishing and established two-step purification procedure for the Sabin 2 strain of polio.

The first step of this method is to run a linear gradient of increasing pH (300 mM sodium phosphorate/borate, pH 5.0–8.2) on a fluoroapatite column to isolate Sabin 2 virus from the cell culture supernatant (Fig. 9.3a). The recovery of the obtained fraction was 94.1%, the protein clearance was 91.87%, and the dsDNA clearance was 88.15%. The second step is to run a linear gradient of increasing sodium phosphate concentration (10–187 mM sodium phosphate, 1 M NaCl, pH 7.2) in the presence of a high concentration of sodium chloride on a hydroxyapatite column

---

Fig. 9.1 (continued) molecular weight marker; **b**2, 1 mg/mL HSA in PBS; **b**3 and **c**2, complete cell culture medium; **b**4, clarified crude cell culture supernatant; **b**5, samples of combined fraction i–v; **b**6, fraction vi; **b**7, fraction vii; **b**8, fraction viii; **b**9, fraction ix; **c**3, fraction x; **c**4, fraction xi; **c**5, fraction xii; **c**6, fraction xiii. (Reproduced from Kuiper et al. 2002)

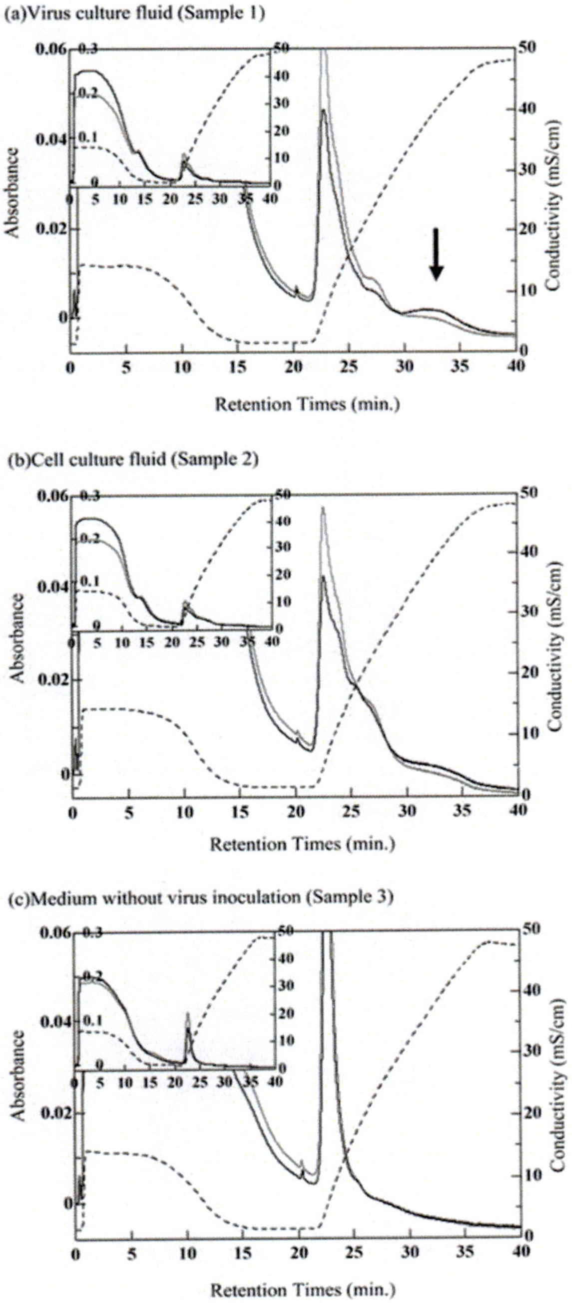

Fig. 9.2 Chromatograms of cell culture and medium on a ceramic hydroxyapatite column. (**a**) C6/36 cell culture medium infected with dengue virus (sample 1); (**b**) C6/36 cell culture medium without virus infection (sample 2); (**c**) medium containing 0.5% FBS without virus (Sample 3). The black line, the UV absorbance at 260 nm; the gray line, the UV absorbance at 280 nm; and the dashed line, the conductivity of the elution buffer. (Reproduced from Kurosawa et al. 2012a)

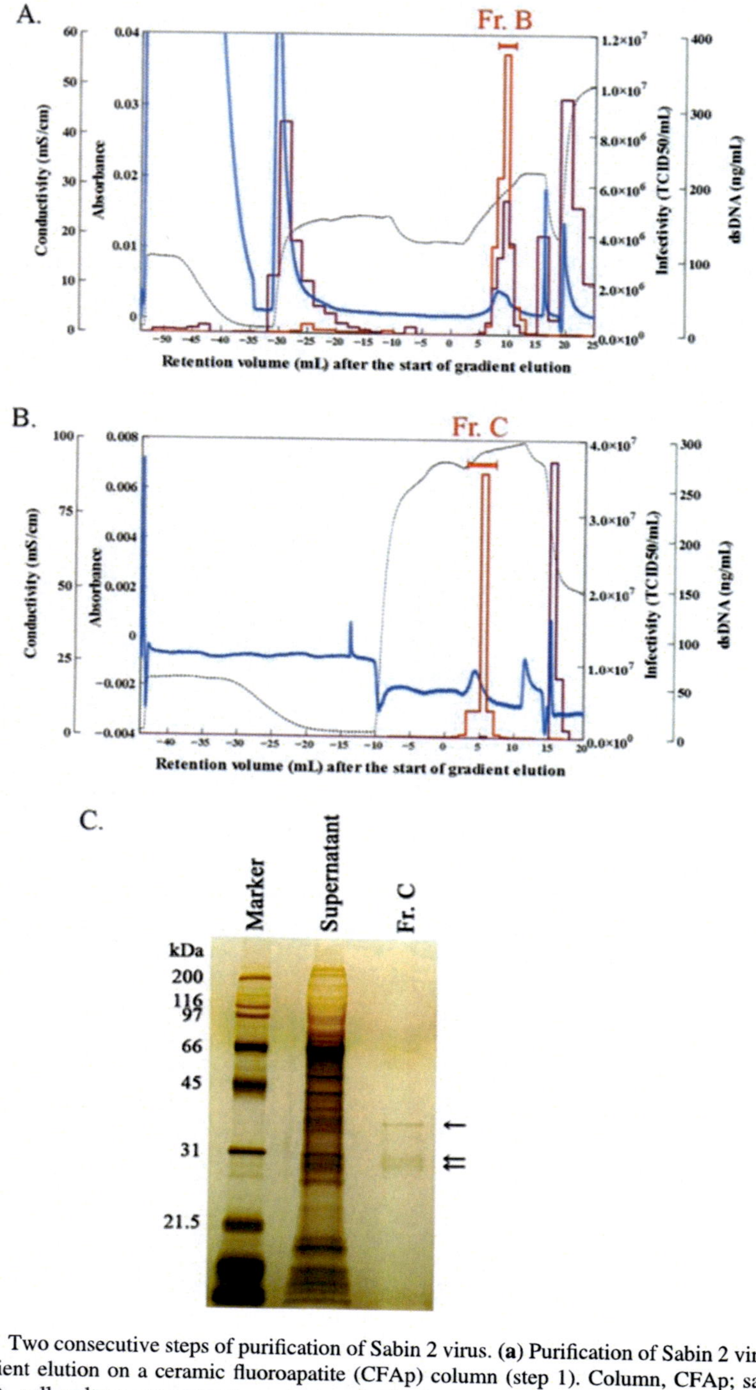

Fig. 9.3 Two consecutive steps of purification of Sabin 2 virus. (**a**) Purification of Sabin 2 virus by pH gradient elution on a ceramic fluoroapatite (CFAp) column (step 1). Column, CFAp; sample (volume), cell culture supernatant containing Sabin 2 virus (10 mL); wash, 10 mM sodium phosphate buffer (NaPB) (pH 6.4, 9 mL) and 300 mM NaPB (pH 6.4, 20 mL); equilibration,

(Fig. 9.3b). In 1 M NaCl solution, the charge repulsion between the phosphate of nucleic acid and the phosphate of hydroxyapatite is cancelled, and the metal coordination with calcium ions is almost unaffected. Therefore, the retention of dsDNA is enhanced, and the separation with the virus is increased. The clearance of dsDNA in this step is 97%. The final purified products including Sabin 2 capsid proteins VP1, VP2, and VP3 with molecular weights of 37, 32, and 30 kDa, respectively, were obtained (Fig. 9.3c).

Purification of phage display virus Filamentous bacteriophages are a family of viruses that infect bacteria (*Inoviridae*). The bacteriophage is named after its filamentous shape, wormlike chain (long, thin, and flexible, reminiscent of a piece of cooked pasta), about 6 nm in diameter and about 1000–2000 nm in length. The outer shell of the virus particle contains five types of viral proteins, which are located in the inner membrane of the host bacteria during the phage assembly process and are added when they are extruded through the membrane into the new virus particles.

Ff filamentous bacteriophages (strains f1, fd, and M13 and their derivatives) infect *E. coli* cells displaying F pili and are the most common vector for phage display. The virion is mainly composed of a tubular array of thousands of major coat protein subunits surrounding a circular single-stranded viral DNA, accounting for only 12% of the particle mass. Each of the four minor coat proteins has five copies covering the ends. Since the net charge is strongly negative, and DNA is completely inaccessible from the outside, the intact particle can be regarded as a long and thin protein-like polyanion physically and chemically.

Smith and Gingrich (2005) used hydroxyapatite (HA) chromatography to purify filovirus particles in one step. The starting virion is fd-tet particles (molecular weight 23.5 MDa; virion length is 1.3 μm; genome size is 9183 bp), which has been partially purified from the cell culture supernatant by two consecutive PEG precipitations. Preliminary experiments have shown that they can be adsorbed on the HA matrix in PBS (0.15 M NaCl, 5 mM NaH_2PO_4, pH 7.0), and will not desorb at NaCl concentrations up to 2.55 M, but desorb at a phosphate concentration of

Fig. 9.3 (continued) 15 mL 300 mM NaPB (pH 5); elution, linear gradient from pH 5 to 8.2 at 300 mM NaPB for 10 mL; wash after separation, 300 mM NaPB (pH 8.2, 5 mL) and 600 mM NaPB (pH 8.2, 10 mL). Fraction B was pooled and further purified in step 2. (**b**) Remove double-stranded DNA (dsDNA) from fraction B. Fraction C was obtained by NaPB elution with 1 M NaCl on a ceramic hydroxyapatite (CHAp) column (step 2). Column, CHAP; sample (volume), fraction B obtained in (**a**) diluted 6.7 times with 0.9% NaCl (17 mL); buffer, pH 7.2; wash, 10 mM NaPB (13 mL); equilibration, 10 mM NaPB + 1 M NaCl (14 mL); elution, in 10 mL and 1 M NaCl, using a linear gradient from 10 mM to 187 mM NaPB; wash, 600 mM NaPB (10 mL). Fraction C was obtained and pooled for further evaluation. (**c**) Sodium dodecyl sulfate-polyacrylamide gel electrophoresis (SDS-PAGE) analysis of the pooled fractions obtained through the two-step purification process. Fraction C and the cell culture supernatant obtained in (**b**) were concentrated 100 times and 10 times by ultrafiltration with a molecular weight of 10,000, respectively. The size is expressed in kDa on the left. The arrow indicates the main protein in the fraction C. $TCID_{50}$, the median infection dose in cell culture. (Reproduced from Kurosawa et al. 2019)

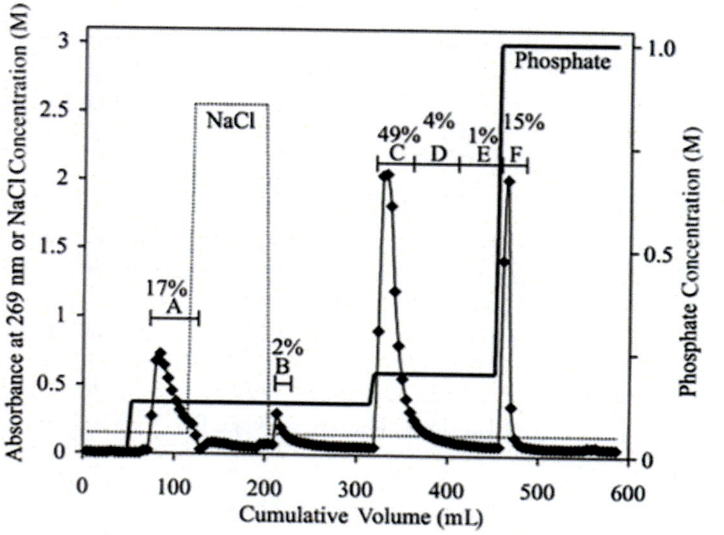

Fig. 9.4 The elution curve of fd-tet virions separated on a hydroxyapatite column. Equilibrium buffer: 0.15 M NaCl, 5 mM NaH_2PO_4, pH 7.0. Washing buffer I: 0.15 M NaCl, 125 mM NaH_2PO_4, pH 7.0. Washing buffer II: 2.55 M NaCl, 125 mM NaH_2PO_4, pH 7.0. Elution buffer solution: 0.15 M NaCl, 200 mM NaH_2PO_4, pH 7.0. Stripping buffer: 0.15 M NaCl, 1 M NaH_2PO_4, pH 7.0. Bars and letters indicate the eluted phage pool. The recovery in each pool is shown. (Reproduced from Smith and Gingrich 2005)

150–200 mM. Therefore, a stepwise gradient chromatographic scheme was designed, in which the virosomes were adsorbed on the matrix and then washed thoroughly with washing buffers I and II, respectively; the buffers I and II were 125 mM phosphate solutions containing 0.15 and 2.55 M NaCl, respectively. The virion was desorbed in a desorption buffer containing 0.15 M NaCl and 200 mM phosphate; then, a stripping buffer containing 0.15 M NaCl and 1 M phosphate was used to "strip" the strongly adsorbed impurities.

Figure 9.4 shows a typical experimental result, in which 8.2×10^{14} virosomes (total mass 32 mg) in 800 mL culture were chromatographically separated on an HA column with an initial bed volume of 10.8 mL. It can be seen from the figure that all the input virus particles are adsorbed on the HA matrix in PBS, thereby concentrating them to 7.6×10^{13} virus particles/mL bed volume. About half of the virions were recovered in the elution fraction pool (pool C). The purified virions did not contain detectable host protein, and their infectivity was not lost.

Purification of human papillomavirus-like particles expressed in silkworm Human papillomavirus (HPV) is a non-enveloped double-stranded DNA virus, of which more than 170 different types are known. HPV infects skin or mucous membrane cells and may cause uncontrolled cell growth, which can lead to warts or malignant growth, penile cancer, or cervical cancer. Therefore, the development of vaccines to prevent or treat human papillomavirus diseases has become a hot spot in the field of biopharmaceuticals. Virus-like particles (VLP) have certain advantages for pharmaceutical applications. Like the virus from which they originate, VLPs have the same capsid as the virus from which they originate or may also have additional membranes, but they do not contain genetic material and are therefore noninfectious. VLP can be used as a vaccine, even if it is not modified, it can cause a strong immune response. If certain antigens are displayed on the surface of the VLP, they can also be used as a multiunit vaccine. VLPs displayed by specific antibodies can target specific types of cells, such as cancer cells. In addition, if the internal voids of VLPs are filled with drugs or conjugated on their surfaces, they have the potential to be used as drug delivery systems.

Minkner et al. (2018) developed a new purification scheme using only chromatographic methods to purify VLP expressed in silkworms. VLP from HPV L1 protein is expressed in silkworm larvae and self-assembled into HPV-LP. The program includes a three-step chromatographic purification. The first step includes an anion exchange (AIEX) system for separating DNA. In the second step, a ceramic hydroxyapatite (CHT) column was used for the separation of baculovirus, because it has the dual principle of metal affinity and cation exchange. Finally, a heparin affinity column was used to concentrate the eluted product in the second stage. Figure 9.5 shows the results of anion exchange chromatography with a large column. VLP eluted before DNA and effective DNA removal is confirmed (Fig. 9.5a). The Western blot results related to this purification are shown in Fig. 9.5b. Comparing the amount of the loaded part and the amount of the eluted part based on the densitometry method, the step recovery of HPV 6b L1 is found to be about 31.4%.

In the second step, CHT chromatography was performed to separate HPV particles from baculovirus using a larger column (about 16 × 35 mm) (Fig. 9.6). Most of the baculovirus protein BmGP64 appeared in the flow-through fraction, and was not bound with CHT. The step recovery of HPV 6b L1 is about 22.7%.

In the third step of purification, heparin affinity chromatography was performed to concentrate HPV 6b L1. BmGP64 protein was not detected in the fraction of HPV 6b L1. The recovery of HPV 6b L1 at this step was 86.9%. The overall recovery is 6% for the three-step purification. In addition, purifications of pepsin-derived monoclonal antibody fragments (Moro et al. 2008) and antibody against Japanese encephalitis virus (Saito et al. 2012) by hydroxyapatite chromatography have also been reported.

9.3 Method Development in Hydrophobic Charge Induction Chromatography

Hydrophobic charge induction chromatography (HCIC) is a unique form of mixed mode chromatography (Li et al. 2019). Its binding mode is similar to hydrophobic interaction chromatography, which adsorbs solutes through hydrophobic interactions with ligands; however, its elution mode is similar to affinity chromatography, which induces electrostatic repulsion between solutes and ligands through a lowering pH step. In traditional hydrophobic interaction chromatography, the density of

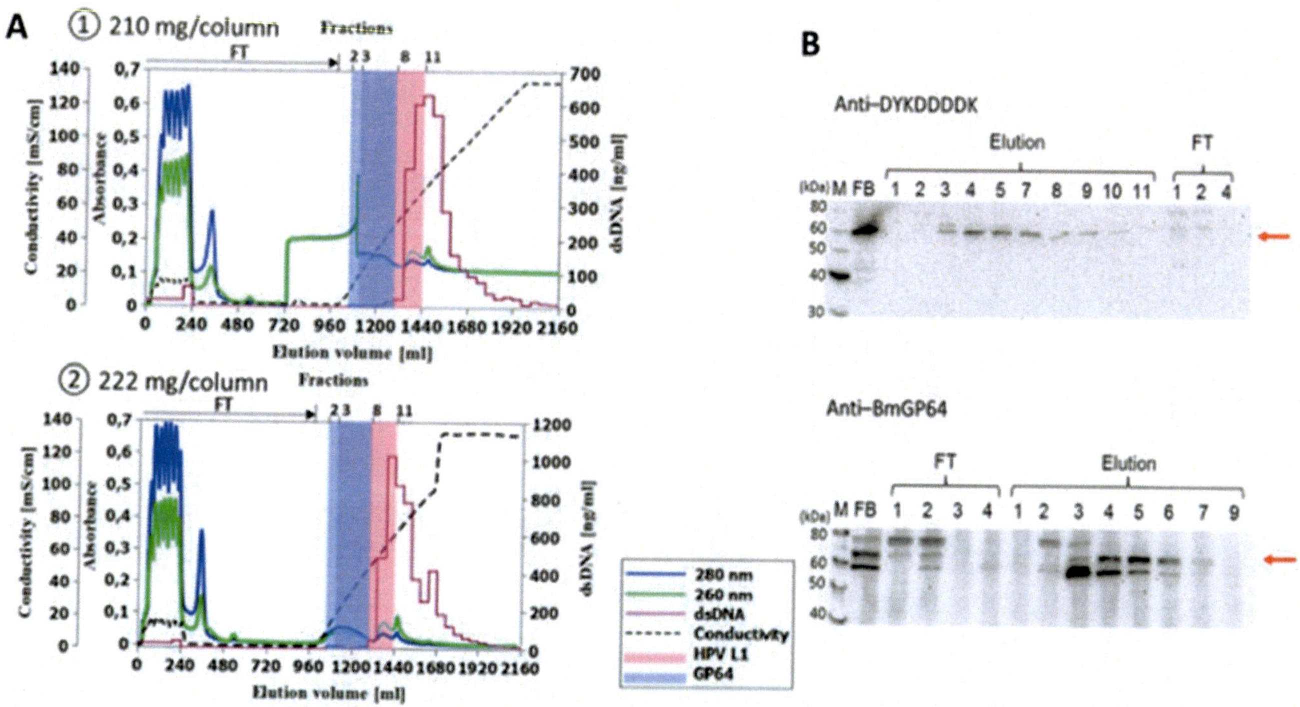

Fig. 9.5 GigaCAP Q-650M column chromatography uses a large column (32 mm × 35 mm) to purify HPV 6b L1 at a flow rate of 1 mL/min. (**a**) GigaCap Q-650M column chromatography. The sample was loaded 7 times in 30 mL batches. In 150 mM sodium phosphate (pH 7.2) buffer, a linear gradient is completed in 960 mL 0–1.5 M NaCl. FT, flow through; blue line, absorbance at 280 nm; green line, absorbance at 260 nm; dashed line, conductivity (mS/cm); purple line, DNA concentration (ng/mL). The pink and purple areas indicate the elution positions of HPV L1 and BmGP64, respectively, which have been confirmed by Western blotting. (**b**) Western blot analysis of each fraction in GigaCap Q-650M column chromatography. The HPV L1 and BmGP64 proteins in each fraction were detected by western blotting using anti-DYKDDDDK tag monoclonal antibody and anti-BmGP64 antibody as primary antibodies, respectively. M, magic mark XP; FB, fat body expressing HPV 6b L1 (9.2 dil + 0.45 μm filter). The red arrows in (**a**) and (**b**) indicate the positions of HPV L1 and BmGP64, respectively. (Reproduced from Minkner et al. 2018)

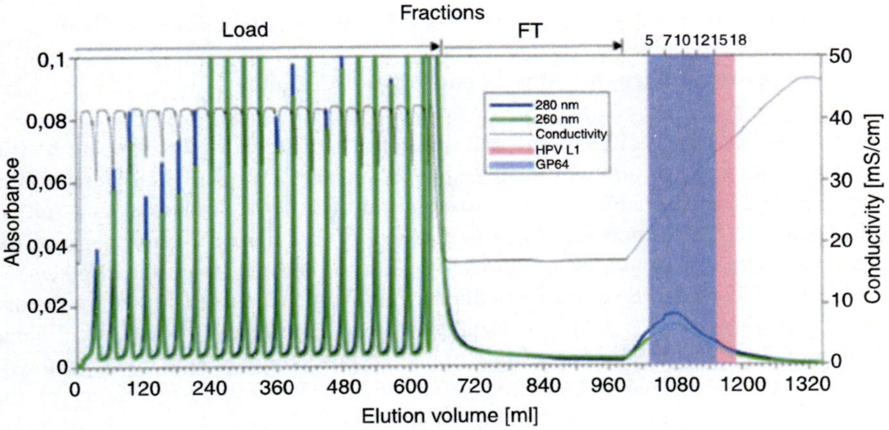

Fig. 9.6 CHT column chromatography purification using a large column (ϕ 16 × 35 mm) (step 2). The chromatogram shows an example of the purification process on the CHT column. The flow rate is 1 mL/min, and the sample is loaded 21 times in 30 mL batches. The linear gradient is completed in 312 mL 150–600 mmol/L sodium phosphate (pH 7.2). FT, flow through. (Reproduced from Minkner et al. 2018)

hydrophobic ligands on the surface of the matrix is low, and solutes can be adsorbed only in the presence of a high concentration of salting-out salt. The solute is eluted through a decreasing salt concentration gradient. In hydrophobic charge induction chromatography, the density of hydrophobic ligands on the surface of the medium is relatively high, and the solutes can be bound without the need to adjust the salt levels in the feed. This simplifies process operations, reduces unit operations (such as percolation or ultrafiltration), and improves process economy. In addition, this method has the advantages of mild elution conditions, high capacity, low price, and versatility and has been widely used in the separation and purification of proteins and antibodies from samples from different sources.

The ligand with a specific structure determines the adsorption performance of a HCIC resin. Generally, HCIC ligands contain at least one hydrophobic moiety and one nitrogen-containing heterocyclic moiety, such as pyridyl, imidazolyl, and indolyl group. A common feature of these ligands is that the proton dissociation constant of their conjugate acid is in the acidic range. Under physiological pH conditions, nitrogen-containing heterocyclic compounds are neutral compounds that adsorb proteins through hydrophobic interactions. However, under acidic conditions, the nitrogen-containing heterocycle dissociates to form a positively charged ion exchange group, which has electrostatic repulsion with the positively charged groups in the protein such as the amino group and the guanidine group on the amino acid residues, thereby speeding up the elution of adsorbed proteins. In addition to nitrogen-containing heterocycles, other factors that affect the adsorption and elution behavior of proteins on hydrophobic charge-induction chromatography ligands include the spacer arm that connects the nitrogen-containing heterocycle to the matrix, the ligand density, and the pH and composition of the mobile phase.

9.3.1 *Factors That Influence Protein Adsorption and Elution*

9.3.1.1 Chromatographic Matrix and Ligand Coupling

Burton and Harding (1998) compared different chemistries used to prepare hydrophobic charge induction chromatography stationary phases. The ligands were coupled to activated cellulose or agarose gel through amino, carboxyl, and sulfhydryl groups. They found that ligands containing thiols are preferred because the activation and ligand attachment chemistry is simple and easy to perform and can be carried out in aqueous solutions without involving the use of organic solvents. However, these thiol ligands are usually expensive, limited in variety, and some require tailored synthesis. The second option is to attach the amine ligand to the *N*, *N*′-carbonyldiimidazole (CDI)-activated substrate. The acid-base titration of the synthesized stationary phases shows that ligand coupling may affect the dissociation constant of the ligand. For example, when the pyridine ring is coupled to the substrate in the para-position, the pK_a value is the highest, and when it is in the ortho-position to the ring nitrogen, the pK_a value is the lowest. The pK_a values are also affected by type of coupling bond. The pK_a values of 2-mercaptopyridine and 2-mercaptobenzimidazole derivatives are lower than the pK_a values of their 2-aminomethyl counterparts. Generally, the bond to the solid phase causes the pK_a of the base to decrease and the pK_a of the acid to increase compared to the typical values of the free molecules. In the process of method development, the influence of the change of the ligand's pK_a value on the protein adsorption and elution behavior should be considered, and the optimal pH value should be determined through screening experiments at different pH values.

Xia et al. (2008) prepared several HCIC adsorbents using macroporous cellulose-tungsten carbide composite beads (Cell-TuC) as a matrix. First, the cellulose beads are activated by allyl bromide (AB) or divinyl sulfone (DVS), and then they are coupled with three mercaptoheterocyclic groups, 4-mercaptoethylpyridine (MEP), 2-mercapto-1-methylimidazole (MMI), and 2-mercaptobenzimidazole (MBI) as HCIC ligands (Fig. 9.7). Four types of HCIC adsorbents were obtained, labeled Cell-TuC-AB-MEP, Cell-TuC-DVS-MEP, Cell-TuC-DVS-MMI, and Cell-TuC-DVS-MBI. Using egg yolk immunoglobulin (IgY, $pI = {\sim}5.0\text{–}5.3$) and bovine serum albumin (BSA, $pI = 7.8$) as model proteins, their adsorption behavior on four HCIC adsorbents was studied. At pH 7, the IgY adsorption capacity on all four adsorbents is 100–138 mg IgY/mL drained gel, and the adsorption capacity of BSA is 17–35 mg BSA/mL drained gel. At pH 4, the adsorption capacity of IgY decreased, but both were higher than one-third of the adsorption capacity at pH 7, which indicated that in order to effectively elute IgY, more acidic elution conditions should be used (e.g., pH 3.0).

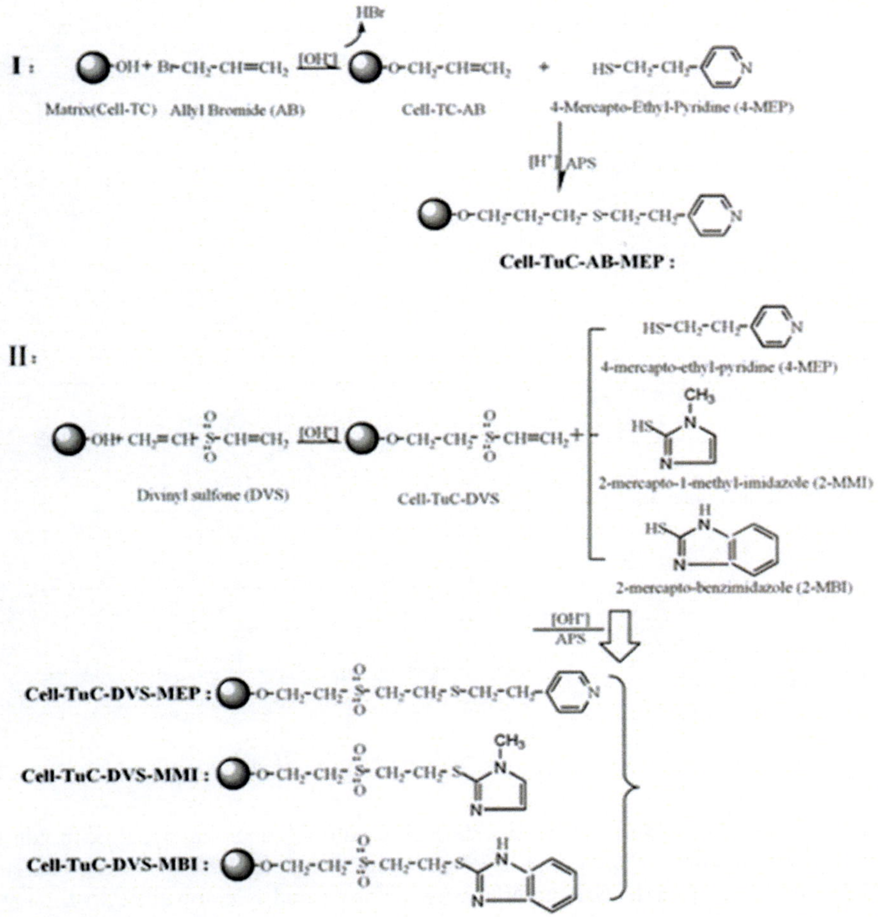

Fig. 9.7 Preparation routes of four HCIC adsorbents. (Reprinted with permission from Xia H-F, Lin D-Q, Wang L-P et al. (2008) Preparation and evaluation of cellulose adsorbents for hydrophobic charge induction chromatography. Ind Eng Chem Res 47:9566–9572. Copyright {2008} American Chemical Society)

9.3.1.2 Ligand Structure

Hydrophobic charge induction chromatography stationary phase usually consists of nitrogen-containing heterocyclic ligands bonded to hydrophilic polysaccharide matrix. So far, the trade name MEP HyperCel is the most widely used hydrophobic charge induction chromatography stationary phase. It is made by covalently attaching the functional ligand 4-mercaptoethylpyridine (MEP) to cellulose microspheres (Table 9.1). The pK_a of 4-mercaptoethylpyridine is 4.8, which can adsorb proteins under neutral conditions. The average diameter of the microspheres is

Table 9.1 Typical functional ligands used in hydrophobic charge induction chromatography

Ligand	Structure	p*K*a
4-Mercaptoethyl-pyridine, MEP		4.8
2-Mercapto-1-methylimidazole, MMI	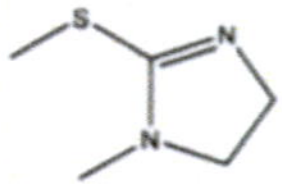	5.3
5-Aminobenzimidazol, ABI	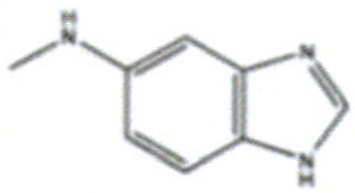	6.5
2-Mercaptobenzimidazole, MBI	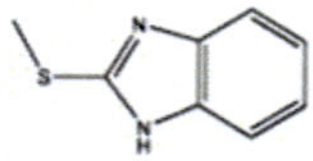	4.1
2-Mercaptoimidazole, MI	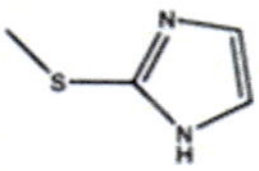	6.5
Histamine	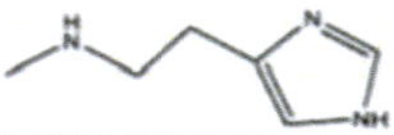	6.04
5-Aminoindole, AI	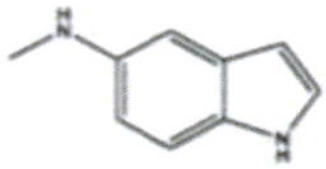	3.9

80–100 μm, the ligand density is 80–125 μmol/mL, and the dynamic binding capacity of monoclonal antibodies is greater than 20 mg/mL. It can achieve excellent flow performance under low column back pressure and is compatible with large-scale production.

In addition to pyridyl, nitrogen-containing heterocyclic compounds such as imidazolyl and indolyl have also been used to prepare stationary phases for hydrophobic charge induction chromatography (Table 9.1). Their behavior in the separation and purification of proteins is expected to be different from that of MEP HyperCel. In the process of method development, the preferred ligands for the target protein can be determined by screening different ligands for optimal balance between purity and yield.

9.3.1.3 Ligand Density

Ligand density plays an important role in protein adsorption and elution. Generally, higher ligand density is good for adsorption but not good for elution. Therefore, in some cases, the ligand density needs to be adjusted to balance the relationship between adsorption and elution. Lu et al. (2013) coupled the MMI ligand to the agarose matrix with different ligand-to-matrix ratios and prepared three resins with different ligand densities. By increasing the density and the number of accessible binding sites on the adsorbent, the adsorption capacity for bovine IgG can be significantly increased.

Zhang et al. (2010a, b) used a coarse-grained pore model to study the effect of ligand density on protein adsorption behavior through molecular dynamics simulation. A self-avoidable chain composed of hydrophobic charged beads and two hydrophobic beads was used as a model of HCIC ligand. The position of coarse beads in the ligand comes from the structure of MEP. Once the pH in the mobile phase drops below the pK_a of the hydrophobic charged bead, it is positively charged. Then, they studied the conformational transition and separation performance of the ligand at 1.474, 1.769, 2.212, and 2.949 μmol/m^2. The fastest adsorption rate was obtained at a ligand density of 2.212 μmol/m^2. The protein-ligand interaction energy diagram, and the free energy diagram show that there is a strong repulsive force at 2.949 μmol/m^2, which reduces the adsorption stability and leads to a decrease in the adsorption rate.

Ren et al. (2009) used MEP-Sepharose to remove anti-double-stranded DNA antibodies and rheumatoid factor from human serum. They found that increasing the density of immobilized MEP can increase the clearance of anti-double-stranded DNA antibodies, while a resin with a MEP density of 98.8 μmol/mL can remove 80% of anti-double-stranded DNA antibodies from human serum. However, further increase in MEP density did not achieve a significant increase in clearance. For rheumatoid factor, compared with anti-double-stranded DNA antibody, effective adsorption can be achieved at a much lower MEP density (46.3 μmol/mL). This difference indicates that the larger size and larger surface area of rheumatoid factor can provide more binding sites and can promote interaction with immobilized MEP in the case of low ligand density adsorbents. The binding of rheumatoid factor on the MEP-Sepharose is limited by the pore size, which is decreased with the increased density of the ligand.

Cheng et al. (2018) designed a new activation method for the effective preparation of hydrophobic charge-induction chromatography resins. Using triphenylphosphine as a catalyst, a resin with an activation density of 157.5 ± 1.2 μmol/g can be obtained within 12 h. The activation density obtained by this method is 1.5 times the activation density obtained by the traditional method. When used for the purification of monoclonal antibodies, the MEP resin prepared by this method showed better performance than MEP HyperCel, the purity of the recovered antibody was 93.5%, and the recovery was 36.9%.

Liu et al. (2014, 2015) prepared HCIC resin by immobilization of a functional ligand, MMI, to a dextran grafted agarose. The resultant resin is denoted as MMI-B-XL. Using hIgG as a model protein, the adsorption isotherms and kinetics of MMI-B-XL resin were studied and compared with the ungrafted HCIC resin MMI-B-6FF. It is found that the dextran graft layer can significantly shorten the mass transfer path, increase the density of the ligand, and increase the dynamic adsorption capacity at high speeds. In addition, they used inverse size exclusion chromatography and confocal laser scanning microscopy to study the structure and functional properties of MMI-B-XL (Liu et al. 2017). They found that the ligands on the grafted dextran can explore much greater space in the pores and thus provide more accessible binding sites for hIgG, resulting in higher adsorption capacity. Due to the relatively small steric hindrance, the diffusion of BSA in the dextran grafted HCIC resin is faster than that of hIgG.

Liu et al. (2016) reported a series of polymer grafted HCIC resins named G-MMI, in which polyglycidyl methacrylate (GMA) was used as the graft polymer and MMI was used as the functional ligand. GMA is grafted onto agarose gel by a surface initiator activator generated by electron transfer atom transfer radical polymerization (AGET ATRP), and poly(GMA) grafted HCIC resin G-MMI was prepared by reacting with the epoxy group of the poly(GMA) chain. The results show that the formation of a three-dimensional binding scaffold with high protein adsorption capacity requires a higher graft density and a suitable ligand density. Compared with the ungrafted resin MMI-L100 and the dextran grafted resin MMI-B-XL-200, the dynamic binding capacity of G140-MMI-L445 is the highest at a linear velocity of 300 cm/h, which is 34.6 mg/g. The stability of the new HCIC resin was studied by performing 50 cycles of separation. The purity of the recovered hIgG was 98.5–99.8%, and the recovery was 90.2–96.9%, indicating the potential for large-scale antibody purification.

9.3.1.4 Mobile Phase Additives

Tong et al. (2013) used sodium caprylate (NaCA) as a selective precipitant for albumin to enhance the adsorption selectivity of MEP HyperCel. The effects of NaCA on the adsorption equilibrium and dynamic binding and displacement behavior of bovine serum immunoglobulin G (IgG) and bovine serum albumin (BSA), as well as the binding and elution behavior in column chromatography, were studied. It is found that adding 50–75 mM NaCA in the eluent can effectively reduce the adsorption of BSA on the MEP resin, but the same treatment has little effect on the adsorption of IgG. By controlling the addition of NaCA in the loading or washing buffer, the process efficiency of purifying IgG from raw materials containing BSA can be improved, and the purity of IgG can reach more than 98%.

Arakawa et al. (2009) found that the elution of antibodies and Fc fusion proteins from MEP HyperCel at neutral pH can be achieved by addition of arginine in the eluent. Neutral elution conditions can minimize conformational changes and help protect protein activity. Arginine interacts well with most amino acid side chains and is a universal solvent additive that can be used to inhibit protein aggregation (Arakawa et al. 2007; Ishibashi et al. 2005; Arakawa and Kita 2014), to facilitate protein elution from chromatographic columns (Hirano et al. 2019). For example, the alkyl chain of the side chain of arginine can interact with the pyridine ring of MEP HyperCel through hydrophobic interactions, thereby reducing the attractive force between MEP and antibody and leading to effective antibody dissociation (Hirano et al. 2014; Pinto and Frey 2015).

In addition, 2-propanol, ethylene glycol, and urea can also be used to wash MEP HyperCel columns (Arakawa et al. 2009, 2010). Ren et al. (2014) used cyclodextrins (CDs) as the mobile phase additives to facilitate desorption of antibodies from HCIC columns. At neutral pH, it forms a reversible non-covalent inclusion compound with the MEP ligand and competes with the antibody for the adsorption site on the ligand, thereby accelerating the elution of the adsorbed hIgG. Using 15 mM β-CD, an average IgG recovery of 87% was obtained.

9.3.2 Examples of Method Development

Purification of antibodies in bovine colostrum The key to establishing a protein purification method based on MEP HyperCel is the choice of adsorption and elution conditions. The main factors affecting protein adsorption are sample concentration and sample loading speed. The main factors affecting the elution are the pH value and salt concentration of the solution. Wu et al. (2015) studied the adsorption behavior of IgG in bovine colostrum on MEP HyperCel. A 40% (wt/vol) ammonium sulfate solution was used to precipitate the bovine colostrum powder solution to obtain a crude IgG solution. After diluting with 50 mM phosphate buffer, it is used as a loading solution and loaded directly on the MEP HyperCel resin. They found that the IgG concentration in the loading solution had no effect on the IgG dynamic binding capacity of the resin, but the linear velocity of the loading solution had an adverse effect on the IgG dynamic binding capacity and recovery. Optimizing the maximum linear velocity to 0.4 cm/min of the loading solution can achieve a 90% IgG recovery. Under these optimized binding conditions, the effects of pH and ionic strength during the elution process were investigated, and it was found that under the elution conditions of pH 4.5 and 0.5 M NaCl, the recovered antibody had better yield (91.5%) and purity (93.9%) (Fig. 9.8). The results show that MEP HyperCel chromatography provides a one-step method for purifying IgG from bovine colostrum.

Purification of monoclonal antibodies in protein-free cell culture supernatant Schwartz et al. (2001) used hydrophobic charge induction chromatography (HCIC) to achieve effective harvesting and recovery of high-purity monoclonal antibodies. First, using human polyclonal antibodies as model proteins, the effects of antibody concentration, solution pH, and ionic strength on dynamic binding capacity were studied. The results show that the dynamic binding capacity of the antibody is not affected by the concentration and ionic strength of the antibody in the solution. However, when the pH increases from 5.0 to 8.0, the dynamic binding capacity increases from 5 to 33 mg IgG/mL, and then it decreases as the pH increases further. Studies on human polyclonal IgG, bovine polyclonal IgG, humanized IgG1, murine IgG1, and murine IgG2 have shown that the dynamic binding capacity of IgG has nothing to do with subclass or species. In addition, it has been observed that the dynamic binding capacity of IgM is less than 20% of that of

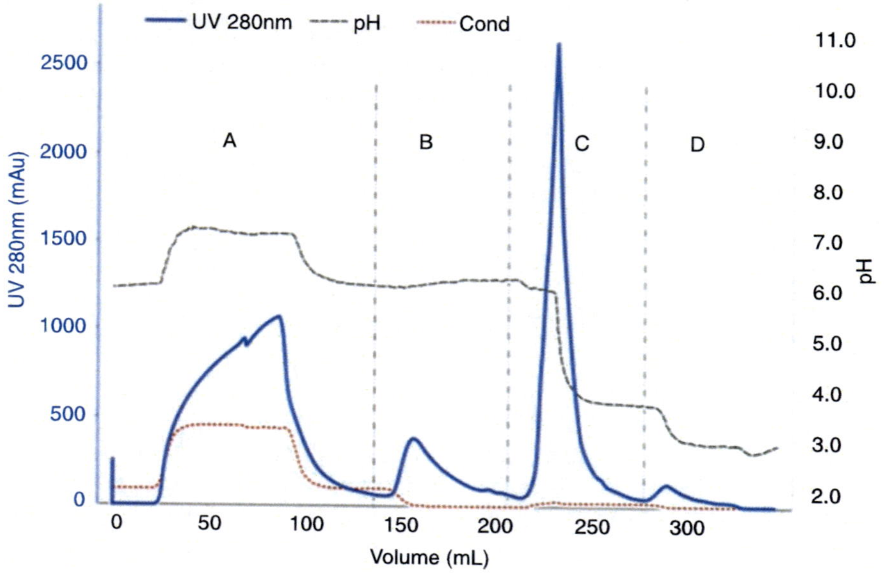

Fig. 9.8 Mercaptoethylpyridine (MEP) HyperCel chromatogram of bovine colostrum. Peak A represents load and elution with 50 mM phosphate buffer pH 7.0. Peak B is deionized water. Peaks C and D are 50 mM sodium acetate solution, pH 4.5, and 200 mM sodium acetate buffer, pH 3.0, respectively. *Cond* the conductivity of the loading solution. (Reproduced from Wu et al. 2015)

IgG, which may be explained by the size exclusion effect of matrix pores on IgM which is five times bigger than IgG. Next, they used MEP HyperCel to purify monoclonal antibodies from protein-free cell culture supernatant (CCS). The CCS containing 200 μg IgG/mL was clarified by microfiltration, but without pH or ionic strength adjustment. After sample loading, the column was washed with equilibration buffer (50 mM Tris HCl buffer, pH 8.0) until the absorbance returns to the baseline. Then the IgG fraction was eluted with 50 mM sodium acetate buffer at pH 4.0 (Fig. 9.9). The purity of the product obtained was greater than 95%, and the recovery was about 92%.

They also evaluated the virus and DNA clearance achieved by the purification of monoclonal antibodies from cell culture supernatants using MEP HyperCel. The results showed that the mouse parvovirus (MVM) was reduced by 4 logs and the DNA content was also greatly reduced. During the loading and equilibration steps, the majority of negatively charged DNA is washed out in the flow-through fraction. In the low pH elution step, the remaining DNA is adsorbed and separated from the eluted antibody product. Strongly adsorbed DNA is removed during the regeneration of the chromatographic column.

Purification of monoclonal antibodies from transgenic goat milk MEP HyperCel is also used for the purification of monoclonal antibodies from transgenic

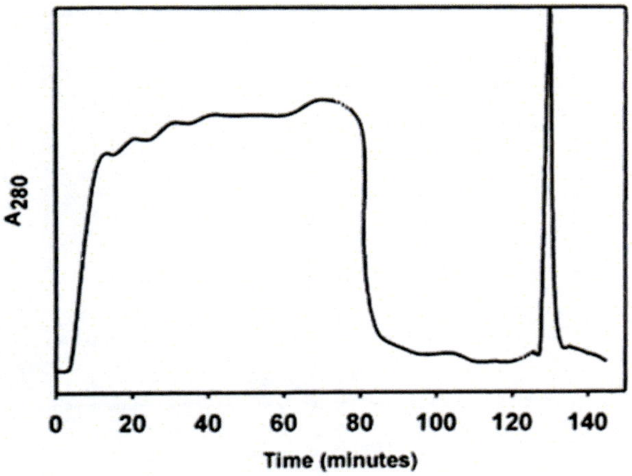

Fig. 9.9 Capture and purification of murine IgG1 from protein-free cell culture supernatant. Chromatography was performed using a 9.0 × 2.5 cm column of MEP HyperCel. The column was rinsed with buffer 50 mM Tris HCl, pH 8.0 at 150 cm/h until equilibration was reached. Thousand milliliter of clarified cell culture supernatant (containing 200 μg IgG/mL) was applied, and then the column was washed with equilibration buffer until the absorbance returns to the baseline. The IgG fraction was then eluted with 50 mM sodium acetate buffer pH 4.0. (Reproduced from Schwartz et al. 2001)

goat milk (Schwartz et al. 2001). Two additional washing steps were included in the purification scheme. After loading and washing with the equilibration buffer, the column was washed with pure water first and then washed with 25 mM sodium caprylate in the equilibration buffer. These washing steps can be used to promote the desorption of impurity components that are less hydrophobic than the target antibody. IgG binding does not depend on salt under a variety of conditions. However, reducing the ionic strength of the mobile phase from 50 mM buffer to pure water can promote the desorption of less hydrophobic proteins, such as α-lactalbumin and β-lactoglobulin. The mild detergent-like properties of sodium caprylate can also promote the desorption of this type of protein. The purity of the product obtained is greater than 95%, which is comparable to the purity of the product obtained by protein A chromatography (Fig. 9.10).

Purification of antibodies from rabbit antiserum MEP HyperCel was used for the purification of recombinant salmon insulin-like growth factor-II antibody from rabbit antiserum (Wilkinson et al. 2004). About 5 mL of antiserum was diluted with 25 mL PBS (pH 7.5), filtered through a 0.22 μm membrane, and then pumped into a 26 × 200 mm column packed with MEP HyperCel at a speed of 2 mL/min. The column was washed using a equilibration buffer PBS (pH 7.5) until a absorbance returned to baseline. A washing step with 150 mL 50 mM Tris (pH 8.0) at a rate of 5 mL/min was performed to remove contaminated proteins. The IgG fraction (a-rsIGF-II) was eluted with 75 mL of 100 mM trisodium citrate (pH 4.0) at a rate

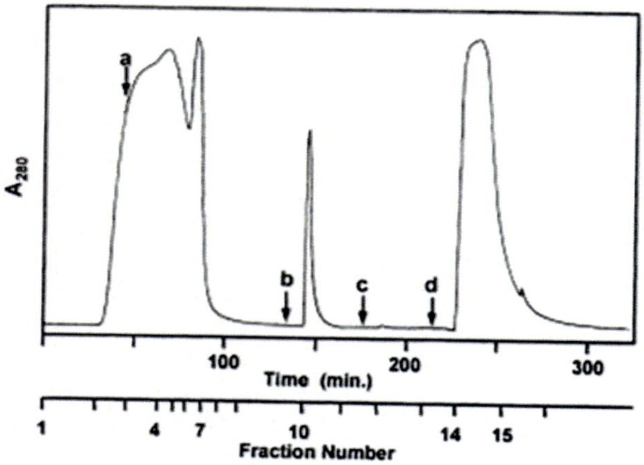

Fig. 9.10 Capture and purification of IgG1 from clarified concentrated transgenic goat milk. Chromatography was performed on a 8.0 × 1.1 cm column packed with MEP HyperCel. During the first 30 min, the column was run at 47 cm/h, and thereafter at 147 cm/h. The column was equilibrated with 50 mM Tris HCl, pH 7.4, and then loaded with feed (15.2 mL, containing 154 mg IgG/mL). Then the column was washed with the following solutions: (a) equilibration buffer; (b) purified water; (c) 25 mM sodium caprylate in equilibration buffer at pH 7.4. Finally, (d) the IgG fraction was eluted with 50 mM sodium citrate buffer (pH 4.0). (Reproduced from Schwartz et al. 2001)

of 2 mL/min. Gel electrophoresis experiments show that in the purified IgG fraction, most of the contaminating proteins have been removed.

Adsorption of high-abundance proteins from saliva Human saliva is rich in protein, and it is easy to obtain potential local and systemic biomarkers from it, which can be used to monitor changes that occur under pathological conditions. However, due to the existence of high-abundance proteins, the identification of low-abundance proteins is very difficult. Ambatipudi et al. (2009) used MEP HyperCel to capture high-abundance proteins from the samples of two age groups. Before loading on the column, the parotid saliva of each group was passed through a 0.22 μm filter to remove particulate matter and clarify it. The MEP HyperCel column was first equilibrated with 10 mL 50 mM Tris HCl pH 8.1, then loaded with saliva, and washed with 10 mL of 50 mM Tris HCl pH 8.1. To elute the adsorbed proteins, the column flow was reversed and a 20 mL gradient elution was applied, changing from 100% 50 mM Tris HCl pH 8.1 buffer to 100% 50 mM sodium acetate pH 4 buffer in 20 min. This was followed by further elution with 24 mL of 50 mM sodium acetate pH 4 within 25 min (Fig. 9.11). The eluted fractions (1 mL each) were collected into a test tube containing 100 μL of 50 mM sodium phosphate pH 9.1. Finally, 40 mL of 50 mM Tris HCl pH 8.1 was used to re-equilibrate the column for subsequent fractionation. Gel electrophoresis experiments showed that in both age groups, most amylases and histones have been successfully separated and

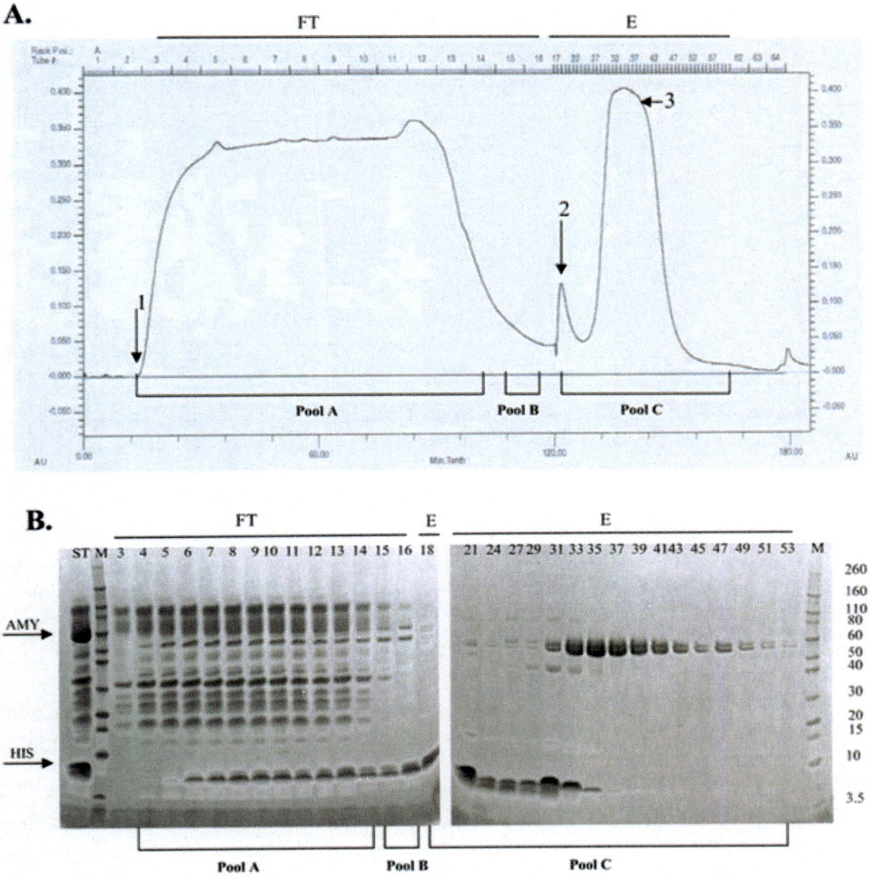

Fig. 9.11 Parotid saliva fractionation and SDS-PAGE analysis. (**a**) The parotid saliva collected from 7 subjects (group 1; age 20–30) was separated by hydrophobic charge induction chromatography (HCIC) on MEP HyperCel. Saliva (76 mg protein) was loaded to the HCIC column (arrow 1) and then washed with 50 mM Tris HCl pH 8.1 until it was close to the baseline. A gradient elution of 50 mM sodium acetate pH 4 was applied from 0% to 100% (arrow 2), followed by further elution with 50 mM sodium acetate pH 4 (arrow 3). Unbound saliva was collected in 5 mL fractions (flow through, fractions 2–16), and the eluate (E, fraction 17–61) was collected as 1 mL fractions. The protein was detected at 214 nm. (**b**) SDS-PAGE analysis of the combined initial saliva (ST lane), size marker (M lane), flow-through fraction (FT, 3–16), and elution fraction (E, 18–53). Amylase and histone are represented by arrows (AMY and HIS) of approximately 55 kDa and ~8 kDa, respectively. The gel was run under reducing conditions and then stained with simple blue. Lane M is a molecular weight marker pre-stained by SeeBlue (Invitrogen), ranging from 3.5 to 260 kDa. (Reprinted with permission from Ambatipudi KS, Lu B, Hagen FK et al. (2009) Quantitative analysis of age specific variation in the abundance of human female parotid salivary proteins. J Proteome Res 8:5093–5102. Copyright {2009} American Chemical Society)

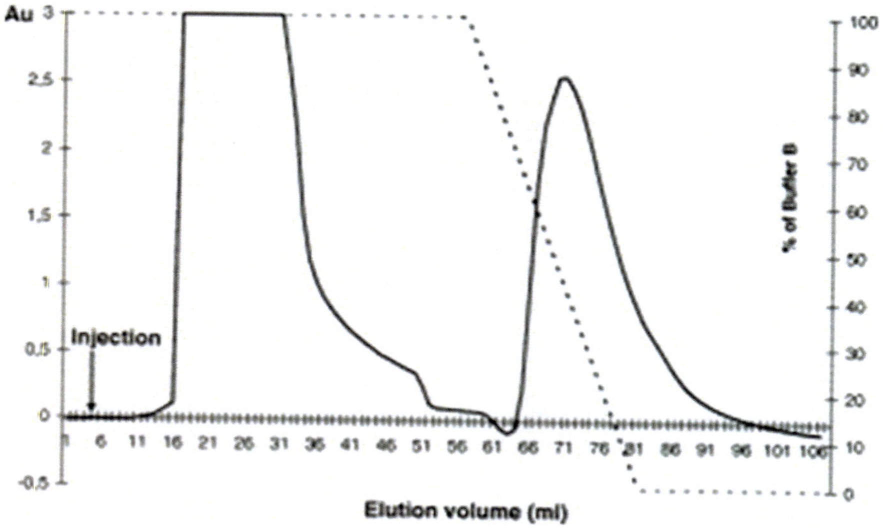

Fig. 9.12 Purification of recombinant penicillin acylase by hydrophobic charge induction chromatography. Column: XK 16/20 (2 mL MEP Hypercel). Sample, crude extract (10 mL); buffer A, 0.5 M $(NH_4)_2SO_4$; 0.02 M sodium phosphate pH 7; buffer B, 50 mM sodium acetate pH 4. Detection at 280 nm. Flow rate, 2 mL/min. (Reproduced from Coulon et al. 2004)

subsequently eluted in fractions 18–53 (pools B and C). In addition to amylase and histones, some faintly stained proteins were also observed in the eluted fraction.

Purification of recombinant penicillin acylase from *E. coli* cell lysate The production of semisynthetic penicillin requires important intermediates, such as 6-aminopenicillic acid (6-APA). The industrial production of 6-APA is through the use of immobilized penicillin acylase. Due to the low recovery of the enzyme purification process, the price of 6-APA remains high. Coulon et al. (2004) developed a method for purifying recombinant penicillin acylase from *E. coli* lysate using MEP HyperCel and evaluated the effect of sample loading rate and salt concentration in the equilibrium solution on the recovery of penicillin acylase activity. In the range of 1–10 mL/min, the recovery decreases as the flow rate increases, but this reduction in recovery can be compensated by using a high concentration of ammonium sulfate. In the optimized process, the specific activity of the enzyme purified from the crude extract is 11 U/mg, the activity recovery is over 90%, and the purification factor is 4.5. A typical chromatogram is shown in Fig. 9.12.

Purification of immunoglobulin from goose serum Lu et al. (2017) prepared a series of 2-mercapto-1-methylimidazole (MMI) resins with different pore sizes and ligand densities to separate IgG from IgG/bovine serum albumin (BSA) protein mixtures. When loaded at pH 7.0 and eluted at pH 4.0, the purity of the IgG monomer was about 95%, and the recovery was about 90%.

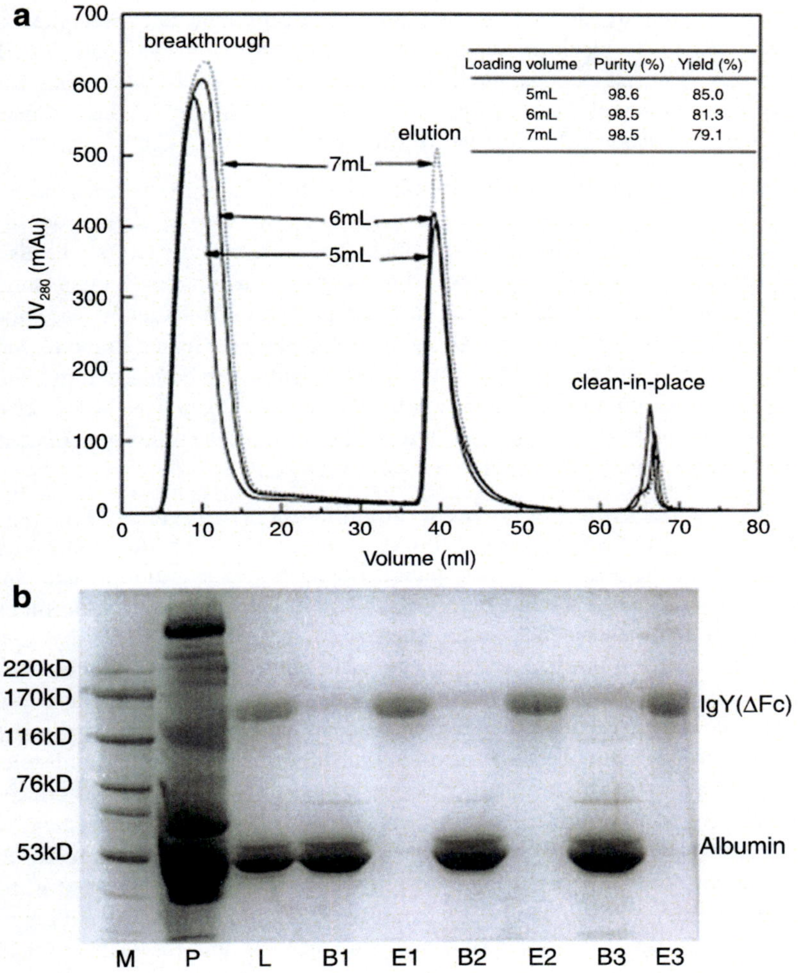

Fig. 9.13 Separation of IgY (ΔFc) from goose plasma by hydrophobic charged induction chromatography on MMI-Bestarose. The sample loading buffer is 20 mM sodium acetate, pH 5.4, and the elution buffer is 20 mM acetic acid, pH 4.0. (**a**) Chromatogram; (**b**) SDS-PAGE. SDS-PAGE lane, protein marker (M); goose plasma (P); loading (L); flow-through fractions with sample loading volume of 5 mL (B1); 6 mL (B2) and 7 mL (B3); elution fractions with sample loading volume of 5 mL (E1); 6 mL (E2) and 7 mL (E3). (Reproduced from Lu et al. 2017)

IgY (ΔFc) is a unique immunoglobulin found in waterfowl. It naturally lacks the CH_3 and CH_4 domains of the heavy chain and has some special therapeutic applications. The synthesized MMI-agarose resin was applied for the purification of goose immunoglobulin IgY (ΔFc) (Fig. 9.13). The adsorption isotherms of IgY (ΔFc) at different pH were studied. The saturation capacity of IgY (ΔFc) at pH 5.0 was found to be 187.5 mg/mL gel, and pH-dependent adsorption behavior was observed. After pretreatment of goose plasma with caprylic acid to precipitate some serum proteins, the resulting supernatant can be directly loaded onto the HCIC column for IgY (ΔFc) separation. The operating conditions, including sample loading pH, elution pH, and sample volume, were optimized. A high separation efficiency was obtained, the final purity was 98.6%, and the yield was 85.0%.

Purification of human immunoglobulin G Yan et al. (2015) prepared a new hydrophobic charge-induction chromatography resin with a ligand density of 195 μmol/mL by immobilizing 5-aminobenzimidazole on polyacrylate beads. The adsorption properties of the synthesized gel was evaluated using human immunoglobulin G and bovine serum albumin as model proteins. The antibody was bound in a binding buffer 0.2 M NaCl pH 8.0, and eluted at pH 5.0. Human immunoglobulin G was purified in one step from the raw material containing bovine serum albumin. The purity of the product obtained was 99.7%. The recovery was 94.6%. Using a similar method to purify monoclonal antibodies from the cell culture supernatant, the antibody purity was 94.9%, and the recovery was 92.5%.

Separation of lysozyme and bovine serum albumin Zhao et al. (2008) reported a 5-aminoindole modified agarose gel (AI-Sepharose) as a hydrophobic charge-induction chromatography stationary phase. The adsorption equilibrium and kinetics of lysozyme and bovine serum albumin (BSA) on AI-Sepharose were determined by batch adsorption experiments under different conditions. Ligand density and salt concentration have a positive effect on the adsorption of the two proteins studied. When the pH decreased from 7 to 3, the adsorption capacity of basic protein lysozyme decreased rapidly due to the increase of electrostatic repulsion, while the acidic protein BSA obtained the maximum adsorption capacity near its isoelectric point. Dynamic adsorption experiments show that under different salt concentrations, the effective pore diffusion coefficient of lysozyme remains constant, while the effective pore diffusion coefficient of BSA decreases with the increase of salt concentration. This is because BSA experiences greater steric hindrance in pores. The high protein recovery can be achieved by adsorption at pH 7.10 and elution at pH 3.0. The purity of the lysate obtained is about 90%, and the recovery is about 90%.

Purification of antibodies in pig plasma Phottraithip et al. (2013) prepared a new hydrophobic charge induction chromatography (HCIC) resin with 2-mercaptoimidazole (MI) as a ligand and used it to purify IgG from pig plasma. The cellulose matrix was activated by divinyl sulfone (DVS) and then coupled with MI ligand to prepare a novel hydrophobic charge-induction chromatography stationary phase with a ligand density of about 100 μmol/mL gel. The prepared resin was used to separate IgG from pig plasma. After the fibrinogen was precipitated by salting-out, the supernatant was loaded onto the chromatographic column at pH 7, and the adsorbed IgG eluted at pH 3.6. The purity of the eluted protein is 89.4%.

Purification of monoclonal antibodies from cell culture supernatant It is well known that hydrophobic charge induction chromatography (HCIC) has the advantages of large capacity, good salt tolerance, and convenient pH controlled elution.

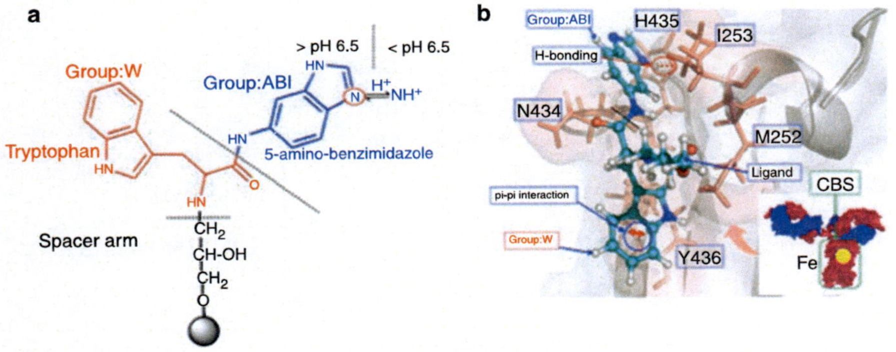

Fig. 9.14 The molecular structure of W-ABI ligand (**a**) and the mode of interaction between W-ABI and IgG (**b**). (Reproduced from Tong et al. 2016)

However, compared with affinity chromatography, its binding specificity is not very good. Tong et al. (2016) proposed to improve binding specificity through multimodal molecular interaction. Therefore, a new chimeric charge-induction chromatography ligand W-ABI that combines tryptophan (W) and 5-aminobenzimidazole (ABI) was synthesized (Fig. 9.14). The indole and benzimidazole groups of the ligand can bind the target IgG at neutral pH, while the imidazole group can induce electrostatic repulsion for effective elution at acidic pH. The W-ABI ligand was successfully coupled to agarose gel for the separation and purification of IgG. The saturated adsorption capacity at pH 7 is 70.4 mg/mL, and the flow rate of the mobile phase has almost no effect on the dynamic binding capacity. In addition, efficient elution can be achieved under mild acid pH (pH 4.5–5.0). W-ABI resin is used to purify mAb from CHO cell culture supernatant. The product purity and recovery are 98.9% and 88.6%, respectively.

Purification of antibodies in human serum Zou et al. (2019) used molecular simulation to design a multimodal hydrophobic charge-induction chromatography ligand, which is composed of phenylalanine (F), tyrosine (Y), glutamic acid (E), and ABI (Fig. 9.15). This new hybrid resin (FYE-ABI) is used for the separation and purification of hIgG from solutions containing human serum albumin (HSA) (Fig. 9.16). The sample was loaded at pH 7 and eluted at pH 4.0. The purity of hIgG obtained was 99.0%, and the recovery was 91.5%. FYE-ABI resin was further applied to separate hIgG from human serum at pH 4.5. The purity of hIgG obtained was 93.9%, and the recovery was 88.9%.

Purification of monoclonal antibody IgM from the culture supernatant of hybridoma cells Luo et al. (2018) studied the adsorption behavior of IgM on four different HCIC resins, including MEP HyperCel, ABI-4FF, MMI-4FF, and W-ABI-4FF. They found that at pH 5.0 ABI-4FF and MMI-4FF have high selectivity for IgM. ABI-4FF was used to capture monoclonal IgM from the culture supernatant of hybridoma cells, with high purity (~99%) and good recovery (80.8%).

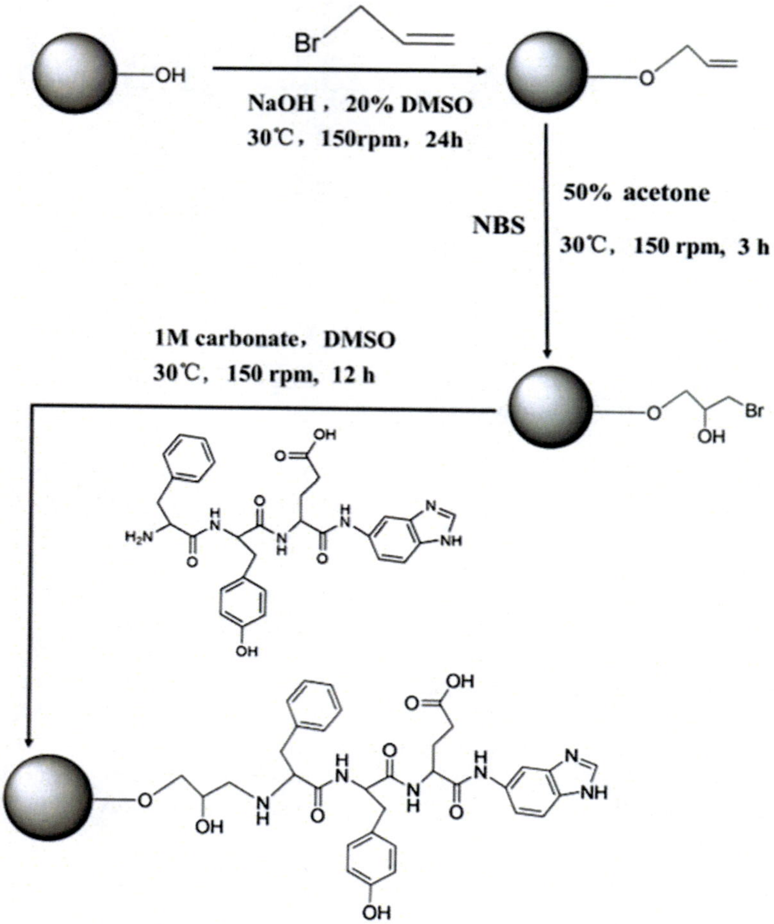

Fig. 9.15 Preparation of FYE-ABI resin. (Reproduced from Zou et al. 2019)

Adsorption behavior of lysozyme on histamine stationary phase Shi et al. (2010) immobilized histamine on Sepharose CL-6B to prepare a new HCIC resin and studied its adsorption behavior on lysozyme. They found that the dehydration of protein and adsorbent promoted the adsorption of lysozyme on the HCIC resin and that the lysozyme molecule undergoes significant conformational changes during protein adsorption. This work provides a theoretical basis for the development of new HCIC ligands. These new HCIC ligands provide a variety of options for researchers in biological separations. However, most ligands have certain limitations and have not yet achieved large-scale industrial applications. However, the discussion and mechanism study of different HCIC ligands may be helpful to the future development of downstream separation processes.

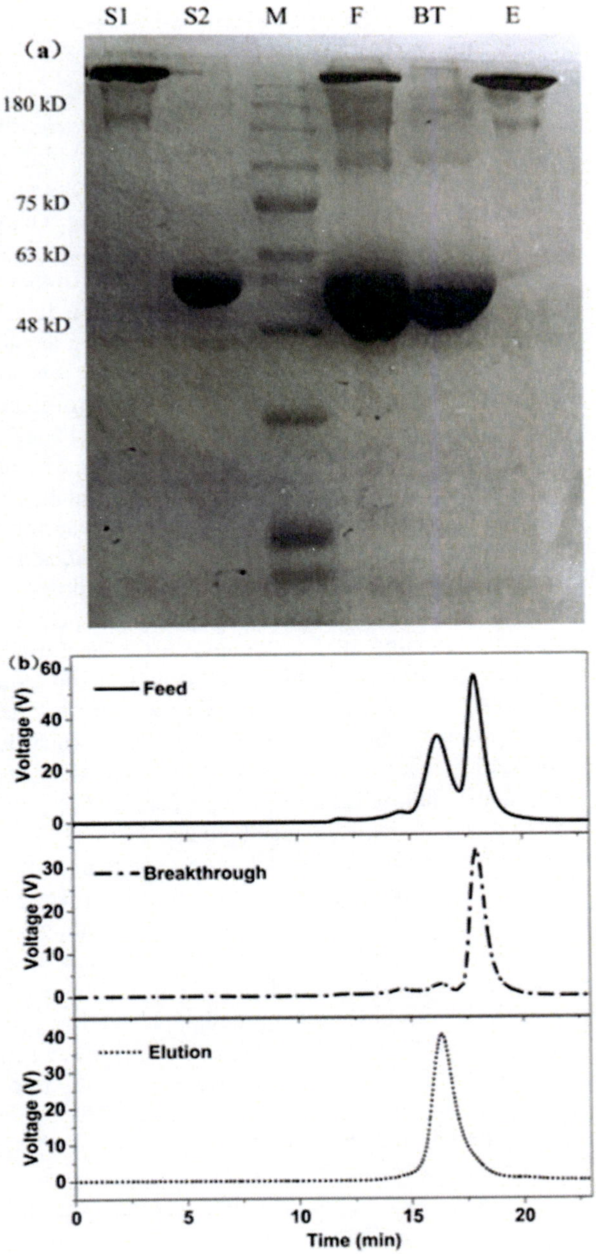

Fig. 9.16 Separation of hIgG (**a**) from the HAS/hIgG protein mixture at pH 7.0. Nonreduced SDS-PAGE results. S1, hIgG standard; S2, HAS standard; M, protein marker; F, protein mixture; BT, flow-through fraction; E, elution fraction (pH 4.0). (**b**) SEC-HPLC analysis of raw materials and elution fractions. (Reproduced from Zou et al. 2019)

9.4 Method Development in Multimodal Ion Exchange Chromatography

Multimodal ion exchange chromatography is the most widely used chromatographic technique in mixed-mode chromatography. It binds solute molecules through a combination of various intermolecular interactions such as electrostatic interactions, hydrophobic interactions, and hydrogen bonding. Therefore, it can identify the local area of the protein surface composed of hydrophobic patches and adjacent charged groups, with unique selectivity different from other chromatographic modes. In addition, due to the superposition of hydrophobic interaction and hydrogen bonding on the ionic interaction, it exhibits obvious salt tolerance for protein adsorption, and the proteins can be adsorbed under high and low ionic strength conditions that traditional ion exchange chromatography or hydrophobic interaction chromatography cannot bind. Therefore, multimodal ion exchange chromatography plays a very important role in the separation and purification of biological macromolecules.

Multimodal ion exchange ligands usually contain a hydrophobic part and an ionic part. The hydrophobic part is composed of a multifunctional group and a linking arm, and its hydrophobicity can be adjusted by adjusting the number of carbon atoms in the alkyl or aryl functional group. The ionic part is composed of dissociable groups such as carboxyl or amino groups or permanent charged groups such as quaternary ammonium. The strength of the electrostatic effect can be adjusted by changing the pH value and ionic strength of the solution. The salt tolerance of multimodal ion exchange ligands is due to the hydrophobic interaction superimposed on the electrostatic interaction, which is equivalent to providing an amplification factor for the electrostatic interaction. In a solution of the same ionic strength, the multimodal ion exchange chromatography ligands show stronger solute affinity than pure ion exchange ligands.

The structures of multimodal ion exchange ligands are diverse. The structures of some typical ligands used in commercial multimodal ion exchange chromatography packing materials are listed in Table 9.2. According to the nature of the group, it can be divided into two major categories: multimodal anion exchange ligands and multimodal cation exchange ligands. Anion exchange ligands contain ammonium groups, which can bind proteins at high pH values and elute at low pH values; cation exchange ligands contain carboxyl groups, which can bind proteins at low pH values and elute at high pH values. In addition, the adsorption and elution of proteins on multimodal ion exchange ligands are also affected by the ionic strength of the solution. Protein retention as a function of the ionic strength of the solution sometimes exhibits a typical U-shaped curve. When the ionic strength of the solution changes from low to high, the solute retention decreases with the increase of ionic strength. After reaching a valley, it shows a rebound trend and increases with the increase of ionic strength. Therefore, solution pH and ionic strength are the main experimental parameters that regulate protein retention and elution. And sometimes they may need to be adjusted simultaneously to optimize binding and elution behavior.

Table 9.2 Typical functional ligands used in multimodal ion exchange chromatography

Ligand	Structure	pK_a
Hexamine (HEA HyperCel)	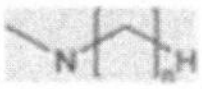 $n = 6$	8
Phenylbutylamine (PBA HyperCel)	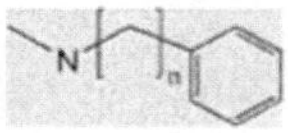 $n = 4$	8
Aminobenzoic acid (CMM HyperCel)	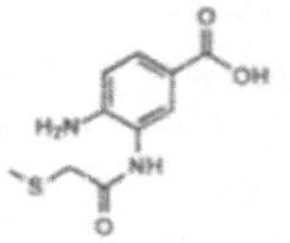	
N-Benzyl-*N*-methyl ethanol amine (Capto adhere)	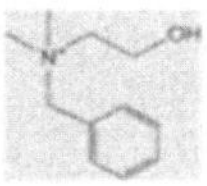	–
2-Benzamido-4-mercaptobutanoic acid (Capto MMC)	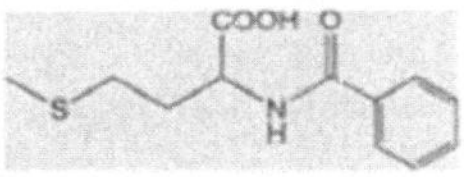	3.3
4-Aminobenzoylglycine (Nuvia cPrime)	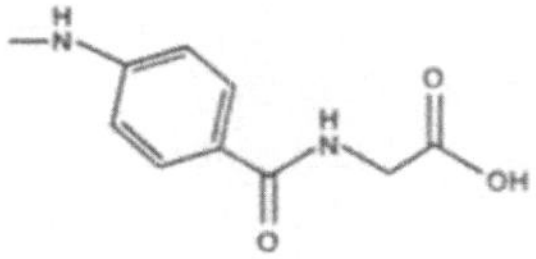	4
Tryptophan (TOYOPEARL MX-Trp)	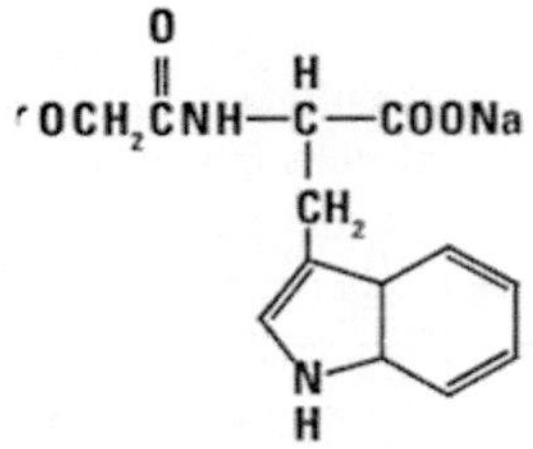	4

The complexity of the interaction between multimodal ion exchange ligands and proteins may bring other challenges to method development in the desorption phase. In most cases, the protein binds to the ligand through multivalent interactions on multiple binding sites. Disrupting any one of these sites or types of interaction alone may not be enough to cause elution. An obvious example is the effect of salt concentration as observed in the U-shaped curve. In the region of high salt concentration, the electrostatic interaction between the protein and the ligand is cancelled, but the hydrophobic interaction between them is strengthened, resulting in an increase in protein retention. Therefore, the most successful eluents so far are based on a mixture of arginine, urea, and salt, which are multipotent in nature, capable of simultaneously modulating two or more interactions between the solute and the ligand.

Salt-tolerant adsorption and unique selectivity are the main advantages of multimode ion exchange materials over single-mode resins. These features enable us to deal with various challenges in the downstream process of biopharmaceuticals, such as capturing proteins from complex raw materials, decolorizing feed streams, purifying difficult proteins, and removing impurities from special products and processes. In most cases, the use of multimodal ligands can greatly reduce the number of unit operations and provide considerable flexibility for the design of effective downstream processes, especially when there is no universal affinity ligand (such as protein A). The complexity of the interaction between the ligand and the target protein may increase the difficulty of method development. However, this problem can be overcome by using a combination of appropriate screening strategies and experimental statistical design, making it possible to optimize complex separation steps in a general and systematic way.

9.4.1 General Strategy

When multimodal ion exchange chromatography is used for the separation and purification of proteins, the method development can be carried out according to the adsorption and elution process. In the adsorption process, the focus is to control the pH of the solution to maximize the ligand's affinity for the solute. In the elution process, a double-gradient elution method can be used. First, the gradient of pH change is run, and then the gradient of salt concentration is applied, or even the gradient of simultaneous changes of pH and salt concentration is used. For multimodal anion exchange chromatography, a gradient with decreasing pH is used. At low pH, the protein and the ligand are both positively charged and the protein will be eluted due to charge repulsion. For multimodal cation exchange chromatography, a gradient with increasing pH is used. At high pH values, the protein and the ligand are both negatively charged, the adsorbed protein will be eluted due to charge repulsion. When an optimal pH for elution is chosen, a gradient of increasing or decreasing salt concentration can be used to optimize protein separation. In the case where the binding of protein and ligand is very strong, and the use of the pH and salt concentration changes alone is not sufficient for elution, one may consider using mobile phase modifiers, such as urea, propylene glycol, guanidine hydrochloride, and arginine, to disrupt any residual affinity.

Due to the presence of aliphatic (i.e., HEA HyperCel™) or aromatic (i.e., PPA HyperCel™) groups and protonated amino groups in the ligand, mixed-mode chromatography works through a combination of multiple interactions. Pezzini et al. (2014, 2015) showed that the retention mechanism of proteins on HEA and PPA HyperCel involves ionic interaction, hydrophobic interaction, and hydrogen bonding. Each of them strongly affects chromatographic separation, and the balance between them is the cornerstone of the original separation. In addition, they

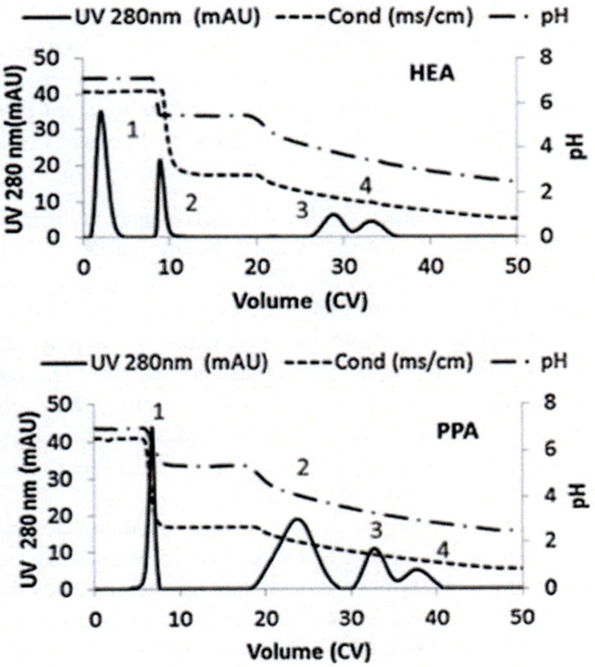

Fig. 9.17 Separation of protein mixtures by adjusting pH on HEA and PPA Hypercel. Column: Tricorn 5/50 (1 mL HEA or PPA HyperCel). Sample: lysozyme, α-chymotrypsinogen A, ovalbumin, BSA (2 mg/mL) in equilibration buffer. Equilibration buffer: 0.2 M NaCl, 0.2 M sodium phosphate, pH 7.4. Elution buffer: 0.2 M citrate phosphate, pH 2.8. Linear gradient over 30 CV. Sample size: 1 mL. Detection at 280 nm. Flow rate: 1 mL/min. (Reproduced from Pezzini et al. 2015)

demonstrated that the distribution of these interactions, especially the hydrophobic and ionic effects (attraction and/or repulsion), directly depends on the physical and chemical environment in which the solute and the ligand are located. The pH value of the buffer and the ionic strength of the eluent are the primary parameters to control solute retention: pH affects the interaction by changing the charge distribution on the protein surface, while electrical conductivity changes the balance between electrostatic and hydrophobic interactions. Finally, adsorption and elution can be obtained in many environments/conditions (pH, conductivity, additives). Several operating modes have also been developed: taking advantage of the charge properties of proteins and chromatographic media by changing the pH and/or conductivity (Fig. 9.17); taking advantage of the hydrophobic properties of proteins and chromatographic media by changing the ionic strength and/or dielectric constant of the mobile phase; and taking advantage of a combination of different attributes by using a combination of specific environments.

9.4.2 Examples of Method Development

Purification of apolipoprotein in plasma Apolipoprotein A1 (ApoA1) is 28 kDa protein that is composed of 244 amino acids, the main protein component of HDL particles in plasma. The chylomicrons secreted from intestinal epithelial cells also contain apoA1, but it is quickly transferred to HDL in the blood. As a component of HDL particles, this protein receives fat from cells (including macrophages in the arterial wall, which have been overloaded by fat ingested by oxidized LDL particles) and transports it to other places through the water outside the cell, including back to low-density lipoprotein particle as liver excretion.

ApoA1 is a cofactor for lecithin cholesterol acyltransferase (LCAT) and is responsible for the formation of most plasma cholesterol esters. ApoA1 has also been isolated as a prostacyclin (PGI2) stabilizing factor, so it may have anticoagulant effects. Defects in the gene encoding it are associated with defects in HDL, including Tangier disease and systemic non-neurological amyloidosis. ApoA1 is often used as a biomarker to predict cardiovascular disease. The ratio of apoB-100/apoA1 (i.e., LDL and larger particles to HDL particles), the ratio of lipoprotein (LDL/HDL) particles measured by NMR is even more so, compared to the old method of measuring lipid transport in water outside the cell; the correlation with the incidence of myocardial infarction is stronger.

ApoA1 plays a key role in preventing atherosclerosis and cardiovascular disease. Purification of ApoA1 from plasma has always been cumbersome, involving multiple steps, resulting in a significant loss of large amounts of plasma and a low yield of purified ApoA1. Arun et al. (2018) developed a two-step method to purify ApoA1 from plasma. First, 60% ammonium sulfate $(NH_4)_2SO_4$ was used to precipitate the plasma, and ApoA1 was enriched in the supernatant. The supernatant was dialyzed into an equilibration buffer containing 50 mM phosphate buffer pH 7.4. Then it was loaded onto the hydrophobic ion exchanger HEA HyperCel™. The bound protein was eluted by gradually lowering the pH value from pH 7.4 to 4.0 and then lowering it to pH 3.5 using 50 mM sodium acetate buffer (Fig. 9.18). Gel electrophoresis showed that the eluted fraction at pH 3.5 contained homogeneous ApoA1, with a purity and yield of 63%. MS analysis confirmed that the purified ApoA1 was a monomer with a mass of 28,079.30 Da.

Purification of β-lactoglobulin in whey In the dairy industry, "ideal whey" is produced by tangential flow microfiltration of milk. It contains soluble protein and does not contain casein and fat. β-lactoglobulin (β-LG) and α-lactalbumin (α-LA) are the main proteins in "ideal whey." β-LG has a special meaning because it is considered to be one of the main allergic components in cow milk. Brochier et al. (2008) compared the performance of three mixed-mode chromatography adsorbents (HEA, PPA, and MEP HyperCel™) in purifying bovine β-lactoglobulin from bovine whey and found that HEA HyperCel showed the best effect. After the bovine whey was directly loaded onto the HEA HyperCel adsorbent, the bovine β-lactoglobulin was partially purified by using a stepwise decreasing pH gradient. Figure 9.19 shows the resulting chromatagrams on three "HEA HyperCel" sorbent packed columns with different volumes. β-LG is completely retained on the chromatographic column and is recovered in a citrate buffer at pH 4.0. Since both bovine β-LG and α-LA are acidic proteins, with the p*I* of 5.4 and of 4.2, respectively, they are negatively charged at pH 7.4 and can interact with HEA ligands through electrostatic and

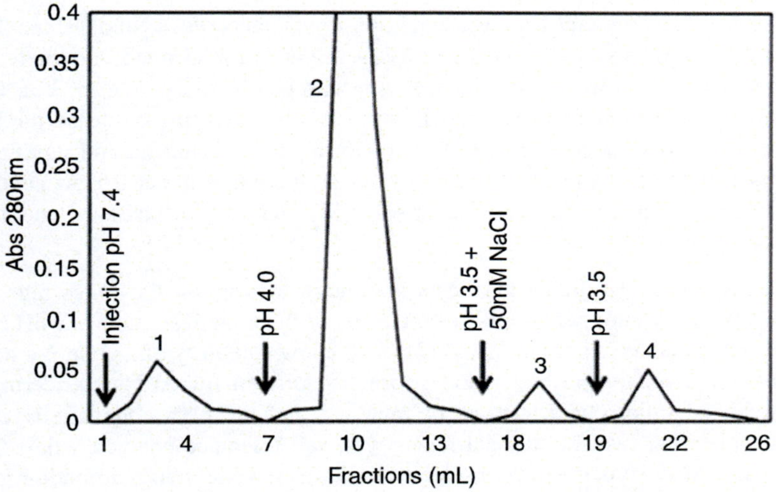

Fig. 9.18 Purification of ApoA1 from plasma using HEA HyperCel™. Column: HEA HyperCel™, 4 mL. Sample: 5 mL of the supernatant fraction from ammonium sulfate precipitation. Equilibration buffer: 50 mM sodium phosphate, pH 7.4. Step 1 elution buffer: 50 mM sodium acetate, pH 4. Step 2 elution buffer: 50 mM NaCl, 50 mM sodium acetate pH 3.5. Step 3 elution buffer: 50 mM sodium acetate pH 3.5. Detect at 280 nm. Flow rate: 1 mL/min. (Reproduced from Arun et al. 2018)

Fig. 9.19 The chromatographic separation of "ideal whey" on three different sized (8 mm ID × 50 mm H; 11.3 mm ID × 50 mm H; 11.3 mm ID × 100 mm H) prepacked cartridges. Injection volume, 4 CV; packing, HEA HyperCel adsorbent, and equilibrated in 20 CV pH 7.4 PBS. The elution step is 5 CV of 50 mM sodium citrate, pH 4.4; pH 4.0; pH 3.0. (Reproduced from Brochier et al. 2008)

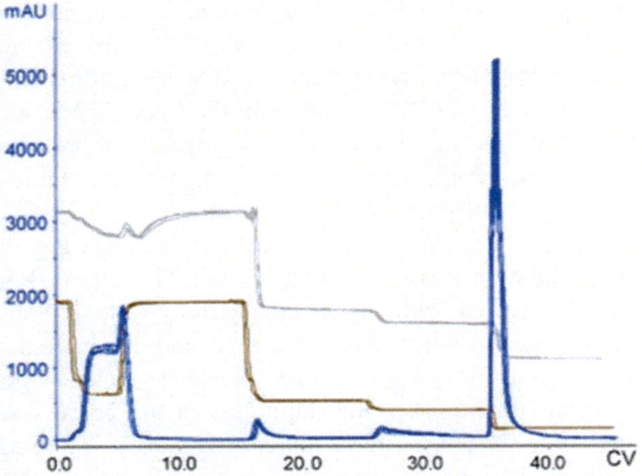

hydrophobic interactions. The adsorbed proteins can be eluted by decreasing the pH of the eluent, with β-LG recovered at the pH of pH 4.0 and α-LA at pH 3.0, respectively.

Purification of recombinant protease inhibitor Multimode ion exchange chromatography has been successfully used to purify several recombinant Kunitz-type protease inhibitor muteins from yeast culture supernatants (Peters and Oehme 2010). Mutein I is expressed in *Saccharomyces cerevisiae* and is secreted into the fermentation supernatant using a complex medium with conductivity >20 mS/cm, which contains a large amount of colored impurities (i.e., melanin). Protease inhibitors (p*I* ~ 6.3) showed reasonable solubility of ~13 mg/mL at pH < 3.2 and significantly decreased to <0.5 mg/mL when the pH was only slightly increased. At neutral or weakly alkaline pH values, acceptable solubility can only be achieved under high salt conditions.

The design of a suitable purification strategy focuses on the following goals: (1) determine the chromatographic step that keeps the protein solution at a pH below 3 or at a high salt concentration, and (2) use the low solubility of mutein I at neutral pH to find a suitable crystal storage form for protein for API. Considering the abnormal solubility characteristics of mutein I and the high conductivity of the starting material, two multimodal resins with salt tolerance characteristics, Capto MMC and HEA HyperCel, were selected to develop a two-step purification for the capture and intermediate purification of the target protein, respectively. The clarified fermentation supernatant (conductivity>20 mS/cm) was directly loaded onto a Capto MMC column with pH 5.5, and the adsorbed protein was eluted under high salt (2 M NaCl) and pH 8.5 conditions. This step effectively removes impurities in the process and products, including colored impurities (melanoids), and the purity of mutein I is >90%. Subsequently, without any conditioning steps, the eluted protein fraction was loaded directly onto the HEA HyperCel column and the adsorbed protein eluted at pH 2.7, with the purified protein kept in a low-salt solution. In this step, the product purity is only slightly improved; only trace amounts of product-derived impurities are separated. However, the HEA HyperCel step allows to remove large amounts of salt from the Capto MMC eluate and immediately lower the pH, which is essential for the preparation of proteins for the crystallization step. Finally, the pH value was increased to above 4 to initiate crystallization to facilitate long-term stable storage of the bulk drug.

Using Capto MMC to capture mutein I is a key chromatographic step in the purification process described above. The clarified fermentation supernatant is directly applied with higher conductivity (>20 mS/cm) without further dilution. The maximum adsorption capacity of the chromatographic column is 17 mg/mL. Although not very high, it is acceptable for the early production process. After this chromatographic step, the impurities in the process and products are effectively removed, and the purity of mutein I is >90%. By using strict washing conditions, high-salt washing conditions with a pH of 5 and low-salt washing steps with a pH of 8.5, the colored impurities (melanoids) are almost completely removed from the fermentation medium, and the Capto MMC eluate is colorless and clean. The decolorization performed by Capto MMC is mainly a pH-driven process. In the process of developing another protease inhibitor mutein (Mutein II), colored impurities from complex yeast media are bound with Capto MMC at a pH of 3–6. It is completely removed when the pH value is greater than 6.5. Therefore, under these conditions, the binding of the required protein to the resin is particularly important for a successful decolorization strategy using Capto MMC.

Purification of recombinant human thioredoxin Human thioredoxin-1 (hTrx) is an acidic protein (p*I* 4.8) with molecular weight of 12 kDa, present in all tissues. Ravi et al. (2020) reported a one-step method for purifying hTrx from *E. coli* lysates using the multimode ion exchange stationary phase HEA HyperCel. Under the conditions of pH 5 and 7, the adsorbed protein can be eluted with a salt gradient to obtain hTrx with a purity of 90% and a yield of 95%.

Purification of recombinant human vascular endothelial growth factor Vascular endothelial growth factor (VEGF) is an important protein that regulates the angiogenesis cascade. Recombinant human vascular endothelial growth factor (rhVEGF) is expressed as inclusion bodies in *E. coli*. The solids are harvested from cell lysates, and rhVEGF is dissolved and refolded at pH 9.8 in the presence of urea and redox reagents. Kaleas et al. (2010) reported the industrial purification process of using Capto MMC to capture recombinant human vascular endothelial growth factor directly from the refolding pool. Due to the multimodal binding properties of Capto MMC, rhVEGF can be completely adsorbed to the multimodal stationary phase under the condition of pH 8.7. However, the development of the elution phase is challenging because the mobile phase components are required to simultaneously disrupt multiple types of interactions between the ligand and the protein in order to recover the active protein. The use an elution gradient of buffers that increase ionic strength and change pH to disrupt ion exchange interactions, or organic solvents that are known to specifically disrupt hydrophobic interactions, has not been successful. Then, a buffer that combines sodium chloride with reagents that disrupt hydrogen bonds or disrupt hydrophobic interactions (such as organic solvents, urea, and L-arginine) proved to be an effective eluent. Under these conditions, the elution of rhVEGF is significantly increased, and it is still biologically active, which indicates that despite the potential denaturing effects of high-concentration solvents, its natural conformation is still retained. The combination of sodium chloride and ethanol or propylene glycol (each of which has very limited effects on its own) improves the elution performance. The shape of the elution peak becomes sharp and the yield increases to 50–60% rhVEGF. A similar effect can be observed by mixing 6 M urea with sodium chloride and high-concentration L-arginine hydrochloride. The last two buffers produced a sharp and compact elution peak, resulting in a yield of over 90%.

After comprehensively considering various factors including process performance (i.e., selectivity and yield) and manufacturability, such as buffer composition, compatibility with downstream chromatography, and cell stability, the composition of the elution buffer was determined. It was found that L-arginine hydrochloride at 0.8–1.0 M can provide consistent and reproducible results. The yield of purified protein is greater than 90%, and the concentration of the host cell proteins is reduced by $\geq$150 times. The preparation of the final elution buffer is simplified by using acid and base conjugates, which can be illustrated by adding 0.85 M L-arginine hydrochloride and 0.05 M L-arginine in 25 mM HEPES (pH 7.5, 41 mS/cm) background buffer without pH adjustment.

The development and optimization of adsorption and elution conditions on a laboratory scale have resulted in robust and scalable chromatographic steps that can provide comparable recovery and purity in pilot and production scales. Under selected conditions, the Capto MMC purification step can reduce the host cell protein level to <1.2%, and the *E. coli* DNA to <10 ng/mg, with a yield of >95%. The purification step was completed in about 13 h, and the volume of the refolding cell was reduced by 75 times, confirming that the unit operation is suitable for commercialization.

Purification of recombinant human fibroblast growth factor Fibroblast growth factor (FGF) is a family of cell signaling proteins produced by macrophages. They participate in various processes and are a key element for the normal development of animal cells. Any irregularities in their function will lead to a series of developmental defects. These growth factors generally act as systemic or locally circulating molecules of extracellular origin that activate cell surface receptors. Recombinant human fibroblast growth factor (rhFGF) is usually purified from *E. coli* lysate using a three-step method of capture by cation exchange chromatography, intermediate purification by heparin affinity chromatography, and polishing by anion exchange chromatography. In order to avoid the use of animal-derived heparin in the purification process, Bolten et al. (2020) reported the use of multimodal cation exchange chromatography instead of heparin chromatography for intermediate purification of rhFGF-2. Figure 9.20 compares the separation of rhFGF-2 captured by cation exchange chromatography on a heparin affinity column, a Nuvia cPrime column, and a Capto MMC column. Buffer A is 25 mM phosphate buffer, 3 mM DTT, 1 mM EDTA, pH 7.5; buffer B is buffer A containing 2 M NaCl. The linear gradient elution is within 10 CV, increasing from 100% buffer A to 100% buffer B. Using heparin affinity chromatography as an intermediate step, two elution peaks were obtained. The target protein rhFGF-2 only exists in the second elution peak, with a purity of 94% and a yield of 90%. Purification of rhFGF-2 using multimodal cation exchange chromatography columns Nuvia cPrime and Capto MMC both produced an elution peak with a purity of 90% and a yield of 85%. It is comparable to the results of heparin chromatography and is suitable for use as a substitute for animal-derived heparin.

Purification of recombinant bispecific antibodies A bispecific monoclonal antibody (BsMAb, BsAb) is an artificial protein that can simultaneously bind to two different types of antigens. Naturally occurring antibodies typically only target one antigen. BsMabs can be manufactured in several structural formats, and current applications have been explored for cancer immunotherapy and drug delivery. There are many formats of bsMabs, but the two main categories are IgG-like and non-IgG-like. The IgG-like retains the traditional monoclonal antibody (mAb) structure, that is, two Fab arms and an Fc region, but the two Fab sites bind different antigens. The most common type is called trifunctional antibodies, because they have three unique binding sites on the antibody: two Fab regions and an Fc region. Each heavy chain

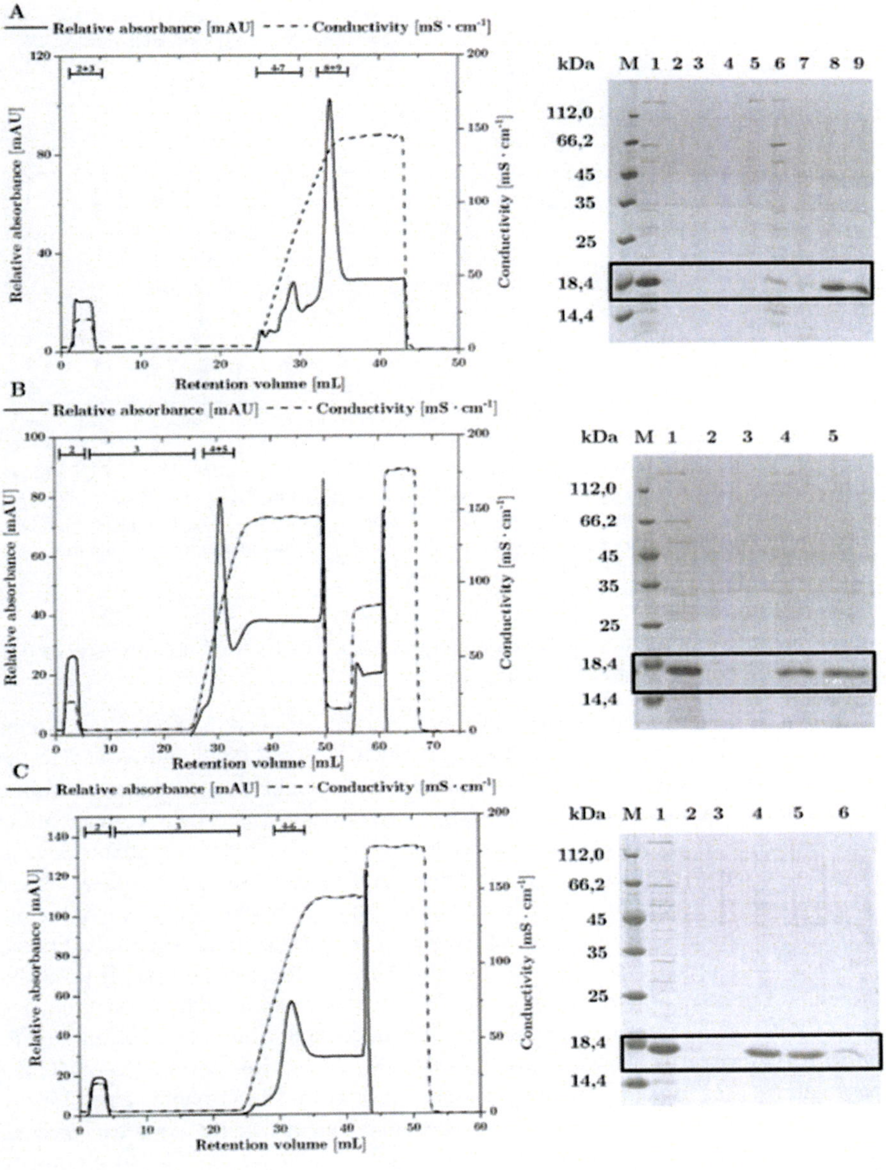

Fig. 9.20 FPLC chromatogram (left) and SDS-PAGE analysis (right) of the intermediate purification step of rhFGF-2 after capture by cation exchange chromatography. The rhFGF-2 band is highlighted at 18 kDa. The binding buffer is 25 mM phosphate buffer, 3 mM DTT, 1 mM EDTA, pH 7.5; at 10 CV, the salt concentration in the binding buffer is increased from 0 to 2.0 M NaCl for elution. (**a**) Heparin affinity chromatography: M, protein ladder; 1, sample; 2 + 3, flow-through fraction; 4–7, fraction of overlapping peaks; 8 + 9, fraction of hFGF-2 elution peak. (**b**) Nuvia cPrime chromatography: M, protein ladder; 1, sample; 2, flow-through fraction; 3, washing step; 4 + 5, rhFGF-2 elution peak fraction. (**c**) Capto MMC chromatogram: M, protein ladder; 1, sample; 2, flow-through fraction; 3, washing step; 4–6, rhFGF-2 elution peak fraction. (Reproduced from Bolten et al. 2020)

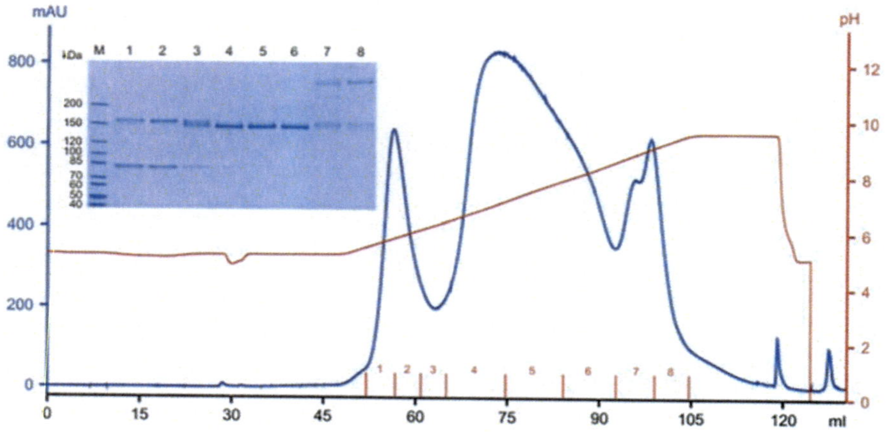

Fig. 9.21 Separation of bispecific antibodies on Capto MMC ImpRes after protein A chromatography. A linear pH (5.5–10) gradient is used for elution. Inset: nonreduced SDS-PAGE analysis of relevant fractions. M, protein marker; lanes 1–8, elution fractions 1–8. The first peak contains a large amount of half antibodies, the second main peak contains intact antibodies, and the third peak contains aggregates. (Reproduced from Tang et al. 2020)

and light chain pair are derived from a unique mAb. The Fc region composed of two heavy chains forms the third binding site.

The IgG-like bsAbs can be manufactured by a recombinant method called "knobs into holes," which relies on introducing a mutation for a large amino acid in the heavy chain from one mAb, and a mutation for a small amino acid in the other mAb's heavy chain. This allows the target heavy chains (and their corresponding light chains) to fit together better and makes bsMab production more reliable. Due to unbalanced chain expression and incorrect chain pairing, various by-products are often observed, including half antibodies, pore homodimers (i.e., hole-hole homodimers instead of knob-hole heterodimers), and light chain deletion antibody, Fab deletion antibody, and aggregates. These by-products cannot be removed by protein A chromatography. Tang et al. (2020) used a multimodal cation exchange resin Capto MMC ImpRes to remove these by-products from bispecific antibodies purified by protein A chromatography. Capto MMC ImpRes and Capto MMC are based on the same ligand, containing cation exchange and hydrophobic parts, but the former has a reduced microsphere diameter and ligand density, so it can provide higher resolution. They initially used linear pH gradient (pH 5.5–10) elution to screen the separation conditions. The chromatogram is shown in Fig. 9.21. Half antibodies are eluted first, then intact antibodies, and finally aggregates (Fig. 9.21, inset). This means that half antibodies can be removed by washing at an elevated pH.

Based on the results of the linear pH gradient, they developed a stepwise elution process that includes a washing step with increased pH (pH 7.4) to remove half antibodies. In addition, it was also observed that the late part of the washing at pH 7.4 mainly contained pore-pore homodimers. The binding strength of the pore-

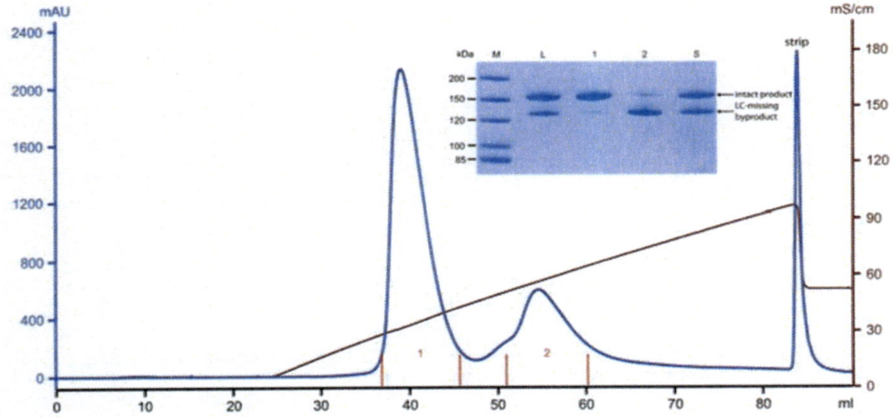

Fig. 9.22 Chromatogram of bsAb 1 on Capto MMC ImpRes eluted with a linear salt gradient of (50 mM sodium acetate, pH 5.5, 0–1 M NaCl). The sample is pooled fractions collected from protein A chromatography of bsAb 1 preparation. Inset: nonreduced SDS-PAGE analysis of relevant fractions. M, protein marker; L, load; lanes 1 and 2, elution fractions 1 (primary peak) and 2 (secondary peak); S, strip fraction. (Reproduced from Wan et al. 2020)

pore homodimer is slightly stronger than that of the half antibody, but much weaker than the intact bsAb. On this basis, they optimized the composition of the washing solution and introduced histidine as the main component of the washing solution. The established washing step can simultaneously remove half antibodies and pore-pore homodimers. In addition to half antibodies and hole-hole homodimers, MMC ImpRes also provides good aggregate clearance. Unlike half antibodies and pore-pore homodimers, which have weaker binding strength than the intact bsAb and can be removed by appropriate washing, the aggregates bind more tightly, so they can be separated by selecting suitable elution conditions for the intact bsAb.

Wan et al. (2020) used Capto MMC ImpRes for the removal of light chain (LC) deletion by-products and aggregates in recombinant asymmetric bispecific antibodies (bsAbs1) and symmetric bispecific antibodies (bsAbs2). In the case of the asymmetric antibody bsAbs1, in a linear gradient with increasing salt concentration (0–1 M NaCl), the intact antibody eluted first, while the LC missing by-products with higher p*I* eluted after (Fig. 9.22). A good separation is achieved with the by-products of LC deletion.

In the case of the symmetrical antibody bsAb 2, the p*I* of the LC deletion by-product is slightly different from that of the intact antibody. Therefore, a linear gradient with increasing pH and salt concentration was tried for elution. As shown in Fig. 9.23, LC deletion by-products are enriched at the tail of the main peak of the intact antibody.

Based on the screening of linear gradient elution, two gradual elution steps of bsAb were established. As shown in Fig. 9.22 for bsAb 1, gradual elution can maintain a good separation between intact bsAb and LC lacking by-products. All

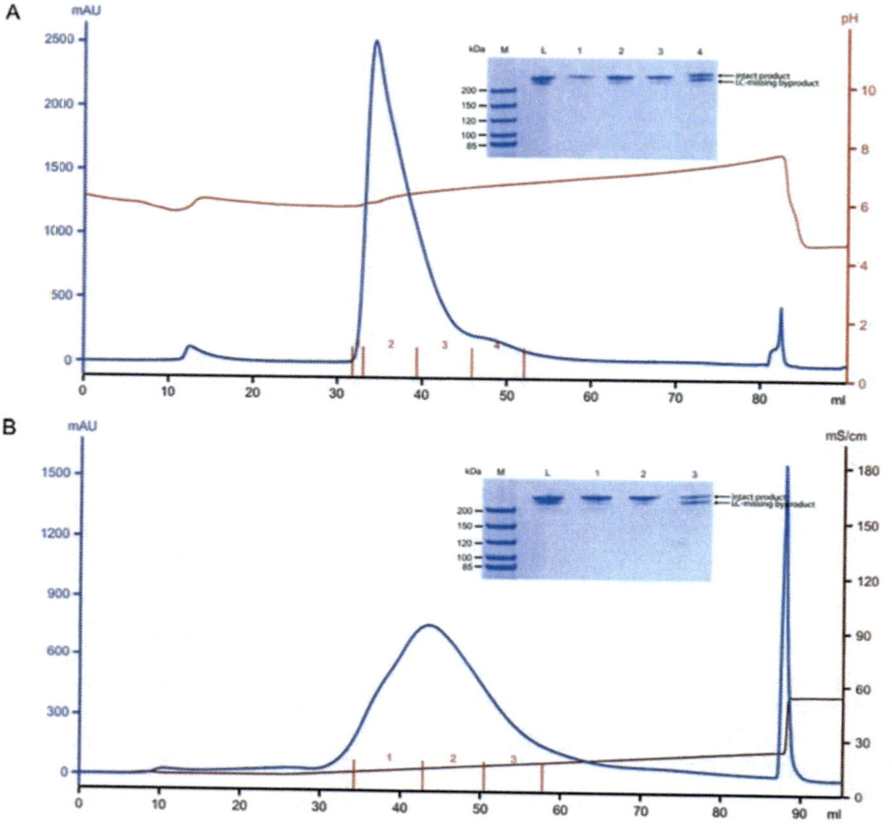

Fig. 9.23 Chromatogram of bsAb 2 on Capto MMC ImpRes under a linear pH gradient elution (20 mM phosphate, pH 6.0–8.0) (**a**) and a salt gradient elution (40 mM phosphate, pH 6.0, 0–150 mM NaCl) (**b**). The sample is pooled fractions collected from protein A chromatography of bsAb 2 preparation. Inset: nonreduced SDS-PAGE analysis of relevant fractions. M, protein marker; L, load; lanes 1–4, eluted fractions 1–4, respectively. (Reproduced from Wan et al. 2020)

eluted fractions contained pure intact products, while LC lacking by-products and aggregates were mainly present in the stripped fractions. The data showed that after purification by MMC ImpRes, the content of LC missing by-products and aggregates decreased from 15.3% and 9.6% to 0.2% and 1.2%, respectively. For bsAb 2, under the gradual elution of pH and salt, a good separation can be achieved between the intact bsAb and the LC missing by-products. Under both conditions, both elution fractions contained pure intact products, while the LC missing by-products were mainly present in the stripped fractions (Fig. 9.24). Between pH and salt stepwise elutions, the latter can provide a higher purity product. After purification by Capto MMC ImpRes, the content of LC missing by-products and aggregates was reduced from 3.7% and 7.8% to 1.2% and 3.5%, respectively.

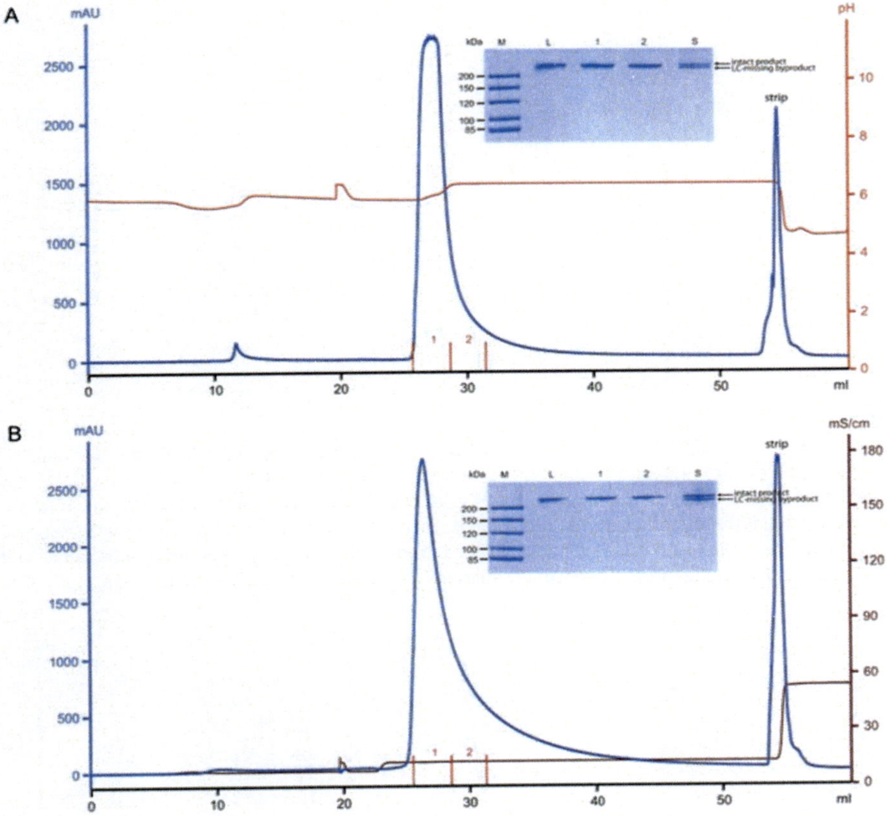

Fig. 9.24 Chromatogram of bsAb 2 obtained on Capto MMC ImpRes using **(a)** pH (20 mM phosphate, pH 6.0–6.5) and **(b)** salt (40 mM phosphate, pH 6.0, 0–47 mM NaCl) stepwise gradient elution. The column loading is pooled fractions collected from purification of bsAb 2 by protein A chromatography. Inset: nonreduced SDS-PAGE analysis of relevant fractions. M, protein marker; L, load; lanes 1 and 2, elution fractions 1 and 2. S, strip fraction. (Reproduced from Wan et al. 2020)

In addition to the half-antibody and LC deletion by-products, in the expression products of asymmetric bispecific antibodies, there is another Fab deletion by-product of, namely, the one-arm by-product. Wan et al. (2021) found that on Capto MMC ImpRes, one-arm by-products cannot be separated from intact antibodies by a single pH or salt concentration increasing gradient. In order to improve the resolution between the intact bsAb and the one-arm by-product, they tried a linear pH-salt double gradient elution (pH 6.0–8.5, 0–250 mM NaCl) (Fig. 9.25). The intact antibody eluted first as the main peak, and the one-arm by-products were enriched at the tail of the main peak. The secondary elution peak appears after the main peak and mainly contains half antibodies, LC missing by-products and aggregates. This means that a salt gradient performed at an increased pH (i.e., higher than the previously used pH 5.5) can provide a higher degree of resolution.

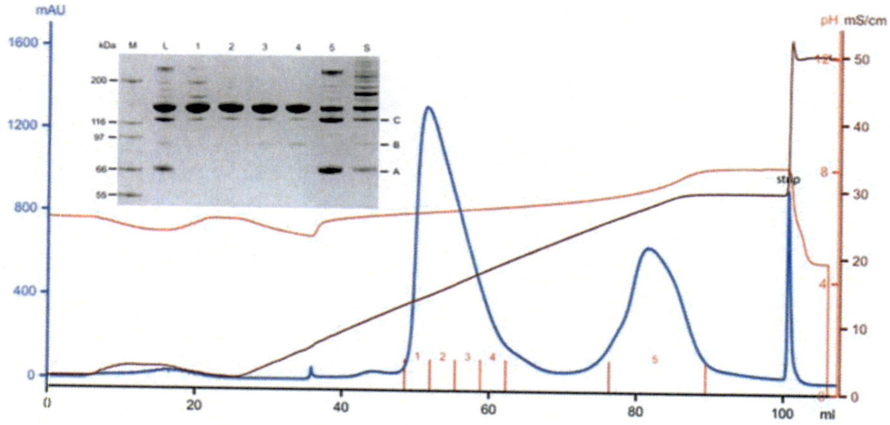

Fig. 9.25 MMC ImpRes chromatogram of bispecific antibody with elution performed under a linear pH-salt dual gradient. For the dual gradient, the pH and salt were increased from 6.0 to 8.5 and 0–250 mM NaCl, simultaneously. Inset, nonreduced SDS-PAGE analysis of relevant fractions. M, protein marker; lanes 1–5, eluted fractions 1–5, respectively; S, stripped fractions. The positions of the bands corresponding to half antibody by-product A, one-arm by-product B, and LC deletion by-product C are respectively marked. (Reproduced from Wan et al. 2021)

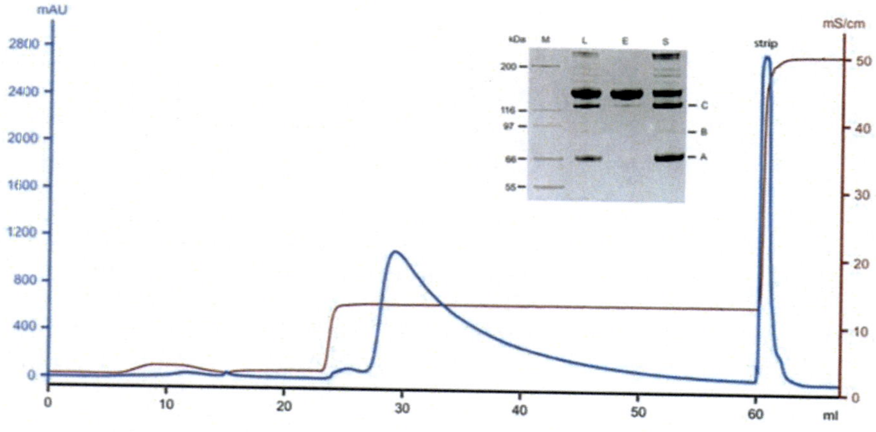

Fig. 9.26 MMC ImpRes chromatogram of bispecific antibody run under stepwise elution. The elution program was developed based on the results of linear pH-salt dual gradient elution and linear salt gradient elution at elevated pH. Equilibration buffer: 20 mM sodium phosphate, pH 6.9. Elution buffer: equilibration buffer + 94 mM NaCl, pH 6.9. Stripping buffer: 50 mM sodium acetate + 1 M arginine, pH 5.0. Inset: nonreducing SDS-PAGE analysis of relevant fractions. M, protein marker; L, load; E, elution fraction; S, stripped fraction. The positions of the bands corresponding to half antibody by-product A, one-arm by-product B, and LC missing by-product C are, respectively, marked. (Reproduced from Wan et al. 2021)

Based on the results of a linear salt gradient elution with a pH of 6.9, they finally developed a stepwise elution to facilitate large-scale production (Fig. 9.26). Elution under optimized conditions can effectively remove most of the low molecular weight by-products and aggregates in the load, and the step yield is 54.8%. The purity of the recovered intact antibody is between 98.1% and 98.8%. The percentages of half antibody by-product A, one-arm by-product B and LC deletion by-product C decreased from 16.7%, 1.8%, and 8.5% to 0.3%, 0.6%, and 0.4%, respectively. The aggregates decreased from 11.1% to 1.2%.

9.5 Summary

The method development process in chromatography mainly involves the selection of chromatographic separation conditions, that is, stationary phase ligand, mobile phase composition, elution method, etc., in order to achieve the key goals of high separation selectivity and high activity recovery. Method development in mixed-mode chromatography is more complicated than in single-mode chromatography. This is because the nature of solute retention is multipotent, that is, the solute can have multiple intermolecular interactions with the chromatography ligand or stationary phase. Canceling one of the interactions by changing the composition of the mobile phase may weaken the retention of the solute, but does not necessarily lead to the complete elution of the solute. Therefore, the separation conditions must be selected according to the characteristics of the specific mixed-mode separation system. In hydroxyapatite chromatography, gradient elution of phosphate solution is usually selected to separate biological macromolecules. Phosphate has a dual role, which can cancel coordination and ion exchange at the same time, achieving the purpose of desorbing solute molecules. In order to improve the selectivity of separation, a gradient of sodium chloride solution in the presence of a low phosphate concentration can also be used to perform the first-dimensional elution, so that the solute bound by the ion exchange site is eluted first. Then a gradient of increasing phosphate concentration is used to elute the strongly retained solutes. It is even possible to use a gradient that simultaneously increases the concentration of sodium chloride and phosphate to optimize retention and elution. In hydrophobic charge induction chromatography, solute molecules bind to neutral ligands mainly through hydrophobic interaction. Then, by changing the pH value of the mobile phase, the ligand and the solute acquire the same charge, causing charge repulsion and elution. In order to improve the selectivity and activity recovery, not only the elution pH value needs to be optimized, but also a washing step can be added before elution to remove adsorbed impurities. Or use mobile phase additives to promote the selective elution of adsorbed proteins. In multimodal ion exchange chromatography, in addition to using a gradient elution that increases the salt concentration, a gradient that changes the pH or the polarity of the solution can also be used for elution. In addition, the use of small molecule displacers is an effective way to improve separation selectivity and recovery. They can disrupt two different types of interactions at the same time, achieving a specific elution similar to the use of substrates or coenzymes in affinity chromatography.

References

Aberin CS, Franklin SG (2002) Plasmid purification using CHT ceramic hydroxyapatite support, Tech Note 2731. Bio-Rad Laboratories, Inc, Hercules, CA. https://www.bio-rad.com/webroot/web/pdf/psd/literature/BULLETIN_2731.pdf

Aberin CS, Franklin SG (2003) Plasmid DNA adsorption behaviour on CHT ceramic hydroxyapatite type II support, Tech Note 2946. Bio-Rad Laboratories, Inc, Hercules, CA. https://www.bio-rad.com/webroot/web/pdf/psd/literature/BULLETIN_2946.pdf

Ambatipudi KS, Lu B, Hagen FK et al (2009) Quantitative analysis of age specific variation in the abundance of human female parotid salivary proteins. J Proteome Res 8:5093–5102

Arakawa T, Kita Y (2014) Multi-faceted arginine: mechanism of the effects of arginine on protein. Curr Protein Pept Sci 15:608–620

Arakawa T, Ejima D, Tsumoto K et al (2007) Suppression of protein interactions by arginine: a proposed mechanism of the arginine effects. Biophys Chem 127:1–8

Arakawa T, Kita Y, Sato H et al (2009) MEP chromatography of antibody and Fc-fusion protein using aqueous arginine solution. Protein Expr Purif 63:158–163

Arakawa T, Futatsumori-Sugai M, Tsumoto K et al (2010) MEP HyperCel chromatography II: binding, washing and elution. Protein Expr Purif 71:168–173

Arun GG, Agamudi SK, Mookambeswaran AV et al (2018) Efficient purification of apolipoprotein A1 (ApoA1) from plasma by HEA HyperCel™: an alternative approach. J Chromatogr B 1073: 104–109

Bolten SN, Knoll A-S, Li Z et al (2020) Purification of the human fibroblast growth factor 2 using novel animal-component free materials. J Chromatogr A 1626:461367

Brochier VB, Schapman A, Santambien P et al (2008) Fast purification process optimization using mixed-mode chromatography sorbents in pre-packed mini-columns. J Chromatogr A 1177:226–233

Burton SC, Harding DRK (1998) Hydrophobic charge induction chromatography: salt independent protein adsorption and facile elution with aqueous buffers. J Chromatogr A 814:71–81

Cheng F, Li M, He W et al (2018) Activation of resin with controllable ligand density via catalytic oxa-Michael addition and application in antibody purification. J Chromatogr A 1570:1–9

Coulon D, Cabanne C, Fitton V et al (2004) Penicillin acylase purification with the aid of hydrophobic charge induction chromatography. J Chromatogr B 808:111–115

Freitag R, Hilbrig F (2012) Isolation and purification of recombinant proteins, antibodies and plasmid DNA with hydroxyapatite chromatography. Biotechnol J 7:90–102

Gagnon P (2008) Improved antibody aggregate removal by hydroxyapatite chromatography in the presence of polyethylene glycol. J Immunol Methods 336:222–228

Gagnon P (2009) Monoclonal antibody purification with hydroxyapatite. New Biotechnol 25:287–293

Gagnon P, Ng P, Zhen J et al (2006) A ceramic hydroxyapatite-based purification platform. Simultaneous removal of leached Protein A, aggregates, DNA, and endotoxins from MABs. BioProcess Int 4:50–60

Hirano A, Maruyama T, Shiraki K et al (2014) Mechanism of protein desorption from 4-mercaptoethylpyridine resins by arginine solutions. J Chromatogr A 1373:141–148

Hirano A, Shiraki K, Kameda T (2019) Effects of arginine on multimodal chromatography: experiments and simulations. Curr Protein Pept Sci 20:40–48

Ishibashi M, Tsumoto K, Tokunaga M et al (2005) Is arginine a protein-denaturant? Protein Expr Purif 42:1–6

Josic D, Loster K, Kuhl R et al (1991) Purification of monoclonal antibodies by hydroxyapatite HPLC and size exclusion HPLC. Biol Chem Hoppe Seylars 372:149–156

Kaleas KA, Schmelzer CH, Pizarro SA (2010) Industrial case study: evaluation of a mixed-mode resin for selective capture of a human growth factor recombinantly expressed in E. coli. J Chromatogr A 1217:235–242

Kuiper M, Sanches RM, Walford JA et al (2002) Purification of a functional gene therapy vector derived from Moloney murine leukaemia virus using membrane filtration and ceramic hydroxyapatite chromatography. Biotechnol Bioeng 80:445–453

Kurosawa Y, Kurane I, Yamamoto A (2009) Observation of Japanese encephalitis virus particles on ceramic hydroxyapatite by scanning electron microscopy. Med Biol 153:607–610

Kurosawa Y, Saito M, Kobayashi S et al (2012a) Purification of dengue virus particles by one-step ceramic hydroxyapatite chromatography. World J Vaccines 2:155–160

Kurosawa Y, Yamamoto A, Kurane I et al (2012b) Development of a purification method for Japanese encephalitis virus particles using ceramic hydroxyapatite chromatography. Med Biol 156:410–416

Kurosawa Y, Saito M, Yoshikawa D et al (2014) Mammalian virus purification using ceramic hydroxyapatite. Tech Note Bulletin 6549. Bio-Lad Laboratories, Inc, Hercules, CA. https://www.bio-rad.com/webroot/web/pdf/lsr/literature/Bulletin_6549.pdf

Kurosawa Y, Sato S, Taoka M (2019) Sequential two-step chromatographic purification of infectious poliovirus using ceramic fluoroapatite and ceramic hydroxyapatite columns. PLoS One 14: e0222199

Li M, Zhang Q, Lin D et al (2019) Development and application of hydrophobic charge-induction chromatography for bioseparation. J Chromatogr B 1134–1135:121850

Liu T, Lin DQ, Lu HL et al (2014) Preparation and evaluation of dextran-grafted agarose resin for hydrophobic charge-induction chromatography. J Chromatogr A 1369:116–124

Liu T, Lin D-Q, Zhang Q-L et al (2015) Characterization of immunoglobulin adsorption on dextran-grafted hydrophobic charge-induction resins: cross-effects of ligand density and pH/salt concentration. J Chromatogr A 1396:45–53

Liu T, Lin D-Q, Wu Q-C et al (2016) A novel polymer-grafted hydrophobic charge-induction chromatographic resin for enhancing protein adsorption capacity. Chem Eng J 304:251–258

Liu T, Angelo JM, Lin D-Q et al (2017) Characterization of dextran-grafted hydrophobic charge-induction resins: structural properties, protein adsorption and transport. J Chromatogr A 1517: 44–53

Lu H-L, Lin D-Q, Gao D et al (2013) Evaluation of immunoglobulin adsorption on the hydrophobic charge-induction resins with different ligand densities and pore sizes. J Chromatogr A 1278:61–68

Lu H-L, Lin D-Q, Zhang Q-L et al (2017) Evaluation on adsorption selectivity of immunoglobulin G with 2-mercapto-1-methyl-imidazole-based hydrophobic charge-induction resins. Biochem Eng J 119:34–41

Luo YD, Zhangn QL, Yao SJ et al (2018) Evaluation of adsorption selectivity of immunoglobulins M, A and G and purification of immunoglobulin M with mixed-mode resins. J Chromatogr A 1533:77–86

Minkner R, Baba R, Kurosaw Y et al (2018) Purification of human papillomavirus-like particles expressed in silkworm using a Bombyx mori nucleopolyhedrovirus bacmid expression system. J Chromatogr B 1096:39–47

Moro A, Yoshitake T, Ogawa T et al (2008) Single-step purification of pepsin-derived monoclonal antibody fragments from crude murine ascitic fluids by ceramic hydroxyapatite high-performance liquid chromatography. J Biochem 144:733–739

Ng P, Cohen A, Gagnon P (2006) Monoclonal antibody purification with CHT. Gen Eng News 26

Peters J, Oehme F (2010) Mixed-mode chromatography in downstream process development. BioPharm International 2010 Supplement

Pezzini J, Cabanne C, Dupuyce J-W et al (2014) A study on the nature of interactions of mixed-mode ligands HEA and PPA HyperCel using phenylglyoxal modified lysozyme. J Chromatogr B 960:209–213

Pezzini J, Cabanne C, Gantierd R et al (2015) A comprehensive evaluation of mixed mode interactions of HEA and PPA HyperCel™ chromatographic media. J Chromatogr B 976–977: 68–77

Phottraithip W, Lin D-Q, Shi F et al (2013) New hydrophobic charge-induction resin with 2-mercaptoimidazole as the ligand and its separation characteristics for porcine IgG. Biotechnol Bioprocess Eng 18:1169–1175

Pinto NDS, Frey DD (2015) Displacement chromatography of proteins using a retained pH front in a hydrophobic charge induction chromatography column. J Chromatogr A 1387:53–59

Ravi A, Foster ER, Perez LM et al (2020) Capture chromatography with mixed-mode resins: a case study with recombinant human thioredoxin from Escherichia coli. J Chromatogr A 1625: 461327

Ren J, Jia L, Xu L et al (2009) Removal of autoantibodies by 4-mercaptoethylpyridine-based adsorbent. J Chromatogr B 877:1200–1204

Ren J, Yao P, Cao Y et al (2014) Application of cyclodextrin-based eluents in hydrophobic charge-induction chromatography: elution of antibody at neutral pH. J Chromatogr A 1352:62–68

Saito M, Kurosawa Y, Okuyama T (2012) Purification of anti-Japanese encephalitis virus monoclonal antibody by ceramic hydroxyapatite chromatography without Proteins A and G. Hybridoma (Larchmt) 31:68–71

Schröder E, Jönsson T, Poole L (2003) Hydroxyapatite chromatography: altering the phosphate – dependent elution profile of protein as function of pH. Anal Biochem 313:176–178

Schwartz W, Judd D, Wysocki M et al (2001) Comparison of hydrophobic charge induction chromatography with affinity chromatography on protein A for harvest and purification of antibodies. J Chromatogr A 908:251–263

Shi QH, Shen FF, Sun S (2010) Studies of lysozyme binding to histamine as a ligand for hydrophobic charge induction chromatography. Biotechnol Prog 26:134–141

Smith GP, Gingrich TR (2005) Hydroxyapatite chromatography of phage-display virions. BioTechniques 39:879–883

Tang J, Zhang X, Chen T et al (2020) Removal of half antibody, hole-hole homodimer and aggregates during bispecific antibody purification using MMC ImpRes mixed-mode chromatography. Protein Expr Purif 167:105529

Tong HF, Lin DQ, Gao D et al (2013) Caprylate as the albumin-selective modifier to improve IgG purification with hydrophobic charge-induction chromatography. J Chromatogr A 1285:88–96

Tong H-F, Lin D-Q, Chu W-U et al (2016) Multimodal charge-induction chromatography for antibody purification. J Chromatogr A 1429:258–264

Wan Y, Zhang T, Wang Y et al (2020) Removing light chain-missing byproducts and aggregates by Capto MMC ImpRes mixed-mode chromatography during the purification of two WuXiBody-based bispecific antibodies. Protein Expr Purif 175:105712

Wan Y, Wang Y, Zhang T et al (2021) Application of pH-salt dual gradient elution in purifying a WuXiBody-based bispecific antibody by MMC ImpRes mixed-mode chromatography. Protein Expr Purif 181:105822

Wilkinson RJ, Elliott P, Hohmann A et al (2004) Development and characterization of a competitive polyclonal antibody enzyme-immunoassay for salmon insulin-like growth factor-II. Comp Biochem Physiol B Biochem Mol Biol 139:193–201

Wu M, Zhang F, Liang Y et al (2015) Isolation and purification of immunoglobulin G from bovine colostrums by hydrophobic charge-induction chromatography. J Dairy Sci 98:2973–2981

Xia H-F, Lin D-Q, Wang L-P et al (2008) Preparation and evaluation of cellulose adsorbents for hydrophobic charge induction chromatography. Ind Eng Chem Res 47:9566–9572

Yan J, Zhang QL, Tong HF et al (2015) Hydrophobic charge-induction resin with 5-aminobenzimidazol as the functional ligand: preparation, protein adsorption and immunoglobulin G purification. J Sep Sci 38:2387–2393

Zhang L, Zhao G, Sun Y (2010a) Effects of ligand density on hydrophobie charge induction chromatography: molecular dynamics simulation. J Phys Chem B 114:2203–2211
Zhang L, Bai S, Sun Y (2010b) Molecular dynamics simulation of the effect of ligand homogeneity on protein behavior in hydrophobic charge induction chromatography. J Mol Graph Model 28: 863–869
Zhao G, Peng G, Li F et al (2008) 5-Aminoindole, a new ligand for hydrophobic charge induction chromatography. J Chromatogr A 1211:90–98
Zou X, Zhang Q, Lu H et al (2019) Development of a hybrid biomimetic ligand with high selectivity and mild elution for antibody purification. Chem Eng J 368:678–686

Chapter 10
Biopharmaceutical Applications

Abstract Except for oligonucleotide drugs that can be produced by chemical synthesis, most biopharmaceuticals are manufactured by recombinant DNA technology. The target gene is inserted into a plasmid, transferred and expressed in bacteria or mammalian cells, and fermented in a bioreactor for large-scale production. The harvested and expressed products are then purified by multistep chromatography to meet the purity requirements of the drug regulatory agency. The separation and purification of biopharmaceuticals plays a key role in the manufacture of biopharmaceuticals, accounting for more than half of the entire production cost. In recent years, mixed-mode chromatography has been increasingly used for the separation and purification of biopharmaceuticals. The reported application examples cover almost all types of biopharmaceutical products, including nucleic acids, antibodies, and enzymes. In this chapter, we will review these developments in detail.

10.1 Introduction

Biopharmaceuticals are drugs with biological activity that are manufactured through biotechnological processes. Compared with traditional drugs, biopharmaceutical products have higher specificity and activity. The beginning of modern biotech pharmaceuticals is marked by the approval of Humulin, a recombinant human insulin used to treat diabetes, in 1982 (Walsh 2003; Rader 2008). Since then, the number and types of biological drugs used to treat diseases such as cancer, inflammation-related diseases, hemophilia, and diabetes have continued to increase. Biopharmaceuticals include a wide range of products, such as antibodies (currently having the largest number of approved products), recombinant enzymes, cell-based therapies, and nucleic acid-based products. In particular, nucleic acid-based products include antisense oligonucleotides, aptamers, modified RNA molecules, small interfering RNA, gene therapy vectors, and DNA or RNA vaccines (Walsh 2018).

In addition to oligonucleotide drugs that can be produced by chemical synthesis, most biopharmaceuticals are manufactured by recombinant DNA technology, using bacteria, mammalian cells, yeast, insect cells, genetically modified plants, and

Q.-H. Wan, *Mixed-Mode Chromatography*,
https://doi.org/10.1007/978-981-16-5485-5_10

genetically modified animals to achieve the expression of biological drugs. After the expression system is established, the bioreactor system is used for large-scale production. Batch, fed-batch and continuous or perfusion culture are used for batch production of recombinant proteins. Continuous bioprocessing has also emerged as a new technology and has been used for upstream and downstream process development of bioprocessing and for the production of therapeutic proteins (Tripathi and Shrivastava 2019). Taking the production of monoclonal antibody drugs as an example, the upstream process includes inoculating the working cell stock solution into a small shake flask culture and then carrying out cultivation in laboratory-scale, pilot-scale, and finally production-scale bioreactors. Downstream manufacturing processes include cell harvesting (using centrifugation and depth filtration), microfiltration, protein A chromatography, virus inactivation, diafiltration, anion exchange chromatography (AEX), virus filtration, hydrophobic interaction (HI) or cation exchange chromatography (CEX), diafiltration, and sterile filtration and formulation.

Affinity chromatography (AC) is the most selective technique for purification of tagged proteins, bispecific antibodies, DNA-based biological products, cell products, viral vectors, and viruses (Zhao et al. 2018; Łacki and Riske 2020). Among them, protein A chromatography is the most widely used method for mAb purification (Gronemeyer et al. 2014; Sifniotis et al. 2019). However, the main problem of this chromatographic method is that the leaching of protein A ligand and the formation of aggregates at the low pH during elution lead to contamination of the products. As an alternative to protein A chromatography, mixed mode or multimodal chromatography can be used in the various steps of purification, such as capture and polishing, providing salt tolerance and high-resolution separations with high binding capacity, and has gradually become an indispensable tool in the separation and purification of biopharmaceuticals. In this chapter, we will focus on the application of mixed-mode chromatography in the production of nucleic acid, plasmid, recombinant protein, and antibody drugs.

10.2 Separation and Purification of Oligonucleotide Drugs

Since the discovery of the targeted gene silencing by synthetic oligonucleotides (Zamecnik and Stephenson 1978; Stephenson and Zamecnik 1978), new therapies based on synthetic oligonucleotides have received widespread attention and have been tried to treat various diseases, such as cancer, viral diseases, skin diseases, and autosomal recessive diseases. Nowadays, research in the field of therapeutic oligonucleotides, including antisense oligonucleotides, aptamers, small interfering RNAs, oligonucleotides targeting microRNAs, and immunostimulatory oligonucleotides, is very active. Oligonucleotide therapeutics face several challenges, such as in vivo stability, targeted delivery, and immunogenicity. These problems can be alleviated to a large extent by chemically modifying the structure of oligonucleotides (Levin 2019).

Oligonucleotides used as therapeutic agents can usually be synthesized using phosphoramidite chemistry. Currently, the process is implemented as solid phase synthesis using phosphoramidite method and phosphoramidite building blocks derived from protected 2′-deoxynucleosides (dA, dC, dG, and dT), ribonucleosides (A, C, G, and U), or chemically modified nucleosides, e.g., LNA or BNA. In order to obtain the desired oligonucleotide, the building blocks are sequentially connected to the growing oligonucleotide chain in the order required by the product sequence. After the chain is assembled, it is released from the solid phase into the solution and deprotected, and the product is collected. The length of the synthetic oligonucleotide (up to about 200 nucleotide residues) is limited by side reactions, because the number of errors accumulates with the length of the synthetic oligonucleotide.

Generally, synthetic oligonucleotides are single-stranded DNA or RNA molecules about 15–25 bases in length. The most common impurities in this type of products are oligonucleotides with deletions or extensions, which are shorter due to coupling of one or more missing nucleotides or longer due to coupling of double and triple nucleotides. Other impurities include incompletely deprotected products, oligonucleotides with missing purine bases, and other degradation products. Anion exchange chromatography, reverse phase chromatography, and ion pair chromatography have all been used for the separation and purification of oligonucleotides. However, there are still challenges in process impurities with similar molecular structures.

In the past 30 years, mixed-mode chromatography has developed into the preferred method for the separation and purification of oligonucleotides (McLaughlin 1989). The multimode ion exchange chromatography column with reversed-phase and ion-exchange characteristics can combine the advantages of two separation methods in a single analysis, which helps to retain and separate hydrophilic ionic compounds. Biba et al. (2013) evaluated the application of mixed-mode chromatographic stationary phases composed of reversed-phase and ion-exchange ligands in the separation of oligonucleotide mixtures. The sample set consists of a family of ribonucleic acid (RNA) oligomers, including a 21-mer parental based on siApoB-15 sense strand, and the consecutive base deletion oligomers prepared from the 5′ end (N-1, N-2 to N-15). These base deletion samples represent the by-products of oligonucleotide synthesis due to the inability of a single nucleotide to attach to the sequence and are one of the main impurity categories. Since automated oligonucleotide synthesis usually proceeds from the 3′ direction to the 5′ direction, the N-X deletion from the 5′ end is most likely to be present in the final product. The molecular weights and total number of charged residues of each member of the sample group are different, and they can be easily resolved by a variety of chromatographic techniques.

In addition to base deletion oligonucleotides, it also includes a sample set consisting of a mixture of "base flipped" isomers. These 21-mer RNA sequences have the same overall composition, charge, and molecular weight, but the order of two adjacent bases within the sequence has been reversed. The chromatographic separation of these "base flip" isomers is expected to be somewhat challenging.

Three commercial reversed-phase/ion exchange mixed mode chromatographic columns are used. The main difference between the three columns is that the ion exchange capacity is increased sequentially. The first column, Scherzo SW-C18, is prepared by mixing two types of bifunctional bonded phase silica particles, one with C18 ligand and strong anion exchange functional group with low exchange capacity, and the other with C18 ligand and strong cation exchange functional group with low exchange capacity. The second column, Scherzo SM-C18, was prepared by mixing particles containing C18 ligand and weak anion exchange group with medium exchange capacity with particles containing C18 ligand and weak cation exchange group with medium exchange capacity. The third column, Scherzo SS-C18, is prepared by mixing particles containing C18 groups and a strong anion exchanger with high exchange capacity with particles containing C18 groups and a strong cation exchanger with high exchange capacity. All columns are 150 mm in length × 4.6 mm in inner diameter and contain silica gel microspheres with a particle size of 3 μm and an average pore diameter of 130 Å.

The linear gradient of increasing NaCl salt was used to elute the oligonucleotide test mixture loaded on the chromatographic column. The resulting separations on Scherzo SW-C18 column are shown in Fig. 10.1. The base deletion standards show a trend that the retention of oligonucleotides increases with the increase of bases, but there are many exceptions. For example, N-3 and N-12 have the strongest retention among the tested compounds, rather than the expected parent 21-mer (Fig. 10.1a). The retention of base flip isomers also showed differences, two of which co-eluted with the parent 21-mer, the retention of two was greater than that of the parent, and one was less than the parent (Fig. 10.1b). These results indicate that the retention of oligonucleotides is not only related to its charge and base number but also to its configuration in solution. The inversion of the base sequence may change the secondary or tertiary structure of the oligonucleotide, thereby changing its surface hydrophobicity. Similar separation can be achieved on Scherzo SM-C18, but the retention time is significantly increased. However, the retention of oligonucleotides on Scherzo SS-C18 was too strong, and even if the salt concentration was doubled, the adsorbed solutes could not be eluted.

Zimmermann et al. (2014) investigated the retention behavior of oligonucleotides in multimodal weak anion exchange chromatography. The stationary phase used is a silica gel matrix bonded phase. The ligand dodecyl-3-aminoquinoline is bonded with silica gel through a sulfide bond. The ligand contains a hydrophobic alkyl chain with embedded polar functional groups (thioether and amide groups) and a weak anion exchange site for quinolinium.

They used a series of 20–23 nucleoside-length oligonucleotides with similar sequences as probes (Table 10.1). The retention of these probes was determined in different elution modes by changing the buffer pH, buffer concentration, and organic modifier, to reveal the relationship between oligonucleotide structure and sequence selectivity. Oligonucleotides 4–18 as molecular probes contain substantially the same sequence as oligonucleotide 1 but show minor structural modifications, such as (1) at the 5′ end, 3′ end, or in the middle of the sequence with a single nucleotide substitution (single nucleotide polymorphism, SNP) (4–12) or (2) a deletion at the 3′

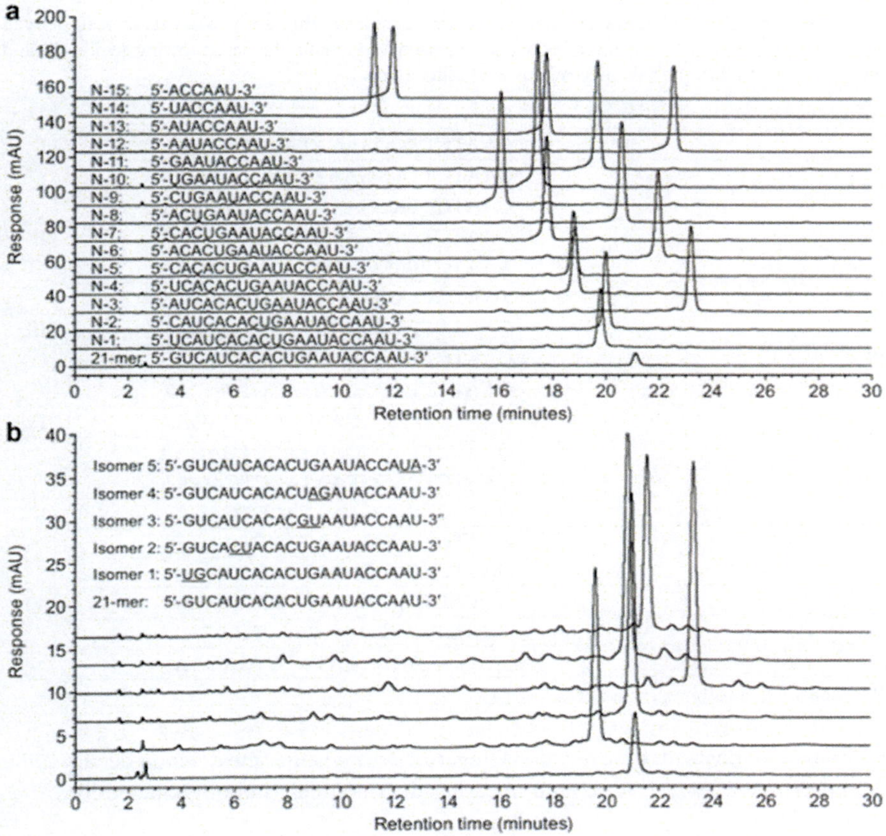

Fig. 10.1 The separation of RNA oligonucleotide samples in mixed mode chromatography using Scherzo SW-C18 column and NaCl gradient elution. (**a**) Separation of N-X deletion sequences and (**b**) separation of "base inversion" isoform standards. Conditions: The mobile phase includes 100 mM Tris pH 7.4 buffer (mobile phase A) and 100 mM Tris pH 7.4 buffer + 1 M NaCl/ acetonitrile (90/10 (v/v%)) (mobile phase B). Use a linear gradient of 35–65% B in 30 min. The flow rate is 1 mL/min, the column temperature is 50 °C, the injection volume is 10 μL, and the UV detection is set to 260 nm. (Reproduced from Biba et al. 2013)

end (13–15) or 5′ end (16–18) of two or three nucleotides (NX; X is 1, 2, or 3) (Table 10.1).

Figure 10.2 shows a representative chromatogram of the oligonucleotide mixture II–VI eluted under the same conditions. The resolution of SNP is related to the position of SNP, decreasing in the order of 3′ end > somewhere in the middle > at the 5′ end. The selectivity of base exchange decreases in the order of C/G > A/G ~ G/T > A/C ~ C/T > A/T.

A representative chromatogram of an oligonucleotide mixture with missing sequences obtained under isocratic elution conditions is shown in Fig. 10.3.

Table 10.1 Synthetic oligonucleotides used as test samples in this study. They have been derived from oligonucleotide 1 by minor sequence variations which might occur during synthesis and represent potential impurities of synthetic oligonucleotide 1

Oligonucleotide abbreviation	Sequence 5′–3′	Mix
1	GAATCTTACGAAATACCTGAGAA	I
2	GAATCTTA**G**GAA**T**TACCTGAGAA	
3	GAATCTTACGAAA_ACCTGAGAA	
4	**C**AATCTTACGAAATACCTGAGAA	II
5	**A**AATCTTACGAAATACCTGAGAA	
6	**T**AATCTTACGAAATACCTGAGAA	
7	GAATCTTACGAAATACCTGAGA**C**	III
8	GAATCTTACGAAATACCTGAGA**G**	
9	GAATCTTACGAAATACCTGAGA**T**	
10	GAATCTTACG**C**AATACCTGAGAA	IV
11	GAATCTTACG**G**AATACCTGAGAA	
12	GAATCTTACG**T**AATACCTGAGAA	
13	GAATCTTACGAAATACCTGAGA_	V
14	GAATCTTACGAAATACCTGAG__	
15	GAATCTTACGAAATACCTGA___	
16	_AATCTTACGAAATACCTGAGAA	VI
17	__ATCTTACGAAATACCTGAGAA	
18	___TCTTACGAAATACCTGAGAA	

Reproduced from Zimmermann et al. (2014)

Oligonucleotides with base deletions, regardless of whether the deletion occurs at the 3′ end or 5′ end, will lead to elution earlier than non-deletion oligonucleotides.

10.3 Separation and Purification of Plasmid DNA

Infectious diseases are currently the disease with the highest fatality rate in the world. Approximately 17 million people die from this type of disease each year, and its fatality rate is higher than that of cardiovascular disease, cancer, or accidental injury. The use of vaccines contributes more to the prevention of infectious diseases than any other medical prevention and treatment methods. Vaccine refers to any biological product that can be inoculated by injection or mucosal route to induce the body to produce specific antibodies or cellular immunity against a specific pathogen, so that the body can obtain the ability to recognize and eliminate the pathogen. Proteins, polysaccharides, nucleic acid, live vector, or infectious agent have all been used as vaccines. Traditional vaccines include inactivated vaccines, live attenuated vaccines, and subunit vaccines made from certain components of natural microorganisms. New vaccines mainly refer to vaccines developed using genetic engineering technology, including recombinant subunit vaccines, gene-deletion live vaccines, genetic engineering live vector vaccines, nucleic acid vaccines, etc. The

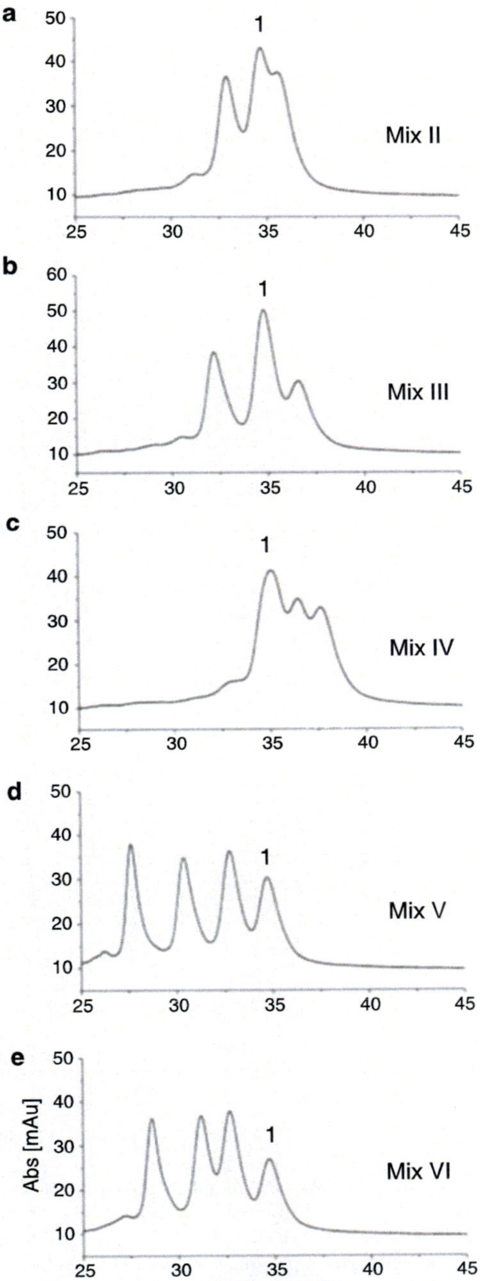

Fig. 10.2 Oligonucleotide mixture II (**a**), III (**b**), IV (**c**), V (**d**), and VI (**e**) on RP/WAX mixed-mode stationary phase using mixed pH/buffer gradient elution separation. HPLC conditions, RP/WAX column; mobile phase, acetonitrile/0.1 M triethylammonium phosphate buffer pH 7.0 (20:80, v/v) (mobile phase A); acetonitrile/0.2 M triethylammonium phosphate buffer pH 8.0 (20:80, v/v) (mobile phase B); gradient, 50–100% B in 25 min, then keep at 100% B until 52 min, then at 100% A re-equilibrate for 8 min. Column temperature, 40 °C; UV detection, 260 nm. (Reproduced from Zimmermann et al. 2014)

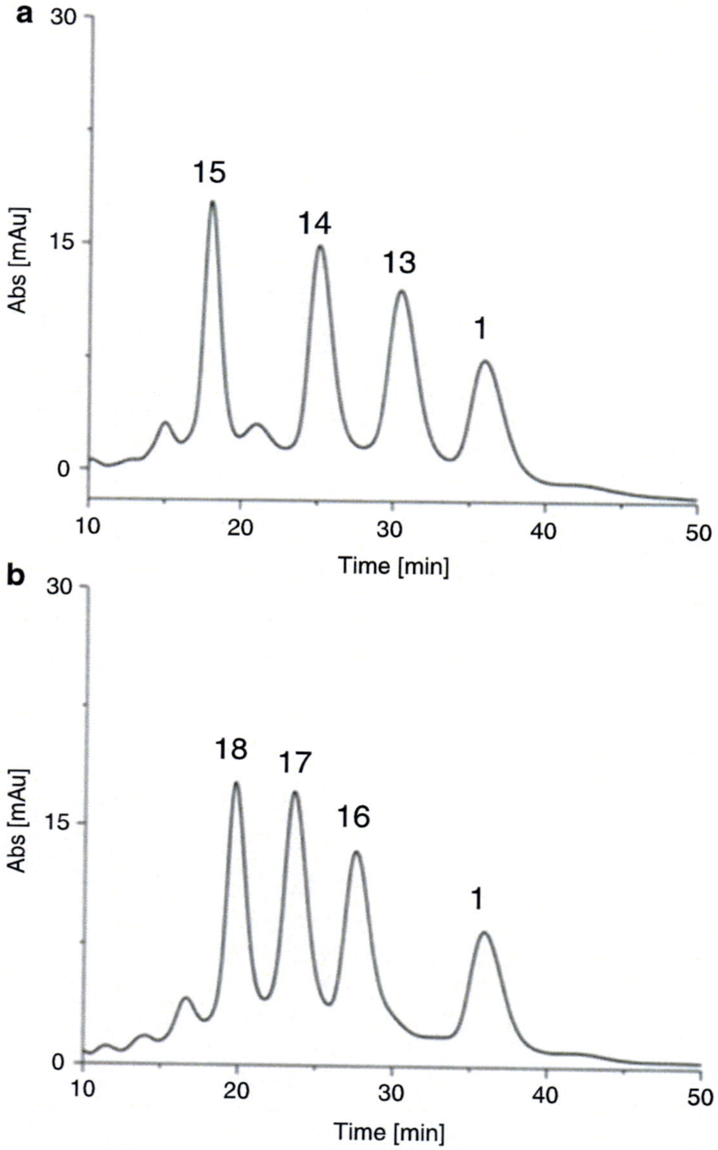

Fig. 10.3 Isocratic separation of oligonucleotide mixtures with similar sequences but different chain lengths (one nt missing in turn) on the RP/WAX mixed mode stationary phase. (**a**) MIx V (1, 13, 14, 15) and (**b**) MIx VI (1, 16, 17, 18). HPLC conditions: mobile phase, acetonitrile/0.2 M triethylammonium phosphate buffer pH 7.5 (30:70, v/v); flow rate, 1.0 mL/min; column temperature, 40 °C; UV detection, 260 nm. (Reproduced from Zimmermann et al. 2014)

development of modern vaccines has evolved from classic virus vaccines and bacterial vaccines to parasite vaccines, T cell vaccines, dendritic cell vaccines, tumor vaccines, contraceptive vaccines, and preventive vaccines to therapeutic vaccines.

Nucleic acid vaccines, also called genetic vaccines or DNA vaccines, refer to the direct introduction of foreign genes encoding a certain antigen protein into animal cells, the expression and synthesis of the antigen protein in the host cell, and a series of immunity similar to vaccination. Responses serve the purpose of preventing and treating diseases (Wolff et al. 1990). Compared with traditional inactivated vaccines, attenuated vaccines, and recombinant subunit vaccines, nucleic acid vaccines have the following characteristics: (1) Nucleic acid vaccines can produce foreign proteins in host cells, and the process of antigen recognition and presentation is very similar to natural infections. Similarly, it induces an immune response that is almost equivalent to the immune response generated after infection with these pathogens or attenuated vaccines and avoids the extraction and purification steps required for the in vitro production of recombinant subunit vaccines and the possible loss or change of epitopes caused by this. (2) After nucleic acid vaccination, protein antigens can directly bind to the histocompatibility complex (MHC) class I and class II molecules in the host cells, causing extensive cellular and humoral immunity, and there is no atavism phenomenon. (3) Since nucleic acid vaccines can rapidly multiply in engineered bacteria as recombinant plasmids, they can be produced on a large scale by using well-established industrial fermentation technology. Therefore, nucleic acid vaccines may cause another revolution in the history of vaccine development.

10.3.1 Topological Structure of Plasmid DNA

Supercoiled DNA The DNA molecules present in natural organisms are very huge, and the size and complexity of DNA molecules of different species are also very different. For example, viruses may only contain a few thousand base pairs, bacteria contain millions of base pairs, and mammalian DNA has billions of base pairs. To accommodate DNA molecules of this length in a small cell nucleus, a tight helical structure must be formed. Therefore, DNA can be further folded into a super helical structure based on the double-helical structure (Fig. 10.4), with the participation of proteins.

The double-helical structure of DNA is in the lowest energy state. If this normal double-helical molecule is rotated a few more or less times, the atoms inside the helix will deviate from the normal position and generate additional tension. If the ends of the double helix are open, the tension generated can be released by the rotation of the chain, and the double helix molecules return to their normal state. But for the plasmid DNA in bacteria, they are a closed circular double helix structure, the tension generated at this time cannot be released. The atoms inside the DNA

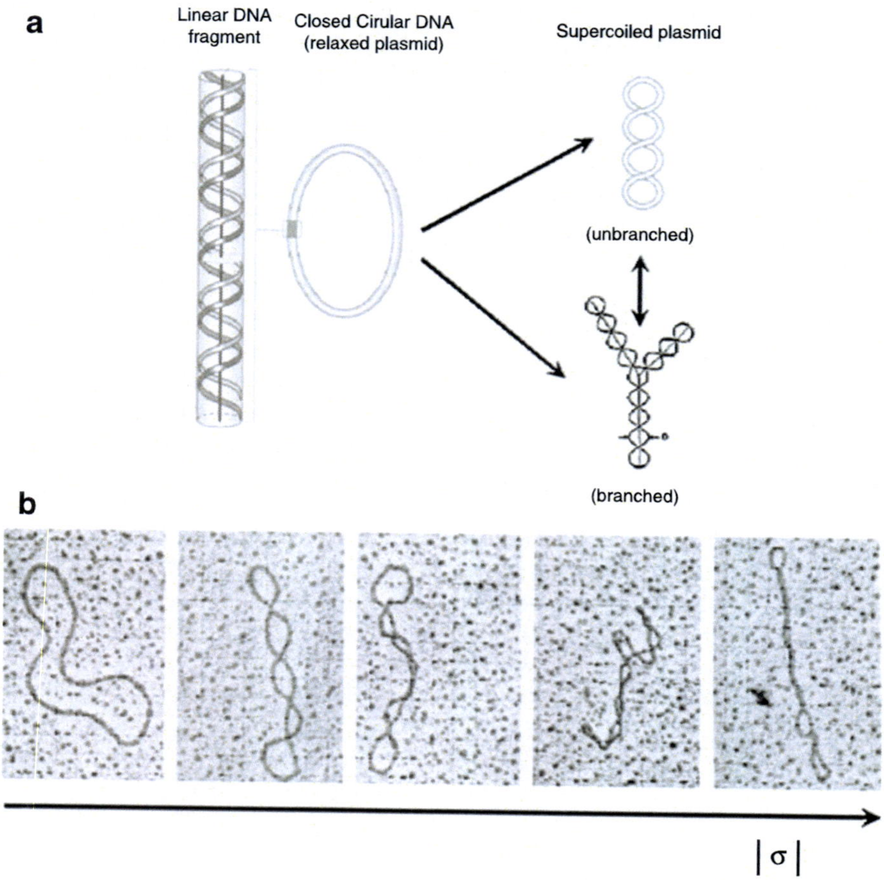

Fig. 10.4 Schematic diagram of DNA structure (**a**) linear fragment, closed loop, and supercoiled topology; (**b**) plasmid DNA supercoiled. (Reproduced from Ferreira 2005)

molecule can be rearranged, which causes the double helix DNA to be further twisted and supercoiled (Fig. 10.4a). Supercoil is the main form of the tertiary structure of DNA. The DNA of most prokaryotes is a covalent closed loop molecule, and this double-helical loop molecule is further spiraled into a supercoiled structure again (Fig. 10.4b).

Compared with linear and open-circlar DNA, supercoiled DNA has a tighter structure, lower viscosity, higher density, faster sedimentation, and faster migration in agarose gel electrophoresis. Supercoiled DNA has a high denaturation temperature. After denaturation, the two strands form a group of random coils that are entangled with each other, which is difficult to separate.

Supercoiled DNA is the most stable form of DNA. Supercoiled plasmids exist in a dense form. When entering cells (whether competent bacteria or eukaryotic cells), the steric resistance is smaller, compared to non-supercoiled plasmids. It is easier to enter the cell; the supercoiled plasmid exists in a stable and dense form and is not easily degraded by nucleases in the body. It can exist in the body for a long time and continuously express the encoded antigen protein.

Nicked (notched), relaxed circular plasmid Plasmid DNA can also exist in the form of open-circular molecules. Nucleic acid replication machines cannot easily access DNA in supercoiled form. In the process of replication, the cell topoisomerase will scribe a strand of the DNA helix and relax the supercoil tension to form a large open circlar DNA, so that the polymerase can approach the DNA.

Linear plasmid Linearized DNA occurs when the DNA helix is cut in two strands at the same position. In gel electrophoresis, linear DNA usually migrates between open and supercoiled forms of DNA. By comparing the uncut plasmid DNA with a plasmid sample linearized with restriction enzymes, the linear DNA form on the agarose gel can be identified.

Circular single-stranded plasmids During the preparation of alkaline cleavage plasmids, the plasmids are denatured because hydrogen bonds are destroyed by alkaline conditions. However, the covalently closed circular chain remains intact and subject to topological constraints. When the pH returns to neutral, hydrogen bonds will re-form, and supercoiled DNA will re-form. However, if the alkaline lysis step is too harsh (e.g., the incubation time is too long), the DNA may be permanently denatured, leaving useless single-stranded closed loops, which precede all other forms of plasmid migration during gel electrophoresis.

10.3.2 Industrial Scale Preparation of Plasmid DNA

Plasmids are usually produced in recombinant *E. coli* hosts through fermentation. The plasmid containing the target gene is introduced into *Escherichia coli*, fermented in a bioreactor, and then alkaline lysis is used to release the plasmid from *Escherichia coli*. In addition to the recombinant plasmid, the bacterial lysate also contains a large amount of cell wall debris, proteins, and nucleic acids and other impurities from the host cell. The recombinant plasmid only accounts for about 3% (w/w) of the *E. coli* extract. In order to obtain pharmaceutical-grade plasmid DNA, an industrial device as shown in Fig. 10.5 must be used for plasmid fermentation, lysis, clarification, concentration, and purification.

The plasmid production process can be divided into upstream and downstream processing processes. The upstream process refers to the fermentation process of *Escherichia coli*, and the downstream process is the treatment after harvesting the bacteria, including operations such as lysis, purification, and concentration. Among them, the purification of plasmids is technically challenging because all products of interest are unstable and therefore require mild processing conditions. In particular, most key pollutants have the characteristics similar to those of plasmids: negative charge (RNA, genomic DNA, endotoxin); the same size (genomic DNA, endotoxin);

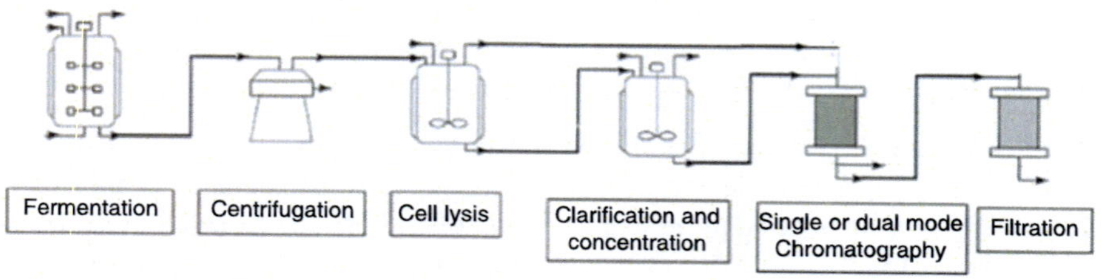

Fig. 10.5 Strategy for purification of supercoiled plasmid DNA. Pure supercoiled plasmid DNA can be obtained by directly transferring from cell lysis to chromatography or by performing clarification and concentration before chromatography. (Reproduced from Ferreira et al. 2000)

and hydrophobicity (endotoxin). However, the number and complexity of the processing steps in the extraction, isolation, purification, and preparation of plasmid DNA can cause a structural stress that may cause damage to the supercoiled plasmid isoforms.

10.3.3 Separation and Purification of Plasmid DNA

DNA vaccines are administered in the form of supercoiled plasmid preparations. Supercoiled DNA is the natural conformation of DNA found in the body. When the DNA helix is cut at the same position on the double strand, linearized DNA is produced. When the DNA helix has a gap in one of the double strands, an open-circular plasmid that is looser than the supercoiled DNA will be produced. Since plasmids exist in multiple topological forms, the isolation and purification of supercoiled plasmids is a challenging task (Ghanem et al. 2013). In addition, the isolation of plasmids from bacterial lysates also faces the challenge of contaminants such as genomic DNA, RNA, and endotoxins. Pharmaceutical-grade plasmid DNA vaccines are composed of more than 90% supercoiled plasmids. The content of residual protein should be lower than the detection limit when the BCA method is used, and the content of RNA should be lower than the detection limit when the 0.8% agarose gel is used. The content of genomic DNA should be <0.05 μg/mg plasmid when PCR method is used, and the content of endotoxin should be <0.1 EU/mg plasmid when LAL method is used.

The purification of pharmaceutical-grade plasmid DNA requires the development of highly reproducible and scalable processing methods that must meet regulatory standards similar to those required for the manufacture of recombinant protein drugs. In order to obtain plasmid products with a purity that meets regulatory requirements, most purification processes can be divided into preliminary isolation, intermediate recovery, and final purification. At present, plasmid purification is done by chromatography. Conventional chromatography, including size exclusion chromatography, anion exchange chromatography, hydrophobic interaction chromatography, metal chelation chromatography, affinity chromatography, and mixed-mode chromatography have all been widely used in plasmid purification.

Mahut et al. (2012a) proposed a unique method based on mixed-mode chromatography for the effective separation of topoisomers of covalently closed circular (ccc) plasmid DNA (Fig. 10.6). This topoisomer is not a topological isomer in the conventional sense, which include linear, open-loop, and supercoiled plasmids, but a series of isomers formed due to the different degrees of tightness of supercoils. The separation is based on the recognition of supercoiled plasmid by the quinine-carbamate ligand on the silica carrier. Under neutral salt gradient elution conditions, the ccc isoforms are resolved into discrete narrow Gaussian peaks, with each corresponding to a discrete supercoiled species. The analysis of the samples collected during the fermentation of *E. coli* using a series of purification steps of size exclusion chromatography and hydrophobic anion chromatography showed that the distribution pattern of topoisomers of the plasmid changed during the upstream processing. Thus, the mixed-mode chromatography provides a useful tool for monitoring the changes of supercoiled plasmid topoisomers during fermentation.

In order to further clarify the separation mechanism of topoisomers of supercoiled plasmids, Mahut et al. (2012b) carried out a structure-selectivity relationship study to find out the basic structural characteristics of the recognition of topoisomers by ligands. By changing the different parts of the selective agent originally derived from cinchona, it is shown that the intercalation of the quinoline part of the ligand is not a necessary feature for the recognition of topoisomers in the chromatography. The key element of selectivity to topoisomers is the presence of rigid weak anion exchange sites and H bond donor sites, which are separated from each other within a defined distance by a spacer of 4-atom length. In addition, the incorporation of weak anion exchange sites into the ring structure provides greater ligand molecular stiffness, which is very advantageous for the separation of topological isomers.

In this work, they prepared 6 sets of stationary phases (SP) with a total of 12 different structures and tested the selectivity of pDNA topoisomers. SP 1, 2, and 5 are based on quinine (QN), while SP 3a–d are based on quincorine, a derivative of QN. SP 4 lacks the quinuclidine ring of SP 1. SP5 and 6 have short bonds between the ligand and the chromatographic support, while all other chromatographic ligands are connected to the support through longer thioether bonds (Fig. 10.7). As shown in Fig. 10.8, 4.9 kbp pMCP1 was separated on these columns using the same mobile phase conditions.

Smrekar et al. (2013) reported a purification process of plasmid DNA based on a hydrophobic ion exchange mixed-mode monolithic column. On the methacrylate matrix, the ligands with hydrophobic and ion exchange functions were introduced into a single monolithic column through the ring-opening reaction of epoxy groups, and three types of methacrylate monolithic columns carrying octylamine, butyl (C4), and diethylaminoethyl (DEAE) group, respectively, were prepared (Fig. 10.9).

The protein adsorption capacity of the synthesized mixed-mode material was measured under the conditions of ion exchange and hydrophobic interaction chromatography, and compared with traditional epoxy resin, C4 and DEAE methacrylate monolithic columns. The monolithic column (gBuMA + DEAE monolithic column)

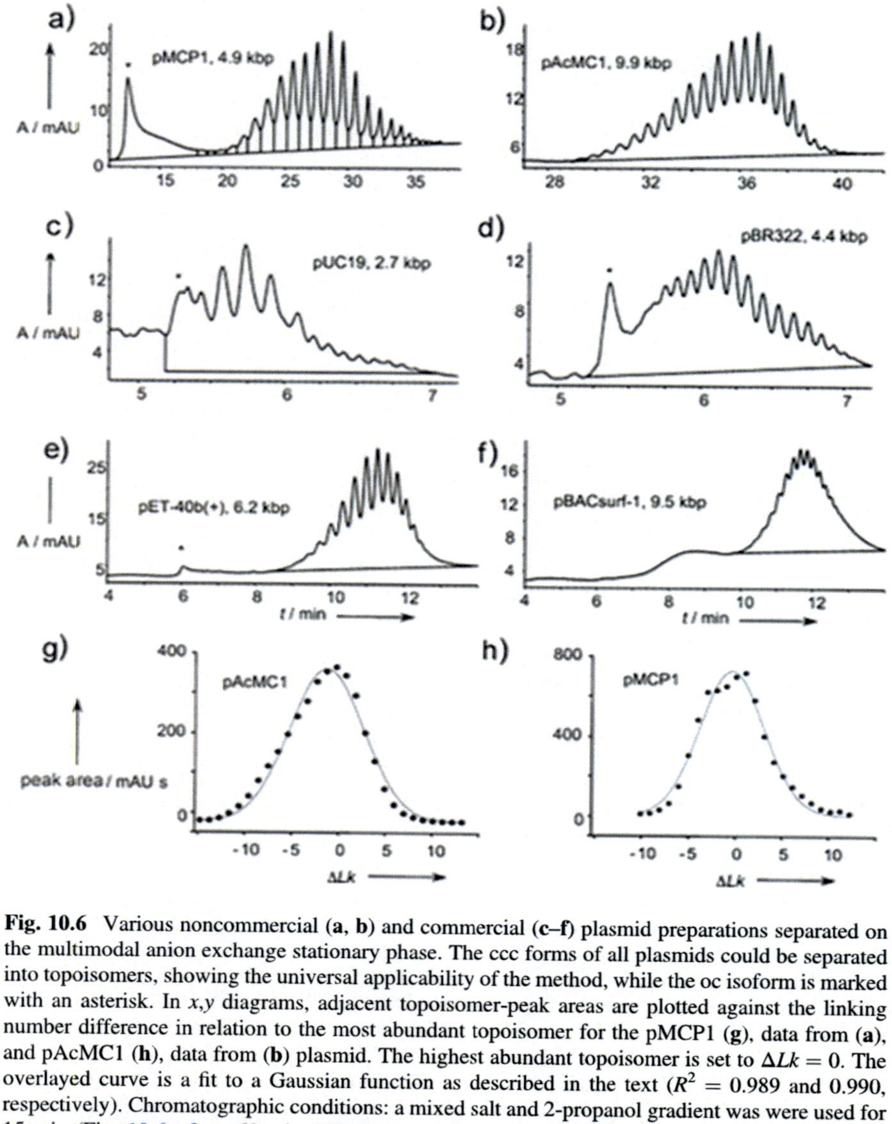

Fig. 10.6 Various noncommercial (**a, b**) and commercial (**c–f**) plasmid preparations separated on the multimodal anion exchange stationary phase. The ccc forms of all plasmids could be separated into topoisomers, showing the universal applicability of the method, while the oc isoform is marked with an asterisk. In *x,y* diagrams, adjacent topoisomer-peak areas are plotted against the linking number difference in relation to the most abundant topoisomer for the pMCP1 (**g**), data from (**a**), and pAcMC1 (**h**), data from (**b**) plasmid. The highest abundant topoisomer is set to $\Delta Lk = 0$. The overlayed curve is a fit to a Gaussian function as described in the text ($R^2 = 0.989$ and 0.990, respectively). Chromatographic conditions: a mixed salt and 2-propanol gradient was were used for 15 min (Fig. 10.6c–f) or 60 min (Fig. 10.6a, b) from 0% to 100% B. Buffer A is composed of 50 mM NaH_2PO_4 pH 7.0. Buffer B is composed of 50 mM NaH_2PO_4, 0.6 M NaCl, pH 7.0, and 10% (v/v) 2-propanol. Between the two analyses, the column was washed with salt solution (inject 50 μL of 3 M NaCl (aqueous)), and then re-equilibrate at 0% B for 5 min. The flow rate is set to 0.7 mL/min, the detection wavelength is set to 258 nm, and the temperature is set to 60 °C (solvent preheating). (Reproduced from Mahut et al. 2012a)

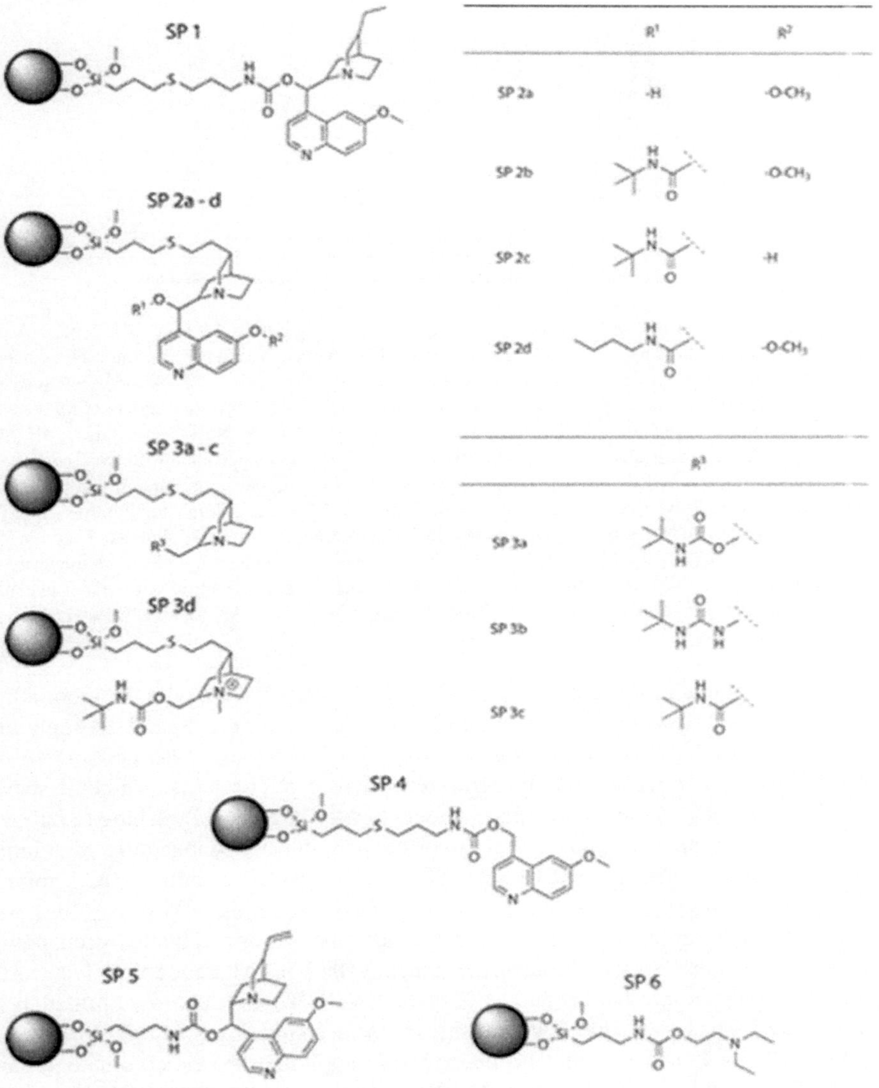

Fig. 10.7 The structure of the chromatographic ligand on the silica gel stationary phase used in this study. (Reprinted with permission from Mahut M, Lindner W, Laemmerhofer M (2012b) Molecular recognition principles and stationary phase characteristics of topoisomer-selective chemoaffinity materials for chromatographic separation of circular plasmid DNA topoisomers. J Am Chem Soc 134:859–862. Copyright {2012} American Chemical Society)

with a combination of C4 and DEAE groups was selected as the packing material for plasmid DNA purification. After repeated experiments, the binding and elution conditions of plasmid DNA on the monolithic column were determined, and the

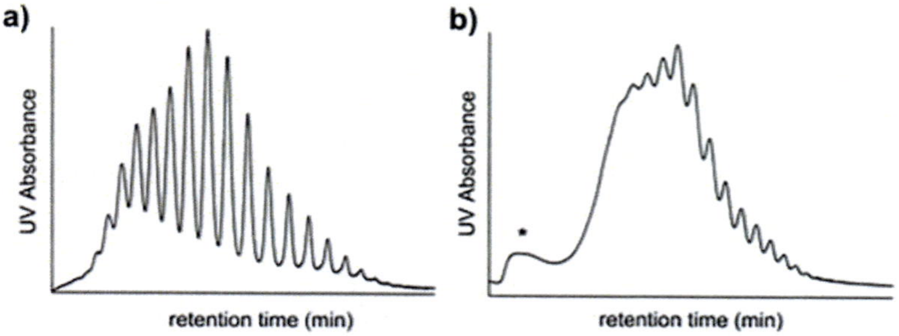

Fig. 10.8 Separation of ccc topoisomers in samples containing native 4.9 kbp pMCP1 pDNA on SP 1 (**a**) and SP 3a (**b**). The most abundant topoisomer has 26 negative supercoils, and each adjacent peak is a topoisomer, which differs by one supercoil, that is, one linking number. (*) Open-circlar plasmid isoforms are marked with an asterisk. Chromatographic conditions: a gradient of mixed salt and isopropanol from 0% to 100% B in 60 min. Buffer A is a 50 mM NaH_2PO_4 solution, pH 7.0. Buffer B is 50 mM NaH_2PO_4, 0.6 M NaCl, pH 7.0, and 10% (v/v) isopropanol. Between different runs, the column is washed with a salt plug (inject 50 μL 3 M NaCl (aq)), and then re-equilibrate at 0% B for 5 min. Flow rate 0.7 mL/min; UV detection at 258 nm; column temperature, 60 °C, solvent preheated to set temperature. (Reprinted with permission from Mahut M, Lindner W, Laemmerhofer M (2012b) Molecular recognition principles and stationary phase characteristics of topoisomer-selective chemoaffinity materials for chromatographic separation of circular plasmid DNA topoisomers. J Am Chem Soc 134:859–862. Copyright {2012} American Chemical Society)

following optimal procedures were developed. pDNA is loaded onto the monolithic column with a low ionic strength buffer and binds to the ligand through the mechanism of IEX. After loading, the monolith containing adsorbed pDNA is washed with low ionic strength buffer to remove host cell proteins. When the stable baseline is achieved, the mobile phase is switched to 0.6 M NaCl washing to remove weakly bound impurities, mainly residual RNA, and further changed to a high ionic strength buffer containing 3 M $(NH_4)_2SO_4$ without NaCl to induce HIC binding. This is followed by washing with a mobile phase containing 3 M $(NH_4)_2SO_4$ and 1 M NaCl to minimize IEX interaction under HIC conditions and finally by changing to 1 M NaCl eluent to elute the adsorbed plasmid DNA from the monolith. It is worth mentioning that if the step containing only the 3 M $(NH_4)_2SO_4$ phase is omitted, part of the pDNA has been eluted when using a buffer containing 3 M $(NH_4)_2SO_4$ and 1 M NaCl. An example of the chromatogram of applying the best chromatographic method and loading 0.6 mg of pure pDNA (the amount observed under IEX and HIC conditions is lower than the binding capacity) is shown in Fig. 10.10. Through this purification process, more than 99% of RNA, host cell proteins, and genomic DNA in the cell lysate are removed. However, this purification process failed to separate the topoisomeric impurities of plasmid DNA.

Černigoj et al. (2013) described the application of multimodal anion exchange chromatography in the separation and purification of plasmid topoisomers. Histamine was connected to a polymethacrylic acid matrix activated by carbodiimide (CDI) to prepare a multimodal chromatographic monolithic column containing imidazole groups (Fig. 10.11). Imidazole is a weak base, and its conjugate acid

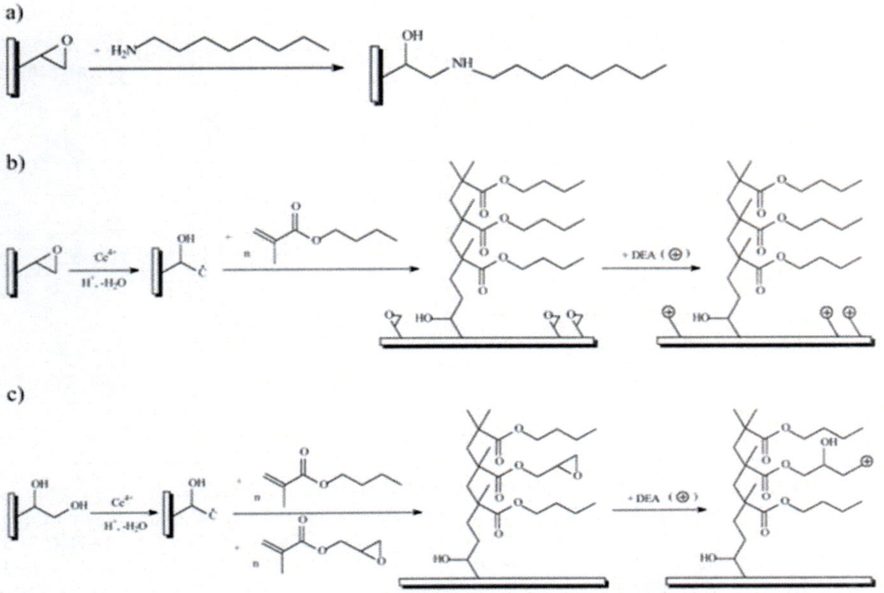

Fig. 10.9 The overall functionalization of methacrylate. (**a**) The reaction between octylamine and epoxy groups, (**b**) butyl methacrylate is grafted onto the epoxy groups, and then the remaining epoxy groups are modified to DEAE groups (gBuMA + DEAE monolithic column), and (**c**) the butyl methacrylate glycidyl ester mixture is grafted onto the hydroxyl group, and then the remaining epoxy groups are modified to diethylaminoethyl (DEAE) functional groups (gBuMA/DEAE monolithic column). (Reproduced from Smrekar et al. 2013)

has a pK_a of 7.05. Therefore, the electrostatic charge of the imidazole ligand depends on the pH of the mobile phase. When the pH is higher than 7, it is neutral and binds to the solute through hydrophobic interaction; when it is below pH 7, it is positively charged and binds to the solute through electrostatic interaction. Therefore, the supercoiled plasmid DNA may be separated from other plasmid isoforms and other process-related impurities by adjusting the salt concentration or pH gradient in the mobile phase.

Under the conditions of hydrophobic interaction chromatography and ion exchange chromatography, respectively, the possibility of separating pDNA isoforms with histamine monolithic column was investigated. In the first experiment, pDNA was dissolved in a solution of 3.0 M $(NH_4)_2SO_4$, 50 mM TRIS, 10 mM EDTA, pH 7.4, to enhance hydrophobic interactions, loaded on the column, and eluted with a linear gradient from 3.0 to 0 M $(NH_4)_2SO_4$. As shown in Fig. 10.12, the oc isomer eluted at 1.67 M $(NH_4)_2SO_4$, while the sc isomer eluted at 1.49 M $(NH_4)_2SO_4$. The sc and oc isomers of plasmid DNA were baseline separated. In order to strengthen electrostatic interaction, the plasmid was dissolved in 50 mM TRIS, 10 mM EDTA, pH 7.4, loaded on the column, and eluted with a linear gradient of 0–3.0 M $(NH_4)_2SO_4$. In this case, $(NH_4)_2SO_4$ is not used as a lyophilic

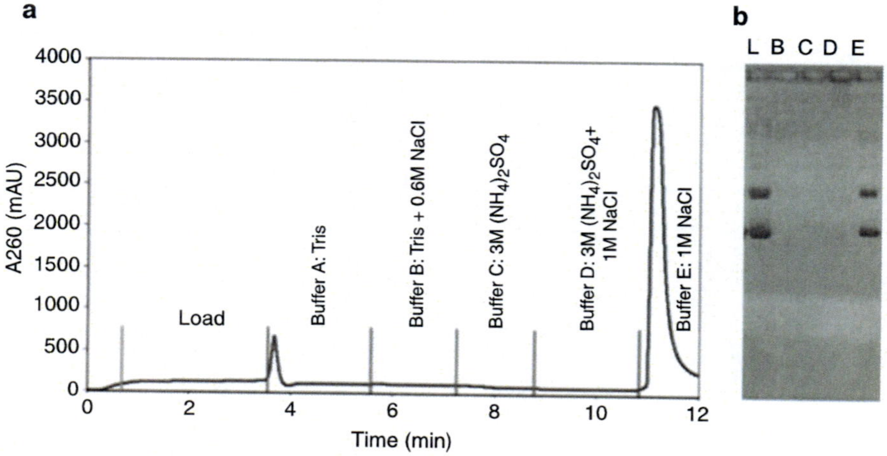

Fig. 10.10 The overall binding of pDNA with gBuMA + DEAE under IEX and HIC conditions. (**a**) Conditions: sample, 15 mL of pure pDNA at a concentration of 40 μg/mL; flow rate, 4 mL/min; stepwise gradient: Buffer A, 10 mM EDTA, 50 mM Tris, pH 7.2; buffer B, buffer 0.6 M NaCl in A; buffer C, 3 M $(NH_4)_2SO_4$ in Buffer A; Buffer D, 3 M $(NH_4)_2SO_4$, 1 M NaCl in Buffer A; Buffer E, 1 M NaCl in Buffer A; UV detection, 260 nm. (**b**) Agarose electrophoresis of the pDNA eluted from the gBuMA + DEAE monolithic column under IEX and HIC conditions. L, loaded plasmid; B, eluted with buffer B; C, eluted with buffer C; D, eluted with buffer D; E, eluted with buffer E. The elution fraction (E) is diluted 1:1 compared to the loaded fraction (L). (Reproduced from Smrekar et al. 2013)

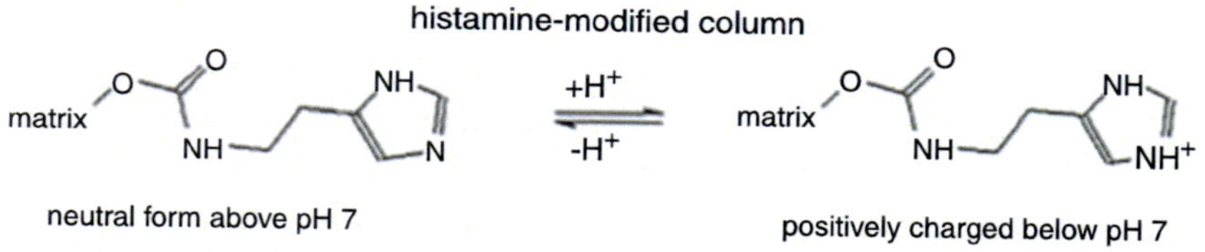

Fig. 10.11 Schematic diagram of the structure of immobilized histamine and its acid-base dissociation equilibrium in aqueous solution. (Reproduced from Černigoj et al. 2013)

agent, but as a modulator to increase the ionic strength of the mobile phase. pDNA eluted at 0.85 M $(NH_4)_2SO_4$, and the isomers were only partially separated. The different elution salt concentrations of pDNA in the two chromatographic modes indicate that pDNA binds to the stationary phase by different mechanisms.

In addition, the authors also found that the oc isoform of the plasmid can be selectively eluted at a constant 3.0 M $(NH_4)_2SO_4$ by changing the pH value of the mobile phase and completely separated from the sc isomer. As shown in Fig. 10.13, at the end of the pH gradient, the chromatographic peak of the oc pDNA isotype was observed, while the sc pDNA was completely retained. In this way, the oc form can be selectively removed from the sc form. To confirm this result, the column was re-equilibrated to the initial conditions (3.0 M $(NH_4)_2SO_4$, 50 mM TRIS, 10 mM

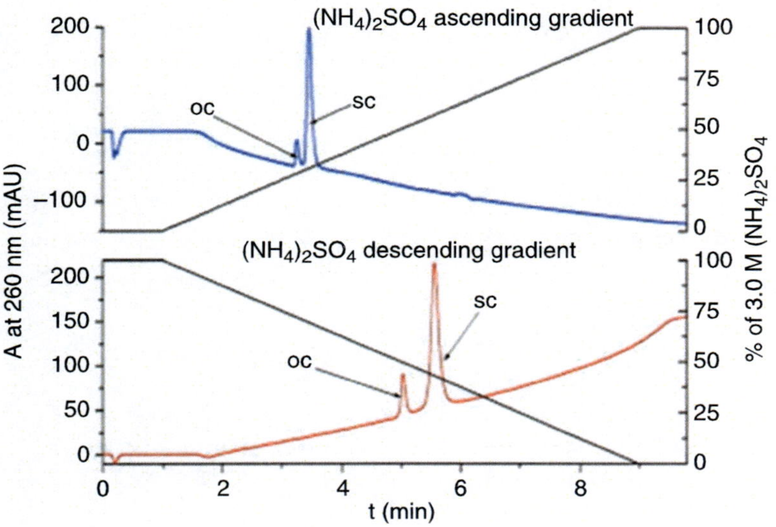

Fig. 10.12 The chromatogram obtained by eluting the pDNA adsorbed on the histamine monolithic column using an ascending or descending $(NH_4)_2SO_4$ gradient. MP A: 3.0 M $(NH_4)_2SO_4$, 50 mM TRIS, 10 mM EDTA, pH 7.4; MP B: 50 mM TRIS, 10 mM EDTA, pH 7.4. Vinj = 100 μL, $\lambda = 260$ nm, $\Phi = 2.0$ mL/min; linear gradient from 0% to 100% MP B and from 100% to 0% MP B in 47 CV. (Reproduced from Černigoj et al. 2013)

EDTA, pH 7.4), and the plasmid fraction recovered from the pH gradient experiment was loaded and eluted with a linear gradient of decreasing $(NH_4)_2SO_4$ concentration (see Fig. 10.13b). The sc peak eluted normally, but the oc peak no longer exists, which proves that the oc form has been effectively removed. In addition, as a control experiment, at a constant pH of 7.4, the original sample was separated in a $(NH_4)_2SO_4$ gradient from 3.0 to 0 M, and baseline separation of the oc and sc peaks was observed, and the sc peak was shown to have the same peak area and retention time as that purified with the pH gradient.

Bicho et al. (2016) reported the use of agmatine monolithic column to purify the influenza virus hemagglutinin-expressing plasmid NTC7482-41H-VA2HA from cell lysates. Agmatine is a cationic biogenic amine derived from arginine, which is fixed on a carbodiimidazole (CDI) activated polymethacrylate monolithic column (Fig. 10.14) to obtain a hydrophobic anion exchange mixed-mode chromatography packing.

The clarified cell lysate was loaded on the monolithic column, and the elution strategies in hydrophobic and ion exchange modes were used for purification. Figure 10.15a shows the chromatogram and agarose gel electrophoresis of the fraction eluted using a stepwise gradient of decreasing $(NH_4)_2SO_4$. The monolithic column was equilibrated at pH 9.6 with 3 M $(NH_4)_2SO_4$ in 50 mM phosphate buffer, and then the salt concentration was lowered to 2.2 M and 1.3 M $(NH_4)_2SO_4$, respectively, to facilitate the elution of the oc isomer (Fig. 10.15a, Lane 2) and sc pDNA (Fig. 10.15a, Lane 3). Finally, after reducing the ionic strength of the buffer

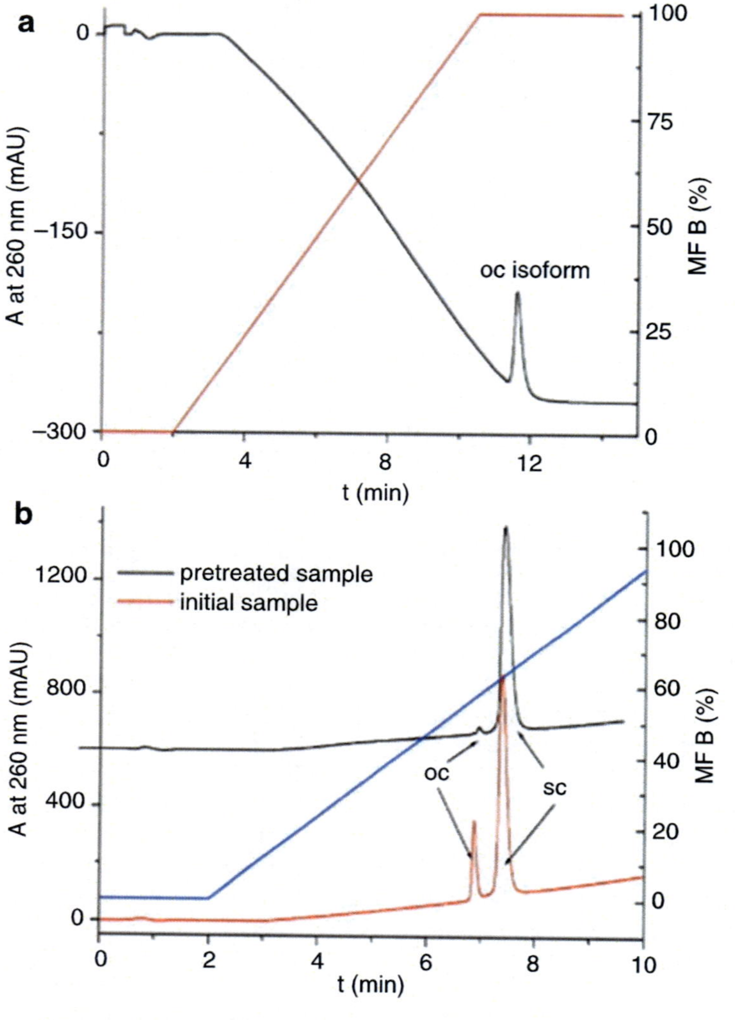

Fig. 10.13 The decreasing pH gradient is used for the elution of adsorbed plasmids. Column: Histamine in monolithic column, $\Phi = 2.0$ mL/min, $\lambda = 260$ nm, $V_{inj} = 500$ μL, pDNA (0.045 mg/mL) dissolved in 3.0 M $(NH_4)_2SO_4$, pH 7.4. (**a**) The elution of oc pDNA in a pH gradient: MF A: 50 mM TRIS, 3.0 M $(NH_4)_2SO_4$, 10 mM EDTA, pH = 7.4; MF B: 3.0 M $(NH_4)_2SO_4$, 50 mM acetate, 10 mM EDTA, pH = 4.5, gradient from 0% to 100% MF B in 50 CV. (**b**) Evidence for selective removal of oc pDNA isomers from the mixture: in 50 mM TRIS, 10 mM EDTA, pH 7.4, linear gradient from 3.0 to 0 M $(NH_4)_2SO_4$; (red line) initial sample; (blue line) the sample is first pretreated according to the procedure in (**a**), and then eluted in a descending gradient of $(NH_4)_2SO_4$. (Reproduced from Černigoj et al. 2013)

to 0 M $(NH_4)_2SO_4$, the strongly adsorbed RNA is stripped from the stationary phase (Fig. 10.15a, lane 4). Figure 10.15b shows the chromatogram of using an ascending NaCl gradient to purify influenza DNA vaccine. Under low ionic strength

Fig. 10.14 Structure of agmatine monolithic column. (Reproduced from Bicho et al. 2016)

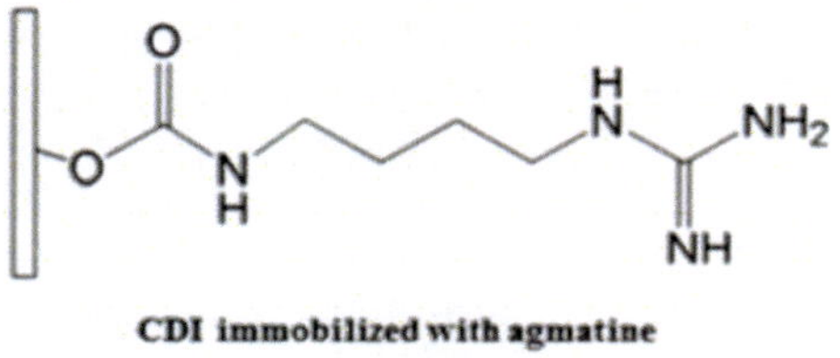

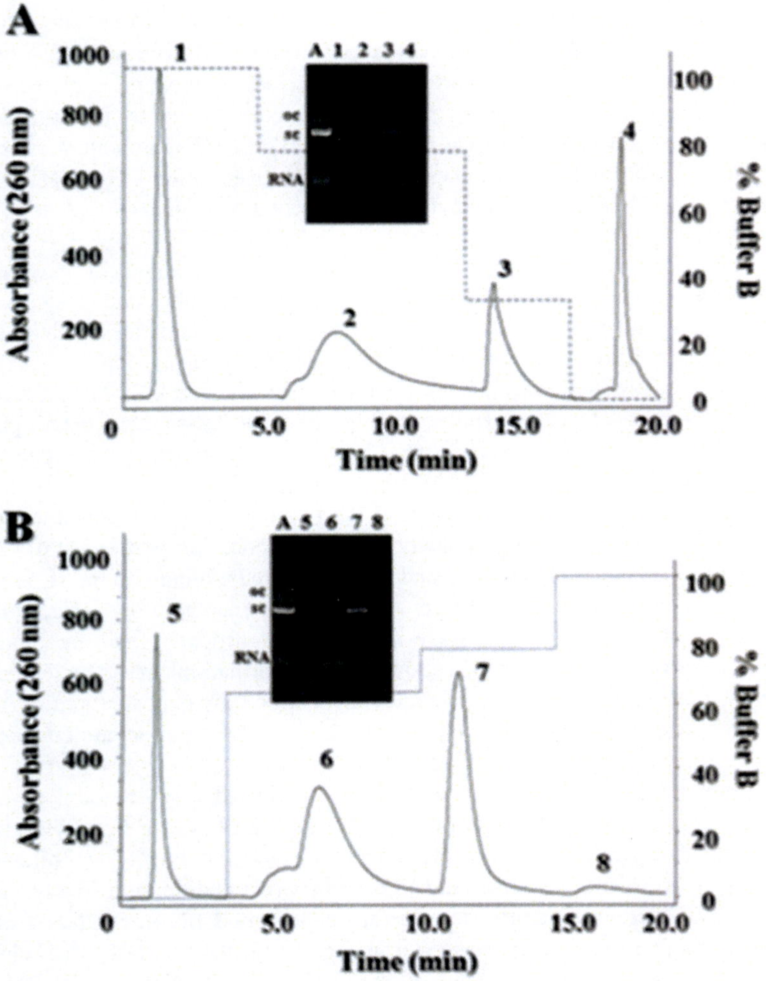

Fig. 10.15 Chromatogram of the purification of sc pDNA isomers from host impurities using agmatine monolith at 1 mL/min, by (**a**) a stepwise decreasing gradient of $(NH_4)_2SO_4$ in 50 mM phosphate buffer pH 9.6 and (**b**) a stepwise increasing gradient of NaCl in 50 mM phosphate buffer pH 8.0. Lane A: lysate feed; lanes 1–8: peaks 1–8; lane 2: oc; lane 3: sc; lanes 4 and 6: RNA; lane 7: oc + sc. (Reproduced from Bicho et al. 2016)

conditions, plasmid DNA is bound to the stationary phase. When the ionic strength of the buffer increases to 1.28 M NaCl, RNA is eluted (Fig. 10.15b, lane 6). The buffer ionic strength was further increased to 1.55 M NaCl to elute the oc and sc isoforms (Fig. 10.15b, track 7). Finally, a 2 M NaCl buffer is applied to elute any ionic impurities bound to the monolithic column (Fig. 10.15b, lane 8). The purity of the plasmids eluted in the hydrophobic and ion exchange modes were 98.3% and 99.6%, respectively; the recoveries were 51.8% and 45.3%, respectively.

Matos et al. (2014) used the multimodal anion exchange properties of Capto adhere to isolate and purify plasmid DNA from crude *E. coli* lysate. The cell lysate was loaded onto a multimodal chromatography column. The plasmid DNA binds to the ligand through electrostatic and hydrophobic interactions and is eluted using a stepwise increasing NaCl gradient to remove protein and endotoxin contamination and obtain purified plasmid DNA (Fig. 10.16). The RNA component is strongly bound to the resin and requires a very high salt concentration (2.0 M NaCl) to be eluted. This method cannot separate the different topoisomers of plasmid DNA (Fig. 10.16).

Silva-Santos et al. (2016) reported the separation of topoisomers of plasmid DNA on Capto adhere, a commercial multimodal chromatography stationary phase. In pH 8 and low-salt buffer, plasmid DNA was adsorbed onto the chromatography carrier. Then, elution was performed by a step increasing gradient of salt. The circular and supercoiled plasmids were eluted at 0.84 M and 0.89 M NaCl, respectively, to produce a baseline separation (Fig. 10.17a). Agarose gel electrophoresis analysis of the corresponding fractions (Fig. 10.17b) confirmed the separation of topoisomers.

The difference in retention behavior of the two nucleic acid topoisomers can be explained based on their charge density and hydrophobic properties. At pH 8, both topoisomers are negatively charged and will therefore be bound on the strong anion exchanger Capto™adhere. However, due to the supercoiling effect, the sc topoisomers are denser, so they have a higher charge density and are capable of stronger electrostatic interaction. In addition, the deformation caused by the inherent torsional strain of the sc pDNA molecule may lead to base exposure and strengthen the hydrophobic interaction with the ligand. Both of these factors are conducive to the retention of supercoiled topoisomers.

Subsequently, Silva-Santos et al. (2017) established a multimodal chromatographic separation and purification method for plasmid DNA in *E. coli* lysate. The method involves *E. coli* culture, cell harvesting and alkaline lysis; followed by isopropanol, ammonium acetate, and PEG-8000 precipitation; and finally, using a stepwise NaCl elution method to separate supercoiled plasmid DNA from the multimodal anion exchanger. As shown in Fig. 10.18, the oc pDNA was eluted in 0.83 M NaCl. Supercoiled sc pDNA was eluted in 0.92 M NaCl, while RNA was eluted in 2 M NaCl. This method provides a baseline separation of plasmid topoisomers and produces a sc-rich fraction (>90%), almost free from contamination by RNA, and with the levels of gDNA ($1.3 \pm 0.3\%$ μg gDNA/μg sc pDNA) and protein (4.97 ± 1.34 μg/mL) impurities within the specifications. Approximately 376 μg sc pDNA was obtained from 200 mL of bacterial cell culture ($OD_{600nm} \approx 19$).

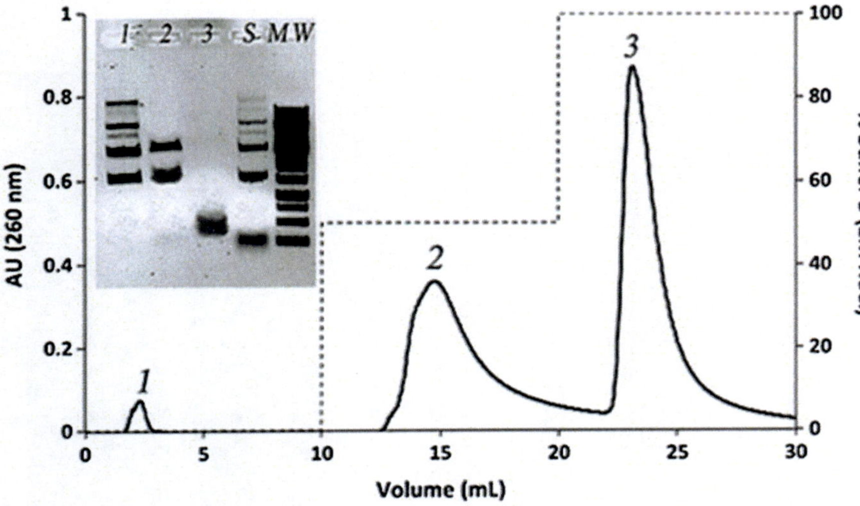

Fig. 10.16 Separation of crude *E. coli* cell extract on Capto Adhere column. The sample concentration is approximately 100 μg/mL, using a volume of 100 μL and a flow rate of 1.0 mL/min. As shown in the figure, the collected fractions are analyzed by agarose gel electrophoresis. (1) Peak 1 is concentrated five times before application on the gel to visualize the presence of different nucleic acids. (2) Peak 2 shows the plasmid DNA of the unconcentrated sample. (3) Peak 3 contains the RNA part of the sample. This part contains 2.0 M NaCl, which partially affects the migration of RNA molecules; S, original sample; MW, molecular weight marker. (Reproduced from Matos et al. 2014)

Using this method to purify plasmids of different sizes (2686, 3696, and 10,410 bp) in cell lysates, similar results were obtained, proving the versatility of the method. Compared with hydrophobic interaction chromatography, which is the preferred chromatographic mode of plasmid DNA isotype, multimode chromatography has advantages because the latter only uses sodium chloride instead of ammonium sulfate, which has less impact on the environment and greatly reduces the cost of production.

10.4 Separation and Purification of Minicircle DNA

Plasmid vectors occupies an important position in gene therapy. In gene therapy, plasmid vectors are used to transfer target genes to eukaryotic cells for expression. Traditional plasmid vectors contain bacterial replication sequences, resistance genes, unmethylated CpG motifs, and some hidden expression signals. These sequences are necessary for plasmid replication, but may cause serious biological safety problems in gene therapy. In addition, resistance genes will spread in the host through horizontal gene transfer, leading to a severe immune response. One of the ways to

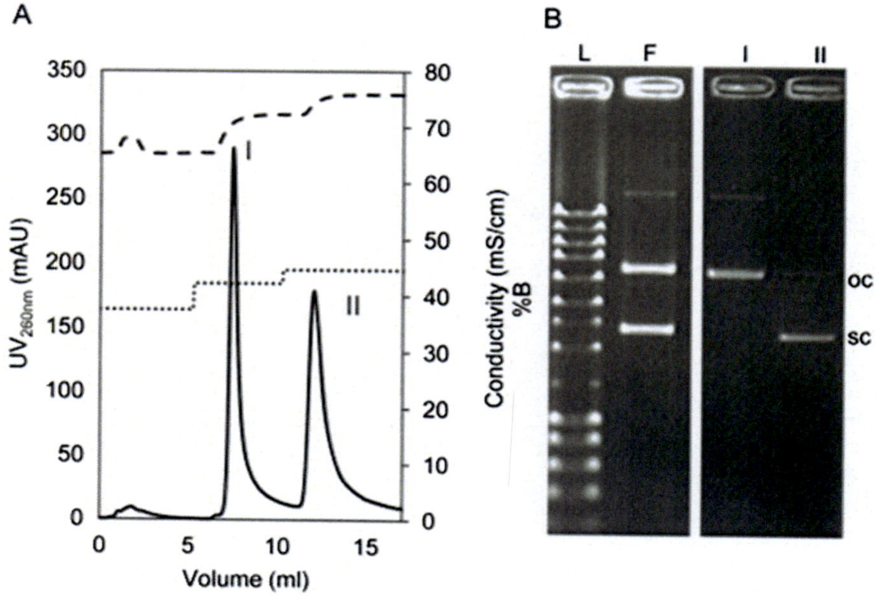

Fig. 10.17 Combined with a stepwise NaCl gradient elution scheme, multimodal chromatography was used to separate plasmid DNA topoisomers. (**a**) Chromatographic separation of topoisomers of supercoiled and open-circular plasmid DNA, 1 mL Capto™ adhesive column (0.7 × 2.5 cm); mobile phase, mixed with low-salt buffer in different proportions in pH 8 TE (A: TE, pH 8) and high-salt buffer (B: 2 M NaCl) prepared; sample, 1 mL purified pDNA (24.7 μg). Equilibration solution: 37.5% B; eluent 1: 42.2% B, eluent 2: 44.6% B. After loading the sample, the column was washed with 2 CV equilibration solution, and then 5 CV elution solution 1 (844 mM, ≈72 mS/cm) and 5 CV elution solution 2 (892 mM, ≈75 mS/cm). Solid line, absorbance at 260 nm; dashed line, conductivity (mS/cm); dashed line, percentage of buffer B (%B). (**b**) Agarose gel electrophoresis analysis of column feed (lane F) and fractions corresponding to peak I and peak II. Lane L, molecular weight marker. Dialysis the sample (15 μL) to desalt before loading. (Reproduced from Silva-Santos et al. 2016)

overcome this problem is to eliminate the bacterial regulatory sequence as a whole during the production of plasmid vectors to obtain minicircle DNA. Minicircle DNA is a small circular supercoiled expression frame obtained by site-specific recombination of traditional plasmids in *E. coli*. It does not contain resistance marker genes, replication origin, and other bacterial sequences, which enhances the safety in clinical applications. In addition, due to the reduced molecular weight of the vector, minicircle DNA also improves the efficiency of transgene expression (Alves et al. 2021).

Minicircle DNA is the product of site-specific recombination in the parent plasmid (Mayrhofer et al. 2008). Recombinase recognition sites are inserted on both sides of the parent plasmid carrying the eukaryotic expression cassette to induce the expression of the relevant recombinase in the body, and the expressed recombinase cuts the DNA sequence in the middle of the recognition site. The parental plasmid is divided into two

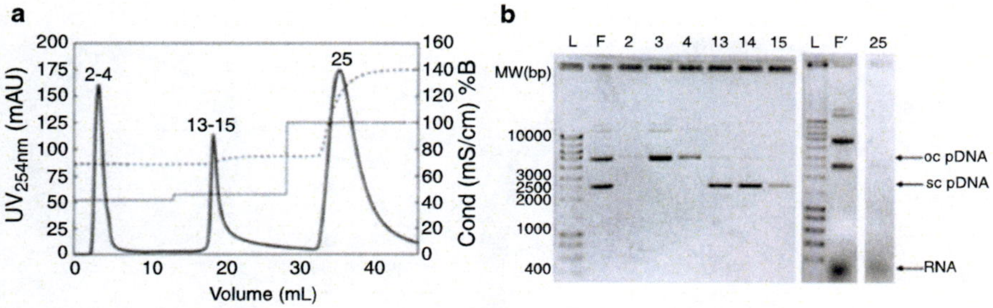

Fig. 10.18 Purification of sc pDNA from a feed stream containing oc pDNA, sc pDNA, and RNA using multimode chromatography combined with a stepwise salt gradient elution method. (**a**) Chromatogram of sc pDNA purification. The sample was prepared by diluting 1 mL of pDNA solution 12 times (75 μg pDNA) and pre-adjusting to 830 mM NaCl, and was injected into a Capto™ adhere column, which was pre-equilibrated with 830 mM NaCl buffer (41.5% buffer B, ≈69 ms/cm). The column was washed with 2 CV of 41.5% B to remove unbound impurities, and eluted with 3 CV of 46% B (≈75 mS/cm) and 3 CV of 100% B (≈140 mS/cm) successively. The number on the peak corresponds to the collected fraction. Solid line, absorbance at 254 nm; dashed line, conductivity (mS/cm); dashed line, percentage of buffer B (%B). (**b**) Agarose gel electrophoresis analysis on the fractions collected during the chromatographic run. Lane L corresponds to the molecular weight marker, and lanes F (2 μL) and F′(3 μL) correspond to the column feed. The numbered lanes correspond to the fractions collected during the chromatographic run (lanes 2–15 are 10 μL samples; lane 25 is 30 μL samples). (Reproduced from Silva-Santos et al. 2017)

supercoiled molecules, namely, the miniplasimid (MP) with replication function and the minicircle (MC) DNA carrying the eukaryotic expression cassette. Figure 10.19 is a schematic diagram of minicircle DNA generation.

The recombinases used to generate microcircle DNA are mainly phi C31, parA, Cre, etc.

Minicircle (mc) DNA is similar to pDNA in that they both contain expression cassettes and can produce high levels of transgene products shortly after delivery in vivo. The difference between them is that mc DNA does not contain all the prokaryotic sequence elements, such as the origin of replication and antibiotic resistance genes. These background pDNA sequences are removed by site-specific recombinases in *E. coli* before the plasmid DNA is isolated. The deletion of these prokaryotic sequence elements also reduces the size of mc DNA, leading to an increase in transfection efficiency. The end result is that mc DNA transfects more cells compared to their full-length pDNA counterparts and allows sustained high levels of transgene expression after delivery in vivo. Therefore, the immune response caused by the mc vaccine is more effective than the response of the full-length pDNA vaccine encoding the same antigen.

The in vivo recombination process produced three different plasmids, namely, minicircle DNA, miniplasmid and residual parental plasmid. Separating the therapeutically effective minicircle DNA from the reaction product faces considerable challenges. So far, the method commonly used in laboratories is to linearize the miniplasmid and parental plasmid with restriction enzymes, and then obtain

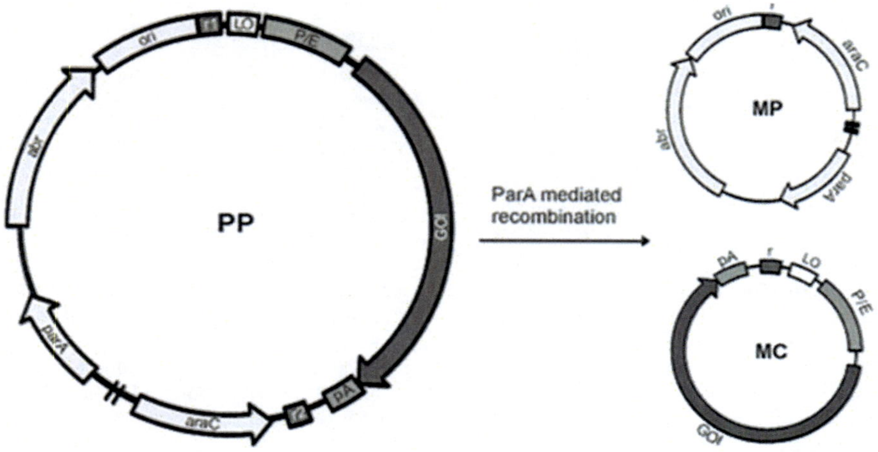

Fig. 10.19 Schematic representation of the parental plasmid for ParA-mediated recombination and the resulting products. Expression of the ParA resolvase, which is under the control of the arabinose promotor/operator system, is induced by addition of L-arabinose. ParA mediates recombination between the resolution sites of the parental plasmid resulting in the generation of a miniplasmid and a minicircle. (Reproduced from Mayrhofer et al. 2008)

minicircle DNA by cesium chloride gradient centrifugation. The main disadvantages of this method are that the enzymes used are expensive, labor-intensive, and the yield of the target product is low, making it difficult to achieve large-scale purification.

One way to solve these problems is to express restriction enzymes in vivo for linearization of small plasmids and parental plasmids. By co-expressing a restriction endonuclease in vivo, the miniplasmid and the parent plasmid are sheared into linear DNA. Then conventional plasmid isolation methods are used to obtain high-purity minicircle DNA. Chen et al. (2005) expressed the I-Sce I endonuclease with the recombinase phi C31 gene in a bicistronic manner. The endonuclease expressed at the same time can recognize the 18 bp residue sequence and act on the small plasmid and parental plasmid to linearize them. Supercoiled MC was then obtained using ion exchange chromatography.

Another method is to use the principle of affinity chromatography to purify minicircle DNA. A recognition sequence that specifically binds to the affinity ligand is inserted into the parent plasmid. After recombination, the sequence is located on the minicircle DNA. The specific binding of the minicircle DNA to the affinity ligand leads to separation from other plasmids. Mayrhofer et al. (2008) used the principle of specific binding of the lactose operon site to the lactose operon repressor (LacI) to construct the lactose operon site into the parent plasmid. After recombination, the minicircle DNA contains the lactose operon site sequence. After loaded on the LacI protein immobilized agarose gel, and separated by affinity chromatography, the minicircle DNA with a purity of 98.5% can be obtained.

Recently, Alves et al. (2016) reported a new purification method. By design, a recognition site with nick endonuclease Nb.BbvCI is inserted into the prokaryotic sequence of the parent plasmid (PP). The minicircle DNA produced by PP recombination in vivo does not contain the recognition site of nuclease. After in vitro enzymatic digestion of the purified mixture of PP, MC, and MP with Nb.BbvCI, supercoiled (sc) MP and unrecombined sc PP were converted into open-circular (oc) isoforms, while sc MC lacked the recognition site of Nb.BbvCI maintains its topological structure. Then sc MC DNA can be separated from oc MP and oc PP by hydrophobic interaction chromatography.

Silva-Santos et al. (2019) described a multimodal chromatography method for purifying minicircles (MC) DNA from *Escherichia coli* culture broth. Minicircle DNAs and miniplasmids (MP) are obtained by intramolecular recombination and division of parental plasmid (PP) in *E. coli*. The minicircle DNA contains only the target gene expression cassette and the miniplasmid DNA contains the prokaryotic gene. The purification process includes alkaline lysis of *Escherichia coli*, tandem precipitation, and endonuclease digestion steps, specifically converting supercoiled (sc) MP into open-circular (oc) MP. The product is then loaded on a multimodal Capto adhere matrix and eluted by gradually increasing the NaCl concentration (Fig. 10.20). The results showed that oc species (MP and MC) eluted first (830 mM NaCl, ≈69 mS/cm), then sc MC (920 mM NaCl, ≈75 mS/cm) and finally RNA and other strongly retained impurities (2 M NaCl, ≈140 mS/cm) eluted. Gel electrophoresis showed that the sc MC fraction was homogeneous (>90%) and free of impurities.

To further elucidate the separation mechanism, the authors also compared the elution behavior of plasmids on multimodal chromatography, anion exchange chromatography, and hydrophobic interaction chromatography (Silva-Santos et al. 2019). Using Sepharose Q and Sepharose Phenyl as reference matrices, control experiments were performed under the same salt gradient as that of multimodal chromatography. The results show that, no matter in anion exchange or hydrophobic interaction chromatography, all nucleic acid species (sc MC, oc and sc MP, RNA) present in the feed stream are eluted in the flow-through as a single peak without retention. Under the same conditions as multimodal chromatography, neither anion exchange (charged amine group) nor the hydrophobic portion (phenyl group) alone can separate nucleic acids. As reported by Alves et al. (2016), the use of high concentration of ammonium sulfate allows the minicircle DNA to be retained on the phenyl Sepharose and eluted with a gradient of decreasing salt concentration.

Nuclease sensitivity determination and bioinformatics analysis further show that due to double-strand instability and B-Z conversion, the base exposure in the sc isomer is more likely to occur. Based on these data, the authors propose the mechanism of nucleic acid separation. The binding of the charged amine group of the matrix with the nucleic acid phosphate backbone produces an asymmetry of charge neutralization, which results in the bending of the nucleic acid chain and the exposure of the base. Therefore, the exposed bases can be used for cation-π, π-π stacking, and H-bonding interactions, thereby enhancing electrostatic interaction and effectively creating a non-covalent bonding network.

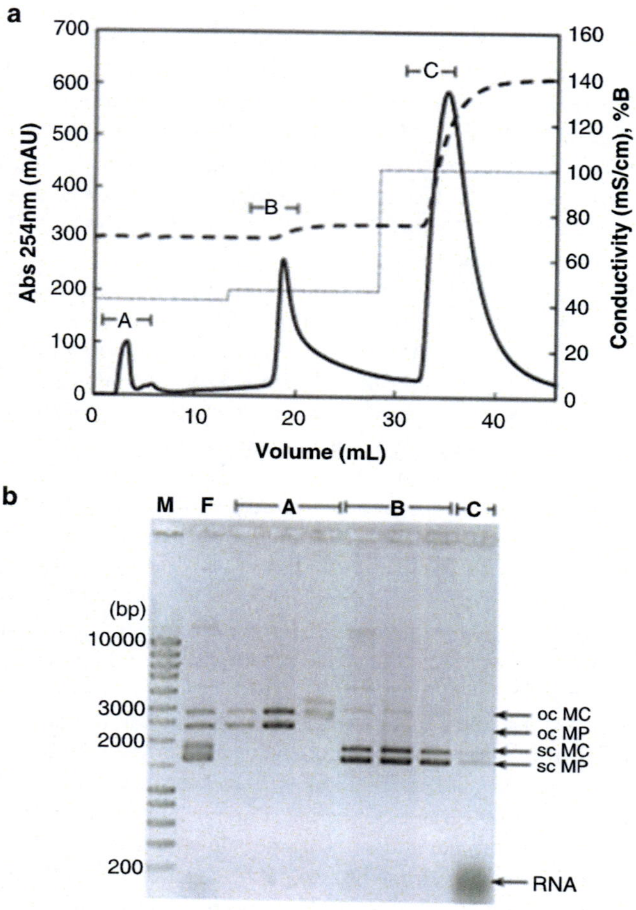

Fig. 10.20 Multimodal chromatography for purification of sc MC DNA and sc MP from a feed stream containing RNA and the oc and sc isoforms of MC DNA and MP. One milliliter of pre-purified sample (pre-adjusted to set the NaCl concentration to 830 mM) was injected into a Capto adhere column, which was equilibrated with 830 mM NaCl buffer (41.5% buffer B, ≈69 mS/cm). The unbound material was washed with 2 CV of 41.5% B and gradually eluted with 3 CV of 46% B (≈75 mS/cm) and 3 CV of 100% B (≈140 mS/cm). (**a**) Chromatogram. Solid line: absorbance at 254 nm; discontinuous dashed line: conductivity (mS/cm); continuous dashed line: percentage of buffer B (%B). (**b**) Agarose gel electrophoresis analysis of the fractions collected during the chromatographic run. M, molecular weight marker; F, feed (3 mL). The fractions collected during the chromatographic run are labeled A (throughput, 10 mL), B (step 1, 10 mL), and C (step 2, 30 mL). (Reproduced from Silva-Santos et al. 2019)

The elution of different substances in the multimodal chromatographic column initiated by NaCl is mainly achieved by the disruption of the ionic interaction between the phosphate group of the DNA backbone and the charged nitrogen of the ligand. As these interactions weaken, the bending of DNA decreases, the exposure of bases decreases, and the corresponding cation-π, π-π stacking, and hydrogen bonding interactions decrease. The increase in salt content also changes the solvation environment and helps reduce the strength of the hydrogen bond between the ligand and the DNA. The observed elution order basically reflects the increase in base exposure in oc DNA, sc DNA, and RNA.

10.5 Separation and Purification of Antibody Drugs

The parts or fragments of antigen molecules that can induce antibody production are called antigenic determinants. An antigenic molecule usually contains multiple antigenic determinants, and each determinant can induce at least one antibody. Therefore, in traditional antisera, there are a variety of different antibodies, respectively, against these different antigenic determinants. These antibodies are called polyclonal antibodies. In principle, appropriate means can be used to separate polyclonal antibodies into different monoclonal antibodies, but the separation method is not efficient and inconvenient to use. Kohler and Milstein (1975) used cell hybridization technology to fuse mouse myeloma cells and mouse spleen cells immunized with sheep red blood cells. The hybridoma cells formed can produce antibodies and proliferate indefinitely, thus creating the hybridoma technology. Monoclonal antibodies can be easily prepared in large quantities. This technology has been widely used in protein function research, gene expression profile analysis, diagnosis, and treatment of clinical diseases.

The first therapeutic monoclonal antibody muromonab-CD3 (Orthoclone OKT3) was approved by the US FDA in 1986. It contains a murine monoclonal antibody that targets T cells expressing CD3. Its function is an immunosuppressive agent and is used to treat acute transplant rejection. However, the commonly used murine monoclonal antibodies also face many obstacles in clinical application. Mouse monoclonal antibodies have strong immunogenicity to humans and can produce strong human anti-mouse antibody responses. In addition, the production cost of murine monoclonal antibody is relatively high, and it is difficult to scale up for widespread applications. With the rapid development of molecular biology, the detailed structure and function of antibodies have gradually been clarified, and recombinant antibody technology has emerged (Fig. 10.21). Morrison et al. (1984) linked the murine monoclonal antibody variable region with the human IgG constant region at the gene level and successfully constructed the first recombinant antibody, the human-mouse chimeric antibody. Jones et al. (1986) replaced the CDR region of human IgG with the complementarity determining region (CDR) of a murine monoclonal antibody and successfully constructed the first humanized modified antibody, also known as CDR grafted antibody. In 1990, Winter et al. (McCafferty et al. 1990) used phage display technology for the discovery and separation of antibodies and established a technology for the preparation of fully humanized antibodies. In 1991, Padlan (1991) proposed to use human antibodies as a reference to transform and replace the surface amino acid residues of murine monoclonal

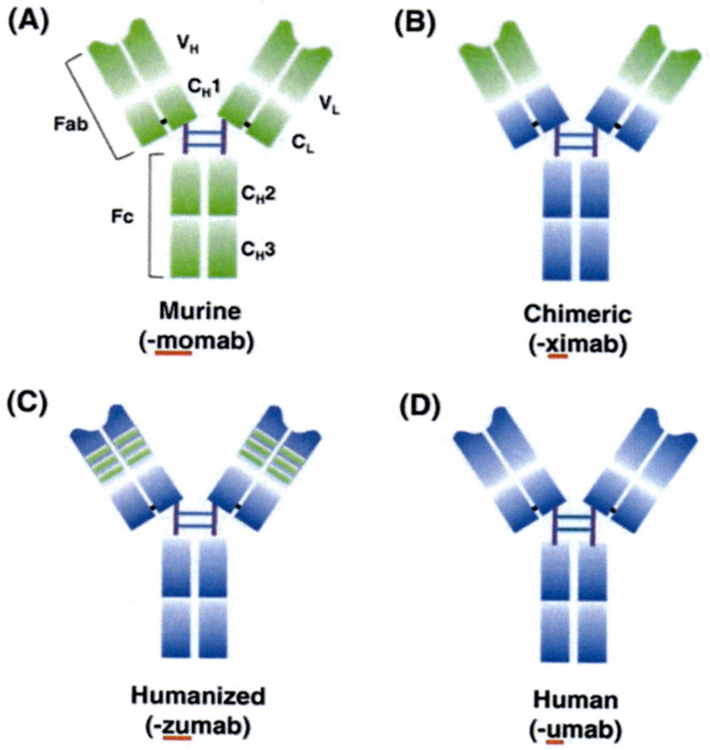

Fig. 10.21 Schematic diagram of antibody humanization from murine antibody (green domain) to fully human antibody (blue domain) and related suffixes. (**a**) Murine monoclonal antibody. (**b**) Chimeric monoclonal antibody: The variable region is of murine origin, and the rest of the chain is of human origin. (**c**) Humanized monoclonal antibody: only includes murine hypervariable fragments. (**d**) Human monoclonal antibody. CH, the domain of the constant region of the heavy chain; CL, the constant domain of the light chain; Fab and Fc, fragments produced by proteolysis; VH, the variable domain of the heavy chain; VL, the variable domain of the light chain. (Reproduced from Lu et al. 2020)

antibodies to obtain veneered antibodies. The first therapeutic chimeric antibody, anti-GPIIb/IIIa antigen-binding fragment (Fab) (abciximab), was approved by the US FDA in 1994 to inhibit platelet aggregation in cardiovascular diseases. The first chimeric antibody with tumor indications, anti-CD20 (rituximab), was approved by the US FDA for non-Hodgkin lymphoma in 1997. The first humanized monoclonal antibody, anti-IL-2 receptor daclizumab, was approved by the US FDA in 1997 to prevent transplant rejection. The first fully humanized therapeutic antibody, adalimumab (trade name Humira), a human antibody against tumor necrosis factor a (TNFa), was approved by the FDA in 2002 for treatment of rheumatoid arthritis.

Up to date, at least 570 therapeutic mAbs have been evaluated in clinical trials worldwide, and 79 therapeutic mAbs have been approved by the US Food and Drug Administration (US FDA) for use to treat cancer, asthma, arthritis, psoriasis, Crohn's disease, transplant rejection, migraine and infectious diseases. Many monoclonal antibody products had annual sales of more than 3 billion U.S. dollars in 2018, and six monoclonal antibody products (adalimumab, nicholizumab, pembrolizumab, trastuzumab, bevacizumab, rituximab) had the annual sales of single products exceed USD 6 billion (Table 10.2). Adalimumab (Humira) sales set a record for biopharmaceutical products, approaching $19.9 billion. Table 10.2 lists the best-selling monoclonal antibody products in 2018. In 2018, the global therapeutic monoclonal antibody market was valued at approximately 115.2 billion U.S. dollars, and it has grown to 150 billion U.S. dollars by the end of 2019. It is expected to generate 300 billion U.S. dollars in revenue by 2025. Therefore, the therapeutic antibody drug market has experienced explosive growth as a new drug. Given that a large number of monoclonal antibodies are still in clinical trials, the therapeutic antibody market still maintains huge room for growth.

10.5.1 Purification of Antibody Drugs

The large-scale antibody drug production process can be roughly divided into three steps. The first is antibody design, gene recombination, cell line construction, and screening optimization; second is the optimization of cell culture and related culture processes; and the last is antibody purification and preparation formulations. The cell expression level of antibodies has been significantly improved with the development of antibody engineering technology, which greatly reduces the cost of large-scale cell culture. However, the complexity of the biological feed and the strict requirements for antibody products make the antibody purification process the key to antibody production, accounting for about 50–80% of the cost of antibody production. A typical purification process is shown in Fig. 10.22. It mainly includes the following parts: (1) harvesting cell culture fluid from the bioreactor, and removing cells and cell debris by centrifugation and filtration; (2) using protein A affinity chromatography to capture antibodies; (3) low pH buffer incubation to remove viruses; (4) a two-step refining process to remove products and process-related impurities; (5) virus filtration, microfiltration, and dialysis to obtain the final antibody product. It can be concluded that the antibody purification process is relatively complicated, and it is necessary to combine a variety of separation methods and design a reasonable purification process to efficiently obtain high-quality antibody products.

Efficient recovery and purification of monoclonal antibodies from cell culture media is a key step of the production process and may account for a large part of the total manufacturing cost. The main consideration in downstream process development is purity. Important product purity attributes include process-related contaminants (such as host cell protein levels, DNA, endotoxin, leached protein A, and certain cell culture additives) and product-related impurities (such as high molecular weight aggregates and low molecular weight deletions or cut-offs). In addition, the process must be able to remove viruses to ensure product safety.

Table 10.2 Top ten best-selling monoclonal antibody drugs in 2018

No.	Drug	Indication (first US FDA approval year)	Company	2018 Revenue (USD)
1	Adalimumab (Humira)	Rheumatoid arthritis (2002) Psoriatic arthritis (2005) Ankylosing spondylitis (2006) Juvenile idiopathic arthritis (2008) Psoriasis (2008) Crohn's disease (2010) Ulcerative colitis (2012) Hidradenitis suppurativa (2015) Uveitis (2018)	AbbVie	$19.9 bn
2	Nivolumab (Opdivo)	Melanoma (2015) Non-small cell lung cancer (2015) Renal cell carcinoma (2015) Head and neck squamous cell (2016)	Bristol-Myers Squibb	$7.6 bn
3	Pembrolizumab (Keytruda)	Melanoma (2014) Head and neck cancer (2016) Non-small cell lung cancer (2015) Lymphoma (2018) Cervical cancer (2018) Microsatellite instability-high cancer (2018)	Merck & Co	$7.2 bn
4	Trastuzumab (Herceptin)	Breast cancer (1998) Gastric cancer (2010)	Roche	$7.0 bn
5	Bevacizumab (Avastin)	Colorectal cancer (2004) Non-small cell lung cancer (2006) Breast ERB2 negative cancer (2008) Renal cell carcinoma (2009) Glioblastoma (2011)	Roche	$6.8 bn
6	Rituximab (Rituxan)	Non-Hodgkin's lymphoma (1997) Chronic lymphocytic leukemia (2010) Rheumatoid arthritis (2006) Pemphigus vulgaris (2018)	Roche	$6.8 bn
7	Infliximab (Remicade)	Crohn's Disease (1998) Rheumatoid arthritis (1999) Ankylosing spondylitis (2004) Ulcerative colitis (2005) Psoriatic arthritis (2005) Psoriasis (2006)	Johnson & Johnson	$5.9 bn
8	Ustekinumab (Stelara)	Psoriasis (2009) Psoriatic arthritis (2013) Crohn's Disease (2016)	Johnson & Johnson	$5.2 bn

(continued)

Table 10.2 (continued)

No.	Drug	Indication (first US FDA approval year)	Company	2018 Revenue (USD)
9	Eculizumab (Soliris)	Paroxysmal nocturnal hemoglobinuria (2007) Atypical hemolytic uremic syndrome (2011) Generalized myasthenia gravis (2017) Neuromyelitis optical spectrum disorder (2019)	Alexion	$3.6 bn
10	Omalizumab (Xolair)	Asthma (2003) Chronic idiopathic urticaria (2014)	Roche	$3.0 bn

Reproduced from Lu et al. (2020)
bn billion

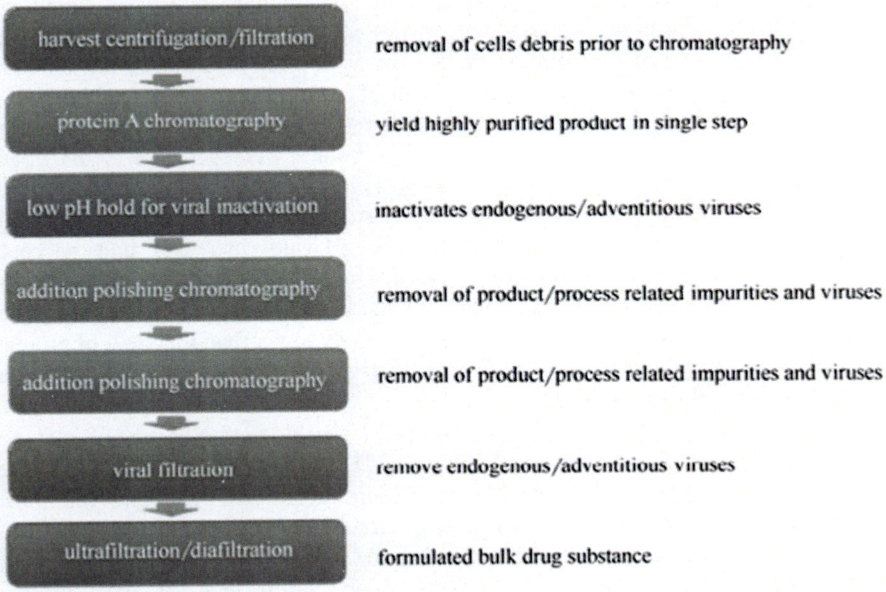

Fig. 10.22 Typical monoclonal antibody purification process

The current typical antibody purification process first uses protein A affinity chromatography to capture the antibody and then further refines the antibody by methods such as cation exchange chromatography and anion exchange chromatography. Protein A affinity chromatography has the characteristics of high separation selectivity. It can directly capture antibodies from complex feeds and achieve concentration at the same time, reducing the burden of subsequent separation. It is considered to be the most important step in the antibody purification process.

However, protein A affinity chromatography has problems such as relatively high media cost, harsh elution conditions leading to production of antibody aggregates, and ligand shedding contributing to contamination of the product. For this reason, research efforts have been focused on seeking alternative technologies, for example, using hydroxyapatite or multimodal ion exchange resins to replace protein A for capture the target antibody from cell culture supernatants or improvements in the polishing step to remove contaminants in the product, including host cell proteins, endotoxin, DNA, leaching protein A, and antibody aggregates.

Freitag and colleagues (Giovannini and Freitag 2001; Schubert and Freitag 2007) evaluated the efficiency of apatite chromatography to capture recombinant monoclonal anti-Rh D antibody (rh-Anti-RhD) directly from the cell culture supernatant and compared it with protein A chromatography. rh-Anti-RhD is a therapeutic agent for neonatal hemolytic disease, expressed in Chinese hamster ovary (CHO) cells. The p*I* of the antibody is 8.3. Under neutral pH conditions, the antibody is positively charged and can interact electrostatically with the P-site of hydroxyapatite. Therefore, the double gradient method is used to elute the antibody with a NaCl gradient and then to elute the strongly retained impurities with a phosphate gradient. First, the influence of the pH of the solution on the adsorption and desorption behavior of antibodies was investigated. At four different pH values, namely, pH 6.0, pH 7.0, pH 7.8, and pH 8.2, the antibodies in the supernatant were completely adsorbed by the hydroxyapatite material, and no active antibodies were detected in the flow-through fraction (Fig. 10.23). The antibody was eluted in the NaCl gradient, while a large number of other proteins are eluted in the subsequent phosphate gradient. The retention time of the antibody decreases as the pH value increases, and the peak shape is also improved, and the elution occurs in a smaller gradient volume. The yield of the product also increased with the increase of pH. At pH 8.2, the yield reached the highest value of 71%, which was slightly lower than the yield of protein A affinity chromatography (86%). The product was homogeneous as shown by a single band in gel electrophoresis.

On fluoroapatite chromatographic materials, similar antibody adsorption and desorption behaviors were observed (Fig. 10.24). The antibody in the cell culture supernatant was bound to fluoroapatite and eluted in the NaCl gradient, while most of the adsorbed protein impurities were eluted in the subsequent phosphate gradient. Antibody retention decreased with increasing pH. In the neutral mobile phase pH, the product yield had the highest value, 90%, which is higher than 86% of the protein A chromatography. Similarly, no protein contamination was observed in the gel electrophoresis of the purified antibody fractions.

As mentioned above, using the double gradient elution procedure of sodium chloride and phosphate, hydroxyapatite chromatography can capture monoclonal antibodies from the cell culture supernatant and remove most of the impurities. However, experiments also found that the purification procedure still has great limitations. IgG binding capacity is easily affected by electrical conductivity of the feed stream. The physiological conductivity of the cell culture supernatant is sufficient to reduce the capacity. In addition, the phosphate used in the cell culture medium will further reduce binding capacity. Therefore, the feed solution must be

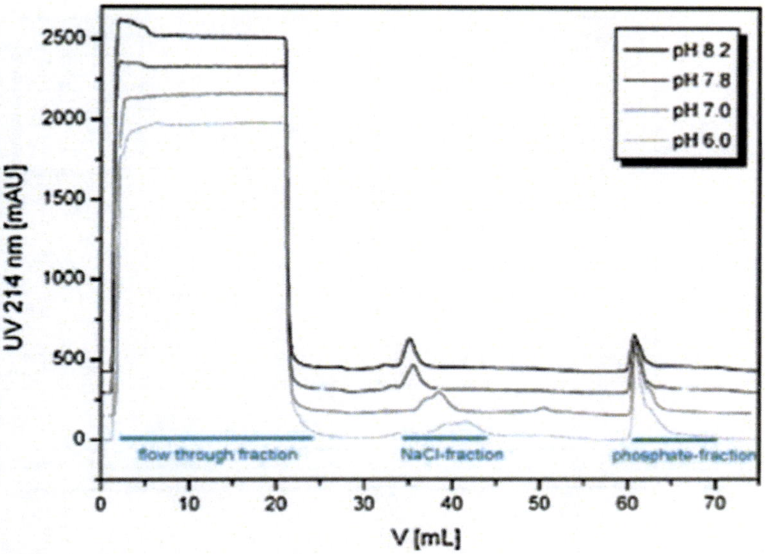

Fig. 10.23 Purification of antibodies by hydroxyapatite chromatography at different mobile phase pH (6.0, 7.0, 7.8, 8.2). Sample: 5 mL of clarified cell culture supernatant diluted 1:4 in binding buffer. Binding buffer (buffer A): 10 mM phosphate, first elution buffer (buffer B): 10 mM phosphate + 1 M NaCl, second elution buffer (buffer C): 400 mM phosphate salt. Elution scheme: 5 CV buffer A, linear gradient from 0% to 100% B (20 CV), linear gradient from 100% to 0% B (4 CV), linear gradient from 0% to 100% C (10 CV), 100% C (5 CV), 100% A (10 CV); the flow rate is 1 mL/min. (Reproduced from Giovannini and Freitag 2001)

diluted or dialyzed to reduce the concentration of salt and phosphate before loading the sample.

Although hydroxyapatite chromatography has greater limitations in the antibody capture step compared with protein A chromatography, it shows unique advantages in the polishing stage after protein A chromatography. First of all, the residual contaminants in the eluate of the protein A chromatography, such as endotoxin, and DNA contain a large number of phosphate groups, which are strongly adsorbed on hydroxyapatite and can only be eluted at higher phosphate concentrations. Secondly, the leached protein A in the eluate of the protein A chromatography is in the form of antibody complexes, and like antibody aggregates, it shows stronger retention than monomers in the hydroxyapatite chromatography. By choosing a suitable elution program, all these contaminants can be removed simultaneously in a single chromatographic run. Gagnon et al. (2006) proposed that in the presence of a certain concentration of phosphate solution in the mobile phase, the gradient elution of NaCl solution with increasing concentration can improve the separation between monomer and protein A-antibody complex or aggregate. In a typical separation, the mobile phase always contains 5 mM phosphate. A salt gradient increasing from 0 to 1 M NaCl is used to elute the adsorbed antibody. Natural antibodies are eluted in 0.3 M NaCl, aggregates are eluted in 0.5 M NaCl, and the strongly retained

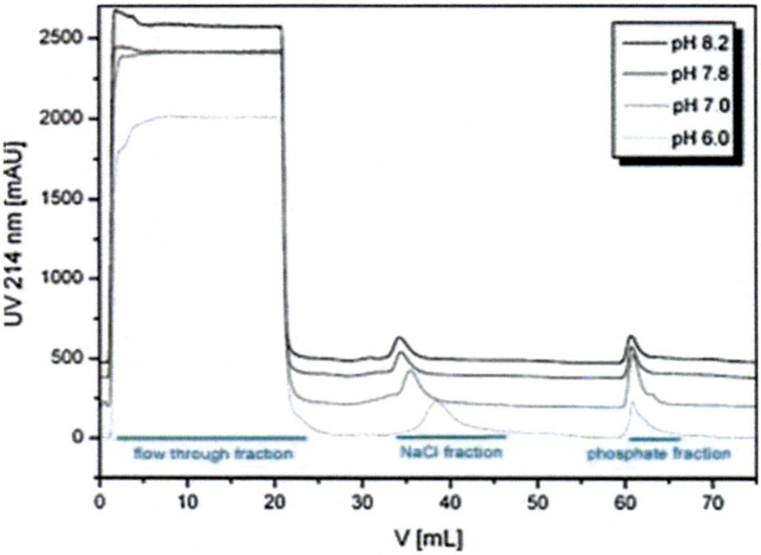

Fig. 10.24 Purification of antibodies by fluoroapatite chromatography at various mobile phase pH (6.0, 7.0, 7.8, 8.2). Sample: 5 mL of clarified cell culture supernatant diluted 1:4 in binding buffer. Binding buffer (buffer A), 10 mM phosphate; first elution buffer (buffer B), 10 mM phosphate + 1 M NaCl; second elution buffer (buffer C), 400 mM phosphate salt. Elution scheme: 5 CV buffer A, linear gradient from 0% to 100% B (20 CV), linear gradient from 100% to 0% B (4 CV), linear gradient from 0% to 100% C (10 CV), 100% C (5 CV), 100% A (10 CV); the flow rate is 1 mL/min. (Reproduced from Giovannini and Freitag 2001)

endotoxin, DNA, and protein A antibody complexes are stripped from the column in 0.5 M phosphate. Most monoclonal antibodies can be eluted in a 1.5 M NaCl gradient. In the exceptional case that antibodies cannot be eluted by simply increasing NaCl, the concentration of the background phosphate can be increased to 10 mM to facilitate the elution of the antibody at a lower NaCl concentration. However, as the concentration of phosphate in the mobile phase increases, the retention of contaminants decreases, resulting in a decrease in the removal efficiency of aggregates, leached protein A, DNA, and endotoxins.

Pezzini et al. (2011) evaluated the ability of four mixed-mode chromatography resins (PPA HyperCel, HEA HyperCel, MEP HyperCel, and Capto adhere) to capture recombinant antibodies from CHO cell culture supernatants. Using a batch adsorption experiment on a 96-well filter plate combined with DoE modeling, the effects of the pH and conductivity of the solution in the washing and elution stages on the clearance of HCP and the recovery of monoclonal antibodies were studied. Figure 10.25 shows the contour plot of HCP content and antibody recovery after washing and elution phases.

According to the balance between the minimum HCP content and the maximum antibody recovery, four mixed-mode chromatography resins are used to determine

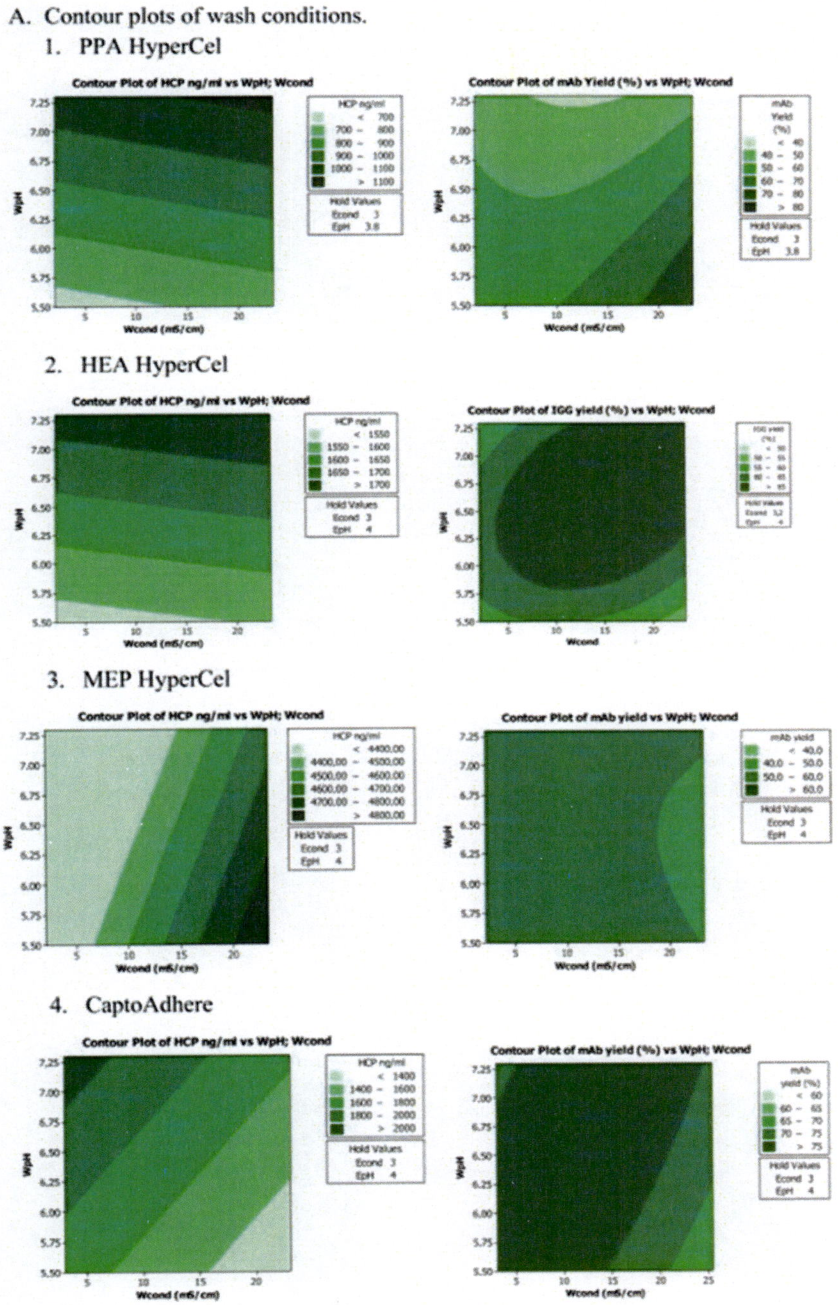

Fig. 10.25 Contour maps obtained after the washing (**a**) and elution (**b**) conditions of PPA HyperCel, HEA HyperCel, MEP HyperCel, and Capto adhere were screened with the pH and conductivity of the solution as variables. On the left is a contour map showing the contents of the HCP. On the right, a contour plot of antibody production is shown. The value corresponding to the color of the contour map is displayed on the scale on the right side of each panel. (Reproduced from Pezzini et al. 2011)

B. Contour plots of elution conditions.

1. PPA HyperCel

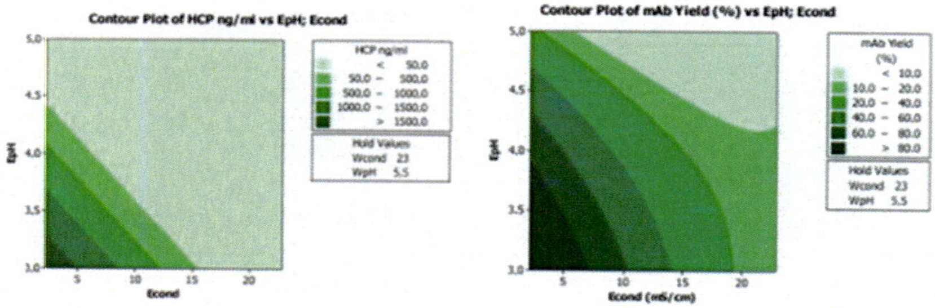

2. HEA HyperCel

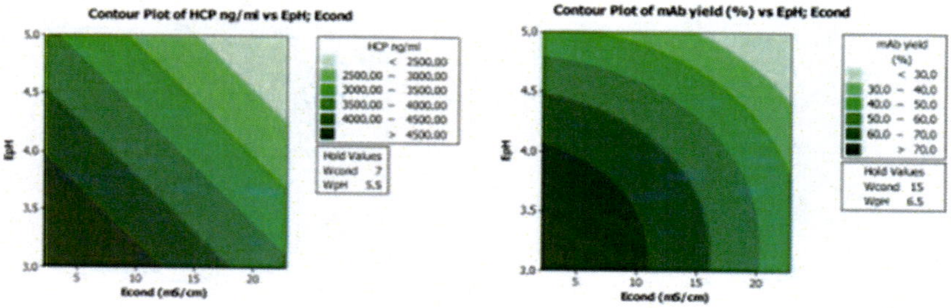

3. MEP HyperCel

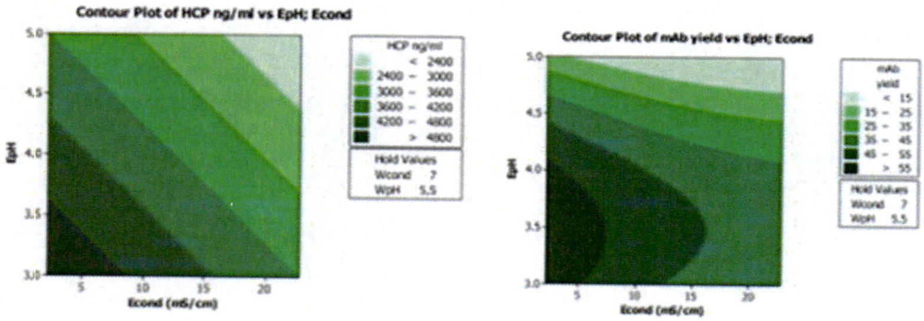

4. CaptoAdhere

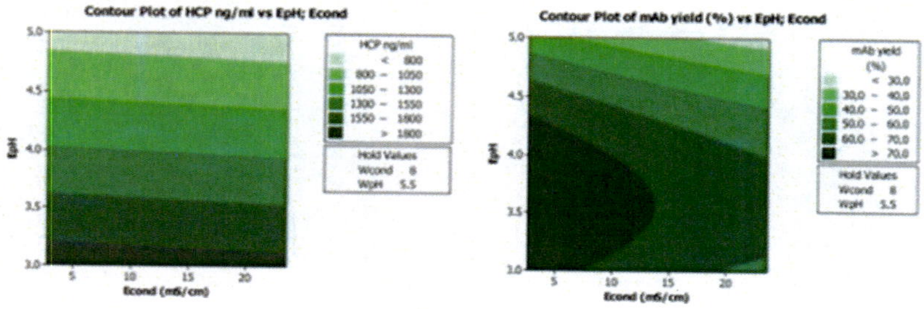

Fig. 10.25 (continued)

Table 10.3 Optimal conditions obtained with the model

Packing	Binding	Wash	Elution
MEP HyperCel	pH 7.3	pH 5.5	pH 4
	13 mS/cm	7 mS/cm	3 mS/cm
PPA HyperCel	pH 7.3	pH 5.5	pH 3.8
	13 mS/cm	23 mS/cm	3 mS/cm
HEA HyperCel	pH 7.3	pH 6.5	pH 4
	13 mS/cm	15 mS/cm	3 mS/cm
Capto adhere	pH 7.3	pH 5.5	pH 4
	13 mS/cm	8 mS/cm	3 mS/cm

Reproduced from Pezzini et al. (2011)

the best conditions for washing and elution after capturing monoclonal antibodies from the cell culture supernatant, as shown in Table 10.3.

The optimal conditions defined by the model were directly applied to a laboratory-scale chromatographic column, and the antibody yield and HCP clearance equivalent to the predicted value were obtained. These four resins can provide good yield (72–93%) and low HCP content (430–2400 ppm) for antibody purification. Under the condition of the crude material, PPA HyperCel seems to be the most effective, showing the highest yield (93%) and the lowest HCP content (430 ppm). The results show that the screening with this method is very effective for designing purification steps using mixed-mode resins. It allows to explore various conditions within a limited time and limited sample consumption, and quickly establish antibody purification methods.

Kaleas et al. (2014) evaluated a method for selectively capturing and purifying mAbs from harvested mammalian cell culture fluid (HCCF) using multimodal chromatography with Capto MMC resin. The determination of the dynamic binding capacity (DBC) of 12 monoclonal antibodies shows that it is feasible to directly load HCCF on the multimodal resin for the binding of monoclonal antibodies, but it may be necessary to adjust the pH of the HCCF to increase the DBC. The raw material conductivity in the range of 9–16 mS/cm does not show much influence on DBC. The optimization of the loading pH conditions makes 75% of the raw materials reach a binding capacity of >30 g/L. It is determined that the load pH is about 3 pH units lower than the mAb p*I* and/or the charge number of the Fab is 13 will produce the best binding capacity.

The elution strategies for recovering monoclonal antibodies from multimodal chromatography columns include pH gradient, sodium chloride gradient, sodium chloride gradient in the presence of urea, and L-arginine HCl gradient. Organic solvents and buffers that are known to specifically disrupt hydrophobic and/or hydrogen bond interactions, such as 20% propylene glycol, 3 M urea or water, were tested separately using stepwise elution, and none of them produced elution peaks, indicating mAb binding is not driven by hydrophobic interactions. However, changes in the pH of the solution or the concentration of the salt solution caused the antibody to be almost completely eluted. As shown in Fig. 10.26, at pH 7.0, the ionic strength gradient from 0 to 300 mM sodium chloride produced a yield of 93–98%,

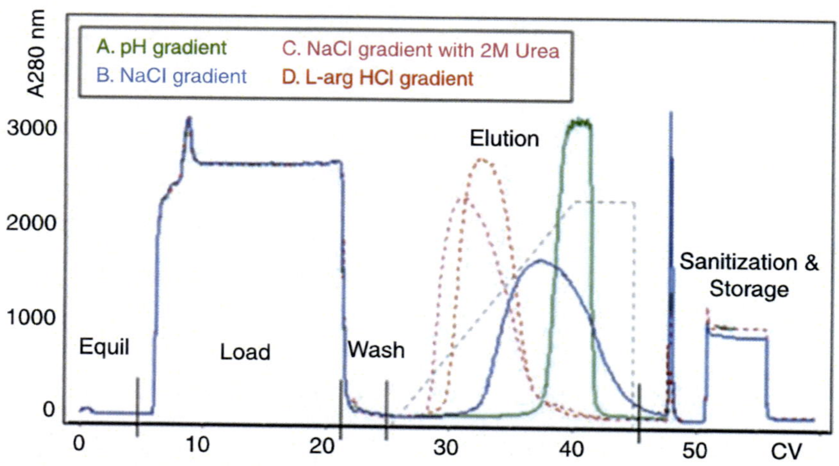

Fig. 10.26 Capto MMC chromatograms of the four elution strategies. The pH value of all elution buffers is 7.0, and each buffer uses a 15 CV gradient and then maintains 5 CV at 100% B, as shown by the gray dashed line. The raw material shown is HCCF with unadjusted mAb2 at a load of ΔpH 3.5 and/or Fab charge of 13.4. (Reproduced from Kaleas et al. 2014)

while at a constant conductivity of 4.5 mS/cm, a pH gradient of from pH 7.0 to 10 produced a yield of 94–100%. Similar elution recovery was observed using the combination of 2 M urea and 300 mM sodium chloride at pH 7.0 and 300 mM L-arginine HCl at pH 7.0. The yields of antibodies produced by the two elution strategies ranged from 90% to 100%. These results indicate that the binding of antibodies on the multimodal Capto MMC resin is mainly through electrostatic interactions, and hydrophobic interactions only play an auxiliary role. Desorption buffers that can simultaneously disrupt electrostatic and/or hydrophobic interactions can selectively recover mAbs and can be applied to other mAb raw materials with minimal optimization.

Then, two mAb HCCF raw materials were tested in a process comparison experiment. The capture step was Capto MMC or protein A affinity chromatography, and both pooled fractions were polished by anion exchange chromatography in flow-through mode (Fig. 10.27). The antibody yields of different capture columns, elution strategies, and raw materials are comparable, with all values ranging from 90% to 100%. Regardless of the elution strategy used, the HCP clearance of Capto MMC is lower than that of protein A. In order to achieve lower HCP levels, the multimode elution method can be enhanced by elution optimization and/or implementation of a washing stage, similar to the washing stage in the affinity chromatography process, to further remove adsorbed HCP. The results show that multimodal chromatography is an effective capture tool. When combined with appropriate polishing steps, it is possible to produce products with a purity comparable to the affinity capture process.

Multimodal ion exchange chromatography can also be effectively used as a polishing step to remove various product-related impurities such as aggregates,

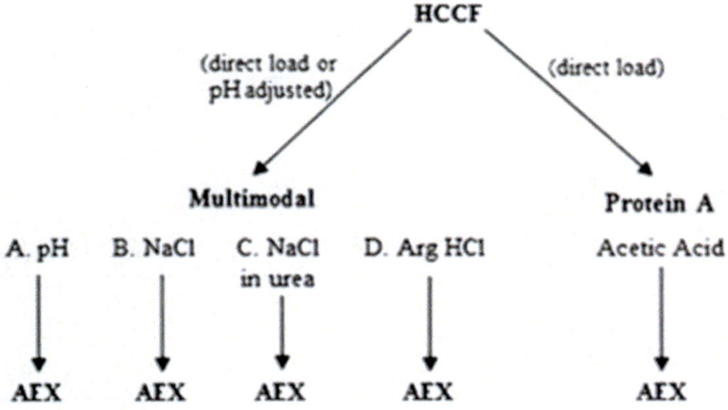

Fig. 10.27 Process flow of multimodal chromatography and affinity chromatography as the initial purification capture step. Four elution strategies (A–D) were evaluated using gradient elution on Capto MMC multimodal resin. Use acetic acid for gradual elution on ProSep-vA affinity resin. In all cases, anion exchange (AEX) chromatography in flow-through mode was used as the second purification step. (Reproduced from Kaleas et al. 2014)

HCP, DNA, endotoxins, and leached protein A. Guo et al. (2013) used three mixed-mode chromatography resins, namely, commercial Capto adhere and self-made agarose resin with benzylamine and butylamine as functional ligands, to remove mAb aggregates from protein A affinity chromatography streams in a flow-through mode and compared with the traditional cation exchange and hydrophobic interaction chromatography for the removal of aggregates. The results show that under the sample load of 50 mg/mL resin, both Capto adhere and benzylamine resins can significantly reduce the aggregate level from 20.5% to 2.6% and 2.4%, respectively, while the butylamine and traditional resins are less effective in removing aggregates. The ability of mixed-mode chromatography to effectively remove aggregates is attributed to the synergistic effect of ionic and hydrophobic interactions between ligands and aggregates.

Chmielowski et al. (2014) studied the use of Capto Adhere™, a multimodal anion exchange chromatography resin, to remove monoclonal antibody aggregates in the protein A affinity chromatography process stream. The characterization of the aggregates showed that the aggregates consisted of dimers with acidic charge distribution and higher-order oligomers with basic charge distribution. In order to take advantage of the ability of multimodal ligands to distinguish monomers and aggregates, they adopted a multivariate experimental design combined with high-throughput screening strategies, with monomer yield and aggregate reduction percentage as the main response for rapid evaluation of the experimental parameters for mixed-mode chromatography, for example, the feed pH in the range of 7.1–7.3, the feed conductivity in the range of 11–13 mS/cm, and the protein load of $\leq$140 g/L resin. The established multimodal chromatographic process can meet the goals of $\geq$80% yield and $\geq$70% percentage reduction of heterogeneous charged aggregates. After being verified by packed column experiments, the multimodal

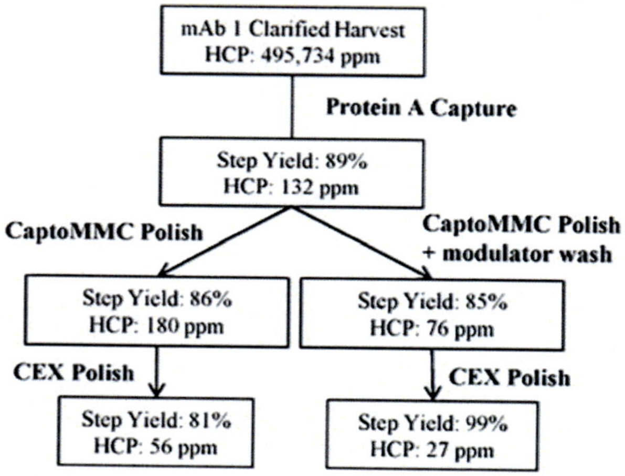

Fig. 10.28 HCP removal during decontamination. The process flow diagram depicts the downstream purification process to check the HCP removal rate, including and excluding conditioning agent cleaning. 25 mM Tris, 50 mM arginine, 0.1 M NaCl, pH 7.0, are used as the conditioning agent washing buffer. The step yield and HCP level after each step are reported in the box above the HCP level of each process intermediate. (Reproduced from Wolfe et al. 2014)

chromatography process was successfully scaled up ≥1700 times and yielded a step yield of 89–92% and a percentage reduction of aggregates of 75–85%. Compared with single-mode cation exchange and hydrophobic interaction chromatography, the protein load of the multimodal chromatography process is increased by 4–5 times, and the step yield is increased by 12–20%.

Wolfe et al. (2014) reported that after the monoclonal antibody is bound with the multimodal cation exchange resin Capto MMC, selective intermediate washing with a buffer containing a mobile phase modulator can further reduce the HCP content in the product. They first studied the effect of a single mobile phase modulator (such as ethylene glycol, arginine, sodium thiocyanate, urea, and ammonium sulfate) on reducing the HCP content in the product elution pool, from which they picked up the modulators that show significant effects. These modulators were used in the intermediate washing before elution by mixing two or three reagents. The most effective mobile phase modulator in reducing HCP includes 0.1 M NaCl and a combination of one or two of the following reagents: 5% ethylene glycol, 50 mM arginine, and 1 M urea. Figure 10.28 compares the Capto MMC chromatographic results of selective washing with and without 25 mM Tris, 0.1 M NaCl, 50 mM arginine, 5% ethylene glycol, and pH 7.0. In this downstream process sequence, the mAb product is captured by protein A chromatography, then bound/eluted and purified on Capto MMC, mainly to remove HCP, and then proceed to the cation exchange polishing step in flow-through mode to remove aggregates and residual DNA. It can be seen from the figure that the selective cleaning step can reduce the HCP level in the product by more than onefold.

In addition to using the multimodal chromatography as a polishing step in the mAb process, multimodal chromatography can also be used as a capture step. When the mobile phase modifier is incorporated into the wash buffer of the Capto MMC capture chromatography, the HCP level can also be reduced while maintaining the step yield. The data of Wolfe et al. (2014) showed that the combination of 50 mM arginine and 50 mM NaSCN was the most effective in reducing HCP levels. However, 50 mM arginine containing 1 M urea or 0.1 M NaCl can also significantly reduce HCP levels. Although the total HCP level did not decrease to the same level (132 ppm) as with protein A chromatography, when the optimal washing method was added to this process step, the HCP level was reduced by more than 30 times (from ~300,000 to ~8000 ppm). In contrast, when the same Capto MMC step was used for product capture without selective washing, the HCP content was much higher, about 14,000 ppm.

10.6 Separation and Purification of Antibody Fragments

Single-chain fragment variable (scFv) antibody fragments consist of the variable regions of the antibody heavy (V_H) and light (V_L) chains fused together via a short flexible linker to form a polypeptide chain that carries a complete antigen-binding site. In addition, they have a monomeric structure and remain stable even at low concentrations and physiological temperatures. Due to its small molecular size, scFv has better tumor penetration characteristics and less immune response and is used for radioimmunoimaging and specific delivery of cytotoxic agents (Damasceno et al. 2004). Multimodal chromatography resins (such as Capto MMC) have been used to capture the single-chain variable fragment scFv hu225 targeting epidermal growth factor receptor (EGFR) expressed in the periplasmic space of *E. coli* (Lindner et al. 2014). Using rha-BAD expression system and high cell density culture technology of *E. coli*, a specific product concentration of up to 20 mg scFv/g CDW was obtained. By combining freeze-thaw, osmotic shock, and pH-induced precipitation of host cell proteins, nearly 70% of the product was extracted from biomass. In the multistep chromatographic purification method, the multimodal cation exchange resin Capto MMC chromatography is used as the capture and desalting step. Under the conditions of pH 5.5 and salt concentration of 225 mM NaCl, the cleared sample solution is directly loaded onto the chromatographic column. Then the washing step is used for impurity removal and desalination, and elution is performed by a gradient of rising pH (Fig. 10.29). The step recovery is 85%, and the product purity is 80%. The polishing step including ion exchange chromatography and hydroxyapatite chromatography further increased the purity of the product to 99.4%, and the recovery of the entire downstream process was 50%.

10.7 Purification of Recombinant Protein Drugs

Kaleas et al. (2010) used Capto MMC to selectively capture recombinant human vascular endothelial growth factor (rhVEGF) from alkaline raw materials. VEGF is an important protein that regulates the angiogenesis cascade. It is an effective and

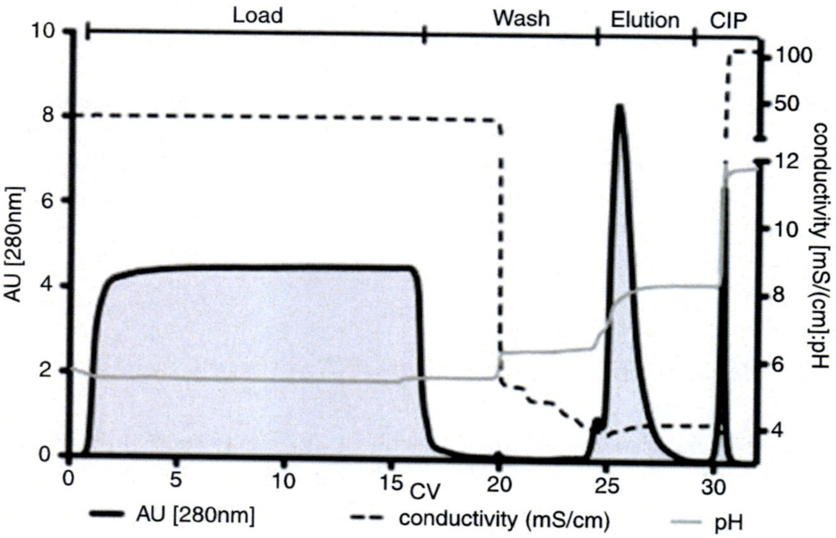

Fig. 10.29 Separation of single-chain antibody scFv hu225 extract on mixed-mode Capto MMC column. Load the sample solution containing scFv on a 110 mL Capto MMC (35 mm × 115 mm) column with a residence time of 2–3 min. Initially the column was equilibrated with 5 column volumes (CV) of binding buffer (40 mM disodium hydrogen phosphate, adjusted to pH 5.5 with 2 M citrate stock solution; 225 mM NaCl; 0.05% Tween20). After loading the sample, the column was washed and desalted with 5 CV washing buffer (30 mM sodium phosphate buffer; 10 mM NaCl, pH 7.0). By raising the pH above 7.4, the bound protein was eluted (30 mM sodium phosphate buffer; 10 mM NaCl, pH 8.0). The black dashed line indicates the conductivity, the gray line indicates the pH, and the gray area together with the black line indicates the absorption at 280 nm. (Reproduced from Lindner et al. 2014)

specific endothelial cell mitogen, which participates in the development of the vascular system and the differentiation of endothelial cells, so it has the therapeutic potential to accelerate wound healing. It is a homodimeric growth factor. Its most abundant and active form consists of two 165 amino acid monomers, each of which contains 7 pairs of intrachain disulfide bonds and is covalently connected through 2 interchain disulfide bonds. The molecular weight of VEGF165 is about 38 kDa and the theoretical p*I* is 8.6. It can be cleaved by plasmin to produce two domains: (1) receptor binding domain with affinity to the kinase domain receptor (KDR) and the fms-like tyrosine kinase receptor (FLT-1) (residues 1–110, disulfide bond-linked homodimer) and (2) heparin binding domain (residues 111–165). The high positive charges of the heparin binding domain indicates that it should have a strong affinity to negatively charged ligands, and most of the hydrophobicity of the receptor binding site can be further utilized to enhance the selectivity of mixed-mode resins.

rhVEGF is expressed in *E. coli* as inclusion bodies and then refolded to restore its biological activity. The active ingredients of the refolding buffer consist of 2 mM DTT, 15 mM L-cysteine, and 1 M urea. 10 mM CHES and 100 mM L-arginine are present to maintain the target pH of the buffer. The refolding step is performed at pH 9.8 and downstream processing is performed within 48 h. Before loading on the column, the refolding pool was conditioned with 1% (v/v) Triton X-100 and adjusted to the required load pH with acetic acid. The ionic strength of this material is relatively low (≤5 mS/cm).

The refolding pool contains 1 M urea at pH 9.8 and has low ionic strength (≤5 mS/cm). Therefore, without major adjustments to the refolding pool, directly using traditional chromatography resins for capture faces some challenges. The first attempt was made to capture with a cation exchange resin, but it was unsuccessful because the protein precipitated on the chromatographic column and was not eluted. Then consideration was given to anion exchange resins. However, when the pH of the refolding cell was adjusted to below the p*I* of the protein, a large amount of rhVEGF precipitated, excluding the use of ion exchange resins. The low ionic strength of the refolding cell discourages the use of hydrophobic interaction chromatography. In contrast to single-mode chromatography, mixed-mode chromatography uses a combination of multiple interaction types to adsorb proteins and shows certain advantages in challenging target protein capture. Especially the multimodal cation exchange resin, Capto MMC, contains a number of different functional groups (carboxyl, phenyl, amide, and thioether), which can interact with proteins by electrostatic, hydrophobic, hydrogen bonding, and thiophilic interactions, and has the potential as the preferred adsorbent for capture of rhVEGF from refolding cell.

Once Capto MMC was selected as the capture adsorbent for rhVEGF, the focus of method development was to explore protein binding and elution characteristics to maximize resin adsorption capacity and active protein yield. In particular, the dynamic binding capacity (DBC) of refolded rhVEGF as a function of residence time and load pH was studied. When the clarified pH 9.8 refolding pool was applied to the column, 11 g of rhVEGF/L resin DBC was obtained within a residence time of 1 min. DBC increases with the increase of residence time. Using a residence time of 1 min, DBC was studied as a function of load pH, and it was observed that DBC increased as the load pH decreased. The maximum DBC was reached at the lowest pH value of 8.5.

The variation trend of DBC of mixed-mode resin with the pH value of the solution can be understood by the combination of multiple interactions between ligand and solute. First, the local charge of rhVEGF can lead to electrostatic interaction with the ligand. There is a positively charged pocket near the carboxyl terminal region for binding heparin. Although both the protein and the ligand have a net negative charge at a pH of 8.5–10, the local positive charge of the protein may still promote the local electrostatic interaction with the ligand. The heparin binding domain is considered a complete pocket, with a local positive charge at most pH values. Peptide mapping analysis of the carboxyl domain indicated that it is very basic in nature, with a p*I* of about 11.8 and an amino acid sequence consisting of about 22% of arginine and lysine residues (Keck et al. 1997). Second, hydrophobic and hydrogen bonding interactions may promote rhVEGF-ligand binding. The crystal structure of the protein shows that the binding surface to the VEGF receptor Flt-1 (also known as VEGFR-1) domain-2 is quite flat, consisting of a β-sheet with a high percentage of leucine and isoleucine residues (Wiesmann et al. 1997). The intermolecular contact at this interface is mainly driven by hydrophobic interactions.

Molecular dynamics simulations were used to determine four direct polar interactions between VEGF and Flt-1: Flt-1-Arg224 and VEGF-Asp63, Flt-1-Lys171 and VEGF-Glu103, Flt-1-Glu141 and VEGF-Lys16, and Flt-1-Thr38 and VEGF-Lys48 (Horta et al. 2008). Electrostatic, hydrophobic and hydrogen bonding interactions dominate the molecular recognition between proteins and receptors and also drive the binding interactions between rhVEGF and Capto MMC.

The multiple nature of protein-ligand interactions poses challenges for method development in the desorption phase. In order to elute the adsorbed protein, it is necessary to find a suitable buffer composition in order to simultaneously adjust two or more interactions between the solute and the stationary phase. In hydroxyapatite chromatography, it has been observed that phosphate is a multipotent buffer component, which can simultaneously inhibit calcium coordination affinity and phosphate ion interaction and can elute adsorbed monoclonal antibodies (Gagnon 2008). In hydrophobic charge induction chromatography, it has been observed that under neutral pH conditions, buffers (such as 1 M sodium chloride or 20% ethanol) are not sufficient to successfully elute bound proteins from mixed-mode resins, while urea or L-arginine HCl buffer is effective (Arakawa et al. 2009). Therefore, during the method development of the elution phase of rhVEGF, different eluent compositions were tried to improve the recovery.

Table 10.4 lists the salt-based buffers and organic buffers tested and the recovery of rhVEGF obtained. From the data in the table, it can be seen that no single-component buffer can completely desorb the protein. The use of elution gradients that increase ionic strength and change pH to disrupt ion exchange interactions, or organic solvents that are known to specifically disrupt hydrophobic interactions, has not been successful. The salt-based buffer produced only a small elution peak and was severely tailed off. The recovery yield of rhVEGF was less than 10%, indicating that protein retention was not dominated by ionic interaction. Like in HIC, the strong retention at high salt concentrations may reflect enhanced hydrophobic interactions, thereby preventing the release of rhVEGF from Capto MMC. Buffers prepared from organic solvents, urea or L-arginine are tested individually using a step-by-step elution method. Each buffer produces few or no elution peaks, and the recovery of rhVEGF is less than 5%. It shows that the binding of protein is not dominated by hydrophobic interaction either.

Finally, the multipotent buffer composed of sodium chloride and reagents that disrupt hydrogen bonds or disrupt hydrophobic interactions (such as organic solvents, urea, and L-arginine) successfully achieved the elution of adsorbed proteins. Under these conditions, the elution of rhVEGF is significantly increased, and it still has biological activity, which indicates that the natural conformation of the protein is still retained in the presence of a high concentration of denaturant. The combination of sodium chloride and ethanol or propylene glycol (each of which has very limited effects on its own) improves the elution performance. The shape of the elution peak becomes sharp, and the yield of rhVEGF increases to 50–60%. Mixing 6 M urea with sodium chloride and a high concentration of L-arginine HCl further improved the elution behavior of the protein, resulting in a sharp and compact elution peak and a yield of over 90%. These results indicate that the latter two buffers have the ability

Table 10.4 Evaluation of elution buffers for rhVEGF desorption on the Capto MMC

Mechanism	Elution buffer component (s)[a]	pH range	Conductivity (mmho)	rhVEGF yield (%)
IEX	1 M sodium acetate	6.5–12	50	0.00
	1 M TMAC + 1 M sodium chloride	8.0	85	
	1 M ammonium chloride	9.5	100	<10
	1.5 M sodium chloride	7.0–9.5	120	
HIC	20% propylene glycol	8.0	<2	<5
	20% ethanol	8.0	<2	
	6 M urea	6.5–7.0	<2	
	0.75 M L-arginine	7.5	10	
IEX + HIC	20% propylene glycol + 1.5 M sodium chloride	8.0	65	50–60
	20% ethanol + 1.5 M sodium chloride	8.0	70	
	6 M urea + 1.5 M sodium chloride	6.5–7.0	78	>90
	1 M L-arginine HCl	6.5–7.5	43	
	0.75 M L-arginine + 0.45–0.75 M sodium chloride	9.0–9.5	43–58	

The column was equilibrated with 5 CV of HEPES buffer, pH 7.5. Clarified refold pool was loaded to approximately 20 mg rhVEGF/mL of resin. The elution was completed using a step for the HIC buffers and gradient elution for all others. The residence time was 2.3 min
Reproduced from Kaleas et al. (2010)
[a]All elution buffers contain 25 mM HEPES

to simultaneously disrupt multiple interaction components between rhVEGF and Capto MMC ligand and successfully release the protein. Compared with propylene glycol and ethanol, urea, and arginine have higher elution efficiency, which may be related to their excellent hydrogen bond and hydrophobic potency in inhibiting protein aggregation.

Through the stepwise gradient elution of the latter two elution buffers, the composition of the elution buffer was optimized to improve the recovery yield of the target protein. The L-arginine HCl buffer experiment consists of five steps from 0 to 1.25 M, carried out in 0.25 M increments (Fig. 10.30a), while in the experiment of urea/sodium chloride, the urea concentration maintains a constant at 6 M, the concentration of sodium chloride increases in 0.25 M increments from 0.25 to 1.5 M (Fig. 10.31a). The protein recovery yield, host cell protein content, and protein quality of each peak were evaluated by rpHPLC (Figs. 10.30b and 10.31b). In Figs. 10.30a and 10.31a, five different elution peaks were observed in the two elution buffers. Peaks A–E eluted from 0.25 to 1.25 M during the elution of L-arginine HCl, and peaks F–J eluted from 0 to 1.0 M NaCl during the urea/salt elution process. Most rhVEGF was detected in two peaks with each elution buffer (B and C;

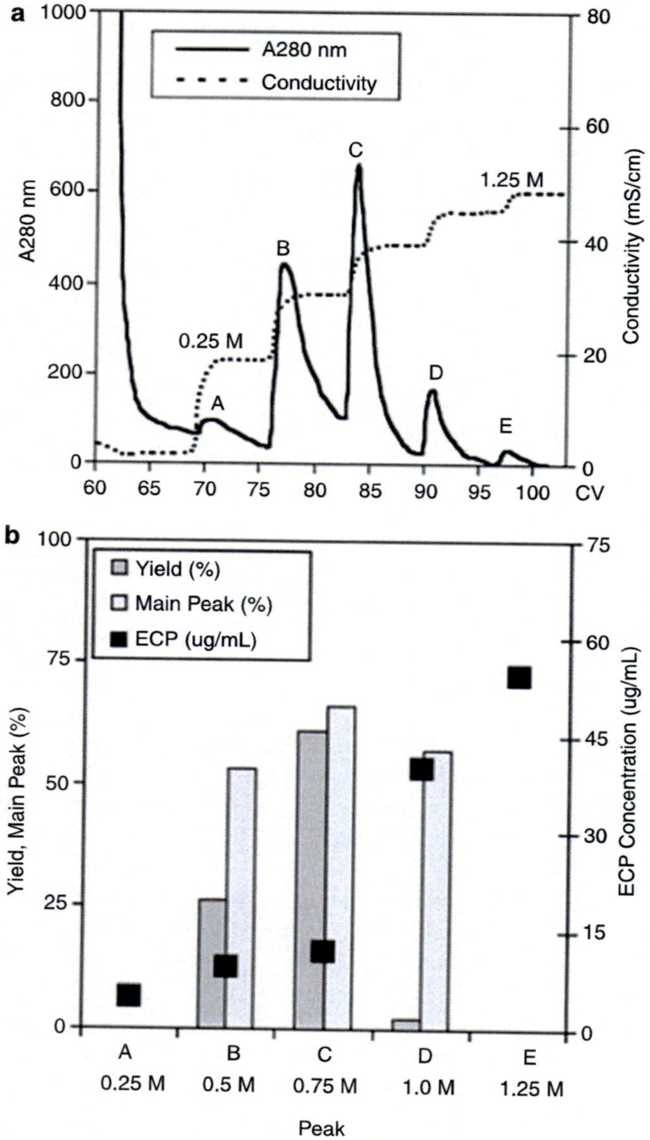

Fig. 10.30 (**a**) Chromatographic separation chart and (**b**) peak analysis of rhVEGF eluted from the Capto MMC column. The elution phase includes a step of L-arginine HCl (pH 7.5) from 0 to 1.25 M in 0.25 M increments. (Reproduced from Kaleas et al. 2010)

H and I). It should be noted that compared with urea/salt buffer, L-arginine hydrochloride can better separate host cell proteins and rhVEGF. After most of the rhVEGF eluted, higher E. coli proteins (ECP) concentrations were obtained in

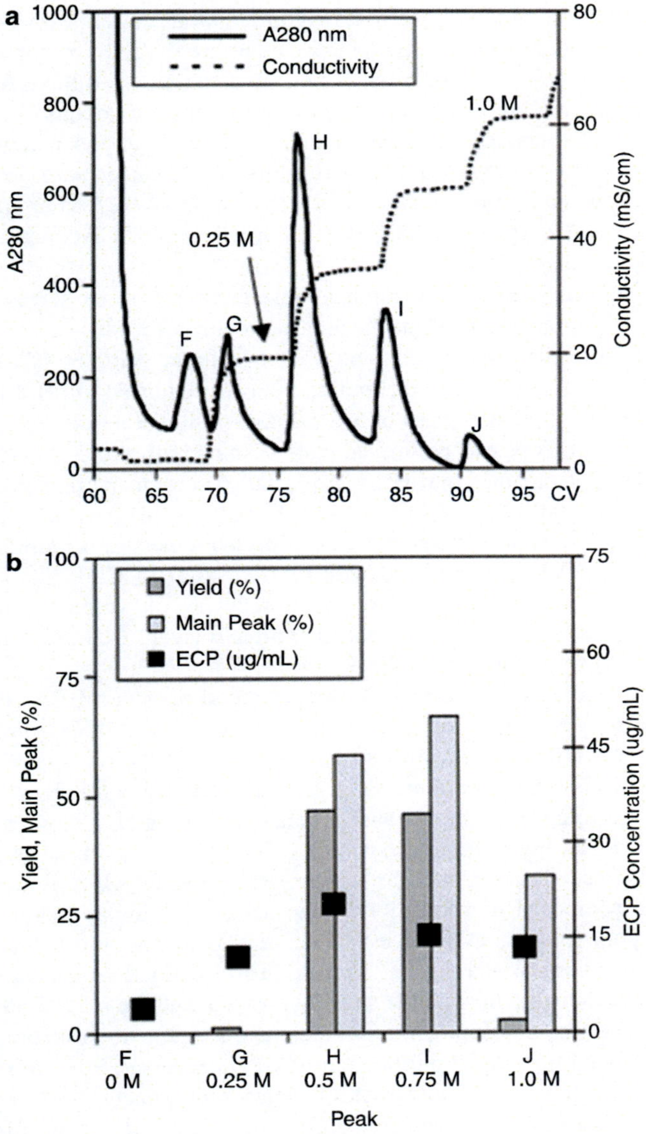

Fig. 10.31 (**a**) Chromatographic separation and (**b**) peak analysis of rhVEGF eluted from the Capto MMC column. During the elution phase, a constant 6 M urea concentration was maintained, while the sodium chloride concentration was increased to 1.5 M (pH 7.5) in 0.25 M increments. No elution peaks were observed with >1.0 M NaCl. (Reproduced from Kaleas et al. 2010)

peaks D and E (≥1 M L-arginine hydrochloride), while in all peaks F–J, ECP and rhVEGF co-eluted. Analysis of the peaks containing rhVEGF by rpHPLC showed that the quality of the refolded product remained constant and the type or concentration of the elution buffer did not introduce new product variants.

With arginine selected as the best eluent component, the conditions of the washing stage after sample loading were further investigated in order to improve the purity of the target protein. The final wash buffer of 0.25 M arginine, 0.20 M arginine HCl, and 25 mM HEPES (pH 9.0) can reduce the ECP concentration in the pool by approximately 25%.

Combining the method development results of the three stages of loading, washing and elution, the final purification protocol is established as follows. The Capto MMC column was equilibrated with 5 column volumes (CV) of 25 mM CHES (pH 8.7). The clarified renaturation pool was adjusted to pH 8.7 with acetic acid. After loading, the first wash of 5 CV with equilibration buffer was performed. A second wash of 5 CV was performed with washing buffer (0.20 M L-arginine HCl and 0.25 M L-arginine in 25 mM HEPES, pH 9.0). Then, an elution buffer (0.85 M L-arginine HCl and 0.05 M L-arginine in 25 mM HEPES, pH 7.5) was used for 5 CV elution. The column was regenerated with 2 M guanidine hydrochloride in 100 mM Tris, pH 10 (3 CV), and sterilized with 0.5 M NaOH (5 CV). The Capto MMC purification step is completed in about 13 h, reducing the volume of the refolding pool by 75 times, and reducing the host cell protein level to <1.2%, and the *E. coli* DNA to <10 ng/mg, with yield >95%.

Guo et al. (2019) established a two-step method based on HEA HyperCel and Capto Q to separate and purify recombinant osteopontin (OPN) from *E. coli* cell lysates. Osteopontin is an essential structural protein in the process of biomineralization, and it plays an important role in bone remodeling and kidney stone inhibition. Due to its role in cell signal transduction, adhesion and migration under normal conditions and during pressure response, OPN can also be used as a molecular marker for tissue diseases such as cancer. Recombinant OPN is a single-chain polypeptide composed of about 300 amino acids. It is highly acidic and has an isoelectric point of 4.3. OPN is usually classified as an intrinsically disordered protein with biological activity but lacks a clearly defined secondary and tertiary structure. As a potential therapeutic and diagnostic agent, the large-scale production of OPN requires the development of efficient and scalable downstream processes.

The traditional method of purification of OPN expressed in *E. coli* is based on fractional precipitation of cell lysates, buffer replacement, and multistep ion exchange processes. The process is lengthy and the yield is low. Since OPN is unstable in high concentrations of ammonium sulfate, hydrophobic interaction chromatography is excluded. Due to the salt tolerance of mixed-mode chromatography, it provides the possibility to capture OPN directly from cell lysates. Since solute retention in mixed-mode chromatography involves complex interactions, it is necessary to screen protein binding and elution conditions to comprehensively evaluate ligand-protein interactions and determine the design space for process development. The authors first used a batch adsorption experiment on a 96-well filter plate to evaluate the adsorption and elution behavior of OPN and total protein

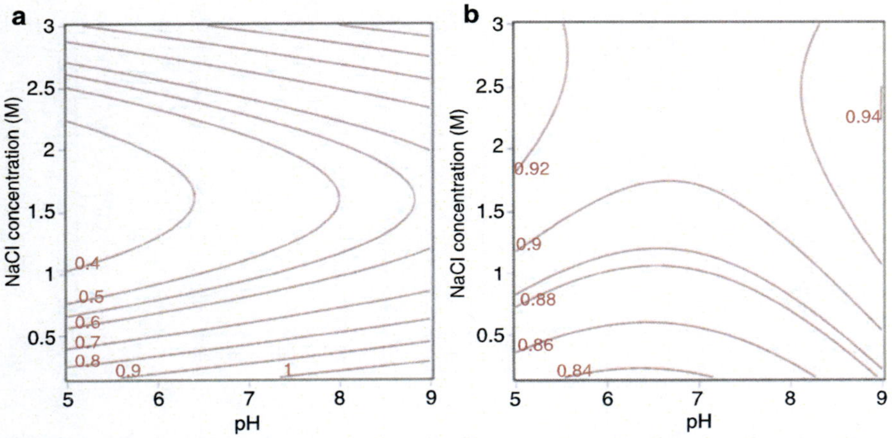

Fig. 10.32 Contour plots of OPN (**a**) and total protein (**b**) bound to HEA HyperCel. The number on the contour grid represents the fraction of protein in the lysate adsorbed to the HEA HyperCel. (Reproduced from Guo et al. 2019)

(TP) on the hydrophobic anion exchanger HEA HyperCel. In the range of pH 5–9 and 0–3 M NaCl, the equilibrium adsorption capacity of OPN and TP on the resin was measured, and the results are shown in Fig. 10.32.

The adsorption contours of OPN show that the adsorption amount of protein increases monotonously with increasing pH. However, the adsorption capacity presents a typical U-shaped curve with the increase of the salt concentration, that is, it first decreases and then increases, with a minimum value at 1.5 M. In the lower salt concentration range (0–0.3 M), the binding between protein and ligand is dominated by electrostatic interaction. As the salt concentration increases to 1 M NaCl, these interactions are significantly weakened, resulting in a reduction in OPN adsorption by as much as 50%. In regions with high salt concentration (NaCl concentration >2.7 M), protein adsorption is dominated by hydrophobic interactions. Because the binding level of host cell proteins in the entire pH range exceeds 80%, changes in pH and NaCl concentration have little effect on total protein (TP) adsorption.

According to the above results, low-salt (0.15 M NaCl) and high-salt (3 M NaCl) adsorption conditions were selected for the study of elution conditions. A combination of pH and salt concentration changes was used to gradually weaken the hydrophobicity and charge interactions and ultimately lead to electrostatic repulsion and complete desorption of adsorbed proteins. The elution step consists of 1 M NaCl, 50 mM Tris buffer (pH 7), then 1 M NaCl, 0.1 M sodium acetate (pH 5), and 0.1 M sodium citrate (pH 3) without NaCl.

A very small amount of OPN (5%) and more than 40% of *E. coli* TP were detected in the flow-through. After washing the resin with binding buffer, the salt concentration was increased for the first elution. Most of the bound OPN (76%) and a small amount of host cell proteins (6%) were desorbed. This step alone purified

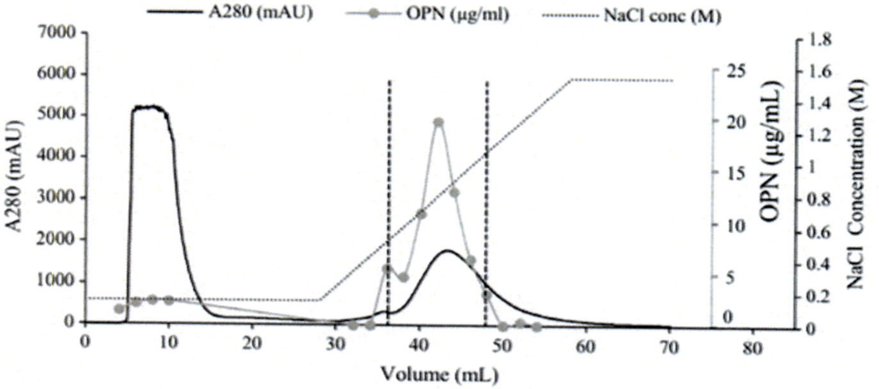

Fig. 10.33 Chromatogram of the low salt binding/elution of OPN on the HEA HyperCel column. The clarified lysate was loaded onto the equilibrated PRC HEA HyperCel column (1 mL) at a rate of 0.2 mL/min to ensure a residence time of 5 min. The column was washed with 20 CV of binding buffer and then eluted through a linear NaCl gradient (0.15–1.5 M NaCl, over 20 CV). OPN eluted in the peaks near 1 M NaCl, shown between the vertical dark dashed lines. (Reproduced from Guo et al. 2019)

OPN 14 times. The remaining adsorbed OPN was recovered in the subsequent elution step.

Under high-salt conditions, more than 80% of the total *E. coli* proteins and 92% of the OPN are bound with the HEA resin. After washing the resin with 10 CV binding buffer, the salt concentration was reduced for the first elution. Most of the bound OPN (74%) and a small amount of TP (10%) were desorbed, achieving a sevenfold purification of OPN. The remaining bound OPN was recovered in the next two elution steps.

In order to confirm the validity of the results of the batch adsorption experiment, the same feed composition and loading volume were used to repeat the batch adsorption and elution process in the packed column (1 mL pre-packed PRC HEA HyperCel). The adsorbed OPN was eluted by using a linear gradient instead of stepwise elution. The resulting chromatogram is shown in Fig. 10.33. The elution salt concentration of OPN spans the concentration range from 0.6 to 1.1 M NaCl, and the resulting fraction (36–50 mL) contains 97% of the loaded OPN, resulting in a 16-fold purification.

The pressure rises due to the precipitation of OPN on the HEA HyperCel under high salt (3 M NaCl) conditions, which exceeds the maximum pressure resistance range of the resin (0.3 MPa). Therefore, the development of OPN purification under high salt (3 M NaCl) binding conditions was abandoned.

After removing the salt by dialysis, the eluted OPN pool was loaded onto a HiTrap Capto Q column (1 mL). A linear salt gradient (0–1.0 M NaCl) was used to achieve additional separation of OPN and *E. coli* proteins. The bound protein eluted in the three main peaks (Fig. 10.34), and OPN was recovered in the second peak (fraction 32.3–33.8 mL).

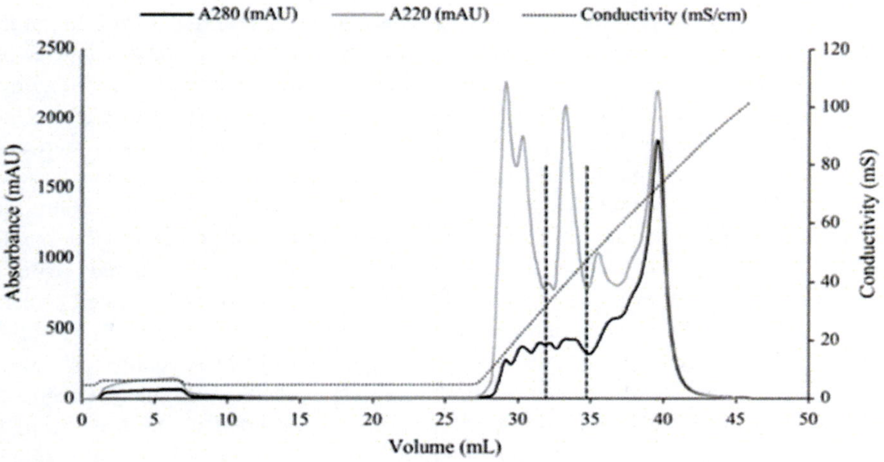

Fig. 10.34 Chromatogram of OPN purification on Capto Q after the HEA HyperCel capture step. After the salt was removed by dialysis, the eluted OPN pool was loaded onto the HiTrap Capto Q column (1 mL) at a flow rate of 0.2 mL/min. A linearly increasing salt gradient (0–1.0 M NaCl) was used to further purify OPN from *E. coli* proteins. OPN eluted in the peaks between the vertical dark dotted lines (32.3–33.8 mL). (Reproduced from Guo et al. 2019)

The HEA HyperCel capture step increases the OPN purity in the crude lysate from 2% to 29%, resulting in a purification factor of 15.6 and a recovery yield of 81%. The second purification step, the Capto Q packed bed column, provides the necessary orthogonality for HEA and an additional threefold OPN purification with a recovery of 90%. Overall, the two-step purification process produced 44% recovery and 95% purity. Compared with previously reported processes that rely on continuous precipitation steps and non-orthogonal chromatography steps, the integration of orthogonal methods achieves higher purity protein purification with fewer unit operations.

Ravi et al. (2020) described the use of mixed-mode chromatography to directly capture recombinant human thioredoxin (hTrx) from cell lysates and the selective elution of adsorbed proteins. Human thioredoxin-1 (hTrx) is a small 12 kDa protein found in all tissues. It has redox active sites and is responsible for maintaining a reduced cellular environment. hTrx has a highly ordered and conserved 4 alpha helix, 5 beta sheet secondary structure, and has thermal stability and pH stability. It is an acidic protein with a theoretical p*I* of 4.8, but a lower net charge of −6.8, and a hydropathic index of −0.096 at pH 8. Potential applications include reducing the mucin network to alleviate the symptoms of cystic fibrosis. The administration of hTrx by aerosol inhalation requires a large dose of 1.5–150 mmol/kg of patients, which is equivalent to approximately 65 g per patient per year. Therefore, a scalable and high-yielding method will be needed to provide hTrx to patients with cystic fibrosis at a reasonable cost.

The purification process of hTrx in *Escherichia coli* reported in the literature mostly undergoes lysate pretreatment, capture, and refinement steps (Stein et al. 1995; Harris et al. 2012; Wollman et al. 1988). The process is tedious and lengthy. Using the salt-tolerant properties of mixed-mode chromatography to capture the target protein directly from the cell lysate can bypass the pretreatment step and conditioning of the feed solution, simplifying the purification process.

In order to select the ideal resin for capture chromatography, the adsorption performance of hTrx on three commercial mixed-mode chromatography media Capto adhere, HEA HyperCel, and Toyopearl NH_2-750F were compared. Using a batch adsorption experiment in a 96-well filter plate, the hTrx and host cell protein (HCP) bound to the resin were tested under different conditions of pH (6, 7, and 8) and electrical conductivity (adjusted by NaCl to 2–10 mS). In the entire test range, Toyopearl NH_2-750F showed weak binding of hTrx. Even at a higher pH value of 8.0 and a conductivity of 2 mS, 80% of hTrx was detected in the flow-through fraction. Capto adhere can completely adsorb hTrx at pH 7.0 and 8.0 and conductivity greater than 5.0 mS. When the pH is lower than 7.0, the hTrx binding will be affected by the conductivity and decrease. If the conductivity is higher than 5.0 mS, the adsorption will be incomplete. HEA HyperCel shows salt-tolerant binding to hTrx in the range of pH 6.0–8.0, which is not affected by pH. Under the optimal lysate conditions of pH 5.0 and 5 mS, HEA HyperCel resin showed complete binding to hTrx.

Experiments have shown that simply using a salt-increasing gradient cannot completely elute the bound hTrx from the HEA HyperCel resin. Three organic molecules (arginine, guanidine, and lysine) were explored as buffer modifiers to reverse the interaction of hTrx-HEA ligands. The effect of 50 mM Tris containing arginine HCl, guanidine HCl and lysine HCl, 5 mM DTT, pH 7.0, on the recovery of hTrx eluted from HEA HyperCel resin was tested by a batch adsorption experiment in a 96-well filter plate, as shown in Fig. 10.35a. Using 1.0 M NaCl, 0.25 M arginine, guanidine, and lysine buffers, the obtainable hTrx recoveries were 33%, 35%, 47%, and 43%, respectively, indicating that each buffer can disrupt the specific interaction between the protein and HEA, thus contributing to the desorption of hTrx. For example, sodium chloride can inhibit electrostatic interactions, and arginine, guanidine, and lysine can promote protein desorption by competing with ligands for hydrophobic interaction and hydrogen bonding.

Therefore, using 0.1 M arginine in combination with 1.0 M NaCl to elute proteins from HEA HyperCel can greatly increase the recovery of hTrx from less than 50% when used alone to 100% when used in combination (Fig. 10.35b). Using 0.1 M guanidine hydrochloride or lysine in combination with 1.0 M NaCl, the same results were obtained, indicating that each buffer modifier (0.1 M concentration) and 1.0 M NaCl act in a cooperative manner. In order to further simplify the elution process, a binding buffer (50 mM sodium acetate buffer, 5 mM DTT, pH 5.0) containing 0.1 M arginine and 1.0 M NaCl was used to elute the bound hTrx, resulting in 100% recovery.

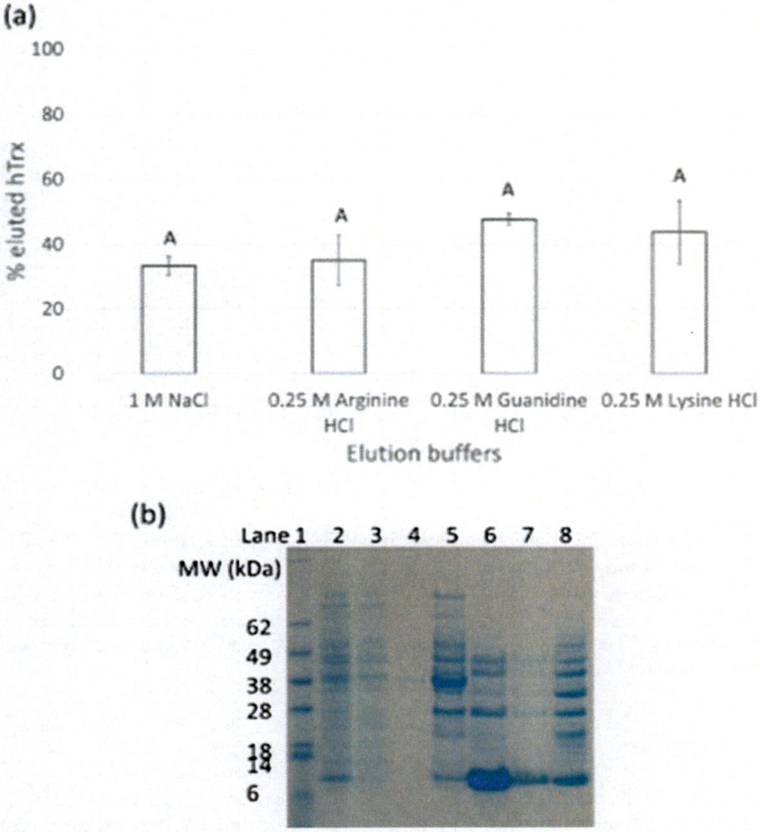

Fig. 10.35 (**a**) Recovery (%) of hTrx eluted from HEA HyperCel resin with different elution buffers. The effects of a 50 mM Tris buffer containing 0.25 M arginine, lysine, or guanidine with the same buffer containing 1.0 M NaCl were compared. (**b**) SDS-PAGE of the binding and elution curve of *E. coli* lysate on HEA HyperCel resin. All samples were diluted to <1 mg/mL TSP. Lane 1, MW marker; Lane 2, clarified lysate; Lane 3, flow-through; Lane 4, wash (3 CV); Lane 5, elution with 0.1 M NaCl; Lane 6, elution with 1.0 M NaCl; Lane 7, elution with 0.25 M arginine HCl (0–2 CV); Lane 8, elution with 0.25 M arginine HCl (2–10 CV). (Reproduced from Ravi et al. 2020)

Under optimized binding and elution conditions, the influence of protein loading conditions was further investigated. As shown in Fig. 10.36, 10 mL of clarified lysate (pH 5.0, 5 mS/cm) was directly loaded onto a 1 mL HEA HyperCel column at a flow rate of 0.25 mL/min without any other adjustments. After the washing step, a 20 CV elution buffer gradient was used, from 50 mM sodium acetate, 5 mM DTT, pH 5.0 to 50 mM sodium acetate, 5 mM DTT, 1.0 M NaCl, 0.1 M arginine HCl, pH 5.0. hTrx was eluted from the resin at a conductivity of 30 mS/cm, with a total volume of 8.4 mL. This capture step resulted in 95% recovery and 90% purity.

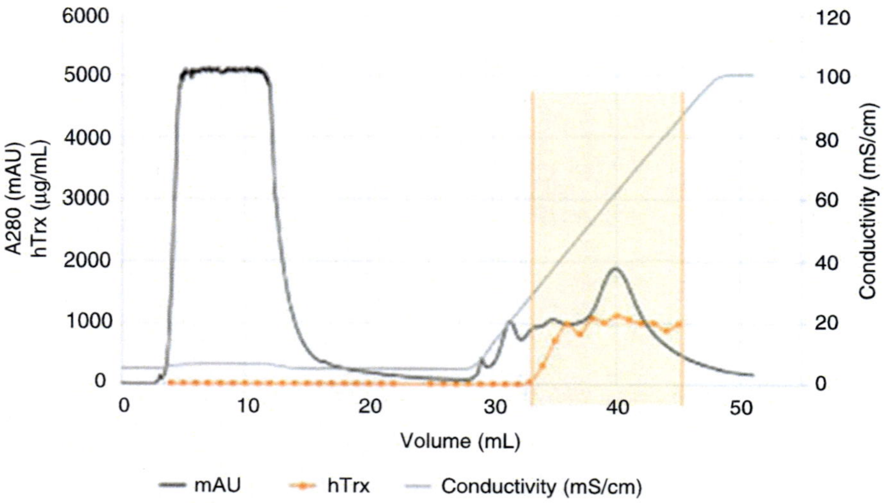

Fig. 10.36 Chromatogram of hTrx purification on HEA HyperCel resin. The clarified lysate (10 mL) was loaded onto each packed bed column (1 mL) at a flow rate of 0.25 mL/min. At a flow rate of 0.5 mL/min, a linear gradient of 20 CV of elution buffer (50 mM Tris, 5 mM DTT, 0–1 M NaCl, 0.1 M arginine HCl) was applied, and 1 mL fractions were collected. The elution fractions containing hTrx are shown as orange shaded areas and are combined and analyzed. (Reproduced from Ravi et al. 2020)

10.8 Summary

The modern pharmaceutical industry is mainly supported by two pillars: synthetic drugs and biological drugs. In the past 40 years, biotechnology pharmaceuticals based on genetic recombination technology have developed by leaps and bounds. Monoclonal antibodies have become the most promising drug category in the pharmaceutical industry, occupying most of the top ten best-selling drugs. The production of biotechnology drugs mostly undergoes the process of expression of recombinant genes and purification of expression products. Chromatography plays an indispensable role in the purification process of biotechnology drugs, and the cost of the chromatographic process accounts for more than half of the entire production cost. At present, protein A chromatography is the most important step in the platform technology of monoclonal antibody purification. However, protein A chromatography also has some problems. For example, protein A is easily cleaved by the protease in the cell lysate; monoclonal antibodies tend to form aggregates when they are desorbed under low pH conditions. These leached protein A and aggregates contaminate the product and must be removed through additional steps to meet the purity requirements of the drug regulatory authority. In addition, protein A chromatographic materials are expensive and have a short service life, which makes the cost of monoclonal antibody drugs high and affects the availability of antibody drugs.

As an alternative to affinity chromatography, mixed-mode chromatography is increasingly used in the manufacture of biotechnology pharmaceuticals. In the separation and purification of nucleic acid drugs, reversed-phase ion exchange mixed-mode chromatography has shown certain potential for the separation and purification of oligonucleotides. The coupling of reversed-phase and ion exchange mechanisms makes this mode of chromatographic separation show higher selectivity than reversed-phase or ion-exchange single-mode chromatography and has advantages in the fine separation of impurities with missing bases and sequence reversals. Multimodal ion exchange chromatography shows advantages in the separation and purification of plasmids and minicircle DNAs. On the multimodal chromatographic stationary phase, the supercoiled plasmid retains more than the open-circular plasmid, so the separation of different topoisomers is easy to achieve. In terms of the separation and purification of monoclonal antibodies, hydroxyapatite chromatography can achieve simultaneous removal of impurities and antibody aggregates in host cells. Both hydrophobic charge induction chromatography and multimodal ion exchange chromatography can provide salt-tolerant adsorption and efficient separation of proteins. In flow-through mode, multimodal anion exchange media can be used to remove host DNA and antibody aggregates. In the purification of recombinant enzyme drugs, mixed-mode chromatography can also provide different selectivity than other mode chromatography. In view of the problem of strong adsorption of proteins on mixed-mode chromatographic media that affect activity, the recovery of active proteins can be improved by selecting suitable mobile phase additives to weaken the various interactions between solutes and ligands.

References

Alves CPA, Simcikova M, Brito L et al (2016) Development of a nicking endonuclease-assisted method for the purification of minicircles. J Chromatogr A 1443:136–144

Alves CPA, Prazeres DMF, Monteiro GA (2021) Minicircle biopharmaceuticals–an overview of purification strategies. Front Chem Eng 2:612594

Arakawa T, Kita Y, Sato H et al (2009) MEP chromatography of antibody and Fc-fusion protein using aqueous arginine solution. Protein Expr Purif 63:158–163

Biba M, Jiang E, Mao B, Zewge D et al (2013) Factors influencing the separation of oligonucleotides using reversed-phase/ion-exchange mixed-mode high performance liquid chromatography columns. J Chromatogr A 1304:69–77

Bicho B, Caramelo-Nunes C, Sousa A et al (2016) Purification of influenza deoxyribonucleic acid-based vaccine using agmatine monolith. J Chromatogr B 1012:153–161

Černigoj U, Vidic U, Barut M et al (2013) A multimodal histamine ligand for chromatographic purification of plasmid DNA. J Chromatogr A 1281:87–93

Chen ZY, He C-Y, Kay MA (2005) Improved production and purification of minicircle DNA vector free of plasmid bacterial sequences and capable of persistent transgene expression in vivo. Hum Gene Ther 16:126–131

Chmielowski RA, Meissner S, Roush D et al (2014) Resolution of heterogeneous charged antibody aggregates via multimodal chromatography: a comparison to conventional approaches. Biotechnol Prog 30:636–645

Damasceno LM, Pla I, Chang HJ et al (2004) An optimized fermentation process for high-level production of a single-chain Fv antibody fragment in Pichia pastoris. Protein Expr Purif 37:18–26

Ferreira GNM (2005) Chromatographic approaches in the purification of plasmid DNA for therapy and vaccination. Chem Eng Technol 28:1285–1294

Ferreira GNM, Monteiro GA, Prazeres DMF et al (2000) Downstream processing of plasmid DNA for gene therapy and DNA vaccine applications. Trends Biotechnol 18:380–388

Gagnon P (2008) Improved antibody aggregate removal by hydroxyapatite chromatography in the presence of polyethylene glycol. J Immunol Methods 336:222–228

Gagnon P, Ng P, Zhen J et al (2006) A ceramic hydroxypatite based purification platform: simultaneous removal of leached protein A, aggregates, DNA, and endotoxins. Bioprocess Int 4:50–60

Ghanem A, Healey R, Adly FG (2013) Current trends in separation of plasmid DNA vaccines: a review. Anal Chim Acta 760:1–15

Giovannini R, Freitag R (2001) Isolation of recombinant antibody from cell culture supernatant: continuous annular versus batch and expanded-bed chromatography. Biotechnol Bioeng 73: 522–529

Gronemeyer P, Ditz R, Strube J et al (2014) Trends in upstream and downstream process development for antibody manufacturing. Bioengineering 1:188–212

Guo D, Wang L-L, Lin D-Q et al (2013) Evaluating antibody monomer separation from associated aggregates using mixed-mode chromatography. J Chromatogr A 1294:70–75

Guo S, Ravi A, Mayfield S et al (2019) Exploring the separation power of mixed-modal resins for purification of recombinant osteopontin from clarified Escherichia coli lysates. Biotechnol Prog 35:e2722

Harris RP, Mattocks J, Green PS et al (2012) Determination and control of low-level amino acid misincorporation in human thioredoxin protein produced in a recombinant Escherichia coli production system. Biotechnol Bioeng 109:1987–1995

Horta BA, Cirino JJ, de Alencastro RB (2008) On the structure, interactions, and dynamics of bound VEGF. J Mol Graph Model 26:1091–1103

Jones PT, Dear PH, Foote J et al (1986) Replacing the complementarity-determining regions in a human antibody with those from a mouse. Nature 321:522–525

Kaleas KA, Schmelzer CH, Pizarro SA (2010) Industrial case study: evaluation of a mixed-mode resin for selective capture of a human growth factor recombinantly expressed in E. coli. J Chromatogr A 1217:235–242

Kaleas KA, Tripodi M, Revelli S et al (2014) Evaluation of a multimodal resin for selective capture of CHO derived monoclonal antibodies directly from harvested cell culture fluid. J Chromatogr B 969:256–263

Keck RG, Berleau L, Harris R et al (1997) Disulfide structure of the heparin binding domain in vascular endothelial growth factor: characterization of posttranslational modifications in VEGF. Arch Biochem Biophys 344:103–113

Köhler G, Milstein C (1975) Continuous cultures of fused cells secreting antibody of predefined specificity. Nature 256:495–497

Łacki KM, Riske F (2020) Affinity chromatography: an enabling technology for large scale bioprocessing. Biotechnol J 15:1800397

Levin AA (2019) Treating disease at the RNA level with oligonucleotides. N Engl J Med 380:57–70

Lindner R, Moosmann A, Dietrich A et al (2014) Process development of periplasmatically produced single chain fragment variable against epidermal growth factor receptor in Escherichia coli. J Biotechnol 192:136–145

Lu RM, Hwang YC, Liu IJ et al (2020) Development of therapeutic antibodies for the treatment of diseases. J Biomed Sci 27:1

Mahut M, Haller E, Ghazidezfuli P et al (2012a) Topology-selective chromatography reveals plasmid supercoiling shifts during fermentation and allows rapid and efficient preparation of topoisomers. Angew Chem Int Ed 51:267–270

Mahut M, Lindner W, Laemmerhofer M (2012b) Molecular recognition principles and stationary-phase characteristics of topoisomer-selective chemoaffinity materials for chromatographic separation of circular plasmid DNA topoisomers. J Am Chem Soc 134:859–862

Matos T, Queiroz JA, Bülow L (2014) Plasmid DNA purification using a multimodal chromatography resin. J Mol Recog 27:184–189

Mayrhofer P, Blaesen M, Schleef M et al (2008) Minicircle-DNA production by site specific recombinant and protein-DNA interaction chromatography. J Gene Med 10:1253–1269

McCafferty J, Griffiths AD, Winter G et al (1990) Phage antibodies: filamentous phage displaying antibody variable domains. Nature 348:552–554

McLaughlin LW (1989) Mixed-mode chromatography of nucleic acids. Chem Rev 89:309–319

Morrison SL, Johnson MJ, Herzenberg LA et al (1984) Chimeric human antibody molecules: mouse antigen-binding domains with human constant region domains. Proc Natl Acad Sci U S A 81:6851–6855

Padlan EA (1991) A possible procedure for reducing the immunogenicity of antibody variable domains while preserving their ligand-binding properties. Mol Immunol 28:489–498

Pezzini J, Joucla G, Gantier R et al (2011) Antibody capture by mixed-mode chromatography: a comprehensive study from determination of optimal purification conditions to identification of contaminating host cell proteins. J Chromatogr A 1218:8197–8208

Rader RA (2008) (Re)defining biopharmaceutical. Nat Biotechnol 26:743–751

Ravi A, Foster ER, Perez LM et al (2020) Capture chromatography with mixed-mode resins: a case study with recombinant human thioredoxin from Escherichia coli. J Chromatogr A 1625: 461327

Sifniotis V, Cruz E, Eroglu B et al (2019) Current advancements in addressing key challenges of therapeutic antibody design, manufacture, and formulation. Antibodies 8:36

Silva-Santos AR, Alves CPA, Prazeres DMF et al (2016) Separation of plasmid DNA topoisomers by multimodal chromatography. Anal Biochem 503:68–70

Silva-Santos AR, Alves CPA, Prazeres DMF et al (2017) A process for supercoiled plasmid DNA purification based on multimodal chromatography. Sep Purif Technol 182:94–100

Silva-Santos AR, Alves CPA, Monteiro G et al (2019) Multimodal chromatography of supercoiled minicircles: a closer look into DNA-ligand interactions. Sep Purif Technol 212:161–170

Smrekar C, Smrekar F, Štrancar A et al (2013) Single step plasmid DNA purification using methacrylate monolith bearing combination of ion-exchange and hydrophobic groups. J Chromatogr A 1276:58–64

Stein M, Jacquot JP, Jeannette E et al (1995) Chlamydomonas reinhardtii thioredoxins: structure of the genes coding for the chloroplastic m and cytosolic h isoforms; expression in Escherichia coli of the recombinant proteins, purification and biochemical properties. Plant Mol Biol 28:487–503

Stephenson ML, Zamecnik PC (1978) Inhibition of Rous sarcoma viral RNA translation by a specific oligodeoxyribonucleotide. Proc Natl Acad Sci U S A 75:285–288

Schubert S, Freitag R (2007) Comparison of ceramic hydroxy- and fluoroapatite versus Protein A/G-based resins in the isolation of a recombinant human antibody from cell culture supernatant. J Chromatogr A 1142:106–113

Tripathi NK, Shrivastava A (2019) Recent developments in bioprocessing of recombinant proteins: expression hosts and process development. Front Bioeng Biotechnol 7:420

Walsh G (2003) Biopharmaceuticals benchmarks 2003. Nat Biotechnol 21:865–870

Walsh G (2018) Biopharmaceuticals benchmarks 2018. Nat Biotechnol 36:1136–1145

Wiesmann C, Fuh G, Christinger HW et al (1997) Crystal structure at 1.7 A resolution of VEGF in complex with domain 2 of the Flt-1 receptor. Cell 91:695–704

Wolfe LS, Barringer CP, Mostafa SS et al (2014) Multimodal chromatography: characterization of protein binding and selectivity enhancement through mobile phase modulators. J Chromatogr A 1340:151–156

Wolff JA, Malone RW, Williams P et al (1990) Direct gene transfer into mouse muscle in vivo. Science 247:1465–1468

Wollman E et al (1988) Cloning and expression of a cDNA for human thioredoxin. J Biol Chem 263:15506–15512

Zamecnik PC, Stephenson ML (1978) Inhibition of Rous sarcoma virus replication and cell transformation by a specific oligodeoxynucleotide. Proc Natl Acad Sci U S A 75:280–284

Zhao M, Vandersluis M, Stout J et al (2018) Affinity chromatography for vaccines manufacturing: finally ready for prime time? Vaccine 37:5491–5503

Zimmermann A, Greco R, Walker I et al (2014) Synthetic oligonucleotide separations by mixed-mode reversed-phase/weak anion-exchange liquid chromatography. J Chromatogr A 1354:43–55